# ION-SELECTIVE ELECTRODE REVIEWS

## VOLUME 1

**Other Related Pergamon Titles of Interest**

ALBAIGES: Analytical Techniques in Environmental Chemistry
BUTLER: The Analysis of Biological Materials
JEFFERY & HUTCHISON: Chemical Methods of Rock Analysis, 3rd Edition
PERRIN *et al:* Purification of Laboratory Chemicals, 2nd Edition
SVEHLA: Automatic Potentiometric Titrations
WANNINEN: Analytical Chemistry

**Related Journals of Interest** (free specimen copy available on request)

*Electrochimica Acta*
*Progress in Analytical Atomic Spectroscopy*
*Talanta*

# ION-SELECTIVE ELECTRODE REVIEWS

## Volume 1

*Editor-in-Chief*
J. D. R. THOMAS
*UWIST, Cardiff, Wales*

**PERGAMON PRESS**
OXFORD · NEW YORK · TORONTO · SYDNEY · PARIS · FRANKFURT

| | |
|---|---|
| U.K. | Pergamon Press Ltd., Headington Hill Hall, Oxford OX3 0BW, England |
| U.S.A. | Pergamon Press Inc., Maxwell House, Fairview Park, Elmsford, New York 10523, U.S.A. |
| CANADA | Pergamon of Canada, Suite 104, 150 Consumers Road, Willowdale, Ontario M2J 1P9, Canada |
| AUSTRALIA | Pergamon Press (Aust.) Pty. Ltd., P.O. Box 544, Potts Point, N.S.W. 2011, Australia |
| FRANCE | Pergamon Press SARL, 24 rue des Ecoles, 75240 Paris, Cedex 05, France |
| FEDERAL REPUBLIC OF GERMANY | Pergamon Press GmbH, Hamerweg 6, Postfach 1305, 6242 Kronberg-Taunus, FRG |



**British Library Cataloguing in Publication Data**
Ion-selective electrode reviews.
Vol.1
I. Thomas, John David Ronald
541'.3724 QD571 80-40914
ISBN 0 08 026044 6

First published in *Ion-Selective Electrode Reviews*, Volume 1, Nos 1 and 2, 1979, and supplied to subscribers as part of their subscription. Also available to non-subscribers.

*Printed in Great Britain by A. Wheaton & Co. Ltd., Exeter*

# CONTENTS

*Ion-Selective Electrode Rev.* Vol.1, pp.1-2.
Pergamon Press Ltd. 1979. Printed in Great Britain.

# EDITORIAL

ION-SELECTIVE ELECTRODE REVIEWS is prompted by strides in the development, theory and applications of selective electrochemical sensors deriving from the description and availability in 1966 of the fluoride solid-state lanthanum fluoride electrode. Progress has been stimulated by the widespread demand for convenient methods of determining ions in various branches of science, medicine and technology.

The present interest in microprocessors and automation focuses even more attention on the need for better and better species sensors, such as ion-selective electrodes. This places increasingly stringent demands on specification regarding selectivity, small size, dependability over long periods, rapid response, etc. The review articles to be published from time to time will have regard to such needs of ion-selective electrodes together with aspects of underlying theory and specific areas of application.

The three articles of the present inaugural issue make a start on a continuing appreciation of the field, helped by the IUPAC Recommendations for Publishing Manuscripts on Ion-Selective Electrodes.

To complement the various reviews, short lists of the more important recent publications will be published from time to time by article title, authors and formal reference. Listings may also be included of forthcoming meetings at which ion-selective electrodes may figure*.

"To travel hopefully is a better thing than to arrive, and the true success is to labour." (R.L. Stevenson).

April 1979

J.D.R. Thomas

J.D.R. THOMAS

* For the present, the following are listed:

| | Information from |
|---|---|
| October 14-19 1979 | |
| 156th Meeting of the Electrochemical Society at Los Angeles, California, USA. Includes a symposium on theoretical and practical aspects of ion-selective electrodes. | Dr. Martin S. Frant, Foxboro Analytical, 78, Blanchard Road, P.O. Box 435, Burlington, Mass. 01803, U.S.A. |
| July 20-26 1980 | |
| SAC 80. An International Conference of The Analytical Division of The Chemical Society at The University of Lancaster, United Kingdom. Includes electro-analysis and ion-selective electrodes. | The Secretary, Analytical Division, The Chemical Society, Burlington House, LONDON, W1V OBN, U.K. |

*Ion-Selective Electrode Rev.* Vol.1, pp.3-30.     0191-5371/79/0601-0003 $05.00/0


# PROGRESS IN DESIGNING CALCIUM ION-SELECTIVE ELECTRODES

G. J. Moody and J. D. R. Thomas
Chemistry Department, UWIST, Cardiff CF1 3NU, Wales

CONTENTS

KEYWORDS: Calcium ion-selective electrodes; Ion-selective electrodes for calcium; neutral carrier ion-sensors for calcium; phosphate ion-sensors for calcium.

## 1. INTRODUCTION

Calcium ion-selective electrodes are among the most widely used of the new generation ion-selective electrodes. This is a direct consequence of the desire and need for determining free calcium ions in biological and cell fluids, but there are also important applications for calcium ion determinations in waters, soil extracts, industrial operations, etc. However, unlike the highly selective fluoride ion-selective electrode with a simple crystal membrane, the successful calcium ion-sensors comprised liquid membranes based on liquid ion-exchangers like calcium bis di-n-decylphosphate sensor dissolved in di-n-octylphenyl-phosphonate solvent mediator [1]. These membranes presented a challenge for electrode design, while the dialkylphosphate sensors themselves, although representing a considerable advance in calcium ion determinations, had several shortcomings - notably restricted pH ranges and interferences from sodium, potassium and magnesium ions. Considerable progress has been made in meeting such challenges and it is timely to briefly assess the progress made.

The most significant milestone in the history of calcium ion determination is undoubtedly the work of Ross [1] who in 1967 described the design and properties of the calcium ion-selective electrode based on calcium bis di-n-decylphosphate sensor with di-n-octylphenylphosphonate solvent mediator. This development was reflected in a commercial version of the electrode [2] and completely overshadowed systems based on paraffin wax and thenoyltrifluoro-acetone/tributylphosphate [3,4]. Nevertheless, the stringent specification demands imposed by the samples to which calcium ion-selective electrodes are applied have stimulated important advances in design, for example, the incorporation of the liquid ion-exchanger in PVC matrix membranes [5-7], the design of microelectrodes [8,9], the design and synthesis of improved sensors based on neutral carrier systems [10,11] and calcium bis-dioctylphenylphosphate [12], and the grafting of ion sensors to polymer matrix membranes [13].

## 2. MECHANICAL DESIGN

The lay-out of calcium ion-selective electrodes is like that of the classical glass pH sensitive electrodes, namely:

| Internal reference element | Internal reference solution | Ion-sensing membrane |
|---|---|---|

The reference element is silver/silver chloride, the reference solution is aqueous calcium chloride (usually 0.1 or 0.01 M), and the sensor membrane is usually a supported liquid or PVC matrix membrane containing the calcium ion-sensor dissolved in a suitable solvent mediator.

Ross [1] used 0.1 M calcium bis di-n-decylphosphate dissolved in di-n-octylphenylphosphonate as the calcium ion-sensing liquid ion-exchanger, while Moody, Oke and Thomas [2] used a 0.2 M solution of the calcium salt in the solvent mediator to form the liquid ion-exchanger trapped in PVC matrix membranes. Although lower concentrations can be employed, a 10 per cent m/m solution of sensor in solvent mediator is a helpful guideline for making up the liquid ion-exchanger.

The ion-selective electrodes are used in test solution systems in conjunction with suitable reference electrodes to give complete potentiometric electrochemical cells in much the same way as glass pH electrodes. The electrochemical response with test solutions depends on the membrane/test solution boundary potential as long as all the other boundary potentials can be assumed to be constant.

## 2.1 Liquid membrane electrodes

The prototype calcium ion-selective electrode consisting of two vertical chambers, with the central one containing the internal reference solution dipping into the liquid ion-exchanger held in an outer tubing sealed at the lower end with cellulose dialysis tubing [1], was developed into a commercial model by a complicated but ingenious piece of fluorocarbon engineering. Here, the water-immiscible ion-exchanger was held within the concentric reservoir compartment by a thin millipore filter membrane [2]. This commercial model has since been modified to give the Orion 93-20 liquid membrane calcium ion-selective electrode with an easily replaceable sensor head in order to overcome the rather tricky recharging procedure and also to avoid the associated waste of liquid ion-exchanger material. The replaceable module is claimed to function for six months under normal laboratory conditions [14].

## 2.2 PVC matrix membrane electrodes

Although other variations of the liquid membrane electrode have been described, a significant advance in calcium ion-selective electrode design was the immobilization of liquid ion-exchanger in PVC matrix membranes [5-7]. These are easily made in the laboratory by pouring a solution of 0.4 g of the commercial calcium liquid ion-exchanger (or alternatively 0.04 g sensor plus 0.36 g solvent mediator) and 0.17 g PVC in 6 $cm^3$ tetrahydrofuran or cyclohexanone into a glass ring resting on a glass plate as in Fig.1(a). The solvent is then allowed to evaporate at room temperature under controlled conditions of the assembly shown in Fig.1(b) over about two days. Discs of appropriate diameter are cut from the "master membrane" and cemented to the flat end of a PVC tubing with an adhesive of PVC dissolved in tetrahydrofuran or cyclohexanone (Fig.2) The other end of the PVC tubing is then connected to an appropriate glass outer casing and assembly of the ion-selective electrode completed as fully illustrated in Fig.3.

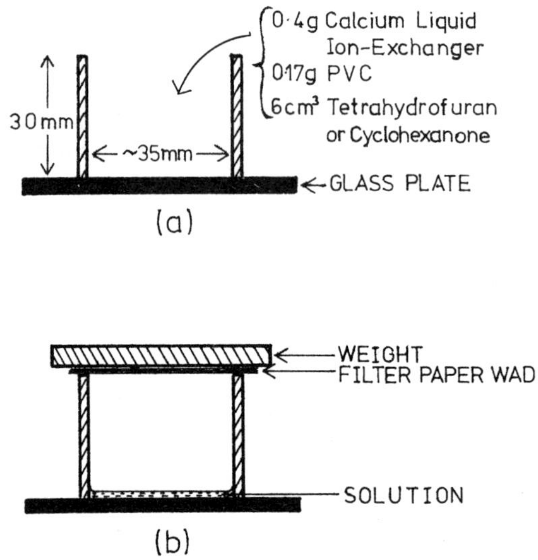


Fig.1. Casting of PVC matrix master membrane

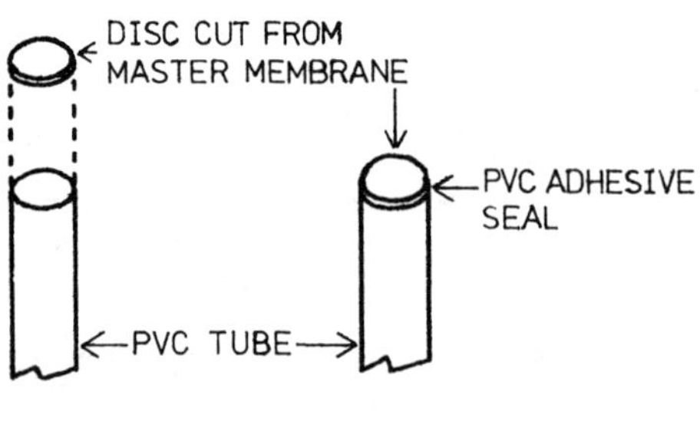


Fig.2. Mounting of PVC matrix membrane

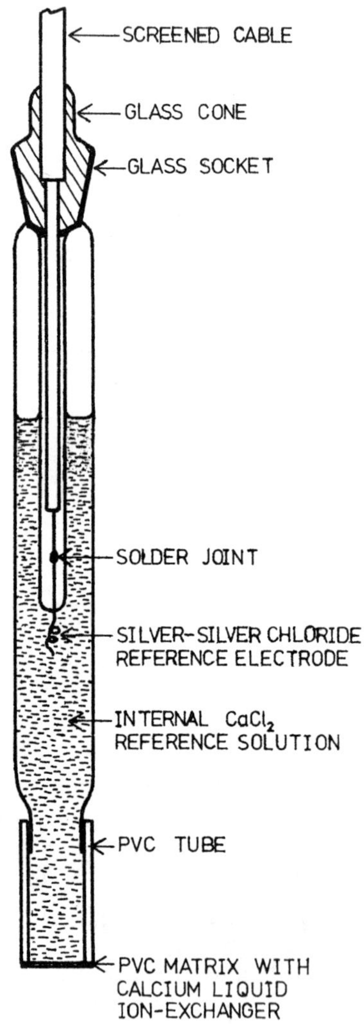


Fig.3. Longitudinal section of a PVC matrix membrane calcium ion-selective electrode

### 2.3 Adaptations and microelectrodes

The various electrodes, once assembled, can be adapted or modified for automatic analysis, for flow-through modes which accommodate samples of just a few hundred $mm^3$ [15], and in flow-injection analysis for about 50 $mm^3$ [16-18]. Small samples, down to 1 $mm^3$ can be handled by sandwiching them between a specially prepared flat surface of a reference electrode fabricated from a glass cone and socket and an ordinary calcium ion-selective electrode of the PVC matrix membrane type [19]. Alternatively, for very small samples and/or intricate measuring situations microelectrodes small enough to measure ionic activities even inside living cells have been designed [8,9]. These can be constructed by putting liquid ion-exchanger selectively sensitive to calcium ions in the tips of micropipettes, either as the liquid ion-exchanger itself [8] or by using the PVC matrix membrane principle when the liquid ion-exchanger in, say, tetrahydrofuran or cyclohexanone is put in the silanised tip of a glass micropipette of 1-2.5 μm tip diameter [9].

Coated-wire type ion-selective electrodes [20,21] dispense with the internal reference solution and consist of a PVC matrix membrane sensing system coated on a platinum or other wire. Although these are small-scale electrodes and can be used for small samples, even the finest wires still render them too large for measuring free calcium ions in cells.

There is some apprehension concerning the apparent absence of a thermodynamically reversible process at the PVC-wire interface in coated-wire electrodes [12,22]. This led to an attempted modification whereby the cleaned silver wire is anodised for one minute at 0.8 mA in 0.1 M hydrochloric acid followed by dip casting of a reference medium membrane of poly(vinyl alcohol) (10 per cent solution in $5 \times 10^{-3}$M potassium chloride) before applying the outer PVC-sensor-mediator or membrane [22]. However, the osmotic gradients caused water to flow between the test solution and the reference medium and breaking the outer membrane. The problem may be lessened by allowing minimal dehydration of the poly(vinyl alcohol) matrix prior to applying the PVC-sensor-mediator membrane [22].

There is now a view [23] that the coated-wire electrodes have an inbuilt reference system and attributed to the permeability of PVC to water and oxygen thus setting up an oxygen electrode at the platinum wire-PVC interface to function as an internal reference. Such an oxygen half-cell is also believed [24] to be responsible for the potential at the solid contact-membrane interface of electrodes where the internal reference solution has been dispensed with in favour of direct PVC membrane to silver [24], graphite [25] or teflonised graphite contacts [24] and which provide better emf stability and analytical sensitivity than silver/silver chloride contacts [24].

## 3. BASIC PRINCIPLES OF SELECTIVITY

With proper operation the calcium ion-selective electrode potential, E, with respect to a suitable reference electrode is given by

$$E = \text{constant} + S \log[a_{Ca} + k^{pot}_{CaB}(a_B)^{2/z_B}] \qquad (1)$$

According to Equation (1) there is selective response to the activity of calcium ions, $a_{Ca}$, in the presence of interfering ions, B, of charge $z_B$ and activity $a_B$; $k^{pot}_{CaB}$ is the selectivity coefficient and S the calibration slope which is usually near to the Nernstian value of 2.303 RT/2F. In the presence of additional interferents as in body and cell fluids there will be further terms within the square parentheses similar in form to the second term.

### 3.1 Selectivity parameters and interference potentials

The use of $k^{pot}_{CaB}$ defined by Equation (1) for assessing the selectivity of calcium ion-selective electrodes is consistent with IUPAC nomenclature recommendations [26], but there are other proposals of selectivity rating [27]. These include a selectivity parameter, $K_M$, derived from a consideration of interference potentials [28] and a more arbitrary parameter, $K_i$ [29].

By measuring the emf, $E_1$, of a solution containing only calcium ions and the emf, $E_2$, of a solution containing calcium ions at the same activity plus interfering ions, it is possible to deduce $k^{pot}_{CaB}$, from

$$E_2 - E_1 = \Delta E = \frac{RT}{2F} \ln[1 + k^{pot}_{Ca,B} \frac{(a_B)^{2/z_B}}{(a_{Ca})}] \qquad (2)$$

by plotting $e^{\Delta E 2F/RT}$ against $\frac{(a_B)^{2/z_B}}{(a_{Ca}}$ . This corresponds to the Srinivasan and Rechnitz [30] method for high $k^{pot}_{Ca,B}$ values (Method IIC of References [15] and [31]). $\Delta E$ may be regarded as an interference potential [27,32].

Another interference potential, $E_M$, has been defined as the difference between an $E_{ideal}$ and an $E_R$ representing, respectively, the theoretical and experimental difference between the emf of a pure calcium chloride solution and that of a mixed calcium chloride/interference monovalent cation chloride or nitrate solution [28]. The two solutions are of equal ionic strength. For the theoretical case, $E_{ideal}$ assumes no interference and also that the single-ion activity coefficient of $Ca^{2+}$ depends only on the total ionic strength, but not on composition. The $E_M$ data obtained in this way for calcium chloride solutions plus sodium chloride or lithium chloride at $<$0.6 M total ionic strength fit the equation [28].

$$E_M = - \frac{RT}{2F} \ln[1 + K_M \frac{a_B}{(a_{Ca})^{\frac{1}{2}}}] \qquad (3)$$

However, both this [28] and later work [32] show that there are several instances where Equation (3) is not obeyed, particularly for bivalent interfering ions where the plots, although usually linear, frequently do not intercept the ordinate at 1.00 and/or give negative slopes and hence negative $K_M$ values.

An arbitrary relation for calculating selectivity arises from misgivings [29] concerning the power term for $a_B$ in Equation (1):

$$E_2 - E_1 = \frac{RT}{2F} \ln[1 + K_i \frac{a_B}{a_{Ca}}] \quad (4)$$

$K_i$ is based [29] on a principle of equal affinity whereby an electrode "has an equal affinity for a primary ion and an interfering ion if the same measured potential is obtained for solutions of the primary ion and the interfering ion which have the same activities, regardless of the charges on the ions".

## 3.2 Critique of interference equations and selectivity parameters

The definitive basis of Equation (4) requires [29] an intersection between the calibration for primary calcium ions and that for interfering B ions, when $K_i = 1$. When the calcium ion calibration line is at more positive emfs than that for the interfering ions, $K_i < 1$; at more negative emfs, $K_i > 1$.

For calibrations of calcium ions and of interfering ions, respectively, which are of equal slope, the condition of equal affinity holds at all points for superposable calibrations and $K_i = 1$ over the complete calibration. This occurs for calcium/magnesium or "water hardness" electrodes based on calcium bis dialkylphosphate sensor with decan-1-ol solvent mediator [33,34]. However, when calibrations of equal slope are not superposable the definitive requirement of Equation (4) cannot be fulfilled; nevertheless, $K_i$ can still be calculated and will be a constant, being <1 for calcium ion calibrations of more positive emfs than those for interferents and >1 for calibrations of more negative emfs.

The power term, $2/z_B$, of $a_B$ in Equation (1) can have a very helpful modifying effect in those instances where $k^{pot}_{Ca,B} > 1$ for monovalent B interfering ions [34], and ensures even in these circumstances a calcium over monovalent B ion selectivity. For example, it can be easily appreciated from the relation

$$a_{Ca} = k^{pot}_{Ca,Na} (a_{Na})^2 \quad (5)$$

used for calculating $k^{pot}_{Ca,Na}$ by the mixed solution Method IIA for References 15 and 31 that as long as $a_{Na} < 1$ electrodes can usefully be used for calcium ion determinations even for $k^{pot}_{Ca,Na} > 1$. Thus, when $k^{pot}_{Ca,Na} = 46$, determined in a $10^{-2}$ M sodium chloride background, the calcium ion-selective electrode still responds down to $4.6 \times 10^{-3}$ M calcium ions with little interference from the background sodium ions [34].

Calibration slopes of an electrode for B ions frequently deviate from Nernstian and calibrations may even be curved as in the case of magnesium ions for the first PVC matrix membrane ion-selective electrode based on calcium bis didecylphosphate sensor with di-n-octylphenylphosphonate solvent mediator [5] where the slope gradually increased from 14.2 to 23.6 mV $decade^{-1}$. No allowance is made for such features in Equation (1) so that $k^{pot}_{Ca,B}$ is not constant, but the significance of this is relatively unimportant in the recommendation [26,35] for quoting background interferent level relating to each $k^{pot}_{Ca,B}$.

There are, however, problems when applying graphical-type relations such as Equations (1-4) over extended ranges. Any fundamental physicochemical deduction made from such equations need to be cognizant of slope changes that must be related to rather more complicated features of membrane/solution interface ion-exchange and membrane diffusion processes than would be the case for constant calibration slopes [27]. Midgley [36] has developed procedures for analysing non-Nernstian responses of ion-selective electrodes arising from the presence of determinant in reagents, the presence of interfering species in test solutions and the solubility of the electrode material. Other factors may also be involved and Midgley recognises that with liquid ion-exchange systems, the time taken to reach equilibrium may be so long in the presence of an interferent that the practical limit of detection is set by kinetic rather than thermodynamic factors. Studies of the permeation of ions through calcium ion-selective membranes seem to confirm this view [37,38] although the permeation studies with the calcium liquid ion-exchanger trapped in PVC were for self-diffusion. The only studies under applied potential are those on 0.01 cm thick PVC membranes containing a neutral carrier calcium ion-sensir and 2-nitrophenyloctylether solvent mediator [39]. Rather high voltages of up to about 25 V were applied to the 0.5 cm diameter membrane discs separating two electrolyte compartments containing calcium chloride, sodium chloride, potassium chloride and potassium thiocyanate as appropriate [39]. In the various current-voltage curves obtained steady-state currents were attained after 5 to 15 min and the associated selectivity in ion transport (permeability selectivity) correlated with potentiometric ion selectivity coefficients [39].

### 3.3 Some practical conclusions

Because of the likelihood of non-Nernstian slopes or curvatures in calibrations of ion-selective electrodes for interfering ions, linear graphs emanating from Equations (2) and (3) are frequently just a coincidence rather than a regular feature. This presents difficulties in presenting data of practical value. In any case, the nature of $k^{pot}_{Ca,B}$ is such that its values for B ions of different valencies are not comparable and, of course, for practical purposes, it is essential for $k^{pot}_{Ca,B}$ data to be accompanied by exact details on their evaluation [26,35]. In this way, $k^{pot}_{Ca,B}$ is a reasonably satisfactory parameter for expressing the practical scope of calcium ion-selective electrodes, although the at-a-glance plots, illustrated by Fig.4 for electrodes containing Orion 92-20-02 calcium liquid ion-exchanger and calcium bis di-4-n-octylphenylphosphate plus di-n-octylphenylphosphonate in PVC matrix membrane are the most instructive [40].

### 3.4 Ionic strength and complexation effects

Although this review is confined to principles of calcium ion-selective electrode design, and an appreciation of the closely related selectivity coefficient parameters, matters concerning the effect of ionic strength and complexation are more marginal and, in any case are deserving of separate treatment. Indeed, the presence

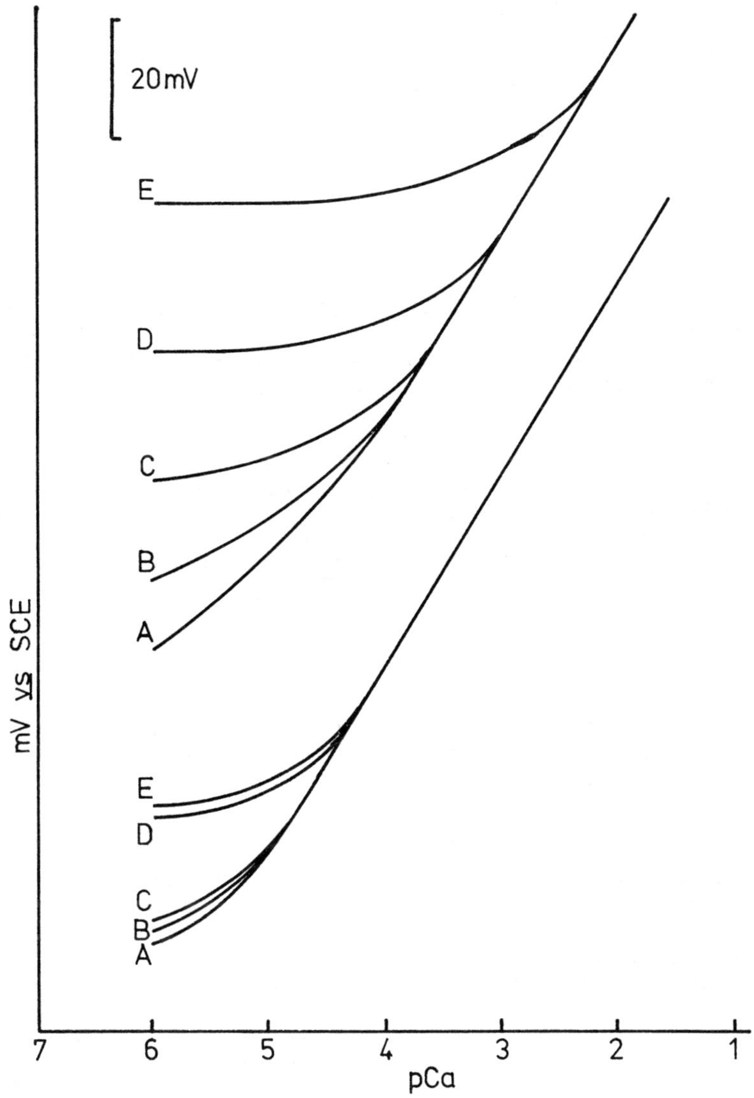


Fig.4. Calibration in calcium chloride standards, of calcium ion-selective electrodes containing Orion 92-20-02 calcium liquid ion-exchanger (top) and calcium bis di-4-n-octylphenylphosphate plus di-n-octylphenylphosphonate (bottom) trapped in PVC membranes. A: calibration with calcium chloride standards; B to E: calibration with calcium chloride standards in background metal chloride solutions of 0.5 M $K^+$ (B), $Na^+$ (C) and $Mg^{2+}$ (D), and of 1 M $Na^+$ (E), respectively. (Data: Craggs, Moody and Thomas)

of complexing agents within the sample, unless introduced either deliberately or unintentionally, is generally of no concern, since the electrodes respond to the activity of free calcium ions, that is, they will not sense bound calcium. It is for this reason that calcium ion-selective electrodes may be used for studying complex formation equilibria [41,42]. Conversely, formation constants of complexes can be used as a basis for deducing the free calcium ion level in calcium ion buffers and for calibrating calcium ion-selective electrodes to very low levels of free calcium ions as discussed briefly in Section 6.

Care must be taken over the effect of ionic strengths of sample and calibrating solutions on the relation between calcium ion activity and concentration [15,43]. Ideally, calcium ion-selective electrodes used for measuring free calcium ions in body and cell fluids, or in any other samples, should be calibrated in standards simulating such fluids.

## 4. SOLVENT MEDIATORS

Considerable effort has been devoted to establishing the relative roles of sensors and solvent mediators in calcium ion-selective electrodes. Clearly, the solvent mediator has a function for dissolving the sensor. Additionally, there are roles in adjusting the

(i) ultimate relative permittivity of the final organic phase;

(ii) mobility of the ion-exchanger/sensor sites according to the viscosity of the solvent mediator; and

(iii) site density by variation of the concentration of the sensor and possibly affecting the tendency towards the dimerisation observed in certain dialkylphosphate sensors [6,12,44].

### 4.1 Solvent mediators for phosphate sensors

Weakly coordinating or non-coordinating solvents do not produce functional liquid-exchanger ion-selective electrodes based on dialkylphosphate sensors [45]; only systems incorporating coordinating solvents such as alcohols, phthalate esters, trialkyl-phosphates and certain phosphonates appear to be functional[34,45,46]. This implies that the solvent mediator not only dissolved the sensor but also partly solvates the metal ions [46]. Thus, the solvent mediator may affect electrode selectivity by influencing the formation constants and partition coefficients which define the ion-exchange between the calcium species in the liquid membrane and the other cations in the sample solution [46,47]. The ability of trialkyl-phosphates to respond to calcium ions in the absence of sensor [47] can well be due to traces of dialkylphosphate, but a more interesting observation is that the longest linear response ranges (to approximately $10^{-5}$M) are obtained for the highest homologues, probably because of greater immiscibility with water [47].

Whatever the underlying reasons for the effect of solvent mediators on calcium ion-selective electrode selectivity, the various liquids are associated with significant differences in selectivity. This is well illustrated by the loss of calcium ion-selectivity by PVC matrix membrane electrodes based on calcium bis di-2-ethylhexyl-phosphate sensor when the di-n-octylphenylphosphonate solvent mediator is replaced by decan-1-ol (Table 1) [34]. The decan-1-ol mediator sensor constitutes the basis of divalent ion or "water hardness" electrodes, but because the alcohol is not a plasticiser the system does not function too well in PVC matrix membranes.

TABLE I. Selectivity coefficient($k^{pot}_{Ca,B}$) data for PVC matrix membrane ion-selective electrodes based on a sensor of calcium bis di-2-ethylhexylphosphate sensor [34]

(Interferent level is 0.001M in all cases)

| Solvent mediator | Di-n-octylphenyl-phosphonate | Di-n-octylphenyl-phosphonate + decan-1-ol(1+1) | Decan-1-ol |
|---|---|---|---|
| $k^{pot}_{Ca,Mg}$ | 0.057 | 0.52 | 1 |
| $k^{pot}_{Ca,Sr}$ | 0.040 | 0.17 | 0.5 |
| $k^{pot}_{Ca,Ba}$ | 0.048 | 0.17 | 0.5 |
| $k^{pot}_{Ca,Mn}$ | 0.28 | 1 | 2 |
| $k^{pot}_{Ca,Cu}$ | 1.7 | 0.66 | 2 |
| $k^{pot}_{Ca,Ni}$ | 0.18 | 0.26 | 2 |

That different solvent mediators tend to produce electrodes with different selectivity properties is demonstrated by the studies of Garbett and Torrance [47] on alcohols and tri-n-alkylphosphates. The phosphates gave calcium ion-selective electrodes, while isomeric octanols showed copper ion-selectivity and the n-alcohols, as in the decan-1-ol data summarised in Table I, showed no predominant cation selectivity.

Occasionally selective ion response is obtained in the absence of solvent mediator. This happens for calcium ion response with the free acid or the calcium salts of di-2-ethylhexylphosphoric acid and di-n-octylphosphoric acid where the materials themselves can plasticise PVC [48]. However, good distribution of the sensor in the membrane is desirable for the best quality electrodes and the solvent mediator provides the vehicle for this purpose [6]. Of the various studies and experiences of possible solvent mediators that can be used with dioctylphenylphosphate and dialkylphosphate sensors for calcium ion-selective electrodes, di-n-octylphenylphosphonate has to be acclaimed as superior, although tri-n-pentylphosphate has recently [49] been shown to be almost as good.

The solvent interactions that cause changes in selectivity are finely balanced. Tri-n-pentylphosphate appears to be the best of the trialkylphosphates [49] and di-n-octylphenylphosphonate is among a long series of dialkylphenylphosphonates recently synthesized [50] for examination [40], where alkyl = pentyl, hexyl, deptyl, octyl, nonyl, decyl and undecyl. Electrodes from the first and last-named degrade slightly more quickly than the others [39]. Both tri-n-pentylphosphate and di-n-octylphenylphosphonate are available commercially in sufficiently pure form for direct use in calcium ion-selective electrode fabrication. Nevertheless, it will be helpful to readers to know that full details for the synthesis of dialkylphenylphosphonates, starting with phenylphosphonic dichloride and the appropriate alkan-1-ol, are described in the literature [50-52]. Both gel permeation chromatography [51] and distillation [52] methods have been described for product purification. The latter is preferred for quantity preparation, but high vacuum must be employed in order to reduce the chance of decomposition.

A possible useful characteristic of di-n-octylphenylphosphonate is its high viscosity (16.7 mN s $m^{-2}$), a factor that appears to be relevant for aromatic nitro solvent mediators in giving long life barium ion-selective electrodes [53]. However, this cannot be the sole reason in view of the utility in calcium ion-selective electrodes of tri-n-pentylphosphate (viscosity = 4.64 mN s $m^{-2}$) [49]. It is, however, interesting that the permittivities of di-n-octylphenylphosphonate ($\varepsilon$ = 6.2) and tri-n-pentylphosphate ($\varepsilon \simeq 7$) are similar and that the permittivities of other suitable trialkylphosphate solvent mediators for calcium ion-selective electrodes based on calcium bis di-[4-(1,1,3,3-tetramethylphenyl)phosphate]sensor fall in the 5.0-8.3 range [49].

No significant advantage accrues on nitrating the phenyl group of di-n-octylphenylphosphonate [54], but see Section 5.1.1.

## 4.2 Solvent mediators for neutral carrier sensors

The general preference of neutral carrier ligands for divalent cations over monovalent cations is associated with solvent mediators of high relative permittivity [58-59]. For example, $k^{pot}_{Ca,Na}$ for a PVC matrix membrane calcium ion-selective electrode based on a neutral carrier sensor containing two ether oxygen atoms between the carbonyl oxygen atoms of two amide groups, similar to the situation in the molecule of Fig. 5, is about 10 in dibutyl sebecate ($\varepsilon \sim 4$) solvent mediator and <0.01 in 4-nitroethylbenzene ($\varepsilon \sim 24$) [55,56]. More recently, 2-nitrophenyl-n-octylether ($\varepsilon \sim 24$) has been used as a high permittivity solvent mediator for promoting divalent over monovalent cation selectivity of synthetic neutral carriers [57-59], although it is possible that 2-nitrophenylphenyl ether ($\varepsilon \sim 28$) will be just as useful.

Fiedler [59] examined the selectivities of PVC calcium ion-selective electrodes comprising the same neutral carrier sensor and various mixtures of two mediators, for example, di(2-ethylhexyl) adipate or dibutylsebacate and 2-nitrophenyl-n-octyl ether with permittivities ranging from 4 to 23.6. For divalent ions, $k^{pot}_{Ca,B}$ decreased with increasing polarity of the mediator solvent mixture.

The pattern concerning the high permittivity solvent mediators could not be established with calcium ion-sensors based on calcium ion complexes with polypropylene glycol (PPG), largely because of poor and erratic electrode response with many of the solvent mediators [60]. However, calcium ion-selectivity was quite prominent when sensors of the PPG complexes were tested with the lower permittivity di-n-octylphenylphosphonate and

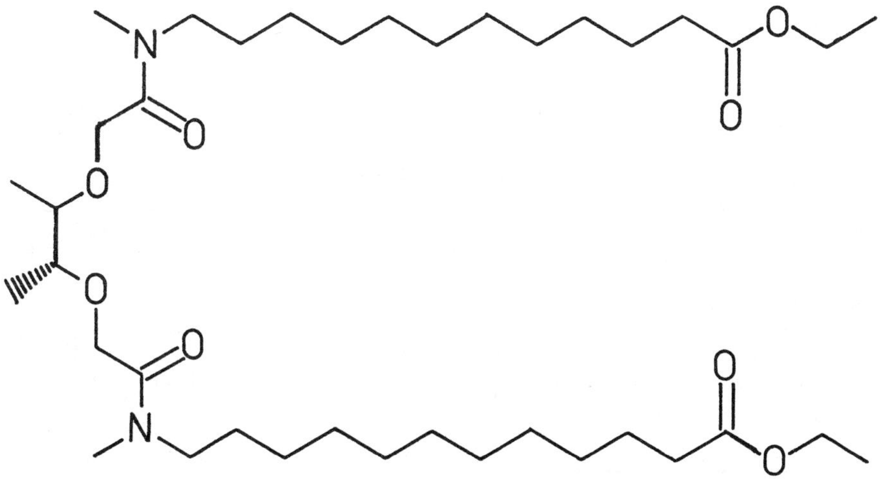

Fig.5 Structure of N,N'-di[(11-ethoxycarbonyl)undecyl]-N,N'-4,5-tetramethyl-3,6-dioxaoctane diamide used as a neutral carrier

tri-n-butylphosphate so that not quite the same rules would seem to apply.

### 4.3 Summarising conclusion

Di-n-octylphenylphosphonate and possibly tri-n-pentylphosphate are the best solvent mediators with calcium bis dialkylphosphate and bis dialkylphenylphosphate sensors, while 2-nitrophenyl-n-octyl ether and maybe 2-nitrophenylphenyl ether are the best with synthetic neutral carrier ligand sensors of the type illustrated in Fig.5.

## 5. SYNTHETIC DESIGN OF CALCIUM ION-SENSORS

Although solvent mediators have a role in calcium ion-selective electrodes, it is the sensors that are the heart and soul of the devices. The calcium bis dialkylphosphate sensors raised the hopes of biochemists and cell biologists, the most notable being calcium bis didecylphosphate. The acid, namely didecylphosphoric acid, may be prepared by reacting phosphorus oxide trichloride with decan-1-ol, followed by alkaline hydrolysis, acidification and extraction stages [51,61]. The calcium salt may then readily be obtained by reacting an aqueous ethanolic solution of the acid with calcium hydroxide at pH >9 [51,62].

## 5.1 Calcium bis di-4-octylphenylphosphate sensors

Despite their usefulness, calcium bis dialkylphosphate sensors had shortcomings including minima or dips [33] in emf *vs* pH curves leading to pH ranges of just *ca* 5.5 to 9, where the electrodes are insensitive to pH changes. The exact range depends on the calcium ion levels in samples with low pCa being associated with an expected reduced pH range. Růžička, Hansen and Tjell [12] suggested that increasing the electrophilic character of the alkyl chain would give a more labile hydrogen atom, thereby shifting the region of the acid form to lower pH values and hence enhancing the pH selectivity range at the acid end. Replacing the alkyl groups of dialkylphosphoric acid with either 4-n-octylphenyl [12] or 4-(1,1,3,3-tertamethylbutyl)phenyl [9,41,49,54] groups does indeed provide calcium ion-selective electrodes for which the characteristic dips in the pH *vs* emf plots shift by approximately one pH unit towards the acid end and leading to useful pH ranges of *ca* 4 to 9.5 [40,50]. However, the real benefit lies in the much improved selectivity, especially for sodium (see Section 5.1.2).

5.1.1 Effect of nitrating 4-octylphenylphosphate. Nitrating the phenyl groups of di-4-octylphenylphosphates leads to some interesting observations, although no further significant advantage is gained for calcium ion-sensing purposes. Thus, calcium bis di-4-(1,1,3,3-tetramethyl)-2-nitrophenylphosphate sensor with di-n-octyl-(3-nitrophenyl)phosphonate solvent mediator does not give a dip in the pH interference curve [27] and neither is the dip observed with the mediator in conjunction with calcium bis di-4-nitrophenylphosphate. However, in the latter case there is a loss of calcium over sodium ion-selectivity and the sensor is water-soluble [54].

The slight dip in the pH curve with the lower permittivity solvent mediators, like di-n-octylphenylphosphonate or tri-n-pentylphosphate in conjunction with the nitrated sensor becomes more prominent with the non-nitrated di-4-(1,1,3,3-tetramethylbutyl)phenyl phosphate sensor [27]. This seems to suggest that the existence of dips in pH interference curves for phosphate ester sensors of calcium ions are related to the diester acid strength. Thus, the sensors of high $pK_a$ with their strong affinity for protons exhibit dips, whilst the sensors of low $pK_a$ are less likely to do so, particularly with solvent mediators of sufficiently high permittivity as in di-n-octyl-3-nitrophenylphosphonate [27].

Jagner and Østergaard-Jensen [63] recorded a rise of about 0.4 pH at the alkaline end for a calcium ion-selective electrode with a sensing membrane of calcium bis di-(4-n-octyl-2-nitrophenyl) phosphate plus di-n-octylphenylphosphonate solvent mediator. Deviations at high pH are in any case rather less evident and despite a small dip at pH *ca* 9, electrodes based on dialkylphosphates and di-4-octylphenylphosphates can be used for calcium ion activity measurements at pH *ca* 11 provided the pH is kept constant [40]. The positions of calibration curves under these conditions will be several mV lower than for the normal 'usable' pH range [40,64].

Dinitration of the phenyl group does not yield the expected phosphorylation to give di-4-(1,1,3,3-tertamethylbutyl)-2,6-dinitrophenylphosphate [62] and purported calcium ion-selectivity of the product [54] is attributable to calcium bis di-4-(1,1,3,3-tetramethylbutyl)-2,6-dinitrophenate [65]. Calcium ion-selectivity of the phenate is slightly inferior to that of the phosphate, especially in the presence of barium ions [65].

Finally, it is interesting that calcium ion-selective electrodes based on calcium bis di-4-(1,1,3,3-tetramethylbutyl)-2-nitrophenylphosphate sensor with nitrated and non-nitrated di-n-octylphenylphosphonate solvent mediator recover quickly compared to the long times required for the non-nitrated sensors after exposure to zinc ions [27]. Unfortunately, however, the calibration ranges are shorter for the nitrated sensor.

5.1.2 Low interference effects. An important additional advantage of calcium bis di-4-octylphenylphosphate sensors in calcium ion-selective electrodes is the low interference from physiologically important cations, like sodium, potassium and magnesium. Table II shows that selectivity coefficients, $k^{pot}_{Ca,B}$, for calcium ion-selective electrodes based on these sensors are much lower than the corresponding coefficients for the more traditional electrodes based on Orion 92-20-02 calcium liquid ion-exchanger. The tremendous advantage of the alternative 4-octylphenylphosphate sensor is clearly demonstrated by Fig.4 (bottom curves). This emphasises that unlike the system illustrated by the top set of curves, these calcium ion-selective electrodes can be used to routinely monitor calcium ion levels at 1 ppm (ca 2.5 x $10^{-5}$ M) or less in the presence of high concentrations of foreign cations such as are encountered in physiological and biological fluids [40,41].

TABLE II. Selectivity coefficient ($k^{pot}_{Ca,B}$) data for PVC matrix membrane calcium ion-selective electrodes [41]

| Electrode Concentration of ion B /M | | Orion 92-20-02 | HDOPP[a] + DOPP[b] | HDTMBPP[c] + DOPP[b] |
|---|---|---|---|---|
| $5 \times 10^{-2}$ | ($Na^+$) | $3.1 \times 10^{-2}$ | $1.1 \times 10^{-3}$ | $1.0 \times 10^{-3}$ |
| 1.0 | ($Na^+$) | $1.0 \times 10^{-3}$ | $9.7 \times 10^{-6}$ | $4.7 \times 10^{-6}$ |
| $5 \times 10^{-2}$ | ($K^+$) | $9.0 \times 10^{-3}$ | $1.0 \times 10^{-3}$ | $4.6 \times 10^{-4}$ |
| $5 \times 10^{-2}$ | ($Mg^{2+}$) | $1.2 \times 10^{-2}$ | $4.9 \times 10^{-4}$ | $3.3 \times 10^{-4}$ |

a HDOPP : di-4-n-octylphenylphosphate

b DOPP : di-n-octylphenylphosphonate

c HDTMBPP : di-4-(1,1,3,3-tetramethylbutyl)phenyl phosphate

The last column of Table II shows that a PVC matrix membrane calcium ion-selective electrode with calcium bis di-4-(1,1,3,3-tetramethylbutyl)phenylphosphate sensor plus di-n-octylphenylphosphonate solvent mediator is comparable to that based on calcium bis di-4-n-octylphenylphosphate with di-n-octylphenylphosphonate as reported elsewhere [9,49,54]. It is noteworthy that di-4-(1,1,3,3-tetramethylbutyl)phenylphosphoric acid is far more accessible than its n-octyl isomer [66], because the starting material, 4-(1,1,3,3-tetramethylbutyl)phenol is much more readily available commercially or can conveniently be synthesised from phenol and 2,4,4-trimethylpent-1-ene [67].

5.1.3 Synthesis. The di-4-(1,1,3,3-tetramethylbutyl)phenylphosphoric acid is prepared by treating a mixture of pyridine and phosphorus oxide trichloride in anhydrous diethyl ether dropwise with 4-(1,1,3,3-tetramethylbutyl)phenol in diethyl ether [66]. Following reflux treatment, there are the separation, alkaline hydrolysis, acidification and extraction stages with calcium salts of the mono- and diesters being obtained by reacting an aqueous methanolic solution of the acid with calcium hydroxide [66]. The diester is separated from monoester by reconverting the mixed calcium salt batch to the free acids with hydrochloric acid, dissolved in benzene and extracted several times with ethane-1,2-diol in order to remove the mono-ester. The benzene layer after washing with water is evaporated to yield the di-4-(1,1,3,3-tetramethylbutyl)-phenylphosphoric acid which may be treated in aqueous methanol with calcium hydroxide in order to obtain the calcium salt [66].

The synthesis of di-4-n-octylphenylphosphoric acid is similar, but as stated above the 4-n-octylphenol starting material is not as readily available as its branched-chain isomer. Synthesis of the phenol is tedious and involves reacting octanoyl chloride and phenol to give phenyl n-octanoate which is then subjected to a Fries rearrangement to 4-hydroxy-n-octanophenone and a Clemmensen reduction in order to obtain the 4-n-octylphenol [66].

## 5.2 Neutral carrier calcium ion-sensors

Calcium ion-sensors involving a macrocyclic siloxane, namely, tetracosamethylcyclododecasiloxane [$(Me_2SiO)_{12}$], usable between pH 1.6 and 5, and the tetraphenylborate salt of a calcium ion complex with propylene glycol, usable between pH 4.5 and >10 have been described by Pickering [68] and Jaber, Moody and Thomas[60], respectively. However, compared with the above calcium bis di-4-octylphenylphosphate sensors these have inferior selectivity to calcium ions over sodium, potassium and magnesium [60,68].

The carboxylic acid antibiotic designated A23187 and extracted from *Streptomyces chartreusensis*, together with nitrobenzene solvent, has been evaluated as a calcium ion sensor but is subject to considerable interferences from even low levels of sodium and potassium [69].

5.2.1 N,N'-di[(11-ethoxycarbonyl)undecyl]-N,N'-4,5-tetramethyl-3,6-dioxooctane diamide calcium ion-sensor. The most tenable of the synthetic neutral carrier calcium ion-sensors are those based on the structure of Fig.5 [10,11,59,70-72]. Indeed, that shown in Fig.5 is included by Morf and Simon [73] in their short-list of the five most attractive synthetic carrier molecules for sensing metal cations, the other four having been selected for their selectivity towards $Li^+$, $Na^+$ and $Ba^{2+}$, as appropriate. The carrier used is a mixture of components of different configurations at the tertiary carbon atoms, namely, 70 mole per cent meso-form and 30 per cent racemate [71,72].

The synthetic carrier (Fig.5) when used in PVC matrix membranes in conjunction with 2-nitrophenyl-n-octyl ether solvent mediator seems to be about as selective a calcium ion-sensor as are the di-4-octylphenylphosphate systems referred to in Table II. Thus, electrodes based on the carrier have $k^{pot}_{Ca,B}$ of $10^{-2}$, $6 \times 10^{-3}$ and $2 \times 10^{-4}$ when B = $Na^+$, $K^+$ and $Mg^{2+}$, respectively, in the presence of backgrounds of $10^{-2}$M aqueous solutions of the metal chlorides [71]. The di-4-octylphenylphosphate type electrodes generally do not pick up any interference arising from $Na^+$ and $K^+$ in the presence of background $10^{-2}$M sodium and potassium chlorides [49,54,64], hence the nearest comparable mixed solution data in Table II are for the much higher background of $5 \times 10^{-2}$M metal chlorides. There are occasions when the phosphate-type electrodes do not even pick up $Na^+$ interference at $5 \times 10^{-2}$M background sodium chloride [54]. Table III summarises some literature $k^{pot}_{Ca,B}$ data for the newer, high quality calcium ion-sensors.

TABLE III. $k^{pot}_{Ca,Na}$ data for calcium ion-selective electrodes

| Sensor | $k^{pot}_{Ca,Na}$ | $Na^+$ level /M | Limit ofuseful $Ca^{2+}$ range (from Equ.5)/M | Reference |
|---|---|---|---|---|
| HDTMPP + DOPP[a] | 0.017 | 0.05 | $4.3 \times 10^{-5}$ | 49 |
| | 0.01 | | $2.5 \times 10^{-5}$ | 27 |
| | 0.001 | | $2.5 \times 10^{-6}$ | 41 |
| | [b] | | $4.6 \times 10^{-6}$ [c] | 54 |
| Neutral carrier sensor[d] + NPOE[e] | 0.01 | 0.01 | $1.0 \times 10^{-6}$ | 71 |

a: HDTMBPP = Di-4-(1,1,3,3-tetramethylbutyl)phenylphosphate
b: Sodium ion interference was not measurable
c: Calibration limit
d: Sensor depicted in Fig.5
e: 2-nitrophenyl-n-octyl ether

An independent attempt has been made [40] to assemble calcium ion-selective electrodes with PVC matrix membranes containing N,N'-di[(11-ethoxycarbonyl)undecyl]-N,N'-4,5-tetramethyl-3,6-dioxaoctane diamide sensor with 2-nitrophenyl-n-octyl ether solvent mediator. Although the electrodes gave good calibration lines of slope 29.7 mV $decade^{-1}$ and good detection limits of $<10^{-6}$M, they quickly deteriorated, especially during pH assessment and could not afterwards be used for selectivity assessments [40]. This is surprising in view of the pH stability over the ~4 to >10 range previously reported [58,72].

In view of the utility of lipophilic anions for generally improving electrode quality and overcoming anion interference [58,73,74], an unexpected observation [40] was the rather noisy electrode response obtained when sodium tetraphenylborate was incorporated in the membranes. The small quantity of sensor in the electrode membranes may be responsible for the disappointing behaviour of this, hitherto, well-praised [58,72-73] neutral carrier calcium ion-sensor. This was enforced by scarcity of the material which is based on a considerably more difficult synthesis procedure than for the apparently equally satisfactory di-4-octylphenyl-phosphate calcium ion-sensors discussed above. On the matter of introducing tetraphenylborate into the membrane, it would be worth-while exploring the facility of preparing the tetraphenylborate salt of the calcium-neutral carrier complex prior to incorporation in the membrane, and as already used for poly(alkyleneoxy) materials [53,60]. However, because of the likely water insolubility of the more lipophilic calcium neutral carrier material the procedure cannot be as straightforward.

5.2.2 Synthesis. Experience in synthesising the neutral carrier sensor of Fig.5 has yielded a product to match a gift from Professor W.Simon regarding various items of specification, including infrared and nmr spectroscopy and thin layer chromatography on silica gel with benzene/acetone (1:2) [40]. The synthesis commences with a multistage procedure of preparing an acid chloride when the appropriate glycol is converted to a diester by treatment with ethyl diazoacetate (stage A), the ester hydrolysed (stage B) and the resulting acid treated with thionyl chloride (stage C) to give the acid chloride [71]:

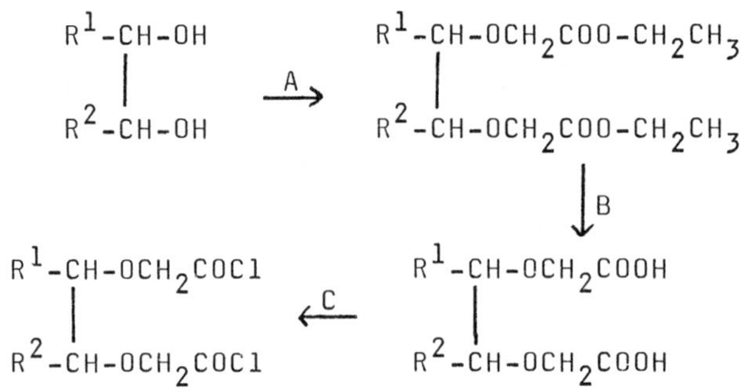


In the next and final stage the acid chloride is converted to a diamide which in the case of the structure in Fig.5 involves treatment with ethyl 12-methylaminolaurate previously prepared from the acid [71]. The crude product is purified, for example, by silica gel column chromatography [40].

## 6. CALIBRATIONS WITH CALCIUM ION BUFFERS

The use of calcium ion buffer solutions for calibrating calcium ion-selective electrodes has been commended by Růžička, Hansen and Tjell [12] and adopted by others [9,40,41,72]. Fig.6 illustrates such a calibration with calcium ion buffers based on NTA and EDTA to *ca* $10^{-8}$M free calcium ions [40] using an electrode based on calcium *bis* di-4-n-octylphenylphosphate sensor. It is noteworthy that the calcium buffer calibration extrapolates to that for serial dilution of calcium ion standards (triangular points). Fig.6 suggests that although the conditional stability constants for a particular system may be unknown, a calcium ion-selective electrode immersed in the solution could give a direct indication of free calcium ion activity. Experiments with carboxylates [41,42] and phosphates [64] confirm this.

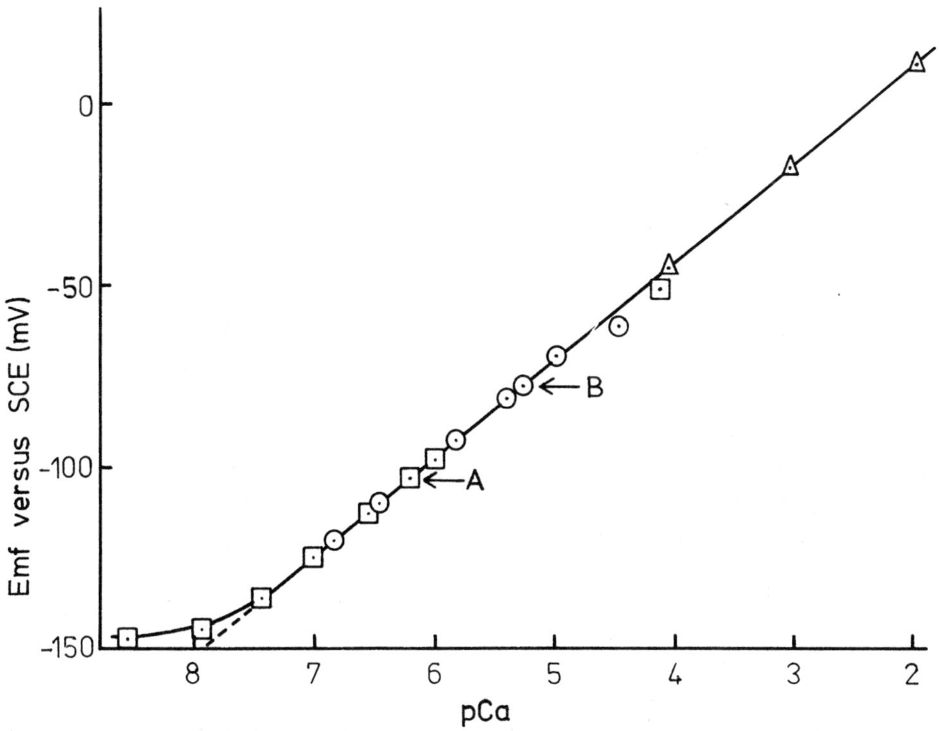


Fig.6. Illustration of a calibration of a calcium ion-selective electrode at 25°C with serial dilution calcium ion standards and calcium ion buffers, each made up to 0.1M ionic strength with sodium chloride. Electrode membrane: calcium bis di-4-n-octylphenylphosphate and di-n-octylphenylphosphonate in PVC. Δ: normal serial dilution of calcium chloride standards: ʘ: $Ca^{2+}$:NTA (1:5) buffered standards for log $\beta_1$ = 6.4; ⊡ : $Ca^{2+}$:EDTA (1:5) buffered standards for log $\beta_1$ = 10.9.

In view of the wide interest in measuring low levels of free calcium ions the method of using calcium ion buffers for electrode calibration is illustrated with reference to Points A and B in Fig.6, based on the buffering solutions with EDTA and NTA, respectively, and for which the relevant data [40] are summarised in Table IV.

TABLE IV. Data for calculating pCa for Points A and B in Fig.6.

| Item | Point A (EDTA) | Point B (NTA) |
|---|---|---|
| $\log \beta_1$ | 10.9 | 6.4 |
| $pK_a$ (proton loss from hydrolysed $Ca^{2+}$) | 12.6 | 12.6 |
| Solution pH | 5.65 | 8.04 |
| $\alpha_M$ | 1 | 1 |
| $pK_1$ | 10.34 | 9.81 |
| $pK_2$ | 6.24 | 2.57 |
| $pK_3$ | 2.75 | 1.97 |
| $pK_4$ | 2.07 | - |
| [Ca] taken/M | $10^{-3}$ | $10^{-3}$ |
| [L] taken/M | $5 \times 10^{-3}$ | $5 \times 10^{-3}$ |
| [CaL] (approx)/M | $10^{-3}$ | $10^{-3}$ |
| $[L]_T$/M | $4 \times 10^{-3}$ | $4 \times 10^{-3}$ |
| $\alpha_L$ | $10^{5.38}$ | $10^{1.77}$ |
| $\log \beta'_1$ | 5.52 | 4.63 |
| Calculated pCa | 6.12 | 5.23 |

To detail the method by which the data of Table IV were obtained it is necessary to consider a solution containing a metal ion, M, in this case calcium, and a chelating agent or ligand, L, in this case EDTA or NTA, with complexes being formed according to

$$M + nL \rightleftharpoons ML_n$$

By expressing each equilibrium in terms of the appropriate stability constant, $\beta_n$, pM may be obtained:

$$pM = \log \beta_n + \log \frac{(aL)^n}{aML_n} \qquad (6)$$

where 'a' denotes the species activity. If the ionic strengths of the solutions do not vary too widely and provided that $\beta_n$ is replaced by the corresponding concentration constant, concentrations may be used instead of activities [75-77]. When the complexing agent is present in excess, almost all of the metal ion is bound in a complex and the concentration of free metal ions is buffered.

Calculation of pM for metal ion buffers containing an excess of ligand is based on the principles of conditional stability constants and side reaction coefficients [75,76]. The most important side reaction is protonation of the ligand to form species with little or no complexing ability. Hence, the pH of the solution and $pK_a$ values of the ligand must be known as well as the stability constants of the metal complexes. If the metal ion undergoes significant hydrolysis the appropriate constants also need inclusion. Thus, "apparent" or "conditional" stability constants, $\beta'_n$, are defined in terms of the equilibrium between a metal complex and its components, except that the free ligand concentration is replaced by the total concentration of all ligand species not bound to the metal and the free metal ions term includes hydrolysed metal ions and metal ions bound to other complexing species. The relation is

$$\beta'_n = \frac{\beta_n}{\alpha_M \cdot (\alpha_L)^n} \qquad (7)$$

where $\alpha_M = ([M] + [MOH] + [M(OH)_2] + \ldots\ldots\ldots\ldots)/[M]$

and $\alpha_L = ([L] + [HL] + [H_2L] + \ldots\ldots\ldots\ldots\ldots\ldots)/[L]$

In the presence of excess strong ligand it is usually sufficient to consider only the formation of mononuclear hydroxy complexes [77]. Hence, $\alpha_M$ reduces to

$$\alpha_M = 1 + 10^{(pH - pK_1)} + 10^{(2pH - pK_1 - pK_2)} + \ldots\ldots \qquad (8)$$

where $pK_1$, $pK_2$ .... are the successive $pK_a$ values for the loss of a proton from a hydrated metal ion. Similarly,

$$\alpha_L = 1 + 10^{(pK_1 - pH)} + 10^{(pK_1 + pK_2 - 2pH)} + \ldots\ldots \qquad (9)$$

where pK values here are for the successive loss of protons from HL, $H_2L$, etc.

For a total metal ion concentration, $[M]_T$, then if most of the metal ions are present as complex, $ML_n$, and the total concentration of ligand species not complexed to the metal is $[L]_T$, it follows from the definition of $\beta'_n$ that

$$\beta'_n \approx \frac{M]_T}{(\alpha_M [M] [L]^n_T)} \qquad (10)$$

which rearranges to

$$pM \approx \log \beta'_n + n \log [L]_T + \log \alpha_M - \log[M]_T \qquad (11)$$

Equation (11), if desired, in conjunction with a simple computer program, can be used for calculating pCa as a function of pH for systems based on NTA and EDTA ligands, values of the ligand protonation constants and metal complex stability constants being obtained from the literature [78] along with the pK value of 12.6 for the mononuclear hydrolysis of calcium ions [77].

Through the use of solutions of constant ionic strength (0.1M for the calibration of Fig.6), in which the activities are directly proportional to the concentration of calcium ions, an operational pCa scale may be defined. With the corresponding emf values for a cell incorporating a calcium ion-selective electrode in solutions of various pCa, calibration points as for the circles and squares of Fig.6 may be obtained.

## 7. CONCLUSION AND RECOMMENDATION

Bearing in mind all factors, including pH range, freedom from interference especially by sodium, potassium and magnesium, and ease of synthesis and availability of materials, a calcium ion-sensor that truly merits attention by scientists and especially by cell biologists, physiologists and biochemists in general is either calcium bis di-4-(1,1,3,3-tetramethylbutyl)phenylphosphate or its isomer calcium bis di-4-n-octylphenylphosphate used in conjunction with di-n-octylphenylphosphonate as solvent mediator. The development of these sensors stems from the work of Růžička, Hansen and Tjell [12] on 4-n-octylphenylphosphate as a calcium ion-sensor, but used at the time in an unpurified form. Fig.7 demonstrates that purification can give improvements in performance to match calcium ion-selective electrodes made from a neutral carrier calcium ion-sensor.

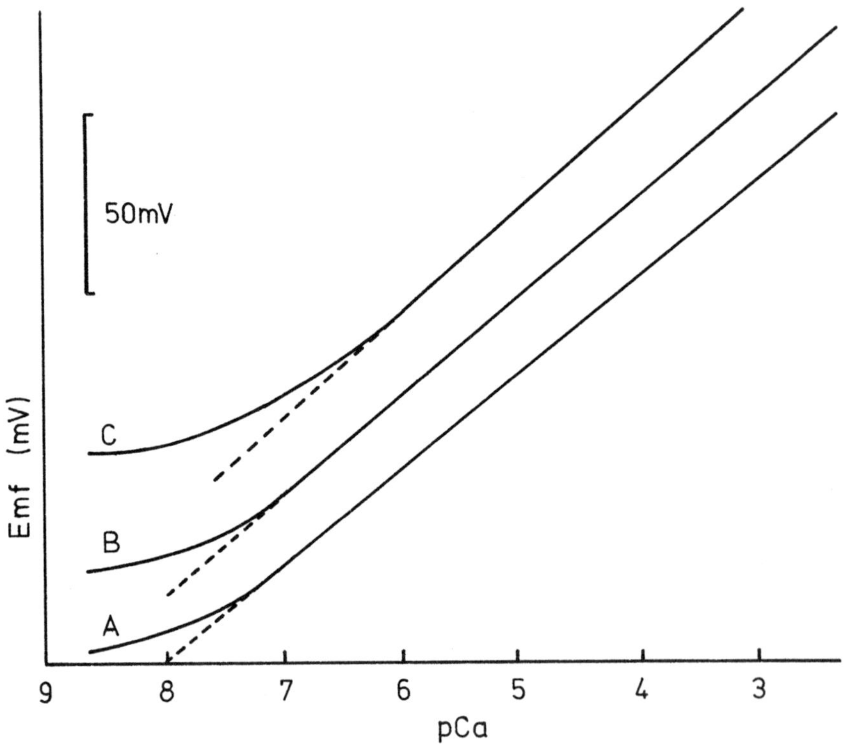


Fig.7. Emf response of cell assemblies in calcium ion-buffered solutions made up to 0.1M ionic strength for various calcium ion-selective electrodes. A: purified di-4-n-octylphenylphosphate [40]; B: neutral carrier material of Fig.5 [72]; and C: impure di-4-n-octylphenylphosphate [12].

Whichever of the three calcium ion-sensing systems is chosen, immobilization in PVC matrix membranes permits use up to 50°C [40]. They can be used in the micro-electrode mode and such neutral carrier calcium ion-selective electrodes have been described [79], but the PVC matrix design described by Brown, Pemberton and Owen [9] also merits attention.

Calcium ion-sensor developments have not reached their nadir. Improvements are still needed to overcome various deficiencies, including restricted electrode lifetime, and reproducibility, but laboratory-made PVC matrix membrane ion-selective electrodes are less likely to fail than commercial non-PVC matrix membrane calcium ion-selective electrodes with a reported 50 per cent malfunction rating [80]. Grafting phosphate sensor to the polymer matrix [13] has not resulted in the hoped-for improvement in electrode lifetime; nevertheless, this and alternative grafting approaches is an avenue that merits further investigation. The products of such researches will, however, have to match the selectivity specifications already achieved by the sensors discussed above, especially with regard to sodium.

Acknowledgements - The authors are grateful for the support of the sponsors to whom acknowledgement has been made in the various primary publications and to their many co-workers. They are also indebted to many willing friends in the ion-selective electrode field who have so readily shared their work in discussion and by providing materials and results - particularly Professors J.Růžička and W.Simon.

## 8. REFERENCES


1. J.W.Ross, *Science*, 156, 1378 (1967).

2. Orion Research Incorporated. *Instruction Manual for Calcium Ion Electrode Model 92-20* (1966).

3. A.Shatkay, *Anal.Chem.*, 39, 1056 (1967).

4. R.Bloch, A.Shatkay and H.A.Saroff, *Biophys.J.*, 7, 865 (1967).

5. G.J.Moody, R.B.Oke and J.D.R.Thomas, *Analyst*, 95, 910 (1970)

6. G.H.Griffiths, G.J.Moody and J.D.R.Thomas, *Analyst*, 97, 420 (1972).

7. A.Craggs, G.J.Moody and J.D.R.Thomas, *J.Chem.Education*, 51, 541 (1974).

8. J.L.Walker, *IUPAC International Symposium on Selective Ion-sensitive Electrodes, UWIST, Cardiff.* Paper 23 (1973).

9. H.M.Brown, J.P.Pemberton and J.D.Owen, *Analytica Chim.Acta*, 85, 261 (1976).

10. D.Ammann, E.Pretsch and W.Simon, *Tetrahedron Letters*, 24, 2473 (1972).

11. D.Ammann, E.Pretsch and W.Simon, *Anal.Letters*, 5, 843 (1972).

12. J.Růžička, E.H.Hansen and J.C.Tjell, *Analytica Chim.Acta*, 67, 155 (1973).

13. L.Keil, G.J.Moody and J.D.R.Thomas, *Analyst*,, 102, 274 (1977).

14. Orion Research Incorporated. *Analytical Methods Guide.* Eighth edition. (1977).

15. G.J.Moody and J.D.R.Thomas, *Selective Ion Sensitive Electrodes*, Merrow, Watford (1971).

16. J.Růžička and E.H.Hansen, *Analytica Chim.Acta*, 78, 145 (1975).

17. J.Růžička, E.H.Hansen and E.A.Zagatto, *Analytica Chim.Acta*, 88, 1 (1977).

18. E.J.Duffield, D.Mason, G.J.Moody and J.D.R.Thomas, *unpublished results.*

19. J.D.R.Thomas, Analytical techniques involving ion-selective electrodes. In: *Ion-Selective Electrodes*, Ed. E.Pungor, Akademiai Kiadó, Budapest (1978). Page 175.

20. R.W.Cattrall and H.Freiser, *Anal.Chem.*, 43, 1905 (1971).

21. G.J.Moody and J.D.R.Thomas, *Lab.Practice*, 27, 285 (1978).

22. M.D.Smith, M.A.Genshaw and J.Greyson, *Anal.Chem.*, 45, 1782 (1973).

23. R.W.Cattrall, D.M.Drew and I.C.Hamilton, *Analytica Chim.Acta*, 76, 269 (1975).

24. A.Hulanicki and M.Trojanowicz, *Analytica Chim.Acta*, 87, 411 (1976).

25. A.Ansaldi and S.I.Epstein, *Anal.Chem.*, 45, 595 (1973).

26. IUPAC. Recommendations for nomenclature of ion-selective electrodes, *Pure and Appl.Chem.*, 48, 127 (1976).

27. G.J.Moody, N.S.Nassory and J.D.R.Thomas, *Proc.Anal.Div.Chem.Soc.*, 16, 32 (1979).

28. J.Bagg, O.Nicholson and R.Vinen, *J.Phys.Chem.*, 75, 2138 (1971).

29. R.W.Cattrall and D.M.Drew, *Analytica Chim.Acta*, 77, 9 (1975).

30. K.Srinivasan and G.A.Rechnitz, *Anal.Chem.*,41, 1203 (1969).

31. G.J.Moody and J.D.R.Thomas, *Lab.Practice*, 20, 307 (1971).

32. G.J.Moody, N.S.Nassory and J.D.R.Thomas, *Talanta*, 26, (1979).

33. J.W.Ross, In: Ion-Selective Electrodes, Ed. R.A.Durst, Special Publication 314, *National Bureau of Standards, Washington, D.C.* (1969) Page 57.

34. A.Craggs, L.Keil, G.J.Moody and J.D.R.Thomas, *Talanta*, 22, 907 (1975).

35. G.J.Moody and J.D.R.Thomas, *Talanta,* 18, 1251 (1971).

36. D.Midgley, *Anal.Chem.,* 49, 1211 (1977).

37. A.Craggs, G.J.Moody, J.D.R.Thomas and A.Willcox, *Talanta,* 23, 799 (1976).

38. A.M.Y.Jaber, G.J.Moody, J.D.R.Thomas and A.Willcox, *Talanta,* 24, 655 (1977).

39. W.E.Morf, P.Wuhrmann and W.Simon, *Anal.Chem.*, 48, 1031 (1976).

40. A.Craggs, G.J.Moody and J.D.R.Thomas, *Analyst,* 104, 412 (1979).

41. B.Birch, A.Craggs, G.J.Moody and J.D.R.Thomas, Detection limits of calcium ion-selective electrodes in relation to ligand-containing systems. In: *Ion-Selective Electrodes,* Ed. E.Pungor, Akademiai Kiadó, Budapest (1978). Page 335.

42. B.Birch, A.Craggs, G.J.Moody and J.D.R.Thomas, *J.Chem.Education,* 55, 740 (1978).

43. G.J.Moody, and J.D.R.Thomas, *Selected Ann.Rev.Anal.Sci.,* 3, 59 (1973).

44. D.L.Manning, J.R.Stokely and D.W.Magowyrk, *Anal.Chem.,* 46, 1116 (1974).

45. K.Garbett and K.Torrance, *IUPAC International Symposium on Selective Ion-sensitive Electrodes, UWIST, Cardiff.* Paper 27 (1973).

46. K.Garbett and K.Torrance, *CERL Laboratory Note No. RD/L/N 153/74.* 12 July1974.

47. K.Garbett and K.Torrance, *CERL Laboratory Note No. RD/L/N 103/75.* 28 May 1975.

48. R.W.Cattrall, D.M.Drew and I.C.Hamilton, *Analytica Chim.Acta,* 76, 269 (1975).

49. G.J.Moody, N.S.Nassory and J.D.R.Thomas, *Analyst,* 103, 68 (1978).

50. A.Craggs, P.G.Delduca, L.Keil, G.J.Moody and J.D.R.Thomas, *J.Inorg.Nucl.Chem.,* 40, 1943 (1978).

51. G.H.Griffiths, G.J.Moody and J.D.R.Thomas, *J.Inorg.Nucl.Chem.,* 34, 3043 (1972).

52. A.Craggs, L.Keil, G.J.Moody and J.D.R.Thomas, *J.Inorg.Nucl. Chem.,* 37, 577 (1975).

53. A.M.Y.Jaber, G.J.Moody and J.D.R.Thomas, *Analyst,* 101, 179 (1976).

54. L.Keil, G.J.Moody and J.D.R.Thomas, *Analytica Chim.Acta.,* 96, 171 (1978).

55. W.Simon, W.E.Morf and P.Ch.Meier, *Structure and Bonding*, 16, 113 (1973).

56. W.E.Morf, D.Ammann, E.Pretsch and W.Simon, *Pure and Appl.Chem.*, 36, 421 (1973).

57. E.Pretsch, D.Ammann and W.Simon, *Research Development*, 25, 20 (1974).

58. W.E.Morf and W.Simon, Ion-selective electrodes based on neutral carriers. In: *Ion-Selective Electrodes in Analytical Chemistry*, Ed. Henry Freiser, Plenum Press, New York and London, Volume 1 (1978). Page 211.

59. U.Fiedler, *Analytica Chim.Acta*, 89, 111 (1977).

60. A.M.Y.Jaber, G.J.Moody and J.D.R.Thomas, *Analyst*, 102, 943, (1977).

61. K.A.Petrov, N.K.Bliznyuk and F.L.Maklyaev, *Zhur.Obshch.Khim.*, 29, 3403 (1959).

62. P.G.Delduca, *M.Sc Thesis, University of Wales* (1978).

63. D.Jagner and J.P.Østergaard-Jensen, *Analytica Chim.Acta*, 80, 9 (1975).

64. A.Craggs, G.J.Moody and J.D.R.Thomas, *unpublished data*.

65. S.Kamata, G.J.Moody and J.D.R.Thomas, *Analytica Chim.Acta*, in the press.

66. A.Craggs, P.G.Delduca, L.Keil, B.J.Key, G.J.Moody and J.D.R.Thomas, *J.Inorg.Nucl.Chem.*, 40, 1483 (1978).

67. B.Loev and J.T.Massengale, *J.Org.Chem.*, 22, 988 (1957).

68. G.R.Pickering, *Ph.D Thesis, CNAA (Brighton Polytechnic)*, (1977).

69. A.K.Covington and N.Kumar, *Analytica Chim.Acta*, 85, 175 (1976).

70. D.Ammann, E.Pretsch and W.Simon, *Helv.Chim.Acta*, 56, 1780 (1973).

71. D.Ammann, R.Bissig, M.Güggi, E.Pretsch, W.Simon, I.J.Borowitz and L.Weiss, *Helv.Chim.Acta*, 58, 1535 (1975).

72. D.Ammann, M.Güggi, E.Pretsch and W.Simon, *Anal.Letters*, 8, 709 (1975).

73. W.E.Morf and W.Simon, *Hungarian Sci.Instruments*, 41, 1 (1977).

74. W.E.Morg, G.Kahr and W.Simon, *Anal.Letters*, 7, 9 (1974).

75. A.Ringbom, *Complexation in Analytical Chemistry, Interscience, New York* (1963).

76. G.Schwarzenbach, *Complexometric Titrations, Interscience*, New York (1957).

77. B.Dempsey and D.D.Perrin, *Buffers for pH and Metal Ion Control, Chapman and Hall, London,* (1974).

78. A.E.Martell and L.G.Sillen, Stability Constants of Metal Complexes, *Special Publication Nos. 17 and 25, Chemical Society, London* (1964) and (1971).

79. M.Oehme, M.Kessler and W.Simon, *Nature,* 258, 754 (1975).

80. S.Madsen and K.Ølgaard, *Clin.Chem.,*23, 690 (1977).

# THE AUTHORS

Dr. G. J. Moody

Dr. J. D. R. Thomas

G. J. Moody is a Senior Lecturer in Chemistry at the University of Wales Institute of Science and Technology and the co-author of *Noble Gases and Their Compounds* (Pergamon); *Dipole Moments in Inorganic Compounds* (Arnold); *Selective Ion Sensitive Electrodes* (Merrow) and *Practical Electrophoresis* (Merrow). He was a Scientific Officer with the United Kingdom Atomic Energy Authority (Industrial Division) in 1955-56 and a Resident Research Associate at the Argonne National Laboratory, Illinois, during 1965-66.

Apart from radiation chemistry and electrophoresis his main research interests cover ion-selective electrodes and ion-exchange separations, and he has published over 150 papers.

J. D. R. Thomas, Editor of *Ion-Selective Electrode Reviews*, is Reader in Chemistry at the University of Wales Institute of Science and Technology (UWIST). He holds the B.Sc., M.Sc. and D.Sc. degrees of the University of Wales. After experience in industry and training as a teacher, he lectured at various colleges of technology before becoming a Senior Lecturer at UWIST prior to his appointment as Reader in 1973. He has published about 200 papers and has co-authored 5 books, namely, *Calculations in Advanced Physical Chemistry* (Arnold); *Noble Gases and their Compounds* (Pergamon); *Dipole Moments in Inorganic Chemistry* (Arnold); *Selective Ion Sensitive Electrodes* (Merrow); and *Practical Electrophoresis* (Merrow). His research interests include coordination chemistry, separations by ion-exchange and electrophoresis, and, of course, ion-selective electrodes, on which he has lectured in many countries in Europe and Asia, and in the U.S.A.

*Ion-Selective Electrode Rev*. Vol.1, pp.31-79. 0191-5371/79/0601-0031 $05.00/0


# ION-SENSITIVE FIELD EFFECT TRANSISTORS

Jiří Janata and Robert J. Huber
Department of Bioengineering and Department of Electrical Engineering, University of Utah, Salt Lake City, UT 84112, U.S.A.

CONTENTS

Keywords: CHEMFET, Chemical sensors, Chemically sensitive semiconductor devices, Field effect transistors, Ion-sensitive field effect transistors Ion sensitive transistors, ISFET, Semiconductor Devices.

## 1. INTRODUCTION

An every increasing number of original papers, reviews, and symposia on the subject of ion-selective electrodes (ISE) attest to the spectacular success of this modern tool of chemical analysis. The interest in, and development of, ion sensitive field effect transistors (ISFET) can be largely attributed to the popularity of ISEs.

The similarity between the two types of sensors is a good starting point for explanation of the operation of ISFET. The evolutionary chain is depicted in Fig. 1, which shows (a) a schematic representation of a conventional measurement with an ISE (with the electroactive membrane M) and a reference electrode, R. The ISE is connected by a wire to an FET input of a pH meter. The readout part of the meter is not shown for convenience. In Fig. 1b the lead connecting the ISE is made shorter until it is completely disposed of, and the membrane M is placed directly on the insulator of the FET. Next, the input transistor is placed in the measured solution (Fig. 1c). In this stage the integration of the ion sensitive membrane and the solid state amplifier is accomplished. Figure 1d shows the details of the ISFET which is very, very small.

Thus, in this first crude approximation, the main difference between the ISE and the ISFET is the length of the conductor connecting the membrane to input of the meter, and the size of the whole assembly. It is interesting to note that all three types corresponding to this diagram exist. There are, of course, many ISEs with solid internal contact; (e.g. coated-wire electrodes). A glass membrane connected with a short length of metallic conductor to the FET amplifier input has been described [1]. Similarly, Nagy *et al.* [2], made a fluoride sensor by bonding a small lanthanum fluoride chip to a FET through a short silver lead. Finally, there is now a rapidly growing family of ISFETs which will be discussed in this review.

It must be realized that the thermodynamics of the cell composed of reference electrode/solution/ion-sensitive device, is not affected by the geometry or size of any of its components (i.e. the length of the conductor between the membrane and the amplifier). In order to fully understand the mechanism of operation of ISFETs it is necessary to perform a complete analysis of the whole measuring circuit, i.e. to move a hypothetical test charge from the semiconductor of the FET through the external circuit, reference electrode, solution, and to the membrane and to describe separately the processes in each phase and at each interface. This will be done in Section 2.4.

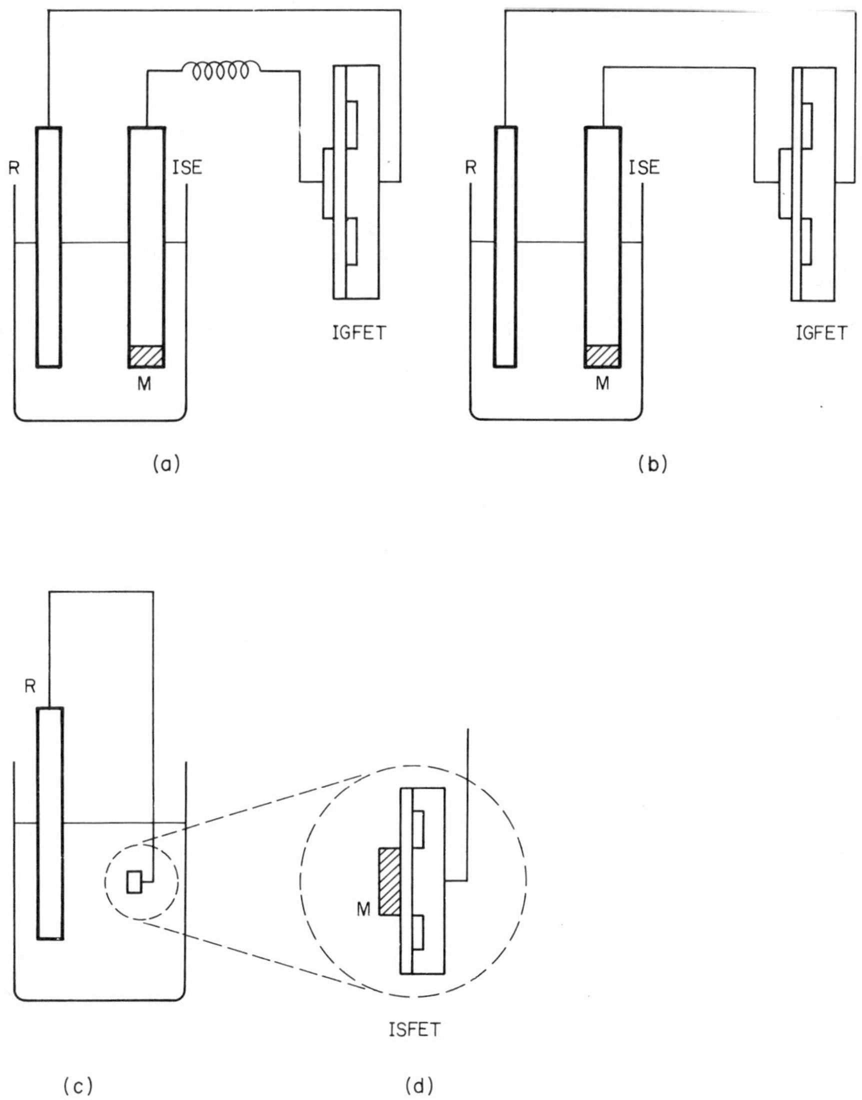


Fig. 1. Evolution of ISFET from ISE and IGFET. R = reference electrode; M = membrane.

### 1.1. Historical Survey

The first report on the ion sensitive field effect transistor is a short communication by Bergveld [3]. Shortly afterwards Matsuo *et al.* [4] reported a similar device in a non-regular Japanese journal. The first full description of the ISFET sensitive to hydrogen and sodium ions was presented by Bergveld 2 years later [5]. He used the device for measuring pH and sodium ion activity as well as for recording physiological transient action potentials. Another pH-sensitive ISFET was described by Matsuo and Wise [6]. There are two major differences between Bergveld's and Matsuo's work which proved to be important for the future development of these devices. Firstly Bergveld did not use an external reference electrode and indeed has claimed that one is not needed for operation of ISFET [5]. Secondly, Matsuo used silicon nitride as the top layer of the gate insulator while Bergveld used silicon dioxide. Although Bergveld later moderated his stand on the reference electrode *(vide infra)* his claim attracted wide attention and stimulated a very useful discussion. The use of silicon nitride proved to be essential for development of ISFETs with ion-sensitive layers deposited over the gate insulator. The first device of this kind was the potassium ion sensitive ISFET [7] in which the ion-sensitive layer was the PVC membrane containing valinomycin which is normally used in conventional potassium-sensitive ISEs. Several other reports describing ISFETs sensitive to pH, $K^+$, and $Ca^{2+}$ have appeared; they will be discussed in the appropriate sections of this review.

The field effect transistor sensitive to molecular hydrogen [8] is just outside the scope of this review. However, it represents a transition from MOSFET to chemically-sensitive FET.

Almost every paper on ISFETs published so far gives at least a rudimentary theory of operation of a field-effect transistor in its introduction. There are, however, several papers surveying the theory and the mechanism of the operation of ISFETs in greater depth. The first review of the area of *chemically sensitive semiconductor devices* was written by Zemel [9] who first used the acronym CSSD. A narrower review focusing on chemically-sensitive field effect transistors (CHEMFETs) was written by Janata and Moss [10], and the mechanism of operation of ISFET was addressed by Kelly [11]. Another article by Zemel [12] more or less reiterates the main points of his previous review and a paper by Revesz [13] deals particularly with the role of surface states in the mechanism of operation of ISFET. A theoretical analysis of the structure, solution/membrane/insulator/semiconductor, supplemented by a thorough experimental study of the system

$$Ag^+ \text{ (or Br-) solution} | AgBr | SiO_2 | Si$$

was done by Buck and Hackleman [14].

The event of major significance for the development of CSSDs was the International Workshop on Theory, Design and Biomedical Applications of Solid State Chemical Sensors held at Case Western Reserve University, Cleveland, Ohio, U.S.A., in March 1977. Many important papers addressing theory, practical aspects, and applications of CSSDs will appear in the symposium proceedings [15].

## 2. THEORY OF IGFETS

There are several kinds of CSSD discussed in the literature, all or which depend in some way on the semiconductor field effect for their operation. The ISFET is no exception. Structurally it is very similar to the IGFET (insulated gate field effect transistor), a device which has been very highly developed as the basis for much of the integrated circuit revolution presently taking place in the field of electronics. The theory of IGFETs is very well developed, although it is generally presented in the language of the electronics engineer or solid state physicist, not the chemists. The purpose of this section is to outline those elements of the theory which are of most use to the reader in understanding and using ISFETs. We begin with a short discussion of the semiconductor field effect.

## 2.1. *The semiconductor field effect*

In an electric field free region of an extrinsic semiconductor containing only one type of dopant, there are equal densities of mobile charges (electrons in *n*-type and holes in *p*-type) and ionized dopant atoms (positively charged for *n*-type and negatively charged for *p*-type). For simplicity we will restrict this discussion to a *p*-type extrinsic semiconductor.

If an electric field from whatever source, is applied to the surface of the semiconductor, the density of the mobile charge carriers (holes) will be either enhanced or depleted depending on the field. If the field enhances the concentration of holes, the surface is said to be accumulated and the semiconductor surface behaves much like a metal in that the excess charge appears at the surface and the electric field does not penetrate. If, on the other hand, the field forces the mobile holes away from the surface, a space charge region consisting of the ionized acceptor atoms, which are fixed in the lattice, forms over an appreciable distance into the semiconductor. The thickness of the surface space charge region is a function of the strength of the field at the surface and the semiconductor doping profile, as is the difference between the surface and bulk potentials of the semiconductor. If the surface potential deviates sufficiently far from the bulk potential, the surface will invert, that is it will contain an excess of mobile electrons. In this case there is an *n*-type conductive channel on the surface separated from the *p*-type bulk by a space charge region. Further increase in the normal surface field will not significantly change the surface potential, but will only change the density of electrons in the *n*-type surface layer, and consequently, the electrical conductivity of this layer. Changes in the surface electric field can be determined by measuring the corresponding changes in the thickness of the surface depletion region if the surface is not inverted, or the conductivity of the surface inversion layer if the surface is inverted.

The surface field can be produced in a number of ways. The semiconductor can be built into a capacitor and an external potential applied (IGFET), or the field can arise from the electrochemical effect between different materials (ISFET).

In both cases, changes in the surface electric field change the density of mobile charge carriers in the surface inversion layer. The physical effect that is measured is the change in electric current carried by the surface inversion layer, the drain current. The devices must be operated under conditions that cause the surface inversion layer to form.

## 2.2. *Current voltage relationships for the IGFET*

Figure 2 shows the typical construction and biasing arrangement for the *n*-channel IGFET. In effect the semiconductor substrate (3) forms one side of a parallel plate capacitor and the metal gate field plate (4) forms the other. In the normal operating mode the gate voltage, $V_G$, is applied between the semiconductor substrate (3) and the gate (4). The polarity and magnitude of $V_G$ are such that the semiconductor field effect gives rise to an inversion layer on the surface of the substrate under the gate. In fact, the surface of the *p*-type substrate (3) becomes *n*-type. This *n*-type inversion layer forms a conducting channel between the source (2) and drain (1) regions. If this inversion layer is not present, application to the drain (1) of a positive voltage $V_D$ with respect to the source (2) and substrate (3) results in no appreciable current flow into drain (1) because the drain-to-substrate *p-n* junction diode is reverse biased. However, if the *n*-type inversion layer exists along the surface of the semiconductor between the source (2) and drain (1), a continuous path for the flow of current, $I_D$, exists. The current, $I_D$, shown in Fig. 2 flows through the channel connecting the drain (1) and source (2). The magnitude of the drain current, $I_D$, will be determined by the electrical resistance of the surface inversions layer and the voltage difference, $V_D$, between the source and drain.

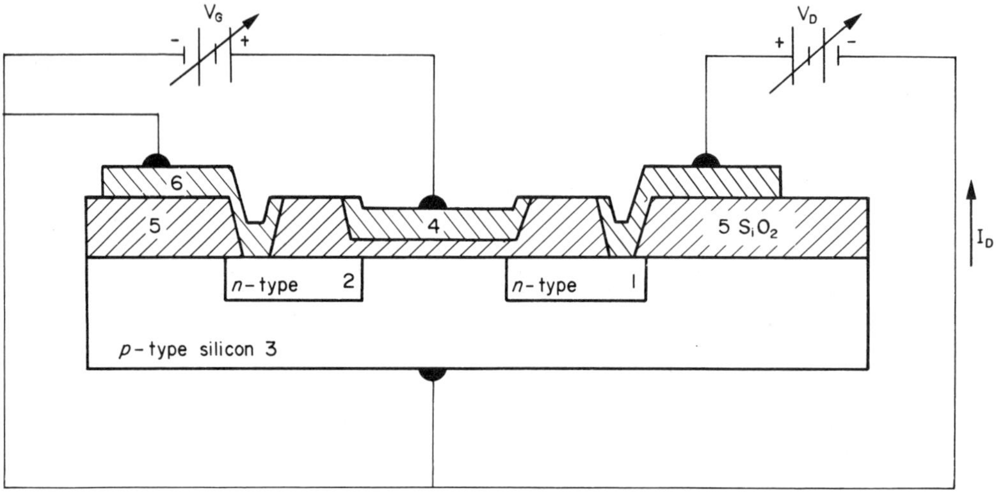


Fig. 2. Schematic diagram of IGFET. 1 = drain, 2 = source, 3 = substrate, 4 = gate, 5 = insulator, 6 = metal contacts.

The basis for the derivation of the current-voltage relationships is the calculation of the density of the mobile electrons in the surface inversion layer as a function of the applied voltages $V_G$ and $V_D$, and position along the channel.

Figure 3 is a schematic representation of the channel region of the IGFET. At a point $y$ along the channel there is a charge density, $Q_n(y)$, per unit area of mobile electrons in the surface inversion layer, and a charge density, $Q_B(y)$, of ionized dopant atoms in the space charge region. These charge densities are both a function of $y$ because of the voltage change along the channel due to the drain current. $Q$s, the sum of $Q_n(y)$ and $Q_B(y)$, is equal and opposite to the charge per unit area on the metal plate which forms the other side of the capacitor (the gate)

$$Q_s = Q_n(y) + Q_B(y). \tag{1}$$

$Q_s$ may be related to the applied gate voltage and the capacitance, $C_o$, of the gate insulator by

$$V_G - V_{FB} = -\frac{Q_s}{C_o} + \phi_s. \qquad (2)$$

$V_{FB}$ is the "flat band" voltage. It accounts for the effects of the work function difference between the gate metal and the silicon and any charges that exist in the gate insulator. $\phi_s$ is the surface potential of the semiconductor.

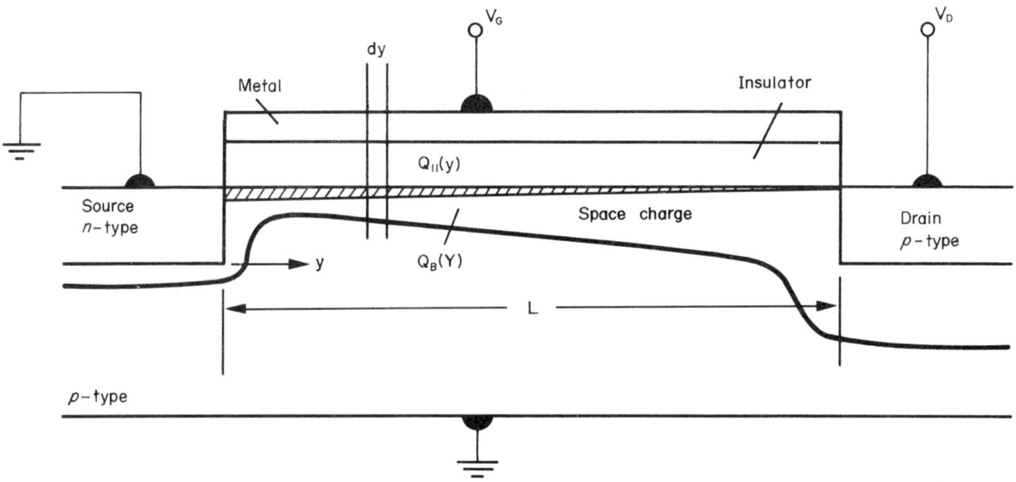


Fig. 3. IGFET channel.

The mobile electron charge in the surface inversion layer is then seen to be given by

$$Q_n(y) = -[V_G - V_{FB} - \phi_s(y)]C_o - Q_B(y). \qquad (3)$$

In writing these equations we assume that a surface inversion layer does in fact exist. At this point in the derivations the assumption is made that the surface potential is given by the conditions of strong inversion

$$\phi_s(y) = V(y) + 2\phi_F. \qquad (4)$$

$V(y)$ is the "reverse bias" applied to the "field effect induced" $p$-$n$ junction composed of the $n$-type surface inversions layer and the $p$-type substrate.

$Q_B(y)$ is the charge in the space charge region. Under the "depletion approximation" which assumes that there are no mobile charges at all in the space charge regions

$$Q_B(y) = -\sqrt{2K_s\varepsilon_o qN_A[V(y) + 2\phi_F]}. \qquad (5)$$

Collecting terms

$$Q_n(y) = -[V_G - V_{FB} - V(y) - 2\phi_F]C_o + (2K_s\varepsilon_o qN_A[V(y) + 2\phi_F])^{1/2}. \qquad (6)$$

Now referring to Fig. 3, the voltage drop along the element of channel, $dy$, can be written as

$$dV = I_D dR \doteq \frac{I_D\, dy}{W\mu_n Q_n(y)}. \tag{7}$$

$W$ is the width of the channel and $\mu_n$ is the electron mobility in the channel.

Substitution of the above expression for $Q_n(y)$ and integrating over the channel length yields the basic current voltage relationship.

$$I_D = \frac{W}{L}\mu_n C_o \left\{\left[V_G - V_{FB} - 2\phi_F - \frac{V_D}{2}\right]V_D - \frac{2}{3}\frac{\sqrt{2K_s\varepsilon_o q N_A}}{C_o}\left[(V_D + 2\phi_F)^{3/2} - (2\phi_F)^{3/2}\right]\right\}. \tag{8}$$

In the derivation of this equation it was assumed that the surface inversion layer actually existed at all points along the channel. In eq. (5) it is seen that $Q_B(y)$ increases with $V(y)$, the voltage drop along the channel. This increase is at the expense of $Q_n(y)$, the mobile channel charge. $V(y)$ must change from 0 at the source to $V_D$ at the drain so if $V_D$ is large enough, the conducting channel will disappear near the drain end of the channel. When this happens eq. (8) is no longer valid and the IGFET is said to become "saturated."

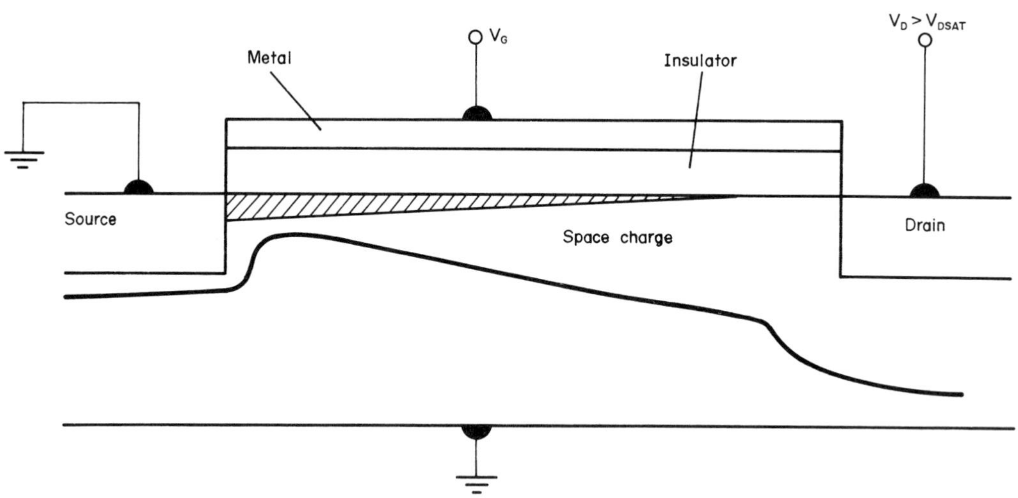


Fig. 4. IGFET in saturation.

Figure 4 is a schematic diagram showing the saturated IGFET. $V_{DSAT}$ is the drain voltage at which the surface channel just disappears at the drain end. Any increase in $V_D$ beyond $V_{DSAT}$ results in a short space charge region between the drain the channel and the drain. Current continues to flow under these conditions because electrons in the channel see no potential barrier restricting flow from the channel across the space charge region to the drain. However, the number of electrons arriving at the drain end of the channel is determined by the voltage $V_{DSAT}$, between the source and the end of the surface channel. It can be shown to be [16]

$$V_{DSAT} = V_G - V_{FB} - 2\phi_F + \frac{K_s\varepsilon_o qN_A}{C_o^2}\left[1 - \left(1 + \frac{2C_o^2(V_G - V_{FB})}{K_s\varepsilon_o qN_A}\right)^{1/2}\right]. \quad (9)$$

For drain voltages greater than $V_{DSAT}$, the drain current, $I_D$ is given by eq. (8) with $V_D$ replaced by $V_{DSAT}$, provided the channel length, $L$, is not appreciably shortened by the space charge region between the end of the channel and the drain.

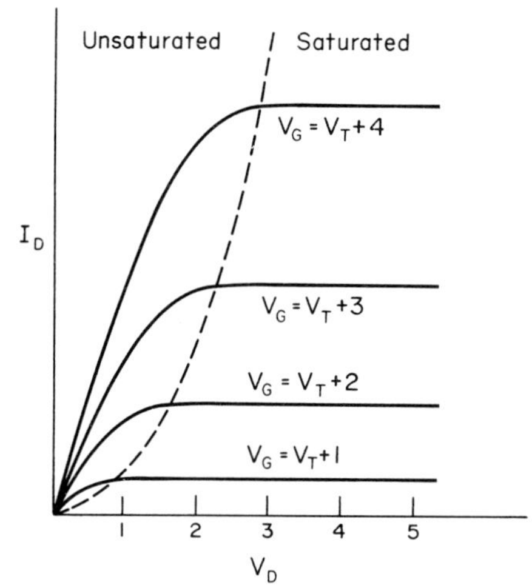


Fig. 5. Change of drain current $I_D$ with drain voltage $V_D$ showing saturation and unsaturation.

The curves in Fig. 5 are divided into the saturated and unsaturated regions by the dotted line. Note that $V_{DSAT}$ is a function of the applied gate voltage.

Even though these eqs (8) and (9) are derived using several simplifying assumptions, they are quite cumbersome because of their complexity. A much simpler set of equations may be derived if the dependence of $Q_B(y)$ (eq. (5)) on $V(y)$ is neglected. In this case

$$Q_B = -\sqrt{2K_s\varepsilon_o qN_A\ 2\phi_F}, \quad (10)$$

and the integrations required to derive the current voltage relationships can be simplified to yield

$$I_D = \mu_n C_c \frac{W}{L}\left\{(V_G - V_T)\ V_D - \frac{V_D^2}{2}\right\}, \quad V_D < V_{DSAT} \quad (11)$$

where

$$V_T = V_{FB} + 2\phi_F - \frac{Q_B}{C_o}$$

and

$$V_{DSAT} = V_G - V_T.$$

If there are no distributed charges in the insulator and no silicon surface states, the flatband voltage, $V_{FB}$, reduces to just two terms

$$V_{FB} = \Phi_{ms} - \frac{Q_{ss}}{C_o}.$$

Equation 11 is the current-voltage relationship for the unsaturated case. For drain voltages greater than $V_{DSAT}$, the current is given by eq.(12) with $V_D$ replaced by $V_{DSAT}$.

$$I_D = \mu_n C_o \frac{W}{L} \frac{(V_G - V_T)^2}{2} \qquad V_D > V_{DSAT}. \tag{12}$$

Equations (11) and (12) are the IGFET current-voltage relations most often quoted in the literature. Their use can lead to significant quantitative errors in calculated currents. They do, however, contain the correct qualitative features of the device, i.e. the saturated and unsaturated regions, the near constant drain current beyond saturation, and the dependence of the saturation voltage on the gate and threshold voltages. Equations (11) and (12) have the distinct advantage of simple form.

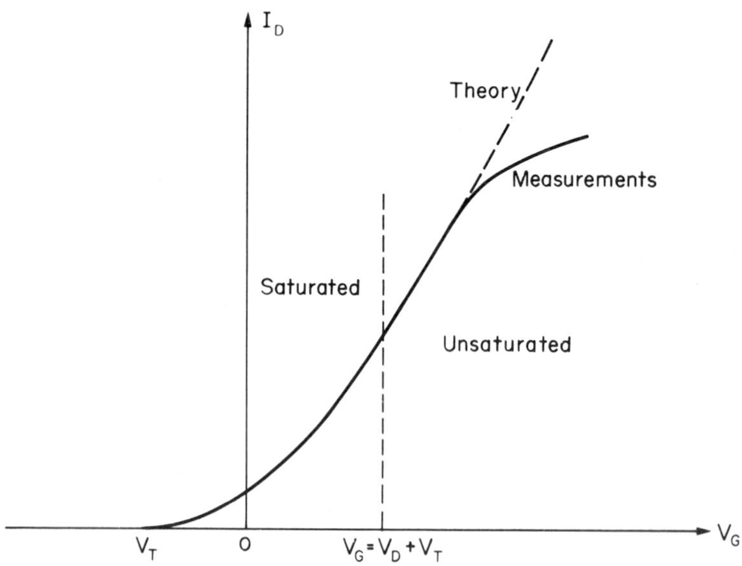


Fig. 6. Change of drain current $I_D$ with gate voltage $V_G$ divided into saturated and unsaturated regions.

Figure 6 is a plot of $I_D$ vs $V_G$ for an $n$-channel depletion mode device. Depletion mode refers to the fact that the device has a conducting channel for zero applied gate voltage. Depletion mode $n$-channel devices are the result of a sufficiently negative flat-band voltage, $V_{FB}$. In the saturated regions, i.e. $V_D > V_G - V_T$, or alternately $V_G < V_D + V_T$, eq. (12) predicts a quadratic dependence of $I_D$ on $V_G$ as shown. In the unsaturated region, i.e. $V_D < V_G - V_T$ or alternately $V_G > V_D + V_T$, eq. (11) predicts a linear dependence of $I_D$ on $V_G$ as shown. Actual data, however show significant departure from linearity in the unsaturated region as indicated in Fig. 6. One cause of this departure from theory is the variation of $\mu_n$, the charge carriers channel mobility, with the electric field normal to the surface of the semiconductor. Surface channel mobility is primarily determined by surface scattering. Strong electric fields that increase the probability of the carrier interacting with the surface reduce the surface mobility. Therefore, as the gate voltage increases, the electric field normal to the surface increases and $\mu_n$ decreases.

A second cause of the departure of the $I_D$ - $V_G$ curve from linearity is series resistance between the end of the channel and the point at which the drain voltage is actually measured. Schematically this is represented as shown in Fig. 7.

The resistive voltage drop actually reduces the effective drain voltage below that applied to the device. The magnitude of this resistance depends on both the individual device geometry and processing parameters, but it may be quite significant. In the case of the ISFETs reported by Janata and Moss [10], the effective $R$ of Fig. 6 above can be 200 Ω. Their experimental data begin to depart from linearity for drain currents of about 2 mA, a value at which the 200 Ω has an appreciable effect.

As a result of the effects of non-ideal structures, second order effects in parameters, and the numerous approximations made in the derivation of the current-voltage equations, eqs (11) and (12) can only serve as a qualitative description of the actual device. Each individual design must be experimentally characterized.

For these reasons, and others to be discussed in Section 3.9, it is advantageous to operate the ISFET in the constant drain current mode.

2.2.1. *Summary of assumptions.* The derivation of eqs (11) and (12) contain the most simplifying assumptions of any of the widely used sets. The most important are discussed below.

(a) It is assumed that there exists a well defined threshold voltage, and that the formation of the surface inversion layer begins suddenly as the gate voltage is increased. It is equivalent to stating that there is a sharply defined semiconductor surface potential dividing surface depletion and surface inversion. In fact, this transition is continuous. In conventional structures this is a good approximation if the gate voltage exceeds the "threshold voltage" by about 0.5 V.

(b) The voltage dropped due to a surface channel current (drain current) flowing has no effect on the thickness of the surface space charge region. This approximation can lead to relatively large errors in the magnitude of the predicted drain current in the saturation region, but the general shape of the drain current vs drain voltage curves is satisfactory, i.e. the qualitative features of the device are not affected by this approximation.

(c) The doping of the semiconductor is constant near the surface where the channel is formed. This is not a good approximation for many ion implanted structures.

(d) The channel length is large compared to the thickness of the depletion region surrounding the $p$-$n$ junctions.

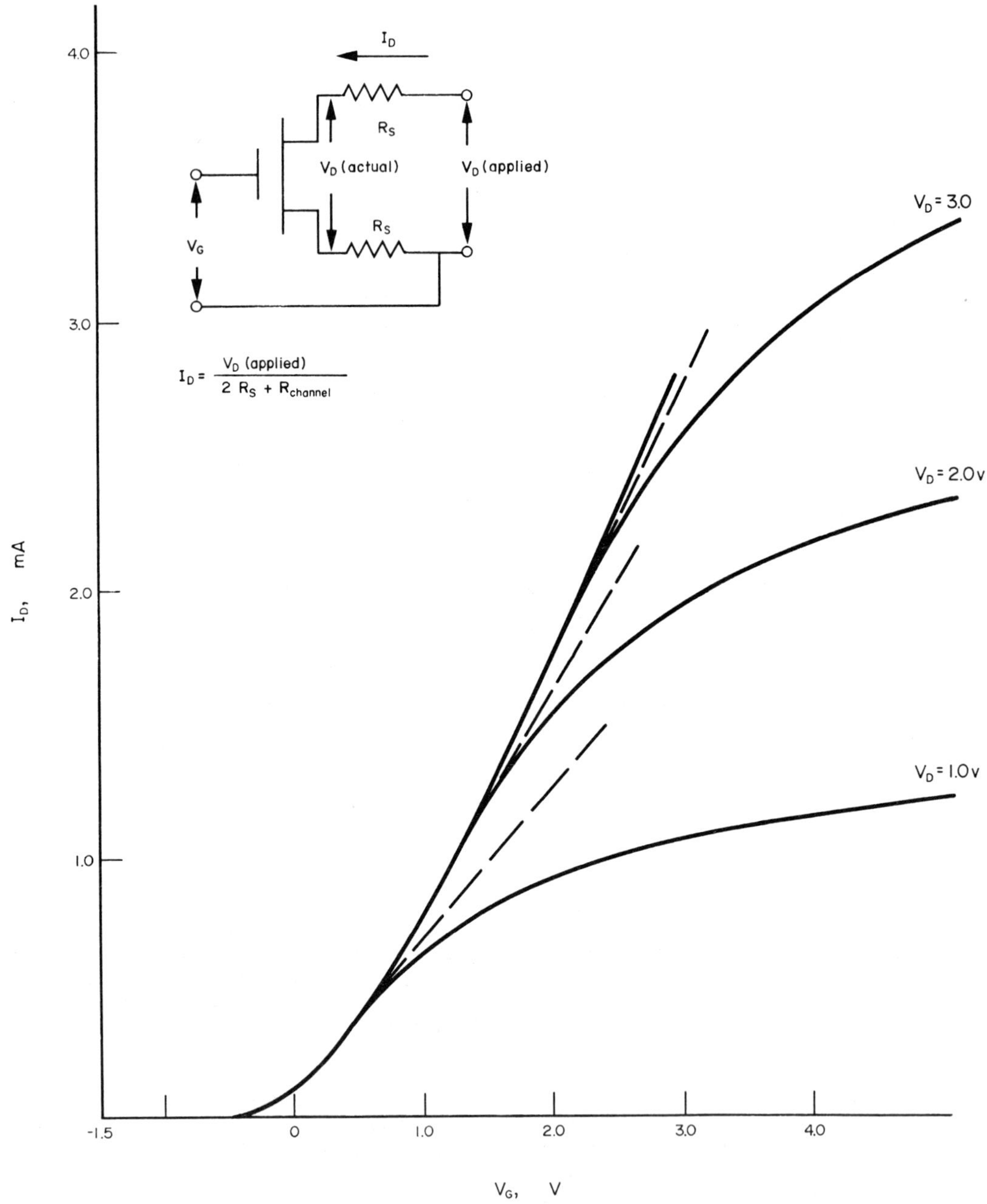


Fig. 7. Effect of series drain current resistance. Theoretical response for $R_S$ = 0 (---).

(e) Both the source and substrate of the device are connected to the same point of the external circuit.

Under assumption (a) above an equation relating the threshold voltage, $V_T$, to the device structure is

$$V_T = \Phi_{ms} - \frac{Q_{ss}}{C_o} + 2\phi_F - \frac{Q_B}{C_o} \quad (13)$$

$\Phi_{ms}$ = The semiconductor-gate metal work function difference.

$\phi_F$ = The electron Fermi potential in the bulk silicon, referenced to the intrinsic semiconductor Fermi potential. The surface potential for the onset of strong inversion is taken to be $2\phi_F$, a fairly arbitrary value.

$Q_B$ = The electric charge contained in the surface depletion layer in the form of ionized dopant atoms at the onset of strong inversion. Assumption (a) above is equivalent to stating that all charge in the surface space region, greater than $Q_B$, is in the form of mobile charges in the inversion layer.

$Q_{ss}$ = The electric charge in a thin layer of the gate insulator lying next to the semiconductor. It is principally a feature of the silicon/silicon dioxide system. As the great majority of IGFET structures employ thermally grown $SiO_2$ as the layer of the gate insulator next to the silicon, $Q_{ss}$ is generally included explicitly in the threshold equation. The charge $Q_{ss}$ is not mobile under voltages and temperatures normally encountered by IGFETs. It cannot communicate with the electrons in the silicon and therefore, is not affected by the silicon surface potential. It is, therefore, not to be confused with surface states on the silicon, usually known as "fast states" which gain a lose charge as a function of the surface potential. In present practice the density of these "fast states" is reduced to negligible levels.
The thermally grown gate oxides in practice are formed under exceptionally clean conditions and contain no other important space charge.

$C_o$ = Capacitance per unit area of the gate insulator. If the insulator is a multilayer structure, $C_o$ is the total capacitance per unit area of all layers.

In the case of the ISFET (Section 2.4) the gate metal is replaced by a more complex structure consisting of an ion-sensitive layer (which may be part of the uppermost layer of the gate insulator), the solution containing the ions of interest, and a reference electrode. Those terms in eq. (13) which are unique to the metal gate must be replaced with quantities which describe the more complex ISFET system. In eq. (13) it is $\Phi_{ms}$ which must be replaced. See Section 2.4 below.

## 2.3. *CHEMFET sensitive to neutral species*

In this review, we adopt a narrow view of the field of CSSDs. We restrict ourselves to a discussion only of the chemically sensitive field effect transistors (CHEMFET). We do not consider the general subject which includes interactions between the semiconductor and its environment other than through the field effect (Section 2.2). We further restrict our discussion to field effect transistors in which the gate dielectric is a perfect electrical insulator, i.e. no charged species can pass through. These restrictions are necessary if the analysis of the CHEMFET is to be based on its similarity with the IGFET.

They are also very practical restrictions. Almost all of the semiconductor technology which makes CHEMFETs practical has evolved in the electronics field. The IGFET is a well characterized and understood device. The art of forming a highly stable near-perfect gate insulator is well established. The practical implementation of devices outside this body of technology will be most difficult. If the field effect transistor contains a perfect electrical insulator, the reference electrode (charge return path) is a requirement of the system (see Section 2.4). The "field effect" (see Section 2.2) results in changes in the conductivity of the surface inversion layer as a function of changes in the electric field at the surface of the semiconductor regardless of the origin of the changes in the field, be it a desired chemical effect or an undesired static charge effect. The only practical way to eliminate the static charge effect (remember we postulate a perfect gate insulator) is to make the device in the form of a parallel plate capacitor. The semiconductor forms one plate. In the case of the IGFET, the other plate is the gate metal. In the case of the ISFET, the other plate is the electrolyte and nonpolarized ion-sensitive membrane. There must be a charge return path between the semiconductor and the other plate.

CHEMFETs sensitive to neutral species can be based on any mechanisms that effects the potential between the semiconductor and the other plate of the capacitor.

One CHEMFET structure sensitive to neutral species is the IGFET with palladium metal gate [8]. Neutral hydrogen molecules, under suitable conditions, are reversibly absorbed by the palladium. The metal semiconductor work function difference, $\phi_{ms}$, (cf. $\phi_s$ in eq. (3) Section 2.2) is a function of the amount of hydrogen in the palladium. Changes in $\phi_{ms}$ result in changes in $V_T$ which lead to changes in the drain current $I_d$ (see eq. (11) Section 2.2).

The ISFET structure could be made sensitive to neutral species in the electrolyte if there is a chemical reaction between the neutral species in the electrolyte and a layer adjacent to the gate insulator that produces a charged species. One mechanism for producing a CSSD sensitive to neutral species often suggested [9,13] is the interaction between the neutral species and semiconductor surface states, i.e. "fast states" which are part of the semiconductor band structure near or at the surface. This is possible in principle, but has great practical problems. First, a whole new manufacturing technology would be needed because current FET production methods effectively reduces these surface states to negligible levels. Second, a very high density of surface states tends to anchor the surface potential [17] at some value. Neutral species interactions that would effect the channel conductivity have to affect the surface state density in a very narrow range of energy and density. The drain current equations, eqs (11) and (12) Section 2.2, are not applicable to this case, because they are derived under the assumption that all semiconductor surface charge resides in either ionized dopant atoms or mobile charges which contribute directly to the channel conductivity.

As we are restricting this discussion to FETs having a perfect gate insulator, neutral species cannot change the distribution or density of fixed charges in the gate insulator, i.e. $Q_{ss}$.

## 2.4. *Ion sensitive field effect transistor, ISFET*

In the CHEMFET (Fig. 8) the metal gate is replaced with a reference electrode (7), solution (8) and a chemically sensitive layer (membrane) (9). The rest of the device is protected by a suitable encapsulant (10). From the electrochemical point of view, there are two types of solution/membrane interfaces: non-polarized and polarized. The non-polarized interface is permeable to at least one type of charged species (electrons or ions). It behaves electrically as a resistor in series with a battery voltage which is described by the Nernst equation. It relates the interfacial potential difference to the activity of the ion in solution. On the other hand, charge cannot cross the polarized interface which is described by the Gibbs-Lippmann equation. Either type of interface can be used with CHEMFET and the thermodynamical analysis of two types of CHEMFETs has been presented [18]. In this review, we shall restrict ourselves to the discussion of CHEMFETs with a non-polarized solution interface - ISFETs.

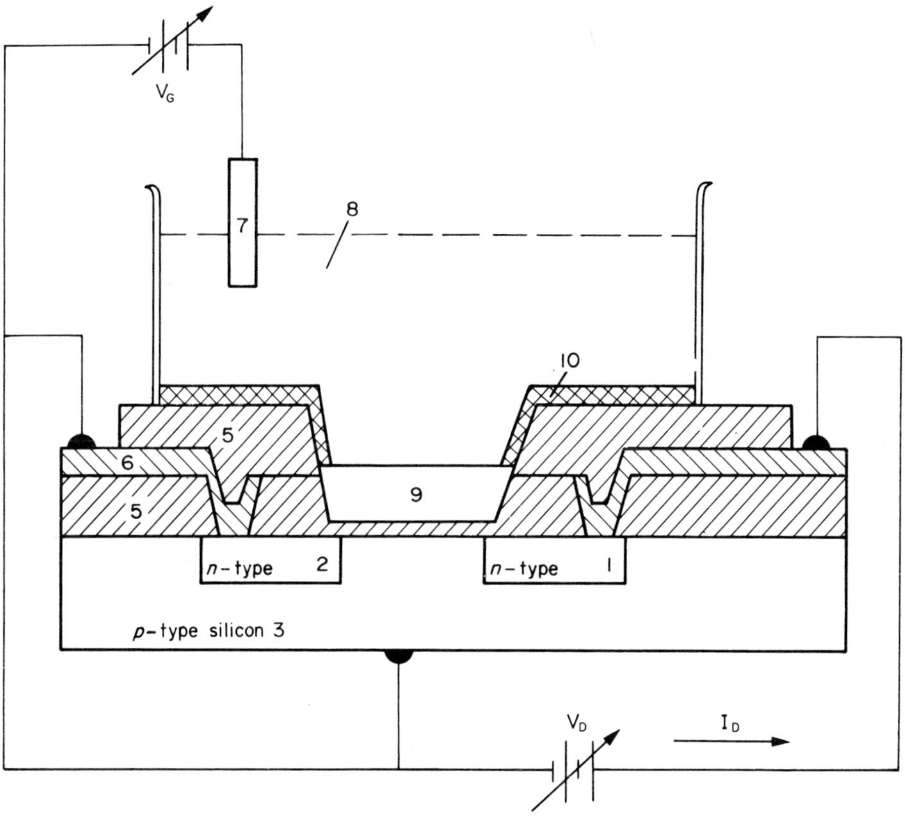


Fig. 8. Schematic diagram of ISFET. 1 = drain, 2 = source, 3 = substrate, 5 = insulator, 6 = metal lead, 7 = reference electrode, 8 = solution, 9 = membrane, 10 = encapsulant. Note that 7, 8, and 9 replace the metal gate 4 of the IGFET (Fig. 2).

The heart of an ISFET is the gate. It has been shown in the preceeding section how the gate voltage $V_G$ controls the drain current $I_D$ in the transistor (eqs. (11) and (12). In order to describe quantitatively the mechanism of operation of the ISFET it is necessary to perform the thermodynamical analysis of the structure shown in Fig. 9, which represents the cross-section through the *whole* measuring circuit including the reference electrode and the connecting leads. It is the simplest case, one where the reference electrode (1) is of the first kind, described by the equilibrium

$$M \rightleftarrows M^+ + e^-.$$

A typical example of such an electrode would be a silver wire immersed in solution of silver ions. Let us further assume that the solution (2) also contains a small amount of ions which can permeate reversibly into and out of the membrane (3) which, therefore, forms a non-polarized interface. A possible example would be a solution

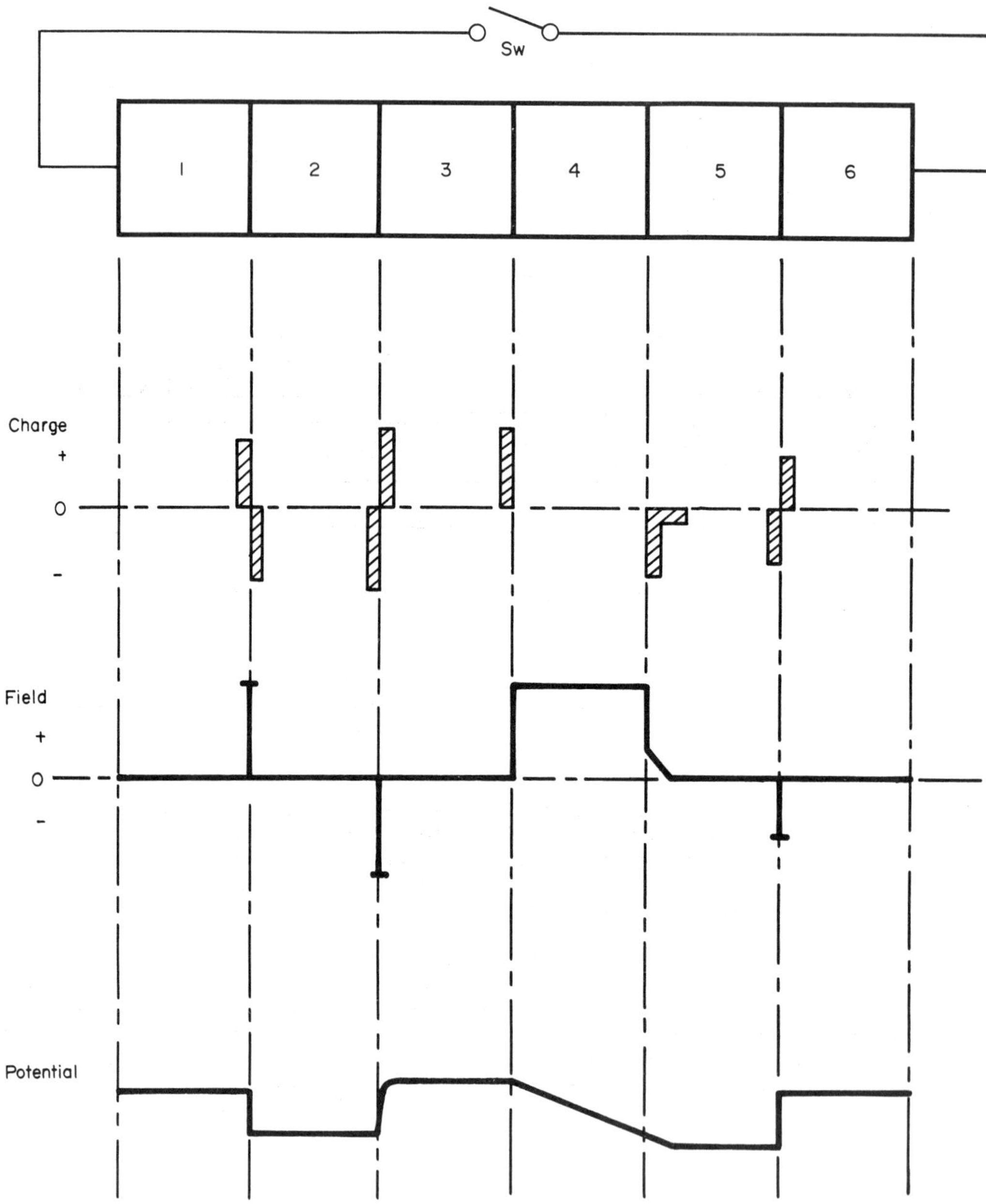


Fig. 9. Charge, field and potential profiles of ISFET.

(2) containing 0.1 M $AgNO_3$ and 0.1 mM $KNO_3$ and the membrane (3) being potassium ion-sensitive such as used in the equivalent ISE. The insulator (4) is assumed to be ideal, i.e. no charge can cross it and it is thicker then electron tunnelling distance ($d \geq 100$ Å). Layer (5) is the transistor semiconductor (such as silicon). The metal (6) will be identical with metal (1). A switch Sw represents operation with (Sw closed), and without (Sw open) reference electrode. The charge, field and potential profiles across this structure are also shown in Fig. 9. Note that this is a very simplified case. A liquid junction, dual layer insulator, trapped charges in the insulator, surface states at the insulator/semiconductor interface, channel doping and a multitude of connecting metals have been omitted for the sake of simplicity. Similar charge, field and potential profiles taking into account some of these elements have been published [10, 11, 14, 18].

From the thermodynamic point of view, this is a multiphase system for which the Gibbs equation must apply at equilibrium

$$\Sigma_i \, dn_i \, \tilde{\mu}_i = 0, \tag{14}$$

where $dn_i$ is the number of moles of species $i$ transported across individual interfaces and $\tilde{\mu}_i$ is the electrochemical potential of species $i$. The equation (14) can be expanded to yield relationship

$$\tilde{\mu}_3 \equiv \tilde{\mu}_2 \equiv \tilde{\mu}_1 \equiv \tilde{\mu}_6 \equiv \tilde{\mu}_5, \tag{15}$$

which simply expresses the fact that in the case of ideal insulator (4) the only way this system can reach equilibrium is through the external pathway (1), (6). If the switch Sw is open, a situation equivalent to operation without a reference electrode (or a signal return), then

$$\tilde{\mu}_3 \equiv \tilde{\mu}_2 \equiv \tilde{\mu}_1 \neq \tilde{\mu}_6 \equiv \tilde{\mu}_5. \tag{16}$$

The inequality of Fermi levels in metals (1) and (6) results in inequality of $\tilde{\mu}_3$ and $\tilde{\mu}_5$. Thus the basic condition for stable operation of ISFET is not satisfied.

As was mentioned in the introduction, it was claimed originally [3, 5, 9, 12, 13] that ISFETs can be operated without a reference electrode as a single ended probe. This point of view has been contested [8, 10, 11, 14] and it is now agreed that a signal return (reference electrode) is needed for a stable, d.c. operation of ISFETs [19, 20]. However, controversy still exists as to whether ISFETs sensitive to electrically neutral species can [19] or cannot [20] be operated without a reference electrode. This point has been addressed in Section 2.3.

Let us now analyze the circuit in Fig. 9. The inner potential in the semiconductor (5) can be expressed as

$$\phi_5 = \frac{1}{F} (\mu_5^e - \tilde{\mu}_5^e), \tag{17}$$

where $\mu_5^e$ is the chemical potential of electron in the semiconductor (the electron-lattice interaction energy) and $\tilde{\mu}_5^e$ is the electrochemical potential of electron in phase 5, normally known as the Fermi level. Similarly, for membrane (3) the inner potential $\phi_3$ is

$$\phi_3 = \frac{1}{z^i F} (\tilde{\mu}_3^i - \mu_3^i), \tag{18}$$

where $z^i$, the number of elementary charges, is positive for cations and negative for anions, and $\tilde{\mu}_3^i$ and $\mu_3^i$ are the electrochemical and chemical potentials of species $i$ in phase 3, respectively. The potential difference across the insulator and the semiconductor space charge region is then

$$\phi_5 - \phi_3 = \frac{1}{F}(\mu_5^e - \tilde{\mu}_5^e) - \frac{1}{z^i F}(\tilde{\mu}_3^i - \mu_3^i). \qquad (19)$$

It is now essential to identify the relationship between species $i$ in the membrane (3) and electron in the semiconductor (5).

We know that ion $i$ can transfer from the solution (2) into the membrane (3), thus according to eq. (15), its electrochemical potentials in the two phases must be equal:

$$\tilde{\mu}_3^i = \tilde{\mu}_2^i = \mu_2^i + z^i F\phi_2, \qquad (20)$$

where $\phi_2$ is the inner potential in the solution. Similarly, Fermi levels in the semiconductor and the metal (6) are equal. Because we defined the metal (6) to be the same as the metal (1) (the reference electrode) we can write

$$\tilde{\mu}_5^e = \tilde{\mu}_1^e = \mu_1^e - F\phi_1. \qquad (21)$$

There is an equilibrium in the metal between the cations and electrons.

$$M_1 \rightleftarrows M_1^+ + e_1^-,$$

for which we can formally write

$$\mu_1^M = \mu_1^{M^+} + \mu_1^e. \qquad (22)$$

Substituting for $\mu_1^e$ in eq. (21) we have

$$\tilde{\mu}_5^e = \mu_1^M - \mu_1^{M^+} - F\phi_1 \qquad (23)$$

Combining eqs (19), (21), (23) and rearranging yields

$$\phi_5 - \phi_3 = \frac{1}{F}(\mu_5^e - \mu_1^M + \mu_1^{M^+}) - \frac{1}{z^i F}(\mu_2^i - \mu_3^i) + (\phi_1 - \phi_2). \qquad (24)$$

The first term on the right-hand side of eq. (24) represents the contact potential between semiconductor and metal

$$\phi_5 - \phi_1 = \frac{1}{F}(\mu_5^e - \mu_1^e) = \Delta\phi_{cont}. \qquad (25)$$

The second term can be related to solution activity of the ion (Nernst equation):

$$\frac{1}{z^i F}(\mu_2^i - \mu_3^i) = E_0^i + \frac{RT}{z^i F}\ln a_2^i, \qquad (26)$$

where $a_3^i$ is assumed to be constant and is included in term $E_0^i$. Finally, $\phi_1 - \phi_2$ is the reference electrode potential $E_{REF}$. Equation (24) can now be written as

$$\Delta\phi_{3/5} = \phi_3 - \phi_5 = \Delta\phi_{\text{cont}} + E_0^i + \frac{RT}{z^i F} \ln a_2^i - E_{\text{REF}}. \tag{27}$$

This equation is essentially identical with that derived by Buck and Hackleman [14]. The voltage across the gate insulator $\Delta\phi_{3/5}$ can be superimposed on externally applied voltage $V_G$ which has the same meaning and function as defined in the theory of operation of IGFET (Section 2.2).

Let us now combine eqs (11) and (27).

$$I_{DS} = \frac{\mu_n C_0 W V_D}{L} \left(V_G + \Delta\phi_{\text{cont}} + E_0^i + \frac{RT}{z^i F} \ln a_2^i - E_{\text{REF}} + \frac{Q_{SS}}{C_0} - 2\phi_F + \frac{Q_B}{C_0} - \frac{V_D}{2}\right). \tag{28}$$

Define $V_T^*$ for ISFET as

$$V_T^* = -\Delta\phi_{\text{cont}} - E_0^i - \frac{Q_{SS}}{C_0} + 2\phi_F - \frac{Q_B}{C_0}. \tag{29}$$

The inclusion of the term $E_0^i$ (but not $E_{\text{REF}}$) in the threshold voltage is rather arbitrary. The reason is that the membrane is *physically* part of the ISFET and thus its standard potential should be included in the constant $V_T^*$. On the other hand, the reference electrode is a completely separate structure. The final equation for the drain current of ISFET sensitive to the activity of ions $i$ is

$$I_D = \frac{\mu_n C_0 W V_D}{L} \left(V_G - V_T^* \pm \frac{RT}{z^i F} \ln a_2^i - E_{\text{REF}} - \frac{V_D}{2}\right) \tag{30}$$

for operation in the non-saturation region and

$$I_D = \frac{\mu_n C_0 W}{2L} \left(V_G - V_T^* \pm \frac{RT}{z^i F} \ln a_2^i - E_{\text{REF}}\right)^2 \tag{31}$$

for operation in the saturation mode.

In the previously derived drain current equations the reference electrode potential was missing [5] and the poorly defined "membrane-semiconductor" [10, 21] or "liquid-semiconductor" [22] work function difference was used instead of the contact potential.

We have frequently encountered the opinion that the thermodynamics of ISFETs is undefined because the "reversible potential cannot be defined at the membrane/insulator interface." Electroanalytical chemists often refer to non-polarized electrodes as "reversible." A polarized electrode is also thermodynamically reversible, but has one more variable (usually charge) than a non-polarized electrode, which means that the interfacial potential can be *measured* directly *only if* the charge density can be kept constant. If it is not, the potential is observed to "drift." This point has been discussed in greater detail elsewhere [18]. What is meant by "non-reversible membrane/insulator interface" is in fact that it is a polarized interface, i.e. that a charged species cannot cross it. It is indeed true because the insulator is considered to be *ideal* and for the voltages involved in these measurements it *behaves like an ideal* insulator. In short, we do not need a non-polarized interface because the charge transport stops at this interface.

## 2.5. *Temperature dependence of the ISFET*

This discussion is based on the strong similarity between the IGFET and the ISFET.

As the drain current in eqs (11) and (12) depends only on the quantities $\mu_n C_o(W/L)$, $V_T$ and applied voltages, any temperature dependence must be through these quantities.

Since $C_o$ is the gate insulator capacitance/unit area

$$C_o = \frac{\varepsilon_{ins}}{d_{ins}}$$

where

$\varepsilon_{ins}$ is the permittivity of the insulator and

$d_{ins}$ is the thickness of the insulator.

These parameters are relatively constant with temperature in the range of interest so any temperature variation must be through $\mu_n$, the charge carrier mobility in the surface inversion layer. This quantity has been experimentally studied as a function of temperature [5, 6]. In the temperature region of practical interest, -55°C to +120°C, the variation of $\mu_n$ with temperature is satisfactorily approximated by a $T^{-1}$ dependence. Therefore, the variation of mobility with absolute temperature, $T$, will introduce a $T^{-1}$ dependence into $I_D$.

Of the terms in the equation for $V_T$, the temperature dependence of $\phi_F$ and $\phi_B$ must be considered. The fixed oxide charge $Q_{ss}$ is observed to be relatively independent of temperature over the temperature range of interest [7]. When dealing with the IGFET, $\Phi_{ms}$ can be considered independent of temperature. However, when the expression for $V_T$ is modified for the ISFET, the terms which replaces $\Phi_{ms}$ in the threshold equation are temperature dependent (see eq. (29), Section 2.4).

The ISFET drain current, eq. (30), may now be differentiated with respect to temperature to show the temperature dependence [10].

$$\frac{dI_D}{dT} = I_D \left[ \frac{1}{\mu_n}\frac{d\mu_n}{dT} - \frac{V_D\left(\frac{dV_T}{dT} - \frac{d(\text{EMF})}{dT}\right)}{\left(V_G - V_T^* \pm \frac{RT}{z^i F}(\ln a_2^i) - E_{REF} - \frac{V_D}{2}\right)} \right] \quad (32)$$

The term $d(\text{EMF})/dT$ is the temperature dependence of the electrochemical cell consisting of the reference electrode, the electrolyte and the ion-sensitive membrane.

The term $dV_T/dT$ in the above equation accounts for the temperature dependence of the semiconductor band gap and bulk Fermi level. It is the same as the temperature variation of the IGFET threshold. It is always negative for *n*-channel structures.

Inspection of eq. (32) shows that given $d\mu_n/dT$ is always negative and the term $(dV_T/dT - (d(EMF)/dT))$ is always negative [10], it is possible to choose an operating point around which $I_D$ is, to first order, independent of temperature. However, it is known that the term $d(EMF)/dT$ varies with the activity of the measured ion and is equal to zero only for one value of activity. The effect of temperature on the drain current can be, therefore, minimized by the proper selection of $V_G$, but not completely eliminated.

## 3. PRACTICE

### 3.1. *ISFET geometry*

Much of the work at the University of Utah has been done using a single silicon chip containing two ISFET devices and two conventional metal gate IGFET devices. The two metal gate devices provide a method of monitoring the basic device parameters independent of the chemically sensitive gate. The two ISFET devices allow for either a multisensor device or a reference device measuring standard conditions.

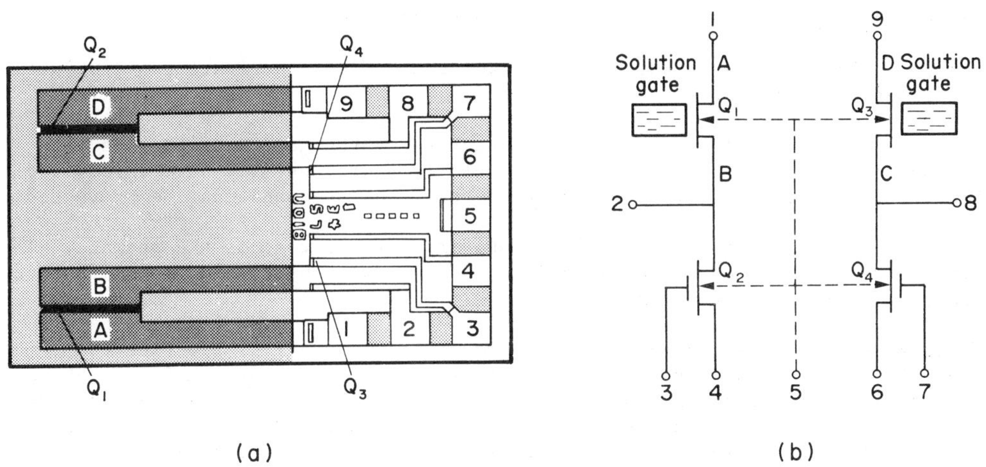


Fig. 10. (a) Layout and (b) circuit schematic of multidevice ISFET chip.

A layout of the dual device chip is shown in Fig. 10a. The overall dimensions of this chip are 1.28 mm x 2.16 mm. There are four field effect transistor (FET) structures on this chip identified as $Q_1$, $Q_2$, $Q_3$, and $Q_4$. Two of them, $Q_2$ and $Q_4$, are conventional metal gate IGFETs and the other two, $Q_1$ and $Q_3$, are devices fabricated without the metal gate. Numbered regions one to nine, are metal contact pads to which wires are bonded for connections to the external circuitry. The chip is so designed that the ISFET gates are well removed from the wire bonding pads to facilitate deposition of the ion-sensitive membranes, and from the area of the chip containing the electrical connections to be coated with an insulating material.

The $n$-type regions which form the source and drain of the two ISFET gates are elongated and form the current carrying paths to the bonding pads for the two

ISFET devices. These elongated *n*-type areas, designated A, B, C, and D in Fig. 10a contain appreciable resistance and form the parasitic resistors shown in Fig. 7, Section 2.2.

Figure 10b shows the circuit diagram of the devices on the ISFET chip. The identified points correspond to the areas of the photograph having the same number. Point 5 is the connection point to the substrate silicon. Other chip designs have been described [5, 22, 27, 29-31, 36], and the original papers should be consulted for details.

### 3.2. *Encapsulation*

Since the beginning of the use of integrated solid state devices, the electronics industry has been trying to provide a better protection against moisture and ionic contaminants. The very nature of ISFETs require that they be used in solutions of electrolytes which from the electronic point of view is a hostile environment. The electrical integrity of the whole ISFET assembly is critical and prerequisite for a successful operation of these devices.

There is no general recipe for encapsulation. The choice of materials is governed by the general device package (e.g. catheter, transistor header, etc.), by the geometry and layout of the chip and by the intended use.

From the point of view of the operation and the device, there are two distinct encapsulation problems:

(a) The encapsulation and electrical integrity of the gate and its immediate surrounding. The electrical leakage in this area will usually have an adverse effect on the electrochemical behavior of the transistor and must be prevented.

(b) The encapsulation and electrical integrity of the rest of the package. This affects the lifetime of the probe and can be an important consideration in electrical safety in medical applications.

The package which has been used by our group is shown in Fig. 11. The top lumen of a 6 French dual lumen PVC catheter (1) is cut off about 5-7 mm from the end. This provides a platform on which the chip (2) is attached with a cyanoacrylate contact adhesive. The connecting laquered copper leads (3) are threaded through the lower lumen (4), anchored with a layer of epoxy (5) and flattened to provide bonding pads. The end of the lower lumen is closed with epoxy. The upper lumen (6) is usually used as a reference electrode compartment, housing a silver/silver chloride wire (7), and a glass capillary (8) which provides a liquid junction. After wirebonding, the chip and the bonding wires are covered with the final encapsulant.

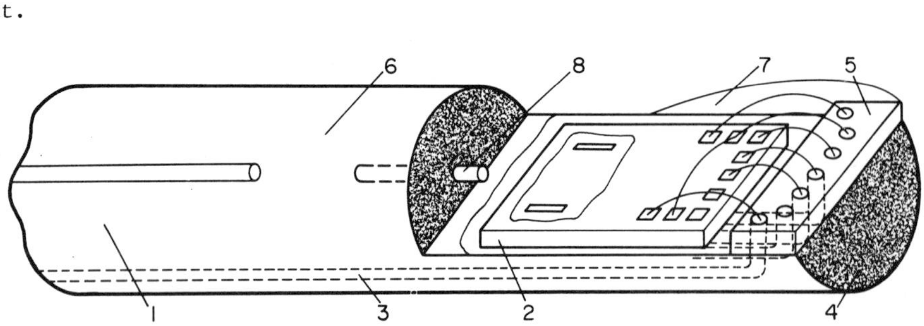


Fig. 11. ISFET chip mounted on dual-lumen 6F catheter (see text for description).

The choice of materials in a package like this has to be based on the following criteria:

(a) low bulk permeability for water and electrolytes;
(b) good adhesion between materials used; and
(c) that the machanical and rheological properties of the non-cured encapsulant are suitable for the fabrication.

The most satisfactory combination of materials used in our package has been found by trial and error over several years. Silicone rubbers and epoxies are generally recognized as good encapsulants for electronic components. Although the uptake of water by silicone rubber is low, its permeability is considerably higher than that of epoxies. Another reason for preference of epoxy as the encapsulant is that silicone rubber does not adhere well to PVC (membrane or catheter). The adhesion of epoxy to the silicon chip coated with silicon nitride can be improved by silanization. It also improves the adhesion of the PVC membrane *(vide infra)* to the gate area of the chip.

A radically different approach to encapsulation has been suggested by Wise and Weissmann [26]. It involves covering of the pre-bonded chip with a thin layer of silicon nitride. Similarly, vapor deposited polymers such as Parylenes, can be used for encapsulation provided that a suitable technique for opening the gate area can be found. Esashi and Matsuo [27] have insulated the sides of the chip by thermally growing silicon dioxide after separation of the chips by anisotropic etching. Bergveld *et al.* [19] suggest the fabrication of devices with leads on the opposite side from the ion-sensitive gates. To our knowledge no report on the device with this geometry has yet been published.

The electrical integrity of the gate is critically important for predictable operation of the ISFETs with membranes (i.e. other than bare silicon dioxide or silicon nitride gate). Any passage of current through the gate "insulator" and the overlaying membrane causes polarization of the latter. This leads to instability of the potential and after prolonged polarization sometimes to the degradation of the membrane. This situation is equivalent to using a low input impedence amplifier to measure voltage across a high resistance ISE membrane.

We have undertaken a detailed study [28] of the electrical integrity of silicon dioxide and silicon nitride prepared by chemical vapor deposition and RF sputtering and found that silicon nitride prepared by either technique offers an excellent insulator. These materials can withstand exposure to 0.15 M sodium chloride for 90 days. On the other hand, thermally grown silicon dioxide lost its insulating property within hours of immersion in the solution. The mechanism of this breakdown appears to be consistent with the formation of microfissures in the top layer of the silicon dioxide which are filled with the solution, and not with the total hydration of the silicon dioxide layer as was originally thought [5, 7, 10]. When a sufficiently high voltage is applied, the electric field at the tip of the fissure is much greater than on the flat surface of the insulator. This leads to a catastrophic breakdown of the remaining insulator beneath the fissure. Depending on the polarity of the applied voltage one or the other of the following reactions then takes place:

silicon negative:

$$H_2O + e^- \rightarrow 1/2\ H_2 + OH^- \qquad (I)$$

silicon positive:

$$Si + 2H_2O \rightarrow SiO_2 + 4H^+ + 4e^- \qquad (II)$$

The first reaction leads to the evolution of hydrogen bubbles which can be observed [5, 7, 10] under a microscope thus aiding in location of the leak. The presence of microcracks in silicon dioxide has been suspected by Esashi and Matsuo [27, 29] and by Schenck [30]. Their presence can also explain why the first ISFET [5] could be operated without a reference electrode, that is without a physically separate structure called "reference electrode." It is likely that the necessary electrical contact to the substrate was inadvertently provided by such a microcrack(s).

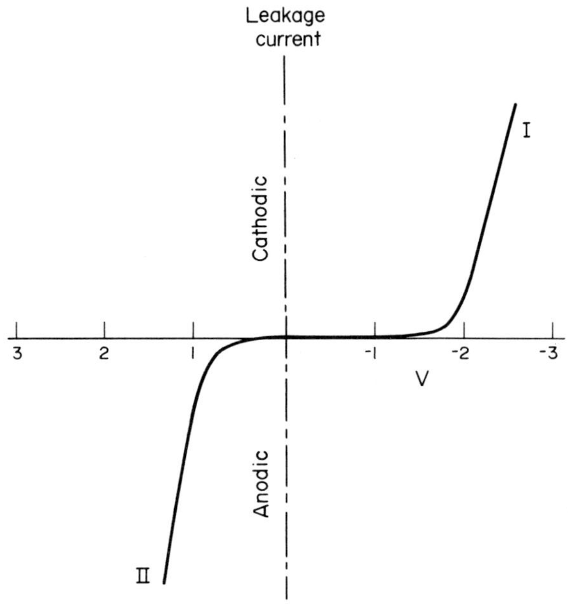


Fig. 12. Leakage current of poorly encapsulated ISFET package. The polarity of the applied voltage is that of the ISFET.

The ultimate test of the electrical integrity of the whole package is to connect all wires of the ISFET (i.e. drain, source and substrate) together, and to polarize the device from +3 to −3V against a low resistance ($R$ < 100 kΩ) reference electrode. The *shape* of the current-voltage curve (Fig. 12) as well as the absolute *value* of the leakage current reveals the possible leak. The increase of leakage current on the cathode side corresponds to evolution of hydrogen (Reaction I). The value of the decomposition potential depends on what material this reaction takes place on (a wire or silicon) and also on the previous polarization history. The anodic branch of the leakage current is due to the Reaction II, if the leak is through silicon. On the metal, the corresponding reaction can be the formation of oxide, insoluble salt or eventually evolution of oxygen. Since all these reactions are pH dependent, it follows that the leak does not make a good reference electrode if the pH of the solution is changing. This can account for inconsistent data obtained with the silicon dioxide gate pH ISFETs. It is also apparent from Fig. 12 that testing between +1 and −1 V cannot tell anything about the encapsulation, a point which is not appreciated by many researchers in this field.

The requirements on the longevity of encapsulation vary from one application to another. With our encapsulation package, described above, the electrical integrity of the encapsulation is maintained for at least 2 months.

### 3.3. *pH ISFET*

The hydrogen ion-sensitive FET with a bare insulator gate is the most widely reported ISFET. Because of this and because of its different mechanism of operation it will be treated separately.

There are two types of bare insulator gate ISFETs: one with only silicon dioxide and the other with silicon dioxide/silicon nitride gate structure.

3.3.1. *Silicon dioxide gate insulator.* The original mechanism of operation of the pH ISFET with a silicon dioxide gate was based on the assumption that the thermally grown silicon dioxide exposed by etching off the aluminum metal gate has ion-exchange properties similar to a pH glass electrode [5]. The observed sensitivity of these devices to changes in concentration of sodium chloride was based on the same assumption. Schenck [30, 31] reported sensitivity values of 47 mV $pH^{-1}$ at pH 7.5 and 0.01 N NaCl and 37 mV $pH^{-1}$ at pH 7.5 for 0.1 M sodium chloride. At the same time it was observed that hydration of the silicon dioxide layer caused a degradation of its insulating properties, so much so that evolution of hydrogen was observed with the ISFET biased sufficiently negative with respect to another electrode. The electrical integrity of silicon dioxide upon immersion in aqueous electrolyte has been questioned by Esashi and Matsuo [27,29] who suggested the existence of pinholes rather than bulk hydration as the cause of the electrical breakdown. We have confirmed [28] the existence of pinholes in thermally grown silicon dioxide as discussed in Section 3.2. It is very likely, that the devices with very thin (∿ 25 Å) silicon dioxide suffered from the same problem but no data on gate leakage current has been given [30].

It is known that quartz [32] as opposed to amorphous silicon dioxide does not hydrate extensively. However, the surface silanol groups dissociate:

$$\equiv SiOH \rightleftarrows SiO^- + H^+$$

and because of this equilibrium the excess charge density on the surface of quartz is pH dependent [33]. Porous silica and silica gel were found to have a surface charge density in excess of the surface density of silanol groups [34] which indicates the formation of a thin hydrated surface layer. This was also suggested by Perram *et al.* [35] who proved that most oxides hydrate to the depth of 40-50 Å. Our capacitance study [28] of the effect of prolonged exposure of 400 Å thick silicon dioxide film failed to detect extensive hydration although the insulation properties of this layer were lost within hours of immersion in solution. The sensitivity of our bridge was, however, insufficient to detect changes below 50 Å.

Based on these facts, and in the absence of a detailed experimental information on silicon dioxide gate ISFETs, it is difficult to formulate a coherent model of operation for these devices. The observed pH response can be partly due to the presence of the reversible double-layer, i.e. the dissociation of the surface silanol groups as supported by the different pH sensitivity for high and low concentrations of inert electrolyte [31]. The possible presence of an alumino silicate layer in Bergveld's devices [5] could explain the weak sodium ion sensitivity although the increased conductivity of the electrolyte could have been also responsible.

Because of the proven poor insulation quality of the thermally grown silicon dioxide and because of the uncertain mechanism of operation, the practical value of pH ISFET with silicon dioxide gate is doubtful.

3.3.2. *Silicon nitride gate insulator.* The mechanism of operation of pH ISFETs with silicon nitride gate seems to be clearer. Properly prepared silicon nitride does not form pinholes and does not hydrate which removes one uncertainty from the interpretation of the data. It has been shown [22] however, that if a "leaky" nitride is prepared the leakage current affects the pH sensitivity. It was found that if the device is biased within the potential range where the silicon/electrolyte interface is polarized (+1.5 - 1.0 vs SCE) the pH sensitivity remains at 50-51 mV $pH^{-1}$. An attempt has been made to explain quantitatively the effect of the leakage current on the pH sensitivity by incorporating the overvoltage term in the drain current equation [22]. However, it is neither necessary nor desirable to complicate the operation of ISFETs with a gate leakage current which introduces a nonequilibrium condition particularly when silicon nitride gate devices with good insulation can be made [6, 7, 21 27, 29 36].

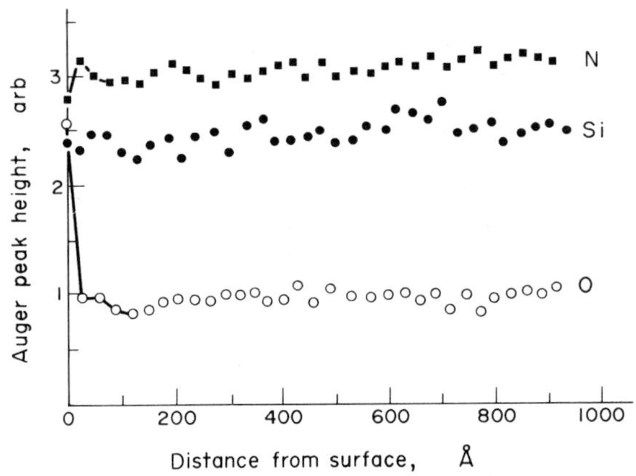


Fig. 13. Compositional profile of the silicon nitride surface (after ageing in water for 14 h). (Reprinted from Ref.29, with permission from IEEE Inc.)

The important contribution to the formulation of the mechanism of operation of the silicon nitride gate ISFET is the excellent paper by Esashi and Matsuo [29]. They have used Auger spectroscopy combined with argon ion sputter etching to determine the concentration profiles of silicon nitride exposed to water for several hours. Their results are in complete agreement with an earlier Auger - argon ion sputtering study of silicon nitride [37, 38] shown in Fig. 13. They have attributed the relatively high level of oxygen in the bulk of the nitride layer to an oxygen leak in their nitride deposition apparatus. We have found exactly the same oxygen profile in our ESCA - argon ion sputtering study of dry silicon nitride [39]. The etching of one of the samples was interrupted after removing the top 200 Å silicon nitride (out of 400 Å) and the sample was then exposed to air for several days. The subsequent ESCA analysis and further etching show the oxygen, nitrogen and silicon profile to be identical to the one obtained originally, i.e. high oxygen content gradually decreasing over 25-30 Å of the new top layer. These results indicate that the identical concentration profiles obtained in the three independent laboratories may represent the equilibrium state of "silicon nitride" which in fact is silicon oxynitride [38]. This agreement could also explain the relatively uniform pH response (50-60 mV) of silicon nitride gate ISFETs prepared in different laboratories and at different times (Table 1).

Table 1

| Reference | Sensitivity mV $pH^{-1}$ | Measured Range pH |
|---|---|---|
| 29 | 54 | 1 - 14 |
| 22 | 50 | 1 - 13 |
| 27 | 60 | 1 - 13 |
| 36 | 59 | 3 - 10 |

If in fact the hydrated layer is represented by a 25-30 Å *interphase* it can be responsible for the pH dependence of the surface charge of silicon nitride (30) in the same way as it is responsible for pH dependence of silicon dioxide [31]. It is also interesting to note that the pH response of a silicon nitride gate ISFET which was oxidized at 1050°C with wet oxygen was reduced to 20-35 mV $pH^{-1}$. This result can be explained by the formation of a surface layer of silicon dioxide which is known to have lower pH response [36]. It should be mentioned here that the pH dependence of surface charge and its effect on the flatband potential of the semiconductor-electrolyte interface has been reported for other oxides. Thus, for example, the flatband potential of a germanium/electrolyte interface was found to shift at the rate of 60 mV $pH^{-1}$ [40,41]. The similar shift of the flatband potential was reported for a titanium dioxide/electrolyte interface while *no* pH dependence was found for a cadmium sulphide/electrolyte over the range of 2-12 pH units, which is in agreement with the absence of hydrogen ion bonding sites on surfaces of cadmium sulphide [42].

The pH equilibrium at the nitride surface can be written as

$$SH \underset{}{\overset{K}{\rightleftarrows}} S^- + H^+$$

where $S^-$ is an unspecified negatively charged site. It follows from the equality of the electrochemical potentials for hydrogen ions in solution and on the surface (SH), that

$$\mu^{H^+}_{sur} + F\phi_{sur} = \mu^{H^+}_{sol} + F\phi_{sol}. \tag{33}$$

Assuming that the surface is stoichiometric and that the dissociation constant $K$ is very small

$$E = \phi_{sol} - \phi_{sur} = \text{const.} + \frac{RT}{F} \ln a_{H^+}. \tag{34}$$

Equation (34) substituted in eq. (29) or (30) gives the response of the drain current to a change of hydrogen ion activity in solution. Although this equation simply describes the observed behavior, it must be realized that the above assumptions may not be justified and that the actual mechanism is probably more complicated.

The wide range of pH sensitivity of silicon nitride may be due to the silanol groups in different strata of the hydrated layer being subjected to different interactions. This would affect their dissociation constant over a broad range. On the other hand, if they were all on the surface in a very similar energy state, then the expected pH response would be centered at ± 2 pH units around the p$K$ value of the dissociation of the surface silanol group. There are three pieces of experimental evidence to support this model:

(a) The surface silanization of a hydrated silicon nitride gate ISFET with amino propyl trimethoxyethyl silane has no measurable effect on the pH response. If only surface silanol groups were responsible for the pH response, then the sensitivity of the drain current to pH should be reduced.

(b) If only one kind of weakly acidic group on the surface of the pH ISFET were present, the pH response of this device should be at ± 2 pH units with respect to the p*K* of that group. This is demonstrated in Fig. 14 which shows the response of an ISFET with immobilized fatty acid (Fig. 14, curve a ) on the surface of a hydrophobic polymer and the pH response of the same polymer without the carboxylic acid (curve b) [43]. The curve a is a typical titration curve corresponding to the titration of a weak acid with a strong base. The lack of significant pH response of an ISFET coated with Parylene film was also reported by Matsuo *et al.* [44] who suggested the use of this device as a reference electrode. Parylene/solution is a blocked interface which could be subject to adsorption of various ions. The suggestion that such an interface can be used as a reference is based on the erroneous assumption that no response to hydrogen ions is synonymous with a stable reference electrode potential.

(c) We have observed that the drain current of the silicon nitride gate ISFET is virtually motion insensitive as would be expected from a non-polarized interface. On the other hand, the transistor described in paragraph (b) above is strongly motion sensitive which suggests the presence of a diffused charge on the solution side of the polymer/solution interface, which is disturbed by the movement of the liquid in the sheer plane.

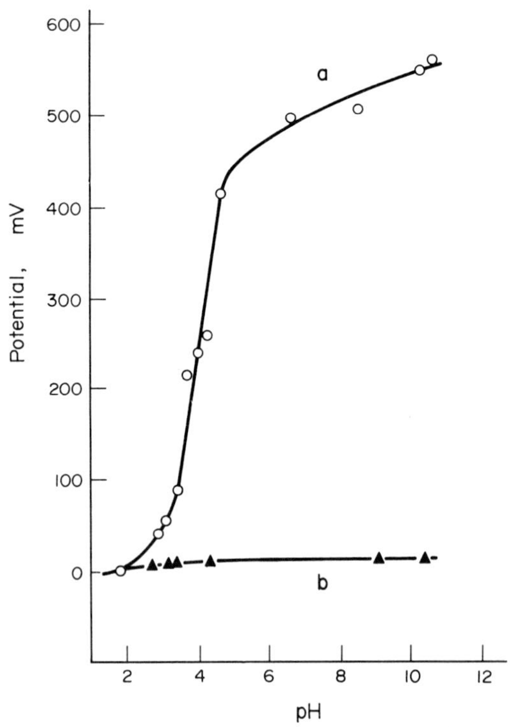


Fig. 14. pH response of hydrophobic polymer (a) with, and (b) without surface carboxyl groups.

The above results amply confirm the earlier suggestions [18, 30] that chemically sensitive transistors can be used as a tool for studying solid/liquid interfaces. This topic is, however, beyond the scope of the present review. The reason for a remarkable hydrogen ion selectively in the presence of $K^+$, $Na^+$, and other ions is not clear and merits further investigation.

### 3.4. *ISFETS with solid state membranes*

The main differences between bare gate insulator pH ISFETs and those which will be discussed in this and the following two sections is the thickness of the ion-sensitive layer. In this case it is much greater than the Debye length. The thermodynamic analysis presented in Section 2.4 applies directly to these devices. Naturally, the mechanism by which the inner potential difference between the solution and the bulk of the membrane is formed is different for each membrane. A detailed discussion of the various mechanisms of operation of ISE membranes can be found, for example, in reviews by Koryta [45] and Buck [46].

The use of solid state membranes with ISFETs is particularly attractive because the deposition techniques, for example, high vacuum evaporation, d.c. or r.f. sputtering and chemical vapor deposition are compatible with integrated circuit fabrication. Although this is a very promising approach, relatively little has been accomplished as yet.

Buck and Hackelman [14] have not made an ISFET, however, the results of their detailed study of the structure silver bromide/silicon dioxide/silicon are directly applicable to the general theory of ISFETs. Esashi and Matsuo [29] describe fabrication, testing, and use of pH and $Na^+$ ISFET. Their sodium ion sensitive layer is either aluminosilicate [47] or borosilicate [48] glass deposited by chemical vapor deposition techniques. The response of the $Na^+$ ISFET with aluminosilicate gate is shown in Fig. 15. The device exhibits equivalent potential change of 55 mV $pNa^{-1}$ in the range pNa 0-3 and measurable response down to pNa 5. Devices with a borosilicate membrane show a similar response.

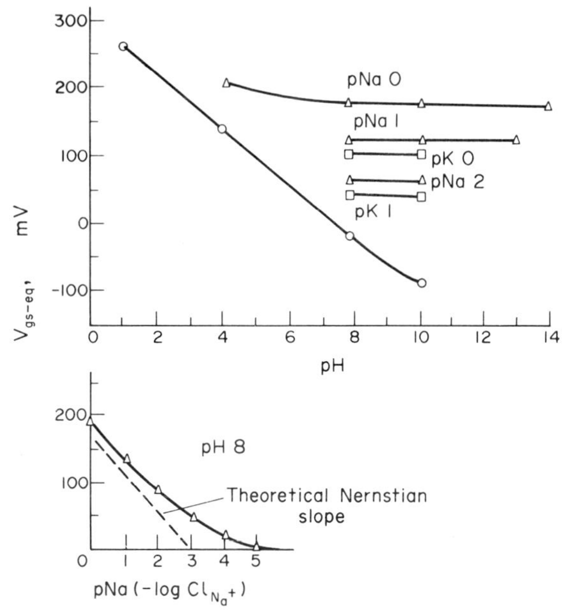


Fig. 15. Measured ion selectivity of the pNa sensor. (Reprinted from Ref. 24, with permission of IEEE Inc.)

Auger electron spectroscopy combined with argon ion sputter etching of fully hydrated alumino silicate has shown that the hydrated layer extends approximately 200 Å from the surface (Fig. 16). This finding is in agreement with the Eisenmann theory for sodium ion selectivity of alumino- and borosilicate glasses [32]. The particularly attractive feature of this sensor is its small size (200 μm tip, 1.5 mm shank) and the fact that it has been fabricated entirely by processes used in integrated circuits production. The only other ISFET with solid state membrane is the F ISFET which was reported in a preliminary form by Moss [49].

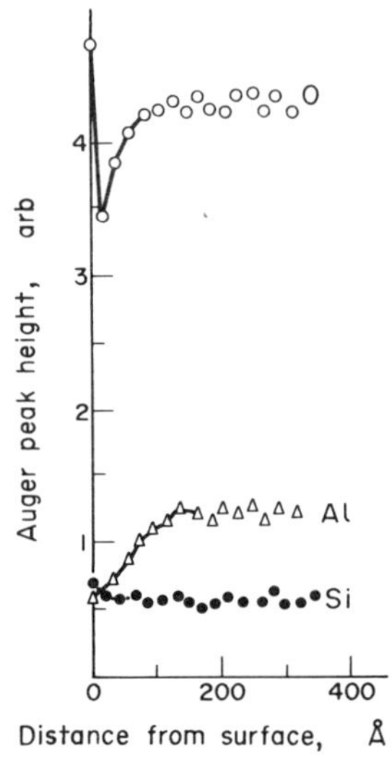


Fig. 16. Compositional profile of the alumino silicate surface. (After ageing in water for 24 h). (Reprinted from Ref. 29, with permission of IEEE Inc.)

There seem to be many problems encountered in the preparation of other ISFETs with solid state membranes. Many materials which form membranes in conventional ISEs cannot be deposited by high vacuum techniques. An example of such a material is silver sulphide used in silver and sulphide ion sensitive electrodes and which is a basic matrix for several other solid state electrodes. Many potentially useful ion-selective materials undergo dissociation and other chemical changes when evaporated.

### 3.5. *ISFETs with homogenous polymeric membranes*

The common denominator of the three types of sensors based on ion-selective field effects transistors (ISFET) described in this section is the mode of application of the ion-selective membrane; solvent casting. It has been pointed out before that the electrical integrity of ISFETs with membranes is a necessary prerequisite for stable operation of these devices. Our recently concluded study has shown that properly prepared silicon nitride provides a stable gate insulator. The surface silanization process combined with an epoxy encapsulant then provides sufficient protection for the transistor chip against the aqueous electrolyte.

In the first device of this type [7] a PVC-valinomycin-dioctyladipate membrane [50] cast out of a tetrahydrofuran solution has been used. Although the principle of operation has been demonstrated with this device, it showed a poor long term stability and relatively short lifetime. These problems have been attributed to the gradual leaching out of the ionophore and the plasticizer from the membrane. It was also estimated that a minimum membrane thickness of approximately 100 μm is required for a reliable performance. Membranes thinner than 50 μm suffer from the presence of pinholes which fill up with the electrolyte and eventually short out the membrane. Similar difficulties have been encountered in the construction [21] of $Ca^{2+}$ sensitive ISFETs utilizing a conventional PVC-based membrane [51]. Further development and the use of better membranes produced $K^+$, $Ca^{2+}$, and $H^+$ ISFETs with considerably improved performance characteristics [52].

3.5.1. *ISFET fabrication.* The general procedure for fabricating ISFETs with polymeric membranes has been published [52]. The outline of this procedure is given below.

The semiconductor chip [7] is eutectically attached to a 1.9 x 3.2 mm Kovar substrate coated with an approximately 0.80 μm layer of vacuum-deposited gold to provide an electrical connection of low resistance between the chip and substrate. This chip-substrate package is silanized with 3-aminopropyltriethyoxysilane in a toluene reflux system, ultrasonically cleaned in toluene and dried in a desiccator. The chip-substrate package is then attached to a specially prepared dual-lumen 6 French PVC catheter with Eastman 910 adhesive and electrical connections are made with an ultrasonic wire-bonder by means of Al-1% Si wires. The devices are encapsulated with epoxy (EPON 825 [Shell] with J230 crosslinking agent [General Mills]) to provide electrical insulation for the bonding wires and exposed silicon regions of the chip as well as to provide a suitable configuration around the chemically sensitive gates for casting membranes. The entire sensor package is finally cured at 60°C for 12-24 h. Potassium and calcium ion-sensitive membranes are cast from a 1:1 (v/v) mixture of tetrahydrofuran and cyclohexanone solution. The valinomycin-based potassium ion-sensitive membrane is prepared according to the formulation developed by Band *et al.* [53]. The calcium ion-sensitive membrane utilizes *t*-HDOPP [di(*p*-1,1,3,3-tetramethylbutylphenyl) phosphoric acid] as the electoractive material [54]. The pH membrane developed by LeBlanc *et al.* [55] is cast from a chlorobenzene-dichloromethylene solution. The membranes are applied to the CHEMFET gates by a sequence of 2-3 solution castings. All three types of membranes are allowed to cure for 8-12 h between membrane applications, and for approximately 24 h before solution testing begins.

3.5.2. *pH ISFET.* The pH response of the pH ISFET was tested by titrating a solution containing 0.1 M sodium acetate and 0.05 M sodium tetraborate with 0.1 M hydrochloric acid. Because of the several dissociation equilibria, the pH changes gradually from approximately pH 12 to pH 2. The results of these measurements are shown in Fig. 17, from which it can be seen that the usable linear range for the General Electric (GE) membrane is between pH 4 and pH 10 with a slope of 47 mV $pH^{-1}$. An average calibration slope between buffer solutions (pH 7 and pH 10) yields a value closer to 55 mV $pH^{-1}$. Bare gate pH sensors, utilizing the solid state silicon nitride gate insulator as a pH membrane, have also shown good response from pH 4 to pH 10; often with a greater slope than those obtained with the General Electric (GE) membrane. In agreement with LeBlanc *et al.* [55], the interference observed from the phthalate ion prevents the use of commercial pH 4 buffer for calibration. This problem can be easily circumvented, however, by using a sodium acetate buffer which does not interfere.

3.5.3. *$K^+$ and $Ca^{2+}$ ISFETs.* Two types of test procedures were used for these sensors; either the sensor with an internal micro-reference electrode was dipped into solutions of known concentrations of $K^+$ or $Ca^{2+}$, or a 1.0 M solution of potassium chloride or calcium chloride was added from a micrometer syringe to 20.0 ml of 0.15 M sodium chloride while the response of the sensor was continously recorded. For values of $Ca^{2+}$ below μM, calcium ion buffers were used [54]. The results of these tests are shown in Fig. 18, and the performance characteristics

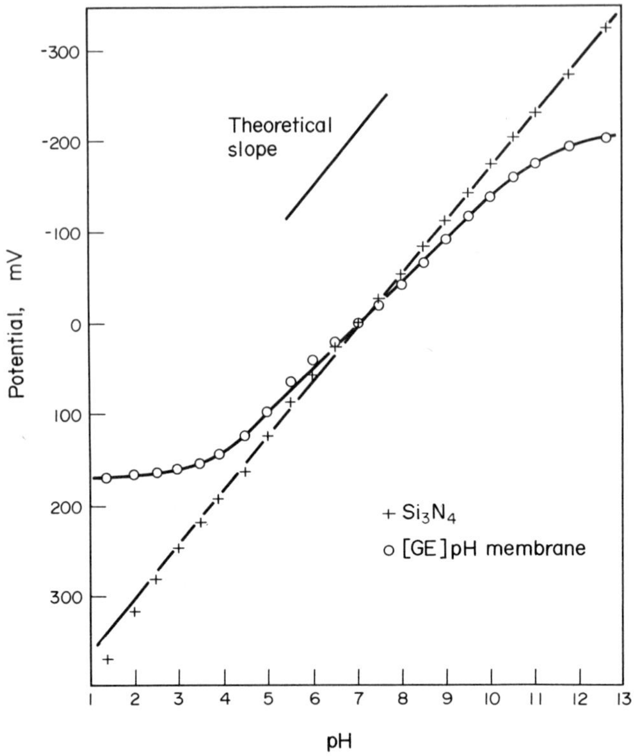


Fig. 17. pH response of pH ISFET with (+) $Si_3N_4$ surface and (O) GE polymeric membrane. (Reprinted from Ref. 52, with permission of Elsevier Scientific Publishing Co.)

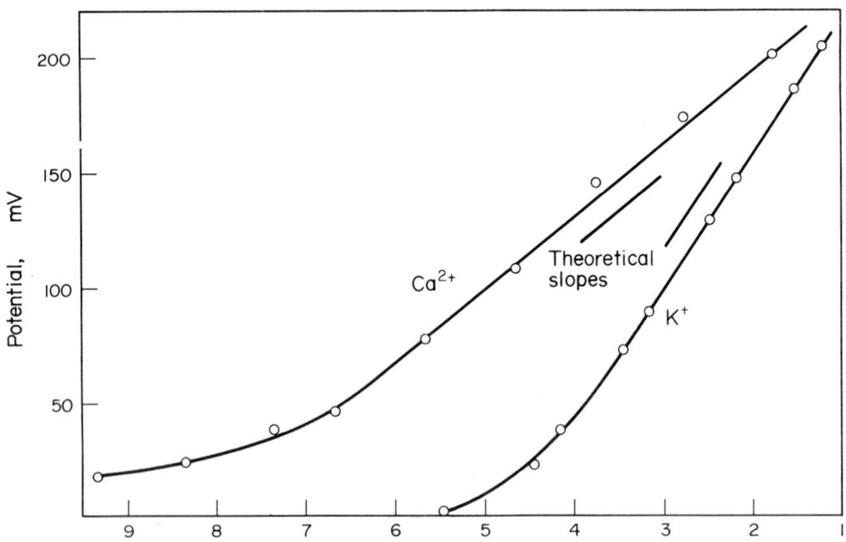


Fig. 18. Response of $Ca^{2+}$ and $K^+$ ISFET. (Reprinted from Ref. 52, with permission of Elsevier Scientific Publishing Co.)

of the three ISFETs are summarized in Table 2. Potassium ISFETs were used continuously for more than 30 days without any detectable degradation of their performance.

The membranes of $Ca^{2+}$ ISFETs, become opaque after approximately 10 days of continuous exposure to the solution with a concurrent loss of response. Recovery of the sensors was observed after the catheter had been dried in a desiccator and a drop of cyclohexanone placed on the membrane itself. Following this recovery, the devices were usable for approximately 5 days.

Table 2

| Membrane | Sensitivity mV $decade^{-1}$ | Range | Stability /mV $h^{-1}$ | Average Lifetime /$day^{-1}$ |
|---|---|---|---|---|
| pH | 56 at 37°C | 3-12 pH | 1.0 | 10-20 |
| $K^+$ | 57 at 25°C<br>59 at 37°C | $10^{-1} \rightarrow 10^{-5}$ M | 0.1 | > 20 |
| $Ca^{2+}$ | 27 at 25°C | $10^{-1} \rightarrow 10^{-9}$ M | 1.0 | ≤ 10 |

The thickness of the membrane has an effect on the response of the device for calcium and potassium ion sensors. Membranes thinner than about 40 μm sometimes produced a substantially lower slope (40 mV $decade^{-1}$ or less). The same pattern was exhibited for calcium. This behavior is attributed to the existence of "pinholes" in the membrane which provide a solution path through the membrane shorting out the potential. With membranes 80-150 μm thick, this phenomenon was not observed, and there was no deterioration of response over periods up to 1 month for potassium ion sensors.

As was discussed previously, a well behaved pH ISFET can be obtained exposing the partially oxidized silicon nitride surface of the gate insulator to solution. The main reason for the study of the membrane-based pH ISFET is the relatively high thrombogenicity of the silicon nitride surface. The linear range of both these types of devices, between pH 4 and 10, makes either type of device suitable for most biomedical applications.

The vastly improved stability of the $K^+$ and $Ca^{2+}$ ISFETs reported here over those described previously [7, 51] is due to better membranes and encapsulation. While the progressive deterioration of the calcium membrane appears to be the limiting factor for the lifetime of this type of sensor, the ultimate lifetime of the pH and $K^+$ ISFETs is determined by the adhesion of the membrane to the surface of the chip and the ability of the encapsulants to protect the chip during long exposures to aqueous media. With the present encapsulation techniques, the average usable lifetime of the sensors (pH and $K^+$) is approximately 20 days. Similarly, the stability and temperature sensitivity of the ISFETs are apparently determined by the transistor itself rather than by the ion-selective membrane.

### 3.6. *Heterogeneous membranes*

The use of heterogeneous membranes in ISEs has been pioneered by Pungor and his co-workers [56]. In general, the membrane consists of a semiconducting electrode material, usually an inorganic salt of low solubility, which is dispersed in a suitable elastomer, e.g. silicon rubber.

Heterogeneous membranes offer a simple solution to the problem of manufacturing ISFETs with silver halides, mixed silver halides-silver sulphides, etc. In order to demonstrate this possibility, we have prepared [57] ISFETs with these materials using silicone rubber (SR) or polyfluorinated phosphazene (PNF) as the elastomer material. Because of the high intrinsic viscosity of silicone rubber it was necessary to dilute it with chloroform. Devices prepared by this method had very sub-Nernstian response (5-27 mV decade$^{-1}$) to either silver or chloride ions.

A considerable improvement was obtained with PNF as a matrix. This polymer [58] dissolves readily in ketones and other common solvents.

3.6.1. *Silver chloride-PNF membrane.* The optimum composition of this membrane was found to be 75% silver chloride, 25% PNF with metyl isobutylketone (MIBK) as the casting solvent. The response to chloride ion was 52 mV decade$^{-1}$ in 0.01 M potassium nitrate with selectivity constants $K^{pot}_{Cl^-,Br^-}$ (at 75 μM $Br^-$) = 1.87, $K^{pot}_{SO_4^{2-},Cl^-}$) (at 10 mM $SO_4^{2-}$) = 0.01 and $K^{pot}_{NO_3^-,Cl}$ (at 0.1 M $NO_3^-$) = 1.3 x $10^{-4}$ and range $10^{-1}$-$10^{-6}$ M. It was irreversibly poisoned by exposure to $I^-$ and by exposure to serum.

3.6.2. *Silver chloride-silver sulphide membrane.* The inorganic salt for this membrane was prepared by co-precipitation of mixtures of chloride and sulphide with silver ions. The optimum molar ratio of $Cl^-$: $S^{2-}$ was found to be 4:1 with 75% of the mixed salt present in the PNF membrane. The casting solvent was again MIBK. The response was 53-58 mV decade$^{-1}$ at 37°C, range $10^{-1}$-$10^{-5}$ M. The selectivity constants determined by the fixed interference method are given in Table 3 and the response is shown in Fig. 19.

Table 3.

| Ion | Concentration M (except pH) | Selectivity constant $K^{pot}_{Cl^-,B}$ |
|---|---|---|
| bromide | 7.5 x $10^{-5}$ | 5.4 |
| iodide | 7.5 x $10^{-5}$ | 0.42 |
| sulphate | 0.01 | 4.5 x $10^{-3}$ |
| nitrate | 0.1 | 5.1 x $10^{-4}$ |
| pH | 5-9 | no sensitivity |

After exposure to iodide ions the sensitivity to chloride ion was reduced but not lost as was observed with the AgCl-PNF membrane, it was, however, poisoned by serum proteins.

One of the possible attractive biomedical applications for the $Cl^-$ ISFET is the chloride analysis in sweat for the detection of cystic fibrosis [59]. We have found that the response of our $Cl^-$ ISFET is not affected by sweat. The sample of sweat was analyzed by a modified [57] Gran plot technique using a known sweat sample as the "solvent" from which the "standard" solution was made (Fig. 20)

3.6.3. *Iodide and cyanide-PNF membrane ISFET.* The membrane for this ISFET was prepared from co-precipitated silver iodide-silver sulphide (6:1) 75% and PNF 25%. The casting solvent was again MIBK. The sensitivity for iodide ions was between 58-62 mV decade$^{-1}$ at 37°C in 0.1 M potassium nitrate solution and 62 mV decade$^{-1}$ for cyanide ions in 10 mM sodium hydroxide solution with a range $10^{-1}$-$10^{-7}$ M for both ions. The presence of 0.15 M sodium chloride in the sodium hydroxide solution somewhat reduces the response to cyanide ions (51 mV decade$^{-1}$). The selectivity

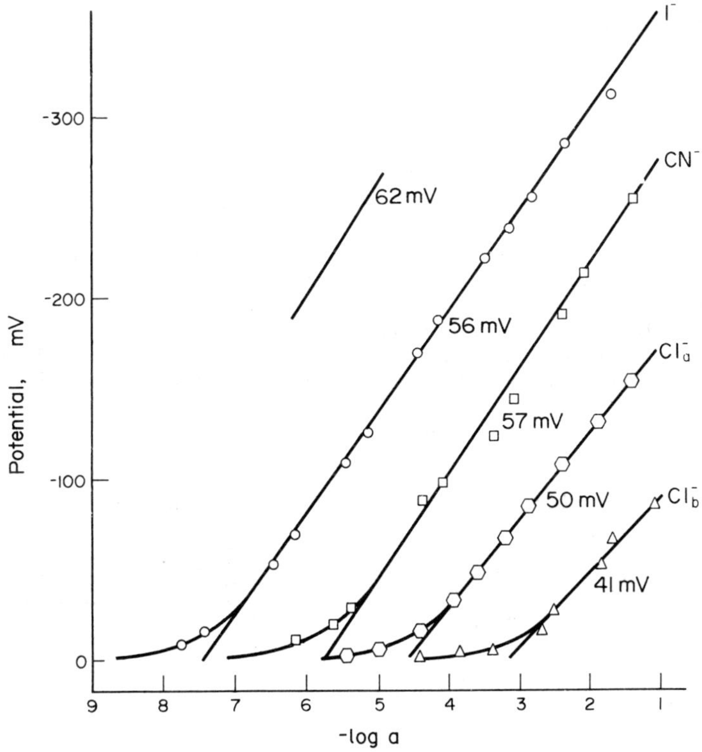


Fig. 19. Response of $Cl^-$ and $I^-$ ($CN^-$) ISFET.

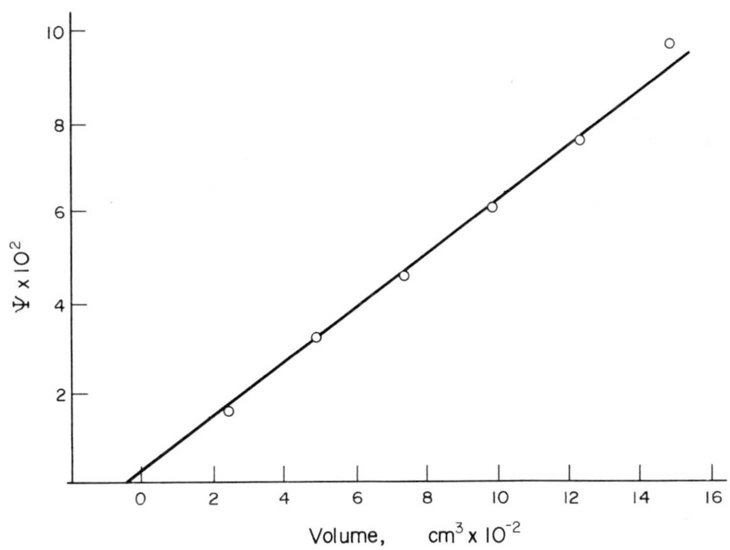


Fig. 20. Gran plot titration of sample of sweat for determination of chloride ions. Standard solution (1.0 + X) M sodium chloride in sweat. $\psi = (V_0 + V_S)\exp(\Delta E/S)$, where $V_0$ and $V_S$ are initial and added volumes, respectively and $S$ is the slope.

constants were $K^{pot}_{I^-,Cl^-} = 1.14 \times 10^{-5}$ and $K^{pot}_{CN^-,Cl^-} = 2.00 \times 10^{-4}$. The cyanide ion response was not affected by 2 h exposures to human serum although the response became slightly sluggish.

### 3.7. *Time response*

It has been postulated [5, 10] that the *in situ* impedance transformation should shorten the response time of an ISFET compared to an equivalent ion-selective electrode. With a typical ISFET gate capacitance of 3 pF and the combined resistance of the reference electrode and membrane of 500 kΩ, the expected time constant would be 1.5 μs compared to the 25 us observed here for the silicon nitride gate. The considerably longer time responses of the present membrane sensors - 40 ms to full response indicates that the response-limiting factor is the formation of a diffusion layer at the membrane surface. The diffusion has been shown to be the limiting factor in the response time for ion-selective electrodes [60, 61]. Thus, it is unlikely that an ISFET can be a faster probe than the corresponding ion-selective electrodes. However, because of their small size, localized, rapid changes of activity can be registered much more easily.

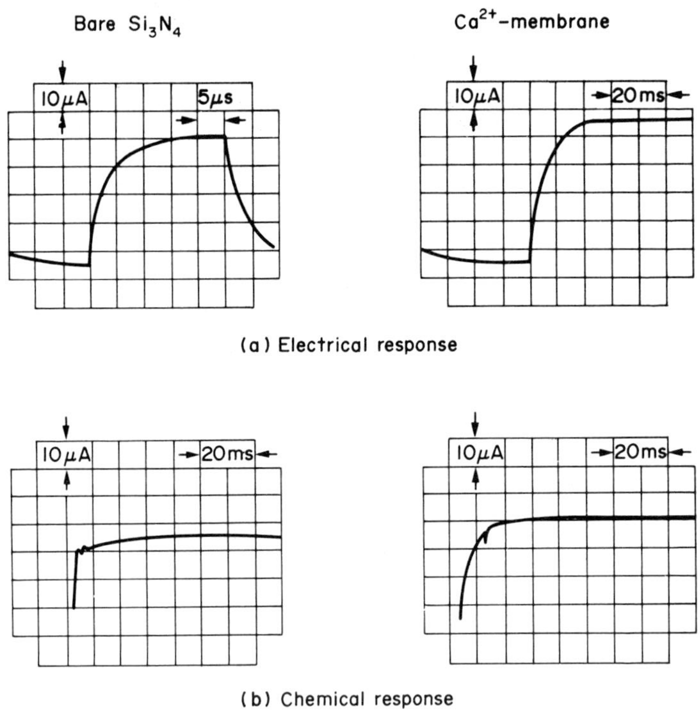


Fig. 21. Time response of silicon nitride and $Ca^{2+}$ membrane ISFET. (Reprinted from Ref. 52, with permission of Elsevier Scientific Publishing Co.)

The dynamic responses of ISFETs with and without a membrane have been tested. The response to a step change in the externally applied gate voltage is shown in Fig. 21a and the time response to a change in solution activity of the relevant ion is shown in Fig. 21b. This experiment was performed by pointing the effluent jet from a syringe containing a higher or lower concentration solution against the active gate of the ISFET. The signal from the sensor was fed into a storage oscilloscope and photographed with a Polaroid oscilloscope camera.

It was interesting to observe that when a jet of solution of the same composition as the background electrolyte was applied against the $Ca^{2+}$ or $K^+$ ISFET, very little change in drain current was observed. This indicates that these two types of ISFETs appear to be relatively insensitive to motion.

## 3.8. *Solid state reference electrode* [62]

Like the ion-selective electrodes, ISFET's require a reference electrode for a predictable operation. The quality of the measurement, therefore, depends on the performance of both half-cells. In addition, it makes sense to develop small, rugged ISFETs only if reference electrodes of the same size, or smaller, can be made. In order to provide a suitable reference electrode for our ISFETs, we have taken two approaches: in the first one we have incorporated a small silver-silver chloride reference electrode in the upper lumen of the catheter [22] which is used as a support for the transistor chip (Fig. 11). The reference electrode compartment was then connected with the sample by a free liquid junction. In the second approach, the pH ISFET was placed in the solution of constant pH which was then connected with the sample solution by a free liquid junction.

3.8.1. *Sensor preparation.* The transistor chips were mounted on the distal end of a dual lumen PVC 6-French catheter and encapsulated with epoxy resin Epon 825. A small well (approximately 500 x 500 x 500 μm) was built around one ISFET gate and a pH-sensitive polymeric membrane [55] was cast on *both* gates. Alternatively, bare silicon-nitride gates were used. After curing the encapsulant a 1% agarose gel prepared in a suitable buffer was packed into the reference gate well and approximately 20 μm glass capillary 200 μm long was inserted into the gel. The reference electrode compartment was then closed with Epon 825. After a short practice the preparation of the reference gate takes between 5 and 10 min. A schematic diagram of the gate and the measuring ISFET is shown in Fig. 22. The drain current of both transistors was measured with a simple differential current follower (Fig. 23).

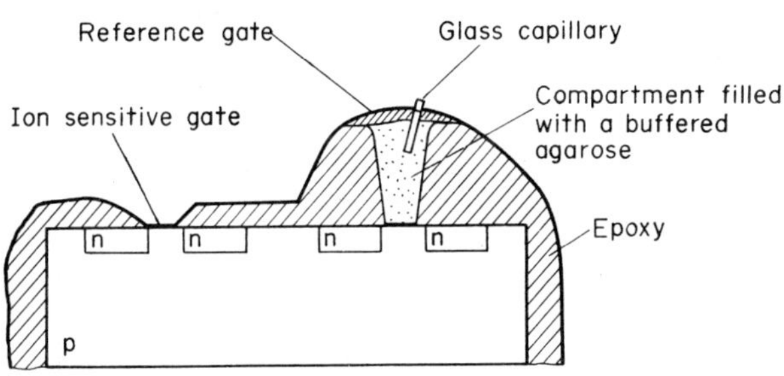


Fig. 22. Reference gate ISFET. (Reprinted from Ref. 62, with permission of Elsevier Scientific Publishing Co.)

3.8.2. *Performance.* The performance of the reference gate has been evaluated according to the following criteria: pH response, life-time, temperature, and noise sensitivity. Figure 24 shows pH response of the pH sensitive gate ($H^+$ - ISFET) curve a, reference gate (curve b, c) and the pH response of their difference (curve d). The useable range, pH 4-11, of the pH-sensitive gate is the same as reported by LeBlanc *et al.* [55]. The change of the reference gate potential with pH during titration of 0.05 M sodium tetraborate is linear, 3 mV $pH^{-1}$ for internal reference with 0.1 M buffer and 1.5 mV $pH^{-1}$ for internal reference with 0.1 M buffer plus

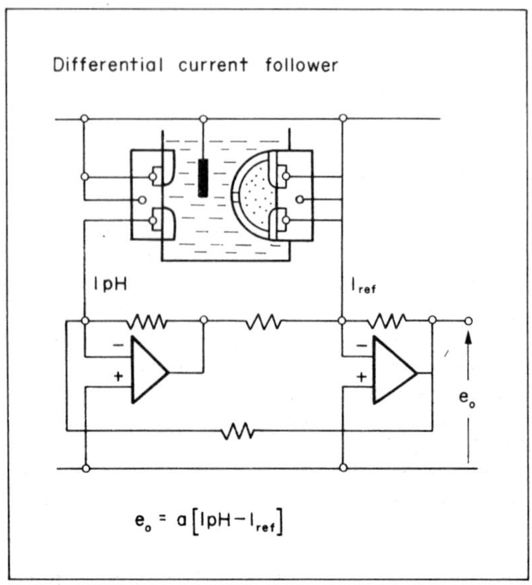


Fig. 23. Differential current follower for measurement of two ISFETs on the same chip. (Reprinted from Ref. 62, with permission of Elsevier Scientific Publishing Co.)

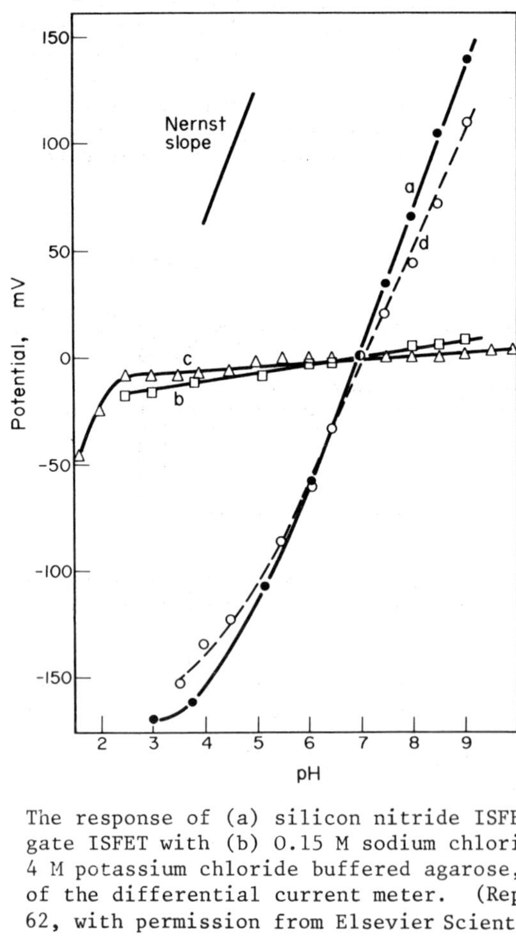


Fig. 24. The response of (a) silicon nitride ISFET, reference gate ISFET with (b) 0.15 M sodium chloride and (c) 4 M potassium chloride buffered agarose, (d) output of the differential current meter. (Reprinted from Ref. 62, with permission from Elsevier Scientific Publishing Co.)

4 M KCl solution. We suspect that the small deviation of the reference gate current at pH lower than 2.5 and higher than pH 10 (Fig. 24, curve c) is due to the presence of the capillary tip pH-potential.

The life-time of the reference gate was tested by leaving the device continuously in the solution and testing the pH response at various intervals. It was found that the performance of the reference gate did not deteriorate upon storage in pH 7.0 buffer solution for a minimum of 4 weeks.

Probably the most attractive feature of the reference gate is its ability to compensate for temperature changes and noise. Figure 25 shows the effect of the change

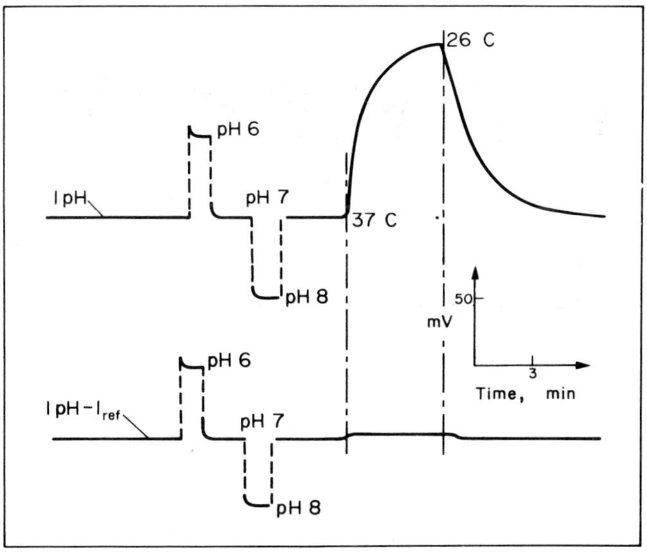


Fig. 25. Effect of temperature on pH ISFET and on the *difference* between pH ISFET and reference ISFET. (Reprinted from Ref. 62, with permission of Elsevier Scientific Publishing Co.)

of pH and temperature on the drain current of a pH ISFET ($I_{pH}$ curve) and the difference of drain currents of the pH and the reference gates ($I_{pH} - I_{REF}$ curve). When the solution contact is made via a proper reference electrode (such as Ag/AgCl, Satd. KCl ‖ sample . . .) the pH response of the pH gate and the difference is almost identical. However, as the temperature is changed from 37°C to 26°C the drain current of the pH gate changes accordingly while the difference remains practically unchanged. It can be shown that the change of the liquid junction potential with temperature is very small. Thus, the small change (0.008 pH/°C) of the drain current difference can be attributed, at least partially, to the difference between the change of the pH in the reference electrode compartment and the sample.

For the measurements discussed up to this point an ordinary stable reference electrode has been used as a contact between the solution and the transistor substrate. However, this is not necessary if we wish to evaluate *only* the differences between the reference and the indicator gate. Provided that an electrical contact is made between the substrate and the solution, the drain current *difference* is unaffected by the floating potential of this contact. Thus, we have used a bare silver wire, an exposed part of the silicon chip, or even a paper clip with no adverse effect on the drain current difference. This point is illustrated in Fig.

26 which shows the response of the drain current difference (lower curve) and the drain current of the pH gate when a noise was "injected" into the solution by touching the solution connection.

We have also experimented with the circuit in which either the reference or the pH gate is operated in a feedback (constant current) mode. Because this is a preferred mode of operation for a single gate measurement we expected a better result than with differential current follower. However, unless the characteristics of the two transistors are perfectly matched (which is rather difficult to achieve) the feedback operation offers no improvement over differential current measurement.

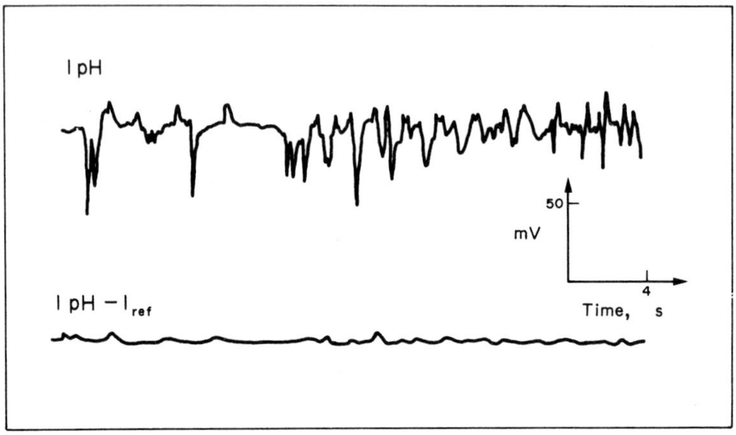


Fig. 26. Noise compensation using reference ISFET. (Reprinted from Ref. 62, with permission of Elsevier Scientific Publishing Co.)

The general applicability of this approach was tested on the probe which incorporated a $K^+$ ISFET and the reference gate. The results obtained with this combination were equivalent to the pH reference gate case.

Because of the small volume of the reference gate compartment it is preferable to use a true buffer solution with high buffer capacity in this compartment rather than a solution of strong electrolyte such as potassium chloride (for a silver/silver chloride electrode). The high concentration of the neutral salt in the reference gate compartment reduces the liquid junction potential. Because the reference gate can be made relatively easily the only disadvantage of this arrangement is that one gate cannot be used for measuring other ions. This is, however, a relatively small price to pay for an excellent temperature and noise compensation.

## 3.9. *Measuring circuitry*

The fundamental operating mechanism of the ISFET is a change in the drain current in response to a change in the activity of the ions in solution as described by eqs (30) and (31), Section 2.4. Maximum sensitivity will occur if the device is operated under conditions that maximize $dI_D/dV_a$ where $V_a = (RT/z^iF)\ \ln a_j^i$. To realize this, the operating conditions must be chosen so that it is biased in strong inversion (Section 2.2) and in the unsaturated mode (see Fig. 4, Section 2.2).

Strong inversion conditions result in all of the change in the semi-conductor surface charge, due to a change in the activity of the ions in solution, appearing as a change in the number of mobile charge carriers in the channel. If biased in weak inversion, only a part of this change affects the channel current. The ISFET may be biased in strong inversions by proper setting of the externally applied voltage $V_G$ as defined in Section 2.4.

For a fixed drain voltage, the sensitivity as defined above is greater when the device is operated in the unsaturated mode (eq. (30), Section 2.4) as can be seen by inspection of Fig. 5, Section 2.2. Unsaturated conditions are realized when

$$V_D < V_G - V_T^* \pm \frac{RT}{z^i F} \ln a_2^i - E_{REF}.$$

However, the operating point must be kept below that at which the sensitivity begins to decrease due to either loss of channel mobility or the effects of parasitic drain resistance as described in Section 2.2.

Once the operating point is chosen, there are two methods of measuring the ISFET response. The applied voltage, $V_G$, may be held constant and the change in drain current measured directly, or the applied voltage, $V_G$, may be changed in such a way that the drain current is held constant following a change in the activity of the ions in solution. As will be seen, this latter method is preferable.

3.9.1. *Constant* $V_G$ *operation.* Conceptually this is the simplest method. All externally applied voltages are held constant and the change in drain current, $\Delta I_D$, is measured following a change in ionic activity.

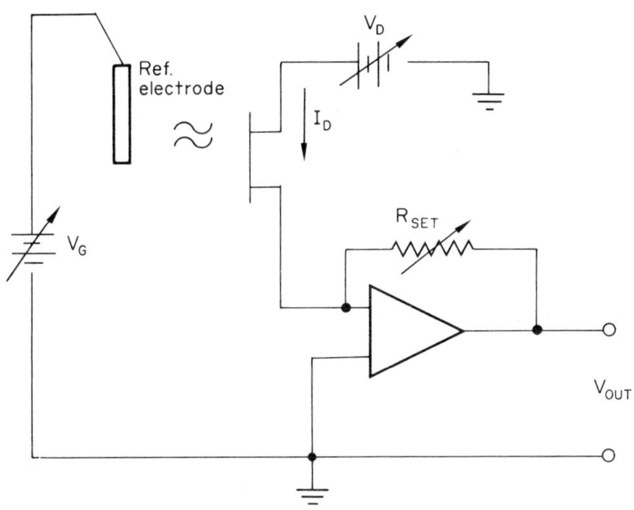


Fig. 27. Schematic of circuit diagram for measurement of drain current at constant $V_G$.

In the circuit of Fig. 27, $V_{OUT} = I_D R_{SET}$. The variable resistor, $R_{SET}$, is a calibration resistor.

While this measuring technique is simple, it suffers from the fact that the ISFET is not an ideal device. The drain current equations do not consider all of the parasitic effects, for example, series resistance in the source and drain. These effects cause the relation between drain current and system voltages to be different than the equations predict. As a result the device must be calibrated over its entire operating range and the calibration will be different for every different design. A better measurement method is constant current operation as described below.

3.9.2. *Constant* $I_D$ *operation.* By using an operational amplifier in a feedback loop as shown in Fig. 28, the operating point of the ISFET may be held constant. Any change in voltage due to a change in ionic activity $i$ appears as a change in the applied potential $V_G$. As the drain current is held constant, the parasitic effects described above are constant and the change in the potential is read out directly.

In the circuit of Fig. 28, the operational amplifier will maintain the voltage $V_{OUT}$ at whatever value is required to make $I_D$ the same as $I_{DSET}$. In response to a change in the activity of the measured ions, it will automatically adjust $V_G$ in eq. (30) or (31) to balance the change in the term $\pm\ (RT/z^iF)$. Since nothing else changes, the readout is directly in mV.

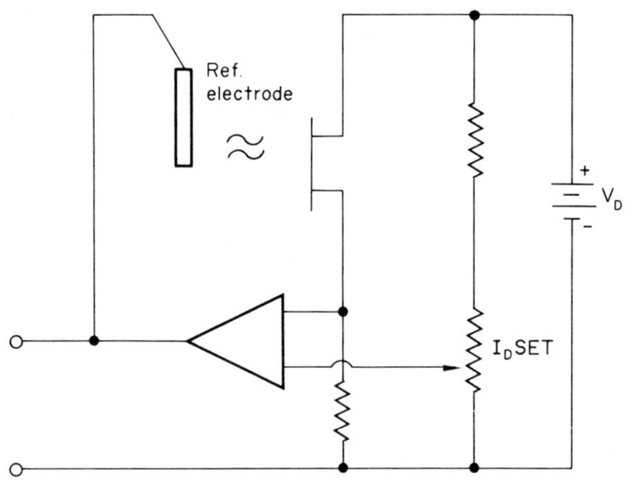


Fig. 28. Schematic of circuit diagram for measurement of ISFET response at constant $I_D$ (feedback).

3.9.3. *Differential measurement.* There are numerous applications in which two ISFET devices on the same chip are used in a differential mode. One device measures the parameter of interest and the other provides a reference signal corresponding to some standard condition. The reference ISFET of Section 3.3. is an example. For these applications, a differential current follower is used as shown in Fig. 23.

## 4. BIOMEDICAL APPLICATIONS

### 4.1. *Measurement of biopotentials*

Historically, the ISFETs have been introduced for various electrophysiological measurements [3-6]; the main reason being small size, excellent signal-to-noise ratio, and wide bandwidth due to *in situ* impedence transformation. Although the initial expectations [3, 5] that the ISFET can be used as a single ended d.c. probe have not materialized it has been used successfully without an *apparent reference electrode* for recording of transient biopotentials, the required signal return being provided either by inadvertent leak through the insulator oxide or by capacitive coupling to either drain or source of the transistor. Thus, the electrophysiological activity of *Taenia coli* has been monitored [5]. The analysis of capacitively coupled ISFETs (in that particular case called "OSFET") has been performed [63] and the device has been used to record action potentials of the *flexor tibialis* muscle of a locust. The important conclusion of this study is the considerably improved bandwidth of this experimental arrangement ($0 \leq f \leq 45$ kHz). Even better frequency response ($0 < f < 100$ kHz) has been reported by Matsuo and Wise [6] who used a conventional reference electrode as a signal return. First reported measurement of hydrogen ion activity in a biological situation came again from Matsuo's group [29]. Figure 29 shows the effect of interrupted respiration on intracranial pH and $pNa^+$ in a rabbit.

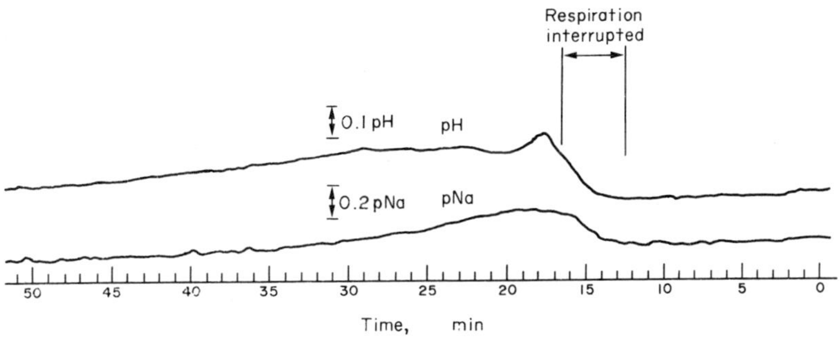


Fig. 29. Measured continuous variations of intracranial pH and pNa in a rabbit when artificial respiration is stopped for 4 min. (Reprinted from Ref.29, with permission of IEEE Inc.)

## 4.2. *Sterilization*

One of the prerequisites for *in vivo* medical application of ISFETs is the sterilization of the probe including the reference electrode. This is particularly important for long term and chronic implants. While the ion-selective electrodes can usually be sterilized by contact (chemical sterilization), the reference electrode presents a problem mainly because the liquid junction, which is by definition an open channel, and the inner compartment can be a potential source of infection. Radiation or high temperature sterilization is impossible because of the damage to the sensor. Thus, the sterilization procedure which is chosen must be both effective in sterilization of cavities and non-destructive with respect to the electronics. One of the accepted sterilization procedures is the treatment with gaseous ethylene oxide (ETO).

The principal advantage of ETO is that it penetrates cavities which would be otherwise inaccessible to other sterilants and that also the sterilization can be done at room temperature. For these reasons ETO sterilization is most suitable for sterilization of delicate medical devices. The main disadvantages of ethylene oxide is that it reacts to form various toxic products [64], such as ethylene glycol, chlorohydrin, etc. These reactions are also responsible for rapid deactivation of ETO as sterilant. In short, it is not possible to use ETO to sterilize objects which contain water and other reactive species.

The following constraints are, therefore, important on ETO sterilizable reference electrode:

(a) the electrode must be dry;
(b) all compartments must be accessible to gaseous ETO;
(c) it must be possible to reconstitute the sterilizable electrode under sterile conditions.

Our sterilizable reference microelectrodes [65] are prepared as described in Section 3.2 (Fig. 4).

One percent solution of hydroxyethyl methacrylate (HEMA) in methanol which is also 0.15 M in sodium chloride is then aspirated into the reference electrode compartment through the capillary by applying a moderate vacuum to the open end of the catheter. The whole assembly is then placed in vacuum oven at 50°C for 12 h to remove methanol. This dry electrode is then sterilized by a standard ETO sterilization/aeration procedure. The sterile electrode is immersed in sterile solution to reconstitute itself. In this stage the capillary force pulls the solution into the reference electrode compartment. The concentration of sodium chloride inside the compartment is initially very high and consequently the osmotic pressure difference forces more water into the compartment. The flow of solvent continues until the osmotic pressure difference vanishes. The electrode is then ready for measurement.

The electrical contact is established within minutes of immersion. The electrode drifts for 2 h and then becomes stable ( $< 0.2$ mV $h^{-1}$). The potential difference against a conventional double junction saturated calomel electrode is -62 mV. The presence of HEMA causes the concentration of sodium chloride to be slightly higher inside the reference electrode compartment thus making the electrode negative by 15 mV with respect to a silver-silver chloride wire immersed in 0.15 M sodium chloride.

The preparation of a sterilizable electrode is very simple. The driving force of the reconstitution process is the osmotic pressure difference, which vanishes as the equilibrium between the two solutions is reached. Consequently, as long as the concentrations of the ions in the sample which determine its osmolality remain constant no net transport of ions through the liquid junction will occur and, therefore, the liquid junction potential should remain zero. In physiological solutions in which the electrode would be used the major component determining the osmotic pressure is sodium chloride which, in the vast majority of applications, remains constant. However, it is recognized that sodium chloride is not the most suitable electrolyte to be used in a liquid junction because of the large difference in the mobilities of sodium and chloride ions. In principle, it is possible to use potassium chloride inside the electrode compartment and sterile potassium chloride as the reconstituting solution. However, most of our applications involve measurement of the activity of potassium ions and thus the presence of a liquid junction which would leak potassium ions is undesirable.

The efficacy of the sterilization procedure has been tested on five electrode/ISFET assemblies. Three were prepared and reconstituted as described above, then cut into 5 cm pieces and incubated in sterile culture broth to test for sterility and pyrogenecity. After 14 days of incubation all five probes were found to be sterile.

Contrary to the finding by Band *et al.* [53], no detrimental effect of ETO on performance of the $K^+$ ISFET has been found. There was also no adverse effect of sterilization on the performance of pH ISFETs (with $Si_3N_4$ or General Electric membrane) or the $Ca^{2+}$ ISFET. We believe that this is due to the absence of the internal aqueous reference solution in ISFETs.

## 5. CONCLUSIONS

Throughout this review, we have been comparing ISFETs and ISEs because of many similarities between the two types of sensors. There is, however, a significant constraint on the future development of ISFETs. While ISEs can be made and studied under very modest conditions, ISFETs can be fabricated only in a well equipped and relatively expensive semiconductor laboratory. There are few Universities in the world which have direct access to such a facility and are actively engaged in ISFET research. Most of the integrated circuit laboratories are owned and operated by industrial firms, which for obvious reasons do not always readily publish the results of their work. This situation is likely to impede the progress in research at least until ISFETs become commercially available.

Inevitably, the question is asked what are the advantages of ISFETs over ISEs? Because ISFETs are still in their infancy while ISEs have been well tested in many analytical situations, the answer is partially speculative.

First of all, the high resistance of the ion-selective membrane is *in situ* transformed into low output impedance. This means that the signal leads do not have to be heavily shielded in order to minimize electrical interferences. In turn, the probe can be made flexible which is important for many medical applications. The completely solid state design, particularly the absence of the internal reference solution, make ISFETs light, rugged, and sterilizable. The actual sensing area is very small compared to conventional ISEs although it may never be as small as glass capillary microelectrodes where the measuring tips are of the order of 1 μm. However, it is well within the state of the art of integrated circuit technology to accommodate tens of transistors on a chip of 1 $mm^2$. Although the membrane deposition techniques have not yet been developed to the point where various membranes can be applied on several gates in a very small area it is only a matter of time before a *multiprobe* can be made without any increase of overall size. The limiting factor in this direction seems to be the number of leads which can be bonded to the chip. The most promising solution to that problem appears to be to include a multiplexor and A/D converter on the same chip. Thus, an integrated data acquisition/processing unit could be made. As was discussed in Section 2.5 the partial temperature compensation can be achieved by tailoring the transistor characteristics and by the proper selection of the operating point. Another possibility for temperature (and light, if necessary) compensation is to build the compensating elements directly on the sensor chip.

The low fabrication cost of conventional integrated circuits could be achieved for manufacturing of ISFETs as soon as the problems of on-line, automatic encapsulation and membrane deposition are solved.

Finally, many ISEs serve as substructures for enzyme electrodes. In direct analogy enzyme FETs (or ENFETs) can be envisaged. The particularly attractive feature of FETs in this application is the small size of the active gate area which could be a significant economical factor in large scale production of these devices.

There are several unsolved problems. The most pressing one is the automatic encapsulation and automatic membrane deposition. IGFETs are vulnerable to static electricity. This problem is mitigated once the ISFET has been placed in a conducting solution, however, considerable anti-static precautions have to be taken during fabrication and initial testing. Every ISFET reported to date has exhibited a drift, usually equivalent to $1 \pm 0.5$ mV $h^{-1}$ change of gate voltage. The exact origin of this drift is not yet known and needs to be further investigated. It is our belief that ISFETs will complement and possibly replace micro ISEs. With the theory of their operation generally understood the remaining problems seem to be technological.

This review covers the literature to October 1978.

Acknowledgement — This work was partly supported by the grant from the National Institutes of Health, Grant No. NIGMS. #22952.

## 6. GLOSSARY OF TERMS

| | |
|---|---|
| $a_j^i$ | Activity of species $i$ in phase $j$. |
| $C_o$ | Capacitance per unit area of the gate insulator. |
| $d$ | Thickness of the insulator. |
| $E_o^i$ | Standard potential of electrode sensitive to species $i$ |
| $E_{REF}$ | Potential of reference electrode. |
| $F$ | Faraday constant |
| $I_D$ | Drain current |
| $K_S$ | Dielectric constant of silicon |
| $L$ | Channel length |
| $N_A$ | Volume density of acceptor atoms |
| $q$ | Charge of electron |
| $Q_B$ | Charge per unit area in semiconductor surface space charge region |
| $Q_n$ | Charge per unit area of electrons in surface inversion layer |
| $Q_S$ | Total semiconductor surface charge per unit area |
| $Q_{SS}$ | Intrinsic $Si/SiO_2$ interface charge |
| $R$ | Gas constant |
| $T$ | Absolute temperature |
| $V_D$ | Drain voltage |
| $V_{DSAT}$ | Saturation voltage |
| $V_{FB}$ | Flat band voltage |
| $V_G$ | Gate voltage |
| $V_T$ | Threshold voltage of IGFET |
| $V_T^*$ | Threshold voltage of ISFET |
| $V(y)$ | Voltage drop along the channel due to $I_D$ |
| $W$ | Channel width |
| $z^i$ | Number of charges of species $i$ |
| $\varepsilon_o$ | Permittivity of free space |
| $\Phi_{ms}$ | Metal-semiconductor work function difference (here = contact potential) |

$\Phi_F$ Fermi level in semiconductor (here equals $\tilde{\mu}_5^e$)

$\Phi_j$ Inner potential in phase *j*

$\Phi_s$ Surface potential

$\mu_n$ Effective electron mobility in channel

$\mu_j^i$ Chemical potential of species *i* in phase *j*

$\tilde{\mu}_j^i$ Electrochemical potential of species *i* in phase *j*

## 7. REFERENCES


1. M. A. Afromowitz and S. S. Yee, *J. Bioeng.* 1 55 (1977).
2. K. Nagy, T. A. Fjeldly, and J. S. Johannessen, *153rd Meet. Electrochem. Soc.* 78 Abstract 108 (1978).
3. P. Bergveld, *IEEE Trans,* BME-17 70 (1970).
4. T. Matsuo, M. Esashi and K. Iinuma, Digest of Joint Meeting of Tohoku Sections of IEEJ, October 1971
5. P. Bergveld, *IEEE Trans.* BME-19 342 (1972).
6. T. Matsuo and K. D. Wise, *IEEE Trans.* BME-21 485 (1974).
7. S. D. Moss, J. Janata and C. C. Johnson, *Anal. Chem.* 47 2238 (1975).
8. I. Lundstrom, M. S. Shivaraman, C. Svensson and L. Lundkvist, *Appl. Phys. Lett.* 26 55 (1975).
9. J. N. Zemel, *Anal. Chem.* 47 255A (1975).
10. J. Janata and S. D. Moss, *Biomed. Eng.* 11 241 (1976).
11. R. G. Kelly, *Electrochim. Acta,* 22 1 (1977).
12. J. N. Zemel, *Res. Devel.* April, 38 (1977).
13. A. G. Revesz, *Thin Solid Films* 41 L43 (1977).
14. R. P. Buck and D. E. Hackleman, *Anal. Chem.* 49 2315 (1977).
15. *Workshop on Theory, Design, and Biomedical Application of Solid State Chemical Sensors,* P. Cheung, D. G. Fleming, W. H. Ko and M. R. Neuman, Eds., CRC Press, 1978.
16. A. S. Grove, *Physics and Technology of Semiconductor Devices,* Wiley, New York, 1967.
17. A. Many, Y. Goldstein and N. B. Grover, *Semiconductor Surfaces,* North Holland, Amsterdam, 1965.
18. J. Janata, in Ref. 15.
19. P. Bergveld, N. F. DeRooij and J. N. Zemel, *Nature* 273 438 (1978).
20. J. Janata, R. J. Huber and S. D. Moss, (unpublished results)
21. S. D. Moss, C. C. Johnson and J. Janata, *IEEE Trans.* BME-25 49 (1978).
22. S. D. Moss, J. B. Smith, P. A. Comte, C. C. Johnson and L. Astle, *J. Bioeng.* 1 11 (1977).
23. O. Leistiko, A. S. Grove and C. T. Sah, *IEEE Trans. Electron. Dev.* ED-12 248 (1965).
24. L. Vadasz and A. S. Grove, *IEEE Trans. Electron. Dev.* ED-13 863 (1966).
25. A. S. Grove, B. E. Deal, E. H. Snow and C. T. Sah, *Solid St. Electron* 8 145 (1965).
26. K. D. Wise and R. H. Weissman, *Med. Biol. Eng.* 9 339 (1971).
27. M. Esashi and T. Matsuo, Proc. 6th Conf. Solid State Dev., Tokyo (1974), Suppl. *J. Jap. Soc. Appl. Phys.* 44 339 (1975).
28. R. M. Cohen, R. J. Huber, J. Janata, R. W. Ure and S. D. Moss, *Thin Solid Films* (1978).
29. M. Esashi and T. Matsuo, *IEEE Trans.* BME-25 184 (1978).
30. J. F. Schenck, in Ref. 15.
31. J. F. Schenck, *J. Colloid Interface Sci.* 61 569 (1977).

32. G. Eisenman, *Glass Electrodes for Hydrogen and Other Cations*, Marcel Dekker, New York (1967).
33. S. M. Ahmed, in *Oxides and Oxide Films*, J. W. Diggle, Ed., Vol. 1 319 (1972).
34. T. F. Tadros and J. Lyklema, *J. Electroanal. Chem.* 17 267 (1968).
35. J. W. Perram, R. J. Hunter and H. J. L. White, *Aust. J. Chem.* 27 461 (1974).
36. D. J. Fung, P. W. Cheung, S. H. Wong, J. A. Topich and W. H. Ko, in Ref. 2, Abstract 81.
37. J. S. Johannessen, W. E. Spicer and Y. E. Strausser, *Thin Sold Films* 32 311 (1976).
38. H. G. Maguire and P. D. Augustus, *J. Electrochem. Soc.* 119 791 (1972).
39. J. Janata, D. Briggs and G. R. Davies, unpublished results.
40. G. Brouwer, *Phys. Lett.* 21 399 (1966).
41. G. Brouwer, *J. Electrochem. Soc.* 114 743 (1967).
42. T. Watanabe, A. Fujishima and K. Honda, *Chem. Lett.* 897 (1974).
43. J. Janata, G. Blackburn and A. Jonkman, to be published.
44. T. Matsuo and M. Esashi, in Ref. 2, Abstract 83.
45. J. Koryta, *Ion Selective Electrodes*, Cambridge University Press, Cambridge, 1975.
46. R. P. Buck, Electroanalytical Chemistry of Membranes *Crit. Rev. Anal. Chem.* 5 (4) 323 (1975).
47. J. A. Aboaf, *J. Electrochem. Soc.* 119 948 (1967).
48. J. Scott and J. Olmstead, *R.C.A. Rev.* 26 357 (1965).
49. S. D. Moss, in Ref. 15.
50. U. Fiedler and J. Růžička, *Anal. Chim. Acta* 67 179 (1973).
51. G. H. Griffiths, G. J. Moody and J. D. R. Thomas, *Analyst* 97 420 (1972).
52. P. T. McBride, J. Janata, P. A. Comte, S. D. Moss and C. C. Johnson, *Anal. Chim. Acta* (1978).
53. D. M. Band, J. Kratochvil and T. Treasure, *J. Physiol.* 265 5P (1977).
54. H. M. Brown, J. P. Pemberton and J. D. Owen, *Anal. Chim. Acta* 85 261 (1976).
55. O. H. LeBlanc, Jr., J. F. Brown, J. F. Klebe, L. W. Niedrach, G. M. J. Slusarczuk and W. M. Stoddard, Jr., *J. Appl. Physiol.* 40 644 (1976).
56. E. Pungor, *Anal. Chem.* 39 28A (1967).
57. B. T. Shiramizu, J. Janata and S. D. Moss, *Anal. Chim. Acta* (1979) in press.
58. H. R. Allcock, *Angew. Chem. Int. Ed.* 16 147 (1977).
59. P. T. Bray, G. C. F. Clark, G. J. Moody and J. D. R. Thomas, *Clin. Chim. Acta* 77 69 (1977).
60. E. Lindner, K. Tóth and E. Pungor, *Anal. Chem.* 48 1071 (1976).
61. W. E. Morf, E. Lindner, and W. Simon, *Anal. Chem.* 48 1596 (1976).
62. P. A. Comte and J. Janata, *Anal. Chim. Acta* (1978).
63. P. Bergveld, J. Wiersma and H. Meertens, *IEEE Trans.* BME-23 136 (1976).
64. Federal Register (Part III), *Ethylene Oxide: Rebuttable Presumption Against Registration; Maximum Residue Limits and Daily Levels of Exposure*, 3800-3815, January 27, 1978.
65. J. Janata and P. T. McBride, *J. Bioeng.* 2 459 (1979).

# THE AUTHORS

Professor J. Janata

Professor R. J. Huber

Jiri Janata obtained his Ph.D. degree in Analytical Chemistry from the Charles University, Prague in 1965. Between 1966 and 1968 he was a Postdoctoral Fellow at the University of Michigan, Ann Arbor, working in the area of organic electrochemistry. In 1968 he joined Imperial Chemical Industries Ltd. in England, where he worked in the analytical section of the Corporate Laboratory until 1976. During sabbatical leave 1973-1975, as Visiting Assistant Professor, he participated in formulation of the CHEMFET program at the University of Utah, where he is now an Associate Professor in the Department of Bioengineering.

Robert J. Huber is a Professor of Electrical Engineering at the University of Utah and director of the Hedco Microelectronics Laboratory there. He received B.S. and Ph.D. degrees in physics from the University of Utah in 1956 and 1961 respectively. He joined Argonne National Laboratory in 1961 where he was engaged in fast reactor studies and fast neutron spectroscopy using semiconductor radiation detectors. In 1967 he joined General Instruments Corporation, Microelectronics Division, in work on MOS integrated circuit development. In 1971 he became a member of the University of Utah Institute for Biomedical Engineering working on the electronics for artificial vision and in 1976 assumed his present position. His present research interests are in the broad area of semiconductor device physics.

*Ion-Selective Electrode Rev.* Vol. 1, pp.81-137. 0191-5371/79/0601-0081 $05.00/0


# INDUSTRIAL APPLICATIONS FOR ION-SENSITIVE ELECTRODES

Peter L. Bailey
Electronic Instruments Limited, 253 Lower Mortlake Road,
Richmond, Surrey TW9 2LL, England

CONTENTS

## 1. INTRODUCTION

Ion-selective electrodes are in many respects ideal sensors for use in the analysis of industrial samples or in the control of a particular constituent in a process stream. Thus, in recent years, the advantages which they offer have led to their adoption in a wide range of applications. Moreover as details of these applications have been published, backed up in several cases by the results of collaborative studies, the methods have been increasingly used by smaller firms and organisations, who lack the resources to develop and prove new analytical methods themselves.

There are several characteristics of the performance of ion-selective electrodes which give them important advantages over other techniques in the field of process control. However, the final choice for a particular application will often be the result of a careful consideration of the relative merits of the techniques available, taking into account such factors as maintenance requirements, the degree of skill needed to run the method, the necessary accuracy and precision, and equipment cost.

The prime advantage of the electrodes is their insensitivity to sample colour, viscosity or suspended solids. Many samples encountered in the effluent control and food industries, as two examples, commonly vary in colour and turbidity from one sample to the next, and the insensitivity of the electrodes to these parameters makes their use preferable to many photometric procedures. Thus, it is unnecessary to have tiresome decolorisation and clarification stages in the pretreatment of samples, which may prevent the development of successful automatic monitoring systems.

Secondly, ion-selective electrodes offer a rapid response to changes in determinand concentration over all but the lowest part of their concentration range. In idealised experimental conditions with a very high sample velocity across the electrode membrane, with the liquid junction of the reference electrode very close, response times of the order of tens of milliseconds have been recorded for electrodes based on inorganic salts. These response times are usually extended to several seconds, or occasionally up to a minute, for measurements on stirred samples in beakers. Consequently in industrial monitors, for example, a fluoride monitor for potable water analysis, the response time for the system will usually be determined not so much by the electrode as by sample pumping rate, the dead volume of the pipework and the design of the measurement cell. However, for the measurement of samples with very low concentration of determinand, as for example in a sodium monitor for boiler feed water analysis, the extended response time of many minutes is due to slow electrode response. Electrodes also respond more slowly in samples containing high levels of interferents or for a period after exposure to such samples.

Furthermore, the electrodes may be used for measurement over a very wide concentration range, often from $10^{-1}$ down to at least $10^{-5}$M, with little or no modification to the experimental method. This can be particularly valuable in the continuous analysis of trade effluents, in which constituents such as cyanide may show major variations in concentration.

Ion-selective electrodes are, with a few exceptions, generally tolerant of small changes in pH. Thus fluoride, nitrate and calcium can all be determined with electrodes in samples within one or two pH units of pH 6; but if alternative procedures for such species are used it may be necessary to buffer samples to a much more definite pH value, which can be inconvenient in process control samples.

A further advantage of analytical methods involving ion-selective electrodes is that they are relatively simple, and cheap to develop, set up and run. The majority of the equipment involved, particularly for laboratory application, is normally readily available in the average analytical or quality control laboratory. The electrodes themselves are not expensive in view of the time saved when using them for analysis. Also they can be used and serviced by junior technical staff unendowed with the skills at handling instrumentation seemingly necessary for some of the more complex techniques; this is in itself a further financial advantage.

On the debit side, ion-selective electrodes are generally unsuitable for analysis or control of species to an accuracy or precision better than $\pm$ 2 percent in terms of concentration, and in many cases, particularly in industrial systems or where temperature variations are occurring, this may slip to $\pm$ 5 percent or $\pm$ 10 percent. This is not surprising when one considers that it is generally accepted that measurements with the relatively well-behaved pH 'glass' electrode are difficult to make to better than $\pm$ 0.01 pH, which corresponds to an uncertainty in hydrogen ion activity of $\pm$ 2 percent. However, by tight control of experimental conditions, often involving the use of flow systems, a precision of 0.5 percent may be achieved with both the pH and other electrodes for direct measurements. In certain circumstances potentiometric titrations or standard addition techniques may be used to improve the precision. Nevertheless, in process control work with dirty samples, an accuracy or precision of $\pm$ 5 percent will frequently be acceptable, particularly at low determinand concentrations, and at least as good as may be readily obtained with other techniques: for higher determinand concentrations, for example at $10^{-2}$M, the comparison with other techniques may not be so favourable.

As a result of the advantages outlined, ion-selective electrodes are now increasingly widely used in process control and reports are appearing of these applications. In the first few years however, enthusiasm for the electrodes was frequently muted because of the mixed success which had been evident in some trials, often attributable to the poor performance of early types of electrode and insufficient understanding of sample pretreatment procedures. Nowadays the mechanical designs of the electrodes are generally much superior to the early models, permitting more reproducible electrode construction and hence performance; moreover the chemical designs of some types of electrode have been developed to give superior selectivity and response range. Also, following many theoretical and application studies, it is now easily possible to find guidance on how to select the best pretreatment procedure for samples to ensure constant ionic strength, minimise interferences and so forth; thus the performance of the electrode and the results of the procedure are optimised.

For many years the obvious analogy between the pH 'glass' electrode and other ion-selective electrodes deterred people from paying sufficient attention to sample pretreatment, as this is almost invariably unnecessary for the glass electrode. Consequently several unsuccessful attempts were made to use ion-selective electrodes without sample pretreatment, for example as in-line sensors, with poor results.

## 2. MEASUREMENT CONDITIONS

The conditions under which process control measurements are made will vary enormously, according to local requirements, from normal laboratory conditions right through to those prevailing at an on-line monitor, with a hot, high pressure sample, in the centre of an industrial plant exposed to an inclement climate. Hence it is clear that no easy generalisation is possible regarding the provision that has to be made for an ion-selective electrode and the system in which it operates for satisfactory measurements to be taken.

### 2.1. Electrode requirements

Whatever the application it is possible to state some objectives to be met if the full potential of ion-selective electrodes, regarding the electrode environment and sample condition, are to be realised.

2.1.1. Temperature. First of all the temperature of both the environment and sample should be kept constant in the short term to $\pm 1^{\circ}C$, or better if high accuracy and precision are required. Also the long term temperature drift should be kept to less than can be allowed for by periodic calibration checks; all cells involving ion-selective electrodes have an appreciable temperature coefficient, and the value of this may be used as a guide to the frequency of calibration necessary to compensate for the effect of the temperature drift experienced. For measurements on samples in beakers in unthermostatted laboratories, the normal diurnal temperature variation of at least $5^{\circ}C$, plus the heating effect of many of the more poorly-designed stirrers, can cause substantial drift of the standard potential of a cell, which will need to be accounted for by frequent re-calibration. It is often also necessary to protect the measurement cell (i.e., the beaker of sample plus the ion-selective electrode and reference electrode) from direct sunlight as this can cause even larger temperature changes. The temperature effect is particularly noticeable when working with, for example, an ammonia probe, which has the relatively large temperature coefficient of 1.5mV $K^{-1}$, and it is easily possible to see the effect on the response as clouds pass across the sun. As a result, it is advantageous to thermostat both the sample and electrode. This is quite straightforward if a continuous flow system is used. Thermostatting is essential for most industrial applications where normally the temperature of both the environment and samples span a wider range than in the laboratory.

2.1.2. Sample pretreatment. The sample must be pretreated by addition of a fixed proportion of a reagent formulated for each type of application; the development of such reagents has been described elsewhere (1) but it is worthwhile to recap the major points. The prime functions of the reagent are to ensure that the sample has the correct pH, a constant ionic strength (or osmotic strength), the determinand is in a state in which it can be measured and is sufficiently free of interferents. Normally the permissible pH range is quite large but limitations arise from both malfunction of the electrode at extremes of pH and the presence of forms of the determinand not sensed by the electrode. For example, the pH of calcium samples must be kept greater than approximately 5 because the ion-exchanger used in many calcium electrodes reacts with hydrogen ions at lower pH, causing the electrode to give an irregular response: at the other end of the scale, at pH > 10, the calcium precipitates as the calcium hydroxide, which is not sensed by the electrode. Similarly for fluoride measurements, the sample must be buffered to pH > 5 to prevent partial complexation of the fluoride ions by hydrogen ions forming protonated species not sensed by the fluoride electrode. The components of the buffer system chosen for the pretreatment reagent must, of course, not interfere with the electrode response.

The pretreatment reagent will usually also contain a concentration of a non-interfering, inert electrolyte, high with respect to the sum of the concentrations of other components of the sample, to ensure that ionic strength variations between successive treated samples and between treated samples and standards are minimised. In that circumstance, the activity coefficient of the determinand may be assumed to remain constant and the activity of the determinand sensed by the electrode, has a constant relationship to its concentration, which is usually the parameter to be measured. Thus, for example, the reagent developed by Hulanicki and Trojanowicz (see section 4.5) for treating calcium samples contains 0.4M potassium nitrate: this is chosen as neither potassium or nitrate ions interfere with the response of the electrode or complex calcium ions. In some cases it may be possible for one reagent to be used to adjust both the pH and the ionic strength. As an example, in the analysis of nitrate in potable waters a reagent containing 0.1M potassium dihydrogen phosphate is used, which when added to samples in the volume ratio 1:10, buffers them to pH 4.5 and fixes the ionic strength.

For measurements of dissolved gases with the gas-sensing probes, instead of making the ionic strength constant, it is the osmotic strength which must be fixed by reagent addition. By this procedure, according to Henry's law, it is possible to obtain a constant relationship between the concentration of the gas to be measured in the sample and the partial pressure of that gas, which is what the probe senses. Because of the way in which these probes operate (2,3), the osmotic strength of the treated sample must not only be kept constant but also approximately equal ($\pm$ ca. 10 percent for probes with microporous membranes, such as the ammonia probe) to that of the probe internal filling solution; otherwise a difference in the partial pressure of water vapour across the probe membrane causes water vapour transfer and hence probe drift. Since osmotic strength is a colligative property of solutions, the reagent is formulated such that the treated sample has the same total concentration of dissolved solids as the internal filling solution. As an example, when using an ammonia probe, containing 0.1M ammonium chloride filling solution, for the measurement of ammonia in effluents, a reagent consisting of 1M sodium hydroxide + 0.1M EDTA is added to samples at a volume ratio of 1:10. Thus, the total concentration of dissolved solids in the treated sample matches that of the probe internal filling solution and, provided that the effluents are relatively dilute, variations in the osmotic strength of successive samples are swamped out. The reagent also serves to raise the sample pH to greater than 12, so that all the ammonium ions in the sample are converted to dissolved ammonia, the species sensed by the probe. In some applications, the sample may already have a high osmotic strength with respect to probe internal filling solution; in the measurement of ammonia in Kjeldahl digests the sample will contain a high concentration of sodium sulphate, and by the time the pH has been adjusted the osmotic strength will be even higher. In such cases, if the sample cannot be sufficiently diluted before measurement, there is no alternative but to modify the internal filling solution (4).

The pretreatment reagent may contain other constituents added to remove species likely to interfere with the measurement, either by complexing the determinand or by interfering with the electrode response. Thus the EDTA, mentioned above as a component of the reagent for treating effluent samples for ammonia measurement, is intended to complex small concentrations of heavy metals, such as copper, which otherwise would complex the ammonia and give rise to a low answer. As a further example, nitrate samples containing an appreciable concentration of nitrite ions are pretreated with a reagent containing sulphamic acid to destroy the nitrite; otherwise if the nitrite ion concentration were sufficiently high, the electrode would give a positive error due to nitrite interference. Occasionally in continuous flow systems an exchange column may be used as an alternative means of removing interferents. Since chloride interferes with the nitrate electrode, nitrate samples containing a lot of chloride may be passed through a column of silver-loaded resin to remove most of the chloride.

2.1.3. Suspended solids. Although ion-selective electrodes usually respond normally to ionic activity in samples containing even quite high concentrations of suspended solids, the electrode response will suffer if the nature of the suspended solids is such that they coat the electrode membrane. Moreover, there is a tendency for the liquid junctions of reference electrodes to become blocked if the sample is very dirty; this tendency can be lessened by increasing the internal pressure of the reference electrode and hence increasing the leakage rate of the junction. Large particulate suspended solids will also be a nuisance in continuous flow systems as they are likely to clog up the tubing and cell and eventually cause the system to stop working. In industrial in-line pH monitors it is customary to clean coated pH electrodes with an automatic mechanical cleaning arm, like a windscreen wiper, or with an ultrasonic cleaner, according to the type of deposit. Unfortunately mechanical cleaning cannot be attempted in monitors based on other ion-selective electrodes and gas-sensing membrane probes, with the possible exception of those with glass membranes, as their membranes scratch very easily or become damaged, with consequent loss of performance. No reports of ultrasonic cleaning of electrodes or probes have been found. Therefore it is usual to filter samples at the in-take of industrial on-line monitors in order to remove particles larger than about 10μm. Unfortunately such advice is easier to give than to act upon, and the design of satisfactory filtration systems for, say, river water is very difficult and a field of study in itself.

2.1.4. Sample velocity. In order to minimise the electrode response time and, for some electrodes, obtain maximum sensitivity at low determinand concentrations it is necessary that the sample velocity across the electrode membrane is rapid and, preferably, constant. For samples in beakers, this is achieved by maintaining brisk stirring, whilst avoiding formation of a vortex. In continuous flow systems, including industrial monitors, the cell design is the critical factor influencing the sample velocity as has been shown by Eckfeldt and Proctor (5). At high flow rates there is also a danger of developing streaming potentials from the pulsing flow produced by the peristaltic pumps; for this reason it is desirable to use a smooth action pump with many rollers. The effects of streaming potentials and methods of minimising them are discussed by Van den Winkel, Mertens and Massart (6). Variations in flow rate can, in adverse circumstances, also affect the response of gas-sensing membrane probes, but such problems can be overcome by careful regulation of both the pump pressure and the pressure of the probe glass electrode on the membrane.

2.1.5. Sample pressure. The ion-selective electrodes with glass or inorganic salt-based membranes can in general tolerate variations in sample pressure. However the electrodes with supported liquid membranes or those having PVC membranes, with liquid or gelled internal electrolyte are, like the gas-sensing membrane probes, unsuitable for use in samples of varying pressure. Electrodes having PVC membranes with solid internal contact, such as the calcium electrode described by Hulanicki and Trojanowicz (7), may be suitable under such conditions.

If the sample pressure does vary, it is essential to pressurise the reference electrode sufficiently to maintain a steady leakage of the bridge solution into the sample; if the sample gets inside the reference electrode, variations in the liquid junction potential and, in some cases, eventual contamination of the reference element will result. To circumvent these and other problems, it is customary to fit industrial monitors with a device, such as a constant head unit, so that the sample inlet is virtually at atmospheric pressure.

From these five points it may readily be seen why it is seldom feasible or desirable to use ion-selective electrodes or gas-sensing membrane probes in in-line, as distinct from on-line, monitors. A further point, which reinforces the argument, is the need for relatively frequent calibration. In pH measurement the precision required is usually much less (in terms of hydrogen ion concentration) than in the measurements performed with the other electrodes and the effect of temperature drift is minimised by use of a temperature compensator with samples and the isopotential setting on the meter. Such temperature drift corrections are normally less feasible with the other electrodes and, moreover, the electrodes themselves may show significant drift in standard potential over a few days (particularly when new): hence frequent recalibration, is desirable, perhaps every six or twelve hours even in a thermostatted monitor, if the required accuracy and precision are to be ensured.

2.1.6. Electrode servicing and replacement. Of the various types of electrode, those with glass membranes such as the sodium electrode, together also with the fluoride electrode, are the longest-lived. The glass electrodes should last for at least a year or two provided the membrane is kept clean and occasionally rejuvenated by a short soak in 0.1M solution of the ion to which the electrode responds; the more dilute the determinand, the more frequently such rejuvenation will be necessary. A glass electrode which fails to revive when so treated may sometimes be saved by a brief dip in dilute hydrofluoric acid, to expose a new glass surface, followed by a thorough wash in distilled water and a prolonged soak in the 0.1M solution. Because of the hazards involved in using hydrofluoric acid, such a procedure is not frequently attempted. The fluoride electrode is also long-lived and resistant to poisoning but its performance will deteriorate if the surface of the lanthanum fluoride crystal forming the membrane becomes scratched. Other causes of eventual electrode failure are leakage through the seal between the membrane and the body and drying out of the internal filling solution; how long an individual electrode lasts before this occurs will depend upon its mechanical design and the conditions in which it is used. The other electrodes with membranes based on inorganic salts, for example, sulphide, chloride and copper electrodes, should also give extended service if the membrane is kept clean and unscratched; regular chemical cleaning or polishing may be advantageous for some electrodes, as described in their instruction manuals.

Electrodes based on organic ion-exchangers and neutral carriers, such as the calcium and potassium electrodes, are more expensive to run as the membranes will usually only last for up to two to three months, or even less if the electrode is exposed to interferents or substances which can poison the membrane.

In many of the new types of commercial electrodes the membrane is included in a disposable capsule which screws onto the end of the electrode body and may be quickly and easily replaced. This represents an advance over old designs which were troublesome to service.

The gas-sensing membrane probes, of which the ammonia probe is met most commonly, also require attention approximately every two months. The usual procedure consists of replacing the gas-permeable membrane and the internal filling solution. After the probe has been in use for a year or so, the probe response may begin to flag because the environment in which the internal pH electrode is operating, a very weakly buffered and near neutral solution, is not ideal for pH-sensing membrane glasses. The response may then be revived by soaking it for 24 hours in 0.1M hydrochloric acid; it may also help to rechloridise the reference electrode (for those probes where this is possible).

The reference electrode which is used in conjunction with all the ion-selective electrodes (excluding the gas-sensing membrane probes) also needs to be kept in good condition. Those which are refillable should last almost indefinitely. They should be topped up or refilled with bridge solution, normally strong potassium chloride solution, every week or, in monitors, whenever the pretreatment reagent tank is refilled; it is essential that the level of the bridge solution should always be above that of the sample stream for the reasons described in section 2.1.5 above. To avoid the necessity of servicing these electrodes, it is common in monitors to use sealed, slurry-filled electrodes; these give as good a performance as the refillable electrodes in the majority of applications. When exhausted these electrodes cause response drift and are thrown away.

## 2.2 User requirements

The next points to consider are those likely to be imposed on the measurement system by the user. These will vary widely depending on whether the measurement is to be carried out manually, semi-automatically in an industrial process control laboratory or fully automatically with an in-line monitor.

2.2.1. Operation and servicing. Where possible, the measurement method or system should be devised so that it may be operated by relatively unskilled staff with the minimum of attention from technical staff. In the laboratory 'direct' measurements of samples are usually preferred, particularly if a pH meter with a direct reading concentration scale or readout is available; a known addition or Gran's plot procedure will only be used if the small improvement in accuracy and precision usually achievable is justified by necessity. It is often preferable to use a continuous flow system, when for example, much sample pretreatment is required, the accuracy or precision need improvement or a large number of samples are to be analysed. Such a flow system accelerates the measurement and imposes standardisation in the way in which samples are treated, thereby reducing operator error. Also, sample preparation and handling must be kept to a minimum as this is a frequent cause of unreliability in measurements, particularly with samples containing low concentrations of determinands.

For on-line monitors all these matters are dealt with automatically and the controls are usually reduced to the minimum 'ON' and 'OFF', with a third option of 'CALIBRATE' to over-ride the automatic calibration sequence. Some makes of monitor take this one stage further and have no external controls at all, the only visible moving part being the measurement display; all controls are reached by using a key to open the front panel. All such monitors are designed to require as little maintenance as possible, usually consisting of filling up the tank of pretreatment reagent, inspection or changing of the pump tubing and a periodic electrode check. The interval between successive services may be set at as long as 30 days in some applications, such as river-water monitoring at remote sites, but frequently a shorter interval may be tolerated; when dirty samples are being analysed the filters often need to be changed or cleaned more often than every 30 days. The system must also be designed so that components of the liquid handling, electrode sensing and electronics units may be easily removed for service or replacement. The frequency at which the electrodes need replacement has been discussed above.

2.2.2. Environmental conditions. No special comments are necessary in respect of laboratory measurement systems. However on-line monitors may often be exposed to varying adverse conditions because of the requirement to install them close to the sampling point of the process stream. The range of environmental conditions for operation of commercial electronic instruments for analytical measurements have been classified by the International Electrotechnical Commission: factors which have been classified include temperature, pressure, humidity, wind, airborne dust, physical shock and exposure to direct sun and rain. It is probable, if current committee considerations reach their likely end, that international and national standards will be issued relating to instruments based on ion-selective electrodes during the next few years. These will require manufacturers, who wish to produce instruments to the standards, to specify the classes of conditions under which their monitors will operate to a stated performance; a considerable list of tests to specify this performance will probably be given. Such tests will naturally apply to the monitor as a whole, including liquid handling, sensing and electronics sections working as a single unit. The provision of such specifications by manufacturers should be valuable to many users when assessing the practicability of using a particular monitor for an application, although the cost of the considerable amount of environmental testing required to produce the specifications will inevitably increase the price of the instruments.

2.2.3. Cost. It is axiomatic that the user requires that the cost of purchasing, operating and maintaining a process control procedure or system is minimised or, ideally, minimal and that the equipment is cost effective. Thus monitors based on ion-selective electrodes are particularly suitable as the initial outlay on a fully automated or, for laboratory use, a semi-automated system based on an ion-selective electrode is small compared with that for a system based on many other techniques, for example spectrophotometry or atomic absorption spectroscopy. Moreover, savings may be made in both the purchase cost and the down time if several monitors are to be operated in a single enclosure by thermostatting the enclosure rather than each monitor separately. For example, in a power station sodium and silica are often measured continuously in several sample streams using monitors based on sodium-selective glass electrodes for sodium and colorimetric analysers for the silica; the monitors are often in installed in a small laboratory or large cubicle near the sampling points. For both types of monitor the liquid handling systems need to be thermostatted and thus part of the monitors are normally fully enclosed and fitted with fans, heaters, temperature sensors, etc.; such monitors exhibit drift due to temperature change whenever opened for servicing. If sufficient of the monitors are in the same cubicle it is more satisfactory to thermostat the whole cubicle and use simpler, more open monitors which are easier to service and remain at thermal equilibrium throughout servicing (8).

The simplicity of the liquid handling system in a monitor based on ion-selective electrodes, coupled with recent advances in electronics, enable the monitor to be made durable and easy to service; thus the purchase cost may be amortised over several years, often five or more. However, the ion-selective electrode itself may need more frequent replacement depending on the type, how well it is designed and made, and also on the amount of trouble which the user is prepared to take to revive a failing electrode, as described previously.

Further contributions to the running cost of the system are the cost of the pretreatment reagent, usually small, and the cost of supplying it to the system. The latter cost is likely to be the more important for on-line monitors remote from the laboratory, hence they are usually designed so that they require topping up no more frequently than once a fortnight or once a month.

Unfortunately it is scarcely worthwhile to outline here actual examples of the cost effectiveness of the use of electrodes or complete monitors based on them in particular applications as local conditions, for example, labour rates and overheads, vary so much from place to place as do other less easily quantifiable factors, such as the importance of the rapid production of the information and the ease of servicing. Moreover cost effectiveness will in many instances only be assessable in the light of a comparison with an established analytical or process control procedure.

2.2.4. Performance. The main performance parameters which the user needs to specify when setting up a process control system, particularly for on-line monitors, are the range of determinand concentrations over which the system must operate, the accuracy and precision required (particularly in respect of any control or set point) and the time lag acceptable between variation of determinand concentration at the sampling point and variation of the reading.

It is one of the advantages of ion-selective electrodes that they give a closely theoretical response over a wide range of determinand concentration, often as much as four or five decades. Hence it is usually possible to use an identical monitor for several different concentration ranges of a particular determinand, often with the same pretreatment reagent. Typically the concentration range of industrial monitors is one or two decades of concentration: unless particularly tight temperature control is used, a more expanded scale to allow greater discrimination in reading is not justified by the performance of the system. As previously mentioned, an uncertainty in pH measurement of $\pm$ 0.01 pH corresponds to an uncertainty in terms of hydrogen ion concentration of $\pm$ 2 percent. Other ions cannot be measured more easily to a better precision or accuracy; indeed $\pm$ 5 percent is more readily attainable and a tighter specification than this is likely to cause a sharp rise in the cost of the system.

The response time of a monitoring system is short if the monitor is close to the sampling point, the flow rate in the monitor is fast and the flow line is short; such factors are normally much more important than the response time of the electrode itself except in applications where the determinand concentration is very low, as in sodium determination in boiler feed-water. Unfortunately most of these factors incur penalties in the design and operation of the system as will be discussed in the next section and it is therefore necessary that, as above, the specification should not be tighter than is really necessary. In practice, a delay of several minutes will be quite acceptable in many applications.

2.2.5. System read-out. In the laboratory, the read-out from the measurement system will normally come from a pH meter, which can be connected to a chart recorder if a permanent record of the measurement is needed. Similarly with an industrial system a recorder may be attached to the monitor; sockets with the various common output ranges for recorders will normally be fitted to commercial monitors. Additionally many commercial monitors are now offering facilities for connection to telemetry systems.

## 3. MEASUREMENT SYSTEMS FOR INDUSTRIAL USE

It is generally the case that analytical procedures for process control are initially developed in the laboratory for the analysis of discrete samples. Subsequently the procedures are adapted for use in laboratory continuous flow apparatus, again usually for discrete samples, and finally the most straightforward and troublefree of these procedures, which have gained the confidence of the plant chemists, are further modified for use in monitors. As yet, reliable ion-selective electrodes have not been available long enough for their use to be at all wide-spread and accepted, particularly in the more conservative industries. However, in some cases, such as in the power generation and water industries, there has been a rapid recognition of the advantages of monitors based on ion-selective electrodes, which has naturally led manufacturers to concentrate on the design of monitors to meet the requirements of these customers. In contrast, in the food industry, although quite a few laboratory applications of electrodes have now been published, there are very few reports of the use of electrodes in continuous flow systems and virtually none relating to on-line monitoring or other plant applicaitons, Thus, most of the automatic systems reviewed will be primarily suitable for the power generation and water industries (including effluent treatment), but none is suitable for the analysis of solid samples, although many examples of the use of electrodes for the non-automated analysis of discrete solid samples have been published.

In this section it is not worthwhile describing the different types of procedure available for laboratory measurements as these have been set out in several books, for example, in Refs. 1 and 9.

The two types of laboratory continuous flow system have been described in several articles, two of which may be particularly commended (10,11); the first deals with the familiar segmented flow type system and the second with a recent type of nonsegmented flow apparatus for carrying out what is known as "Flow Injection Analysis". Both articles deal primarily with the theoretical and practical aspects of the flow systems, with particular emphasis on the degree of dispersion of the sample in the flow, but the sensing systems and some applications are also discussed. In a further paper, by Basson (12), the factors affecting the speed of analysis are considered and a system is described for analysing samples at the rate of 200 per hour (although with a spectrophotometric sensor). One of the critical points is the method for injecting the sample into the carrier stream, which was achieved successfully in this case by use of a micro volume sampling valve. Segmented flow systems, having ion-selective electrodes as the sensor, are discussed from both theoretical and practical points of view by Potts (13).

Laboratory continuous flow systems have been constructed using several electrodes so that several components of the sample may be analysed at the same time. Because the sample pretreatment required for each electrode is usually different, it is not generally possible to mount several ion-selective electrodes in a single flow cell. Instead the sample stream is split into several channels, each of which is treated separately. The signals from each of the electrodes may be fed into a mini-computer, which can be programmed to scan the outputs of the high impedance amplifiers connected to each electrode; the computer may then calculate and print out the required concentrations by use of calibration data previously recorded (14). A mini-computer may also be programmed to determine whether the electrode has reached a steady reading before using that reading for computation and can control standard addition procedures (15); rainwater samples have been analysed in this way for fluoride, chloride and ammonia (16). For those few electrodes which give stable, reproducible results in the presence of relatively large concentrations of interferents, a computer may be used for interference corrections (17).

In the subsequent sections the various parts of both industrial and laboratory continuous flow analysers will be discussed in turn; separate sections will be devoted to flow cells, liquid handling equipment, electronic equipment and the analyser housing. No further mention will be made of computer-linked systems (which are to be reviewed by another author subsequently in this journal) or discontinuous automated analysis systems (e.g. Ref. 18). Also only on-line monitors, analysing a portion of the sample bled off from the main stream, rather than in-line monitors, will be discussed as, for reasons already mentioned, direct immersion of the electrodes in untreated sample streams is very seldom satisfactory.

### 3.1 Flow cells

In order to incorporate ion-selective electrodes or gas-sensing membrane probes into continuous flow systems it is necessary to use a special flow cell for the electrode and reference electrode. This is designed such that the hold-up volume is as small as possible and the velocity of the sample across the ion-selective electrode membrane is very rapid. Also provision is made for a reference electrode mounted downstream from the ion-selective electrode so that the bridge solution leaking from it cannot interfere with the measurements. When this is done in gas-segmented flow systems, it is important to ensure that an uninterrupted stream of the sample is maintained between the two electrodes so that electrical continuity is maintained at all times hence the flowing liquid is usually debubbled immediately before it passes into the cell.

Three main types of cell are in common use with continuous flow systems. Examples of the first type are shown in Fig. 1. The ion-selective electrode is fitted with a small flow cap which allows the sensing membrane to be exposed to the flowing solution, it has been shown by two sets of workers (16,19) that the presence of a stirrer bar inside the cap, motivated by an external magnet, accelerates the electrode response and gives greater potential stability. The liquid junction of the reference electrode is fitted with a similar cap and the two caps are joined by a short length of tubing (the reference electrode is not needed with gas-sensing membrane probes). As an alternative, the sensing membrane of electrodes based on inorganic salts may be drilled out and fitted with tubes so that the sample may pass directly through it (e.g., Ref. 20). Such systems can work well, provided that they can be kept free from bubbles and that streaming potentials are small and constant. The causes of streaming potentials have been investigated by Van den Winkel, Mertens and Massart (6). For these potentials to be minimised the conductivity of the treated sample should be high, pulsations in the flow should be kept to a minimum (by using a good peristaltic pump fitted with pulse suppressors, if necessary) and the tubing between the two electrodes should not be unnecessarily narrow; it may also be advantageous to put a grounding electrode in the flow. A flow system in which streaming potentials have been eliminated is

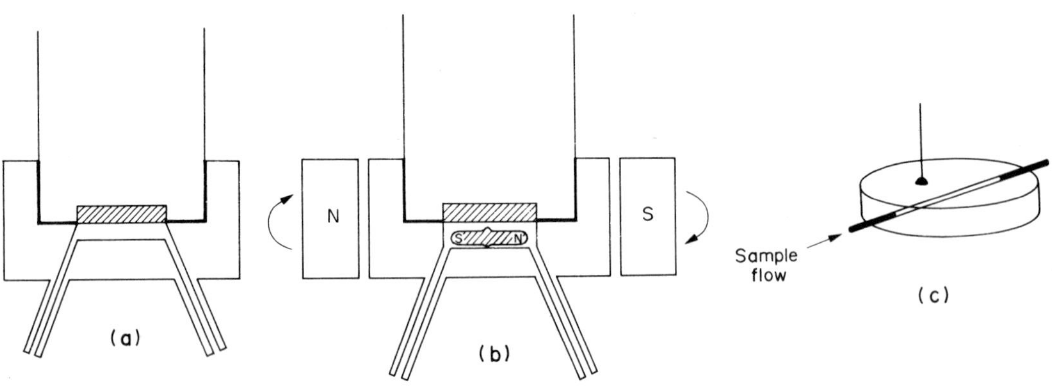


Fig. 1. Flow-through sensors (a) cap fitted to ion-selective electrode
(b) cap with stirrer bar and peripheral rotating magnet
(c) membrane drilled for sample stream.

described in the 'Chloride' measurement section (4.7).

A second type of flow cell is that in which the treated sample is squirted in a jet directly onto the membrane of the ion-selective electrode. In one configuration the liquid junction of the reference electrode is fixed just below the membrane of the ion-selective electrode so that there is a continuous stream of liquid between the two of them. Alternatively the ion-selective electrode is placed so that its membrane is just above the surface of a reservoir of the liquid, which is kept at a constant level, and the tip of the reference electrode is put into the reservoir. These two alternatives, which may be expected to give similar results, are depicted in Fig. 2. That in Fig. 2a is basically a degraded form of that developed by Tóth, Gavallér and Pungor (21), for measurement of the response times of various silicone rubber, precipitate-based electrodes; this is shown in the original form diagrammatically in Ref. 22. Their original apparatus could be simply modified for use in fast continuous flow analysis systems by arranging for successive samples to flow alternately through the two jets and relating the frequency of alternation of the jets to the sampling frequency; the jets would also have to be made smaller to reduce the total flow, although the pump pressure would have to be sufficient to maintain the velocity of the liquid jet. For most applications, such a system would give a faster response than necessary and this at the expense of a high rate of sample and reagent consumption. The modification shown in Fig. 2 has a single stationary jet used in conjunction with conventional sample pretreatment in either a segmented or non-segmented flow system. The form of cell shown in Fig. 2b was developed for use in Flow Injection Analysis (23) but would probably also work well in other systems. Both cells are less sensitive to small variations in sample flow rate than the cells shown in Fig. 1. Both cells also provide optimum conditions for fast electrode response as the hold-up volume of the cell is virtually nil and the sample velocity across the membrane is high; the only problem which is likely to occur is blockage of the jet if the sample contains suspended solids. The cells may conveniently be constructed from a block of a plastic such as Perspex, so machined that the electrodes may be fitted in at the same angle and placed reproducibly with respect to the jet. Until now such cells have only been used in laboratory continuous flow systems, but they could equally well be used in industrial monitors measuring relatively clean or well-filtered samples.

An example of a third type of cell is shown in Fig. 3; such cells are primarily designed for use in industrial monitors but may also be used in laboratory flow systems in situations where a somewhat longer response time may be tolerated. They are designed for use in continuous monitors with non-segmented flow and can be used on samples containing more suspended solids than other types of cell as the tubing bore is larger. They include a stirrer to mix the sample and pretreatment reagent. Small quantities of bubbles in the sample, resulting from the rise in temperature as the sample equilibrates inside the monitor or resulting from the action of the pump, generally have little effect on the measurement as they can escape past the ion-selective electrode, without causing a discontinuity between that and the reference electrode. In cells of this type the hold-up volume is larger than the other types of cell and the sample velocity at the membrane surface is much lower and thus the response time to changes in determinand concentration will be three or four minutes instead of just a few seconds with the others (at determinand concentrations in the middle of the sensing range of the electrode). Nevertheless in very many, and probably the majority of applications involving continuous monitoring, this relatively slow response is not an important disadvantage; indeed robust cells of this type, which permit easy servicing of electrodes, are simple in construction and difficult to block, will usually be preferred.

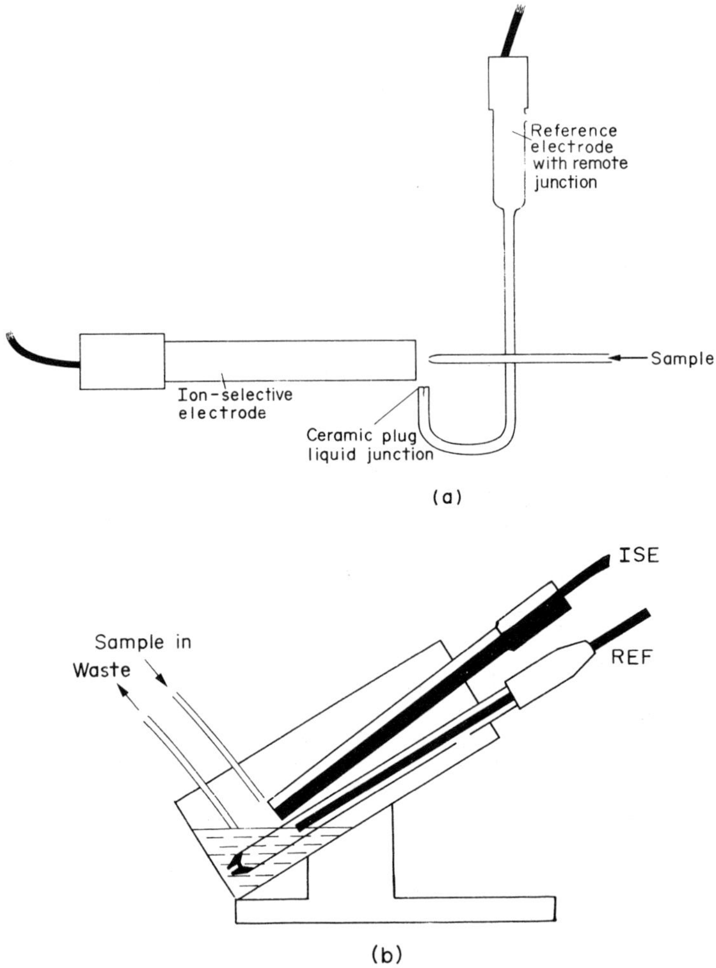


Fig. 2. Flow cells with liquid jets. (Fig. 2b from Ref. 23 by courtesy of The Chemical Society.)

3.2 Liquid handling equipment

In both laboratory and industrial on-line monitors it is now long-established and almost universal practice to use peristaltic pumps for pumping samples and reagents. These pumps are fairly rugged and some makes are easy to adjust and service. They are particularly suited to use in monitors because none of the moving parts of the pump come into contact with the sample stream. The choice of pump for a laboratory application can be quite difficult as the quality of the pump (i.e., its soundness of construction and design) does not appear to be closely related to its price or, in some cases, the renown of its manufacturer. This is not the place to offer a buyers's guide to peristaltic pumps but there are several simple points for which to look. First of all the pump should have sufficient rollers in order to prevent too much pulsing; also the rollers should have as large a diameter as possible and be able to rotate so that they do not unduly shorten the life of the pump tubing. Depending upon the application, it may be desirable for the pump to accommodate tubing of several diameters, in which case the pressure on each channel will need to be independently adjustable. Finally, and unfortunately the only satisfactory way of proving this is by trial and error, the design of the motor and gearing must be such that the pump is capable of continuous operation for prolonged periods without attention.

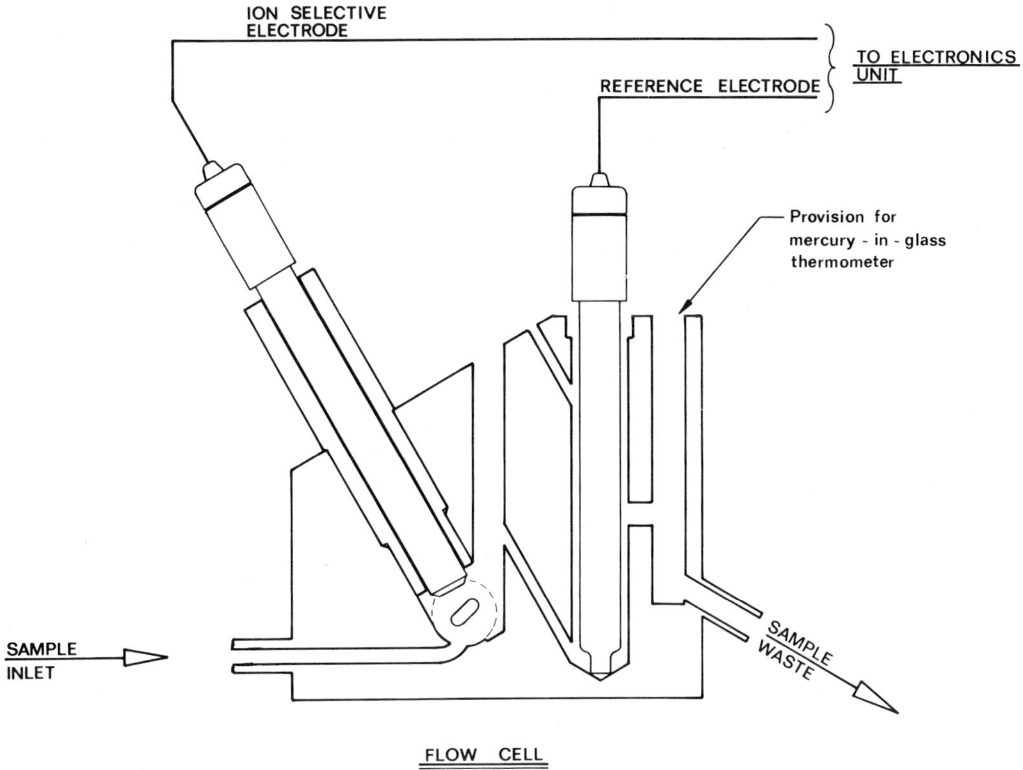


Fig. 3. Flow cell for an industrial monitor.

The use of a pump may occasionally be avoided in those applications where the sample is at a fairly high pressure and so will flow through the monitor unaided; the flow rate will then be set by a pressure regulator on the input. Such a system can only be used if the sample does not need pretreatment (which is rarely the case) or may be treated by other means than pumping, such as by gaseous diffusion. In this latter process the sample is passed through a porous, hydrophobic tube, usually silicone rubber, immersed in a solution of a volatile dissolved gas; the gas diffuses through the tube to pretreat the sample. Several examples may be quoted of monitors using diffusion pretreatment such as the chloride monitor described by Bailey et al. (24), the Orion sodium monitor and, using a variation of the theme in which the sample is diffused into the reagent, the free cyanide analyser of Durst (25).

Other components of the liquid handling systems such as mixing coils, de-bubblers, pulse suppressors, constant inlet pressure regulators and so forth are common to all types of continuous flow system and outside the scope of this article. Another part of the system which is crucial to the successful running of most on-line monitors, is the filtration system on the monitor inlet. In view of recent trends towards the use of monitors with very narrow bore tubing, in order to reduce response time and reagent consumption, there is increasing pressure for the development of satisfactory filters because as the bore of the tubing decreases, so the water needs to be more free of suspended solids to avoid tubing blockage. Unfortunately virtually no information is published on the subject to which the reader can be referred; in any case, the filter required for an application will probably be particular to that application as its design will at least partly depend on the particle size distribution and the composition of the suspended solids present.

3.3 Electronic equipment

Recent advances in electronics, including the development of microprocessors and new types of integrated circuit, have made it possible for manufacturers to produce control modules for analysers which are at once both more advanced in their capabilities and yet more reliable than their immediate antecedents. Moreover these modules are in many cases easier to service as the printed circuit boards contain fewer discrete components; plug-in integrated circuits (ICs) which are easily exchangeable, perform most of the functions of the circuit. Devices such as mechanical cam timers, driven by an electric motor, to operate calibration cycles at preset intervals, are now almost completely replaced by electronic timers, made from a small handful of ICs and other components. Moreover modern types of FET can provide the very high input impedance essential with some types of electrode (gas-sensing membrane probes may have an impedance as high as 3GΩ, thus requiring a module input impedance greater than 1TΩ) without the severe thermal drift formerly associated with such devices.

In the laboratory, with continuous flow systems for the analysis of discrete samples, the control module takes the form of an expanded scale pH or 'pIon' meter, usually in conjunction with a chart recorder. It is also possible to obtain chart recorders fitted with a high impedance input and other pH-metering facilities, such as a slope control and so forth; these are also suitable. Meters which can be calibrated to give a direct concentration reading can save a lot of time and give the answer quickly and unambiguously. The use of such meters, several of which are available, avoids the necessity to construct calibration graphs. However they can be dangerous as usually they use only a double point calibration; the imprecision resulting from this is not as apparent as it otherwise would be and is perhaps concealed behind a display giving readings to three figures. It is, of course, of the greatest importance that such direct-reading concentration scales are only used over ranges of determinand concentration in which the ion-selective electrode gives a linear, Nernstian or near-Nernstian response.

In industrial monitors, designed to run continuously, the electronics must be tailored to fit the requirements of each user and each application; therefore some makes of monitor offer a range of options to the user. For example, high and low alarms, with variable set points, are an option which may be particularly useful in process control application. Also, because ion-selective electrodes have a wide response range, it is usually possible to offer a choice of response ranges for the monitor; in general a monitor may be calibrated over any two decades of determinand concentration in which the electrode gives a linear response. Instruments are also available which automatically calibrate themselves. At a pre-determined interval the sample stream is diverted and replaced by a stream of a standard solution of the determinand; after allowing sufficient time for the signal to reach equilibrium, the reading is corrected for any drift electronically. This option is valuable as it serves to maintain the accuracy of the measurement and can give an indication of when the electrodes begin to age and require servicing. Other options can include a proportional plus integral control module, to operate pumps for reagent dosing: thus a fluoride monitor may be set up to dose fluoride into a water supply to supplement the naturally occurring level of fluoride in areas where this is low. A monitor may also include an amplifier with two high impedance inputs so that more freedom of choice is offered in the selection of electrodes; there may also be electrical advantages in some cases. Finally, the reading itself may be by means of a chart recorder, a periodic print-out or a digital or analogue display: facilities for a BCD output for a computer system or a conventional 4-20mA output (or 0-20mA or one of the several other alternative current or voltage ranges) for a remote recorder may also be provided. Which of these various options will be required for a particular application is up to the user to decide.

No doubt the advent of microprocessors will eventually result in industrial monitors capable of more sophisticated operations designed to improve precision and accuracy. Already microprocessor-based laboratory instruments have been developed for reading and processing the results of standard addition tests and for performing potentiometric titrations; it can only be a matter of time before other profitable applications are exploited.

### 3.4 Analyser housing

Because the physical construction of laboratory instruments is outside the scope of this article, consideration will only be given to the housings of industrial analysers. The main purpose of the analyser housing is to protect the liquid handling system, the flow cell and the electronic equipment from the external environment and provide a durable framework and enclosure for them. The housing is normally constructed from steel, treated to inhibit corrosion and painted with a chip and water resistant paint. The interior of the case is often lined with plastic to inhibit corrosion caused by drips accummulating in the bottom. Frequently the liquid handling and electronics sections of the monitor are housed in separate compartments of the housing, with the electronics modules at the top, to minimise the danger of any accidents with the liquids causing damage to the modules.

In order to keep the flow cell at a constant temperature, it is surrounded by a small thermostatted water jacket; the sample may be pumped through a temperature equilibration coil in the jacket to bring the sample stream to the same temperature as the cell. Alternatively, the whole liquid handling section of the analyser may be thermostatted with a built-in fan heater; this method has the advantages that all the equipment in this whole section should behave in a more stable fashion and that the pretreatment reagent is kept at a constant temperature.

### 3.5 Total systems

Laboratory flow systems have been adequately described in the papers referred to previously. However information on how the factors mentioned in the previous sections are realised by manufacturers of industrial monitors is only available by reading the manufacturers' sales pamphlets. There seems little point in reviewing these as the principles of all remain essentially the same, although shapes, model numbers, options, etc. may change, thus irrevocably dating any comments in a short time; moreover the author cannot claim to be a totally unbiassed critic.

Thus, two examples are given, as representative of the many industrial monitors. The first example (Fig. 4) shows the flow diagram and interrelationships of the electronics modules in a low-level sodium monitor. The sample is dosed with an air stream carrying ammonia vapour to raise the pH of the sample stream to the required level. A picture of such a monitor is given in Fig. 12. The second

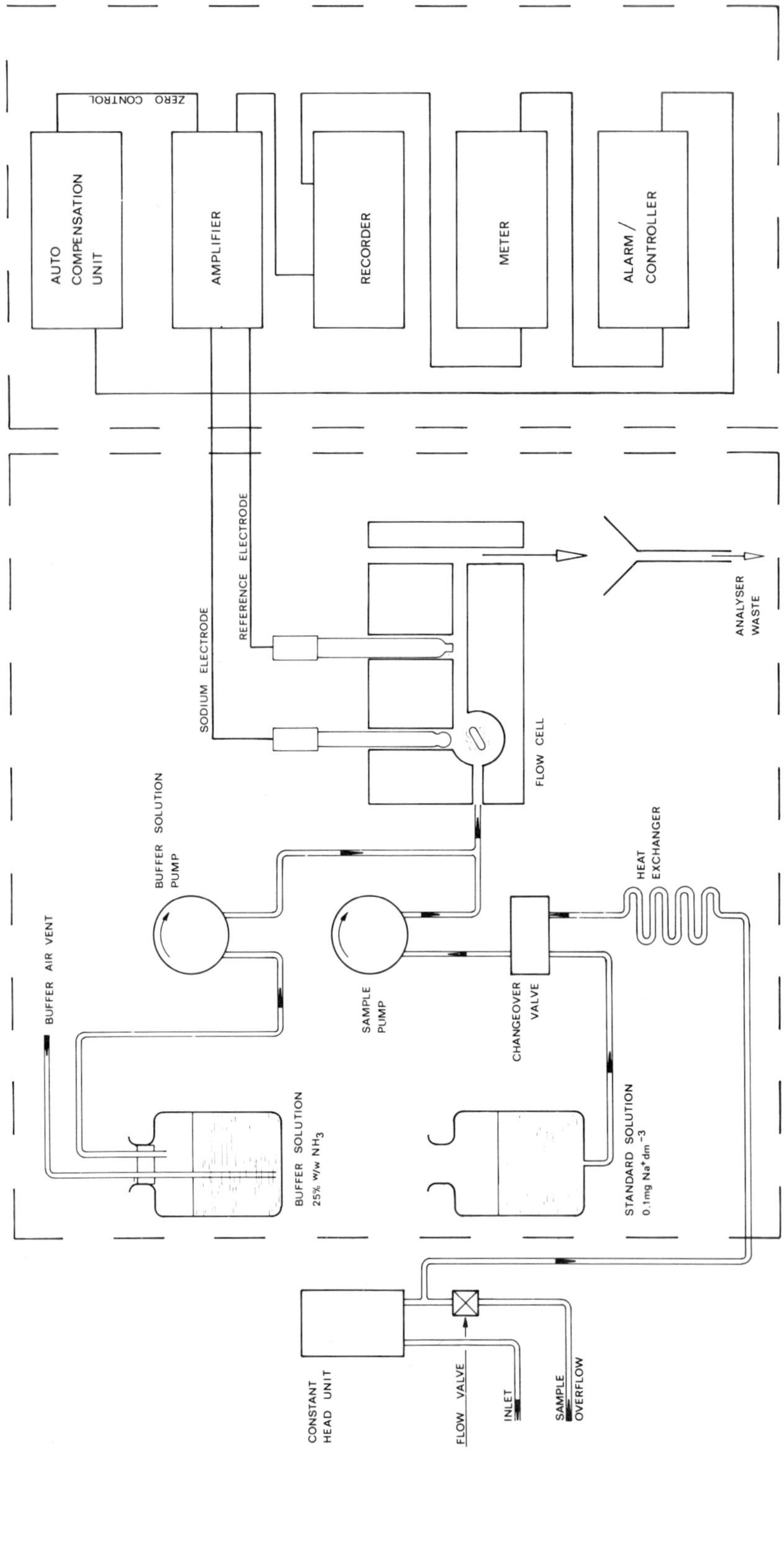


Fig. 4. Diagram of sodium monitor (by courtesy of Electronic Instruments Ltd.)

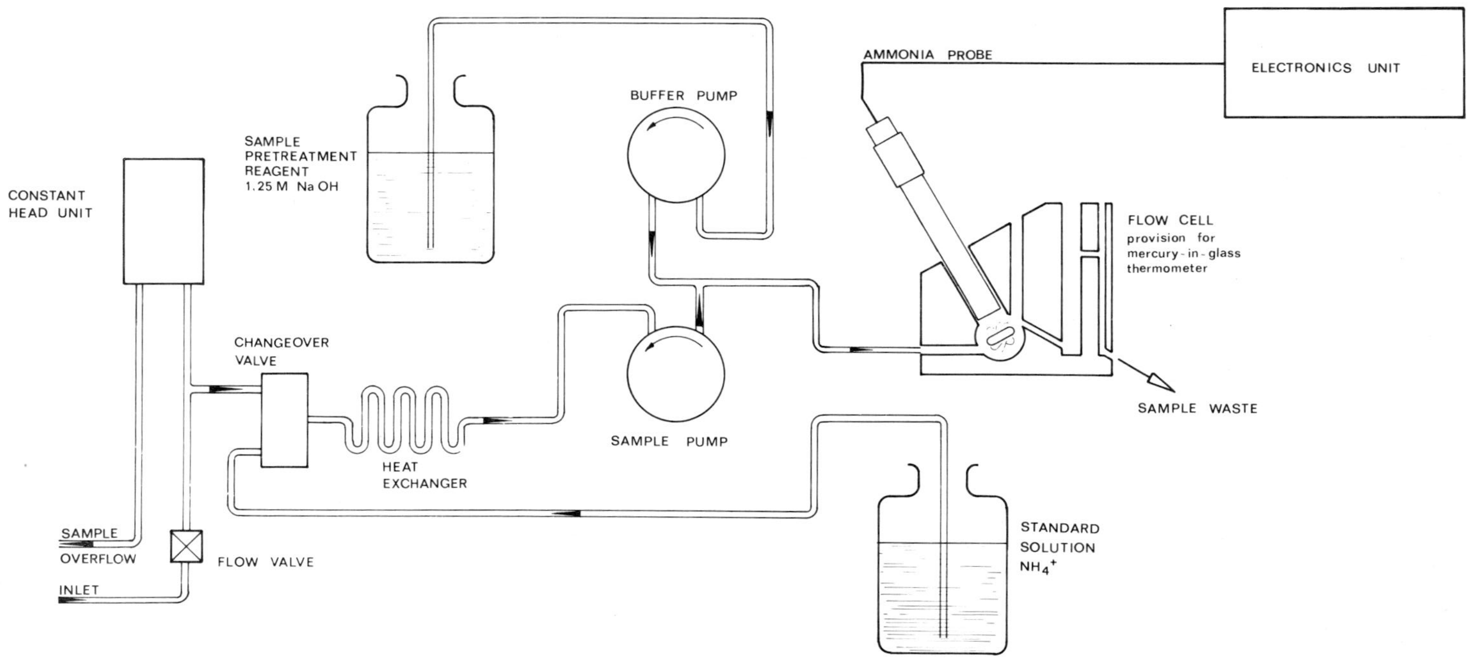


Fig. 5. Diagram of the flow system of an ammonia monitor (by courtesy of Electronic Instruments Ltd.)

example (Fig. 5) shows the flow diagram for the liquid handling unit of an ammonia monitor; the electronics are similar to that of the sodium monitor. In the ammonia monitor it may be noted that no reference electrode is required as the sensor is an ammonia-sensing membrane probe. The sample is treated with dilute sodium hydroxide to provide a constant background for the measurement (with equal osmotic strength to that of the probe internal filling solution) and raise the pH to the required value. In both systems may be seen the change-over valve, which permits the sample to be entirely diverted to waste during a calibration cycle and replaced by a stream of a standard solution for automatic calibration of the sensor. Some types of monitor are designed with provision for an automatic two point calibration; this will decrease the availability (on-line measuring time) of the analyser and will normally only be justified if the electrode being used is suspect or the calibration range and likely variation of determinand concentration are large.

## 4. APPLICATIONS

The applications discussed in this section include uses of ion-selective electrodes in on-line analysers both for direct process control and for continuous monitoring. Also a selection of methods developed for analysis of discrete samples is presented; this includes methods for the related fields of process control and quality control. No attempt has been made to review or cite all such applications which have been published, as many are repetitious or have been subsequently improved. The methods are mostly grouped together according to the determinand.

### 4.1 Aluminium

Aluminium has been determined by potentiometric titration with fluoride using the fluoride electrode as sensor. A good example of an application of this procedure is the following used in the paper industry.

Aluminium salts are added to paper mill waters in order to improve the retention of compounds added at the wet end of the process (26). The aluminium concentration, and the ratio of this to other constituents of the water, need to be carefully monitored. Although colorimetric and other titrimetric methods have been developed these are thought to be either difficult to operate satisfactorily in a plant environment, because of the need for careful timing and pH control, or inherently inaccurate and irreproducible. Homola and James (26) have developed a rapid and accurate procedure in which a 100 $cm^3$ sample of the mill water is treated by addition of 5 $cm^3$ pH 4 acetate buffer and titrated with 0.1M sodium fluoride (also buffered to pH 4) to a constant potential corresponding to approximately $10^{-3}$M free fluoride. The result is corrected for the blank. They have demonstrated that in the pH range 4 to 5.5, for aluminium concentrations up to 3 x $10^{-3}$M, the average ratio of fluoride to aluminium in the complex species formed is sufficiently close to 3.0 for the purposes of the analysis. Interferences caused by calcium and magnesium, which also form fluoride complexes, were investigated: it was found that magnesium, at all realistic concentrations, had no significant effect and that calcium up to 200 mg $dm^{-3}$ could be tolerated. The relative standard deviation of the results of seven replicate determinations of samples containing 10 mg $dm^{-3}$ aluminium was found to be 1.1 percent. Because of the simplicity of the method it could easily be automated, although the results of Radic (27) suggest that several minutes would be needed for the chemical system to reach equilibrium in the region of the end-point.

### 4.2 Ammonia

Three types of ion-selective electrode are available for the measurement of ammonia, a glass electrode, an electrode based on a neutral carrier and a gas-sensing probe. Of these the glass electrode is only suitable for the analysis of waters containing other monovalent cations at concentrations low with respect to the ammonium ion concentration as otherwise interference results. The electrode has found only limited use for this reason. However, Goodfellow and Webber (28) have shown that the electrode is suitable for the on-stream analysis of both boiler feed waters and condensates containing ammonium ion concentrations down to 10 $\mu g\ dm^{-3}$ ($6 \times 10^{-7}$M); they also describe a procedure for the analysis of discrete samples in the laboratory. In order to suppress hydrogen ion interference, the samples are buffered with a triethanolamine-hydrochloric acid buffer. The effect of a range of other interferents, from the alkali metals through to various amines, was tested; although the electrode responded to some extent to nearly all of them, none is likely to be present in normal power station samples in a sufficiently high concentration to affect the ammonium measurement (except in certain circumstances with sodium and n-octadecylamine). The results of the ammonium measurement by electrode were shown to be within 5 percent of those obtained by the indophenol blue method, although the results for a couple of samples needed correction for sodium interference.

A comparison of the neutral carrier electrode and ammonia probe for the same application has been published (29) and it appears that there is little difference between the two or, by turning again to Ref. 28, between the three although each has one disadvantage which may disqualify it for some waters. The glass electrode suffers sodium interference, the neutral carrier electrode has a short life-time (only 2 weeks in continuous use) and the ammonia probe suffers interference from volatile amines. The work of several other authors, although mostly not in direct comparisons, suggests the limit of Nernstian response of the ammonia probe is appreciably lower than that of the neutral carrier electrode (cf. results cited in Refs. 30 and 31). This and the better selectivity of the probe, except in samples containing volatile amines or detergents, may be taken as the reason why very few process control applications of the neutral carrier electrode have been published, although there are many involving the ammonia probe.

The pH of samples to be measured with the ammonia probe must be raised, usually by addition of sodium hydroxide, to at least 11.5 to ensure that virtually all ammonium ions are converted to ammonia, the partial pressure of which is the parameter actually measured by the probe. Also, in order to prevent osmotic drift (3), the total concentration of dissolved solids on both sides of the membrane must be kept the same. This is normally achieved by addition of sodium hydroxide solution so that the final concentration of this in the samples is 0.1M; thus the required pH and osmotic strength are achieved with one reagent. This is only unsatisfactory if the total concentration of dissolved solids is very high (ca. 0.02M or greater), as in sea water samples or Kjeldahl digests. In such cases the usual answer is either to dilute the sample or to change the probe internal filling solution (4).

As previously mentioned, the temperature coefficient of the ammonia probe is high (1.5 $mV\ K^{-1}$) and therefore temperature variations should be minimised by use of a thermostat. A convenient solution for laboratory continuous flow measurements is to immerse as much as possible of the flow system in a water bath maintained at a constant temperature to $\pm 0.1^{\circ}C$ as done, for example, by Buckee (4).

The measured Nernstian range of an ammonia probe appears to vary considerably from one user to another as pointed out by Bailey and Riley (30). Shibata (32) reports that if care is taken to reduce or eliminate absorption of ammonia on the probe body or measurement cell, ammonia volatilisation and leakage of probe internal filling solution, almost Nernstian response may be obtained down to very low ammonia levels with approximately 18 mV between 2 and 5 μg $dm^{-3}$ $NH_3$ -N (1.4 and 3.6 x $10^{-6}$M $NH_3$ -N).

Returning to the analysis of ammonia in boiler feed waters, this is one well investigated and widely used application of the ammonia probe (33,34). Industrial monitors having this probe as sensors are in use in many power stations for the on-line measurement of ammonia. Midgley and Torrance tested the effect of a range of interferents but of these only volatile amines such as cyclohexylamine and octadecylamine produced a major effect and these at levels well above that normally found in power station waters. In continuous determinations of standards, with automatic single point standardisation every 12 hours, a standard deviation of less than 10 percent was obtained from 95 hourly readings; it was also shown that, as expected, the standard deviation increases as the frequency of standardisation decreases. Two probes used for the experiments were still functioning satisfactorily after 4 months and 2 months continuous use. Comparisons were carried out between the results of the probe method and glass electrode method: the sodium content of the waters was determined by flame photometry. The two ammonia methods gave results differing by an average of only 0.3 percent and a maximum of 6 percent (after correcting the glass electrode results from one power station for sodium interference).

Another major application of the ammonia probe is in the laboratory or on-line analysis of waters, effluents and treated or untreated sewage. Among many important papers in this field, those by Evans and Partridge (35) and Beckett and Wilson (36) are particularly thorough. In such applications a pretreatment reagent of 1M sodium hydroxide + 0.1M EDTA is generally suitable; one volume of the reagent is added to ten volumes of sample. The EDTA is added to the reagent so that metals, such as copper, in the sample which might otherwise complex the ammonia, are removed. The EDTA also helps in flow systems by preventing the build-up of hardness deposits and avoiding the formation of metal-hydroxide precipitates which would eventually block the system.

Evans and Partridge (35) used a laboratory manual method for analysing discrete samples ranging from relatively clean samples such as potable water to dirty samples such as airfield drainage runoff, slaughterhouse waste and partially treated sewage. Results were compared to those obtained from a Nessler and an indophenol blue procedure. Swimming pool waters containing chlorine were treated by addition of sodium thiosulphate to prevent chloramines interfering with the measurement. A series of experiments to test the recovery of spikes of ammonia standards from samples of the various waters gave excellent results, generally averaging better than 99 percent with a standard deviation of 3.3 percent except for the lowest ammonia spike (0.1 mg $dm^{-3}$ $NH_3$ -N) which gave a standard deviation of 16 percent. The precision of the analysis was found to be 4 percent for ammoniacal nitrogen concentrations greater than 0.4 mg $dm^{-3}$ and 0.015 mg $dm^{-3}$ for concentrations less than 0.4 mg $dm^{-3}$. Although few, if any, references are available giving details, it is known that many on-line ammonia monitors are in use in sewage works and effluent and water-treatment plants in several countries analysing automatically samples similar to those analysed manually by Evans and Partridge.

In the work by Beckett and Wilson (36) the ammonia probe was shown to be suitable for the analysis of discrete samples of river water. The probe results were compared with those from an automated indophenol blue method for samples in the range 0.6 to 2.4 mg $dm^{-3}$ $NH_3$ -N; on average the probe gave results 1.7 percent lower than the spectrophotometric method. Within-batch relative standard deviations were typically 2-3 percent and between-batch variations 4 percent (for

0.79 mg $dm^{-3}$ $NH_3$ -N). The total analysis time was 5 minutes per sample plus 20 minutes for calibration. The probe membrane was said to need replacement, on average, approximately every 3 months. The authors also recommend rechloridisation of the internal silver/silver chloride reference electrode if the coating wears thin. The authors concluded that the probe was likely to be suitable for the on-line analysis of river waters and, indeed, since that time on-line ammonia monitors have been widely used in river-side monitoring laboratories in many countries. In the Netherlands both floating and river-side laboratories are used to continuously monitor the purity of rivers, such as the Rhine, Lek and Maas (37); ammonia, fluoride, chloride and nitrate are all measured with industrial monitors. The floating laboratories are also equipped for performing other types of analysis, fish studies, and so forth. The servicing interval for the monitors and the filtration system varies between 10 and 28 days, depending on the organic content of the sample water. The inside and outside of one of the laboratories are shown in Figs. 6 and 7.

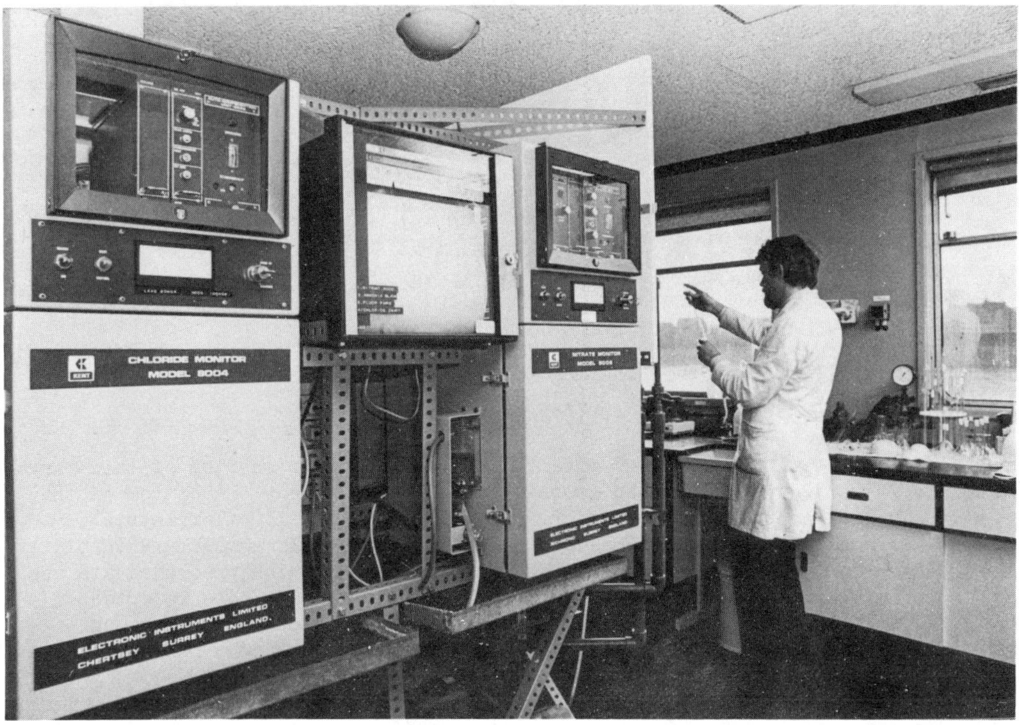


Fig. 6. Internal view of a floating water quality monitoring laboratory on the River Maas at Eysden in the Netherlands (from Ref. 37, by courtesy of George Kent Ltd.).

A laboratory continuous flow analyser for the analysis of discrete water samples at a rate of 20 per hour has been described by Sekerka and Lechner (14); samples were in the concentration range 0.1 to 100 mg $dm^{-3}$ $NH_4^+$. More recently, Philbert et al. (38) have used an ammonia probe in a 'second-generation' Technicon ISE AA II with full scale set at 2.00 mg $dm^{-3}$ N and a sampling rate of 30 per hour. Their traces show good linearity over the range 0.1 to 2.0 mg $dm^{-3}$ N but they had

Fig. 7. External view of the floating laboratory at Eysden. (from Ref. 37, by courtesy of George Kent Ltd.)

some difficulties with carry-over effects when samples near the top of this range immediately preceded samples with ammonia concentrations near the bottom of the range. The method was said to perform 'as well as the AA II colorimetric method' and had the advantage of being insensitive to sample colour and turbidity; the disadvantage was the relatively high detection limit which precluded its use with some natural water samples low in ammonia. Other papers have reported similar continuous flow systems, but with higher ammonia concentrations and operating at the rate of 60 per hour (3,4).

The ammonia probe has been used for the direct potentiometric analysis of several urea ammonium fertilisers (39) such as urea ammonium phosphate and urea ammonium nitrate. It was shown that the presence of the high concentration of urea in the samples did not affect the probe response. However in the analysis of the phosphate fertilisers, which are made from 'wet process' phosphoric acid, impurities of iron caused a precipitate of ferric hydroxide which clogged the probe membrane when the sample solution was made alkaline; this was prevented by the addition of tartaric acid solution to complex the iron. The results obtained with the probe were compared with those from a conventional reduced pressure distillation method and excellent agreement was obtained. For example, nine replicate analyses of samples of urea ammonium phosphate were carried out; the probe gave a result of 11.82 percent ammoniacal nitrogen, range 0.44, standard deviation 0.12, whilst the distillation method gave 11.81 percent, range 0.27, standard deviation 0.09. The measurement with the probe took approximately 5 minutes per sample, whereas the distillation method took 30 minutes.

Another application of the probe with samples containing high concentrations of ammonia is the method developed by Benz and Simko for the analysis of household cleaning solutions (40). Window cleaners, containing 0.05 to 0.15 percent ammonia, were diluted tenfold, made alkaline with sodium hydroxide and analysed by a straightforward standard addition procedure. For analysis of all-purpose cleaners containing more ammonia (0.5 to 1.5 percent) and also some surfactants, the same procedure was used except that samples were diluted a thousand-fold in two stages in order to prevent damage to the probe membrane by the surfactants. As the all-purpose cleaners has specific gravities much higher than 1, and samples were dispensed by volume, the results had to be corrected by dividing by their specific gravity. The precision of the method was tested and relative standard deviations ranged from 1.6 to 4.7 percent for replicate analyses of four different window cleaners and from 1.4 to 2.3 percent for eight different all-purpose cleaners. The accuracy of the method was assessed by comparison with a distillation method. The window cleaners gave the same results with both methods but the probe produced results approximately 2 percent higher with the all-purpose cleaners; this difference is small enough for process control purposes. Typically each probe membrane lasted for 50 to 100 analyses; such a short lifetime is almost certainly due to the presence of so much surfactant in the samples.

A further example of the use of the probe, although not in the direct determination of ammonia, is in the analysis of urea in body lotions (41). 100 - 200 mg sample is boiled in a conical flask with 30 - 40 $cm^3$ water to produce a homogeneous suspension; the suspension is transferred to a volumetric flask and diluted to 100 $cm^3$. To 5 $cm^3$ of the diluted suspension is added 20 - 30 mg of a urease preparation, and the mixture incubated at 40$^\circ$C for half an hour; after this time all the urea should be converted to ammonia. The sample is then made alkaline and the ammonia measured directly. The relative standard deviation from ten replicate analyses was found to be 1.5 percent.

The ammonia content of copper plating solutions may be controlled by use of the ammonia probe. The solution may be sampled either automatically or more usually, manually. Samples are diluted one-hundredfold, which typically brings the ammonium ion concentration into the range 3 to 6 x $10^{-4}$M and the copper concentration down to approximately 5 x $10^{-3}$M, and then treated with the usual sodium hydroxide + EDTA pretreatment reagent. Direct measurement is usually sufficiently accurate for control purposes.

A final example of ammonia determination is that developed by Drawert and Nitsche (42) for beer analysis. They found that the Henry's law constant varied with the concentration of ethanol in the sample; hence it was necessary to match the alcohol content of the standards for calibration of the probe with the samples. An increase in the alcohol content from 1 to 10 percent, at constant ammonia concentration, caused a shift in probe output of about 10mV. A range of nitrogenous compounds were tested for interference, but only volatile amines, such a methylamine and ethylamine, caused any errors. The relative standard deviation of analyses was 5.6 percent with direct potentiometry and 2.0 percent with a standard addition method.

Other applications of sensors measuring ammonia appear in the subsequent section on 'Nitrogen'.

## 4.3 <u>Boron</u>

It is feasible to determine boron as fluoroborate, with a fluoroborate-selective electrode, in a range of materials, although the often complex methods needed for sample preparation make automation difficult. The linear (Nernstian) response range of the electrode is from $10^{-1}$ down to $10^{-4}$M. Only a few papers on the determination of boron with electrodes have been published, but two of them may be used as examples of what would be possible with other samples.

Carlson and Paul (43) developed the fluoroborate electrode and used it for determining boron in waters. The main problems are to isolate the boron and then convert it rapidly into tetrafluoroborate without using such a large excess of fluoride that the fluoride interferes with the electrode response; both problems may be overcome at the same time by use of a boron specific resin, Amberlite XE-243, which has as its active group N-methylglucamine but also contains groups with weak base anion exchange properties. Thus the procedure is to pass the water sample through a column of this resin, which collects the boron. If necessary the column may next be washed with dilute ammonium hydroxide to remove interfering anions. The boron is converted to tetrafluoroborate with a small volume of 10 percent hydrofluoric acid and after 10 minutes surplus acid is removed by rinsing with a small volume of water. The tetrafluoroborate is eluted from the column with dilute sodium hydroxide solution and passed through a cation exchange column to acidify it. It is diluted to a fixed volume for direct determination with the electrode. A plastic bodied reference electrode with a bridge solution of strong potassium chloride is used. A calibration curve for boron concentrations in the range 0.1 to 20 mg $dm^{-3}$ was prepared but this range could be altered to some extent by variation of the procedure. For a sample of tap water containing 0.66 mg $dm^{-3}$ boron, excellent agreement was obtained with the curcurmin method.

As an example of what can be achieved with more intractable samples, Wilde (44) has described a method for determining boron in aluminium oxide-boron carbide matrices. These materials are used as consumable neutron absorbers in nuclear reactors. The method has also been applied by the author to the analysis of borosilicate glass. Samples are fused at 950°C for 1 hour with sodium carbonate and sodium fluoride; after cooling, the residue is leached with 1:1 hydrochloric acid. An aliquot of the leach is titrated with 6M sodium hydroxide to raise the pH to 4.5 and diluted to 100 $cm^3$ with water. The aluminium is present as a precipitate of $Na_3AlF_6$ in the prepared sample. For eight replicate analyses of an NBS borosilicate standard containing 3.96 percent boron, a mean reading of 3.93 percent was obtained with a relative standard deviation of 0.67 percent. Analyses of samples of $Al_2O_3.B_4C$ of known composition both by the electrode method and by a mass spectrographic procedure, showed at a high boron concentration (1.3 percent) the electrode method is more accurate and has a smaller standard deviation (as well as being much more convenient) but at low boron concentrations it may be somewhat less accurate.

## 4.4 Bromide

Bromide-selective electrodes are available which give a Nernstian response down to at least $10^{-5}$M and would, at least in principle, be suitable for process control applications. The electrode suffers interference from iodide, cyanide, sulphide and large excess of chloride; in practice these interferences considerably limit the application of the electrode. Routine determinations of bromide are usually carried out, after appropriate sample treatment, by potentiometric titration with silver nitrate solution using a metallic silver indicator electrode; for example, bromide in glasses is measured in this way (45).

Conacher and Mc Kenzie (46) compared the titration method with a direct bromide electrode determination for the analyses of brominated vegetable oils and found that the titration method was much more accurate and reproducible; the electrode clearly did not work well in the alkaline sample matrix resulting from the fusion of the oil sample in potassium hydroxide. The determination of bromide in wine by a standard addition procedure with a bromide electrode has been reported (47). Components of the wine caused the electrode to have a very long response time and eventually poison the membrane; a response time of 15 minutes was quoted for samples containing 1 mg $dm^{-3}$ $Br^-$. No tests were carried out on the accuracy of the method.

## 4.5 Calcium and water hardness

Two types of calcium electrode are available, electrodes based on ion-exchangers and electrodes based on a neutral carrier. Both have approximately the same Nernstian concentration range, down to approximately $3 \times 10^{-5}$M, but the electrode based on the neutral carrier would appear to offer better selectivity. However as the neutral carrier electrode is relatively new, most reported applications have been developed using the ion-exchange electrode but would probably be just as feasible with the other electrode. A version of the ion-exchange electrode with reduced selectivity, which responds equally to calcium and magnesium, is available for water hardness determinations.

The majority of reported applications of calcium electrodes are in clinical analysis but there are also important examples of the use of the electrodes in other fields. In particular, several papers have appeared on the determination of calcium in waters, which is a very important measurement throughout industry. The paper by Hulanicki and Trojanowicz (48) offers a direct method with an accuracy and precision which will be satisfactory for many purposes. The samples are first de-aerated by bubbling nitrogen through for 10 - 15 minutes at room temperature in order to minimise the carbon dioxide content; failure to do this causes the results to be about 25 percent low due to formation of the calcium-hydrogencarbonate complex. One volume of the deaerated sample is mixed with one volume of a 'Constant Complexation Buffer', and, after allowing two or three minutes for the system to reach equilibrium, the electrodes are put into the solution and the readings taken. The 'Constant Complexation Buffer' consists of 0.4M $KNO_3$ (to give a constant ionic strength), 0.02M disodium iminodiacetate (to weakly complex the calcium to a constant extent, thus preventing errors due to variable amounts of even weaker complexes such as calcium-sulphate), 0.04M acetylacetone (to mask magnesium), 0.02M $NH_3$ and 0.02M $NH_4Cl$ (to buffer to pH 9.1). Despite the complexation of the calcium ions, the electrode gave a linear response for sample calcium concentrations in the range 20 to 800 mg $dm^{-3}$. The error was usually less than 4 percent with the results tending to be high in comparison with titrimetric measurements.

Calcium may also be determined by potentiometric titration, using the calcium electrode as indicator, and tetramethylammonium-EGTA as titrant (49), or, alternatively, using the copper electrode as indicator and EDTA as titrant (50). For analysis of mixtures of calcium and magnesium (51), a two stage procedure has been suggested, whereby the total calcium plus magnesium concentration is determined in a titration with EDTA at pH 9.7 in a glycine buffer and then, using a second sample buffered to pH 12.0 with sodium hydroxide solution, calcium alone is determined by another EDTA titration. These analyses were carried out using an automatic or semi-automatic titration system. Calcium in the concentration range 40 - 320 mg $dm^{-3}$ and magnesium in the range 30 - 200 mg $dm^{-3}$ were determined with average errors of 0.2 percent and 0.3 percent respectively (semi-automatic procedure) or 0.4 percent and 0.7 percent (automatic procedure).

An elegant method for the sequential titration of calcium and magnesium in single samples of drinking waters has been described by Christiansen et al. (52). Their method employs a laboratory automatic titrator controlled by a mini-computer, which is programmed to compute the size of the next aliquot of titrant to be added in the light of the effect produced by the last aliquot and on the assumption that the titration curve will be S-shaped. Increments are reduced as the end-point is approached; also the computer, by potential sampling method, waits for the electrode potential to stabilise before continuing the titration. The end-point is calculated as the point of inflection on the second differential curve. Samples containing no bicarbonate were buffered 1:1 with 0.02M 3,4-dihydroxybenzoic acid plus 0.03M glycine (pH 9.7) in order to get the optimum end-point break for both calcium and magnesium. However for samples containing bicarbonate, a lower pH is necessary to avoid the formation of calcium carbonate which only reaches equilibrium with EDTA very slowly. Consequently for these samples the buffer used was 0.02M acetylacetone plus 0.04M TRIS buffer (pH 8.5). Using this second buffer, replicate samples of drinking water were automatically titrated with a mean standard deviation of 0.25 percent for the calcium determination and 0.23 percent for the calcium plus magnesium determination. A typical trace, showing the excellent curve separation, is shown in Fig. 8. It is interesting to note that this procedure makes use of the response of the electrode down to pCa 8.

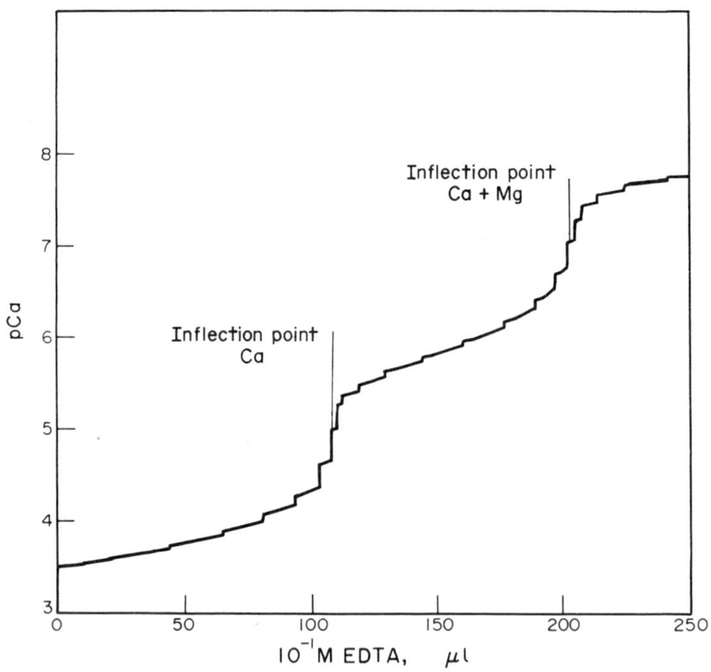


Fig. 8. Digital titration curve for ca. $10^{-3}$M $Ca^{2+}$ plus $10^{-3}$M $Mg^{2+}$ in 0.01M 3,4 - dihydroxybenzoic acid at pH 9.7 (from Ref. 52, by courtesy of the American Chemical Society).

The calcium content of batches of soda-lime for glass-making was analysed by Knupp (53). Samples are digested in hydrochloric acid, neutralised with ammonium hydroxide to pH 9.0 $\pm$ 0.1 and then measured directly with the calcium electrode. The actual results quoted appear not very accurate by comparison with those from a CDTA titration, but this would seem to be due, at least in part, to the poor performance of the electrode used.

## 4.6 Carbonate/carbon dioxide

Gas-sensing membrane probes for carbon dioxide generally have a restricted response range and are primarily suitable for the measurement of carbon dioxide in blood. The use of a carbon dioxide-sensing 'air-gap electrode' for measurement of TIC and TOC in waters has been reported (54). TIC is measured after straightforward acidification. For TOC measurements, samples are initially oxidised by boiling under pressure for 30 minutes with alkaline potassium persulphate solution and then acidified. However preconcentration is required for carbonate concentrations below 2 x $10^{-3}$M by precipitation of the carbonate as lead carbonate, filtration and treatment of the residue with a small volume of acid in the electrode sample container.

The concentration of carbonate impurities in a nickel hydroxide mixture used in NiFe storage batteries has been measured with a carbon dioxide probe (55). The samples consist of about 80 percent nickel hydroxide mixed with small quantities of graphite, barium sulphate and nickel carbonate. A weighed quantity, about 50 mg, of the mixture is treated with 100 $cm^3$ 0.025M hydrochloric acid and the carbon dioxide measured by a procedure involving multiple standard addition of carbonate. Concentrations of hydrochloric acid greater than 0.025M caused interference from hydrogen chloride.

Attempts to measure carbon dioxide in power station waters with a probe continue to be met with little success (56).

## 4.7 Chloride

Chloride is one of the most abundant and widespread anions and in many situations can also be one of the most corrosive. Hence there is a wide range of samples which need to be analysed for chloride from quite pure boiler waters through to food, drink and soil. Three types of electrode are available for chloride determination, silver/silver chloride electrodes, chloride-selective electrodes based on silver salts and, thirdly, chloride-selective electrodes based on mercury salts.

The first type consist simply of a silver wire or silver billet coated with silver chloride. This is an electrode of the second kind and not therefore considered strictly to be an ion-selective electrode. Such an electrode has however been used in several direct chloride analyses, particularly in boiler water analysis (57), and is very widely used as indicator electrode in potentiometric titrations of chloride. Its only disadvantage over the second type of electrode is its response to variations in redox potential; in compensation, the electrode is virtually indestructible, the surface coating may be easily renewed and it gives a faster response to the lowest concentrations of chloride. The limit of Nernstian response of the electrode is $10^{-4}$M (3 mg $dm^{-3}$).

The second type of electrode is the chloride-selective electrode based on silver salts; the membrane usually consists of a mixture of silver chloride and silver sulphide (58). This electrode has the same Nernstian range as the previous electrode. It is the most commonly used electrode of the three for all applications, except for the analysis of low chloride waters, involving direct or standard addition techniques. The electrode responds to chloride in samples over the pH range 2 to 11 and only suffers interference from those species forming more insoluble or stronger silver compounds than chloride, such as sulphide, bromide, iodide and cyanide. These species may usually be removed from the sample by pretreatment prior to measurement.

The third type of chloride electrode is that developed by Sekerka and Lechner (59), which is based on a pressed pellet of mercurous chloride and mercuric sulphide. This electrode works in acid-buffered solutions and is primarily designed for low-level chloride measurements (limit of Nernstian response 2 x $10^{-6}$M, limit of detection 5 x $10^{-7}$M).

In order to prevent corrosion in power station boilers, it is essential to continuously monitor the chloride concentration in the water circuit so that any leakage of chloride into the system or gradual build-up of chloride can be controlled. The water purity which is required depends on the type of boiler concerned, but the chloride concentration is generally less than, and sometimes much less than 50μg $dm^{-3}$. Analytical methods currently in use are mostly based on the silver/silver chloride electrodes, which has been shown to give a linear potential vs. concentration response in the region between the limit of Nernstian response and the limit of detection. The sensivity of the measurement in this range increases as the temperature decreases, and hence measurements are customarily made in a system thermostatted at $10 \pm 0.1^{\circ}C$ (response at $25^{\circ}C$ is 2.26 mV per 100μg $dm^{-3}$ and at $10^{\circ}C$ is 3.54 mV per 100μg $dm^{-3}$). A manual method for carrying out such measurements has been described and subsequently an apparatus has been developed and tested for continuous measurement. The continuous apparatus was calibrated over the range 0 to 150 μg $dm^{-3}$ and was shown to be capable of detecting changes in chloride concentration of less than 1μg $dm^{-3}$. None of the substances likely to interfere with the electrode response is normally present in a sufficiently high concentration to cause errors (57).

Over the last few years there has been continuing work to develop methods, and eventually monitors, based on the chloride electrode with mercury salts in the membrane. Use of this electrode should offer much greater sensitivity in the lower concentration ranges and a decreased limit of detection. Sekerka, Lechner and Wales (60) report a direct manual method and an automated method for discrete samples; the pretreatment reagent, which served to keep the ionic strength and pH constant, was saturated potassium nitrate solution adjusted to pH 2 with nitric acid. They analysed samples of boiler water, river-, lake- and waste-water containing chloride from 0.05 to 100 mg $dm^{-3}$ with satisfactory results in comparison with a standard colorimetric method.

A version of this electrode has also been used by Marshall and Midgley (61) for the analysis of boiler water and the results compared favourably with those from a mercury (II) thiocyanate colorimetric method in the range 0.1 to 1.0 mg $dm^{-3}$; the electrode was found to be suitable for measurements down to 10μg $dm^{-3}$, but the spectrophotometric method was insufficiently sensitive for measurement in this region. The standard deviations at 500, 100, 50 and 10 μg $dm^{-3}$ were found to be 25, 4, 5 and 15 μg $dm^{-3}$ respectively; the value of 4μg $dm^{-3}$ for 100μg $dm^{-3}$ samples may be compared with the standard deviation for 40μg $dm^{-3}$ previously obtained with the silver/silver chloride electrode, thus showing the much greater precision obtainable in measurements at low concentrations with the mercury-based electrode. The electrode method is also more precise than the spectrophotometric method for samples with chloride concentrations less than 1 mg $dm^{-3}$. The interference produced by a range of possible sample constituents was measured and found to be insignificant; only iron (III) caused an important error (100μg $dm^{-3}$ $Fe^{3+}$ in a sample containing 100μg $dm^{-3}$ $Cl^{-}$ caused the electrode to respond as if to 66μg $dm^{-3}$ $Cl^{-}$) and this interference could be removed by addition of fluoride to the buffer.

A prototype continuous monitor for low-level chloride based on this electrode has been described (24). Samples are buffered to pH 3 by passage through a length of hydrophobic gas-permeable tubing immersed in glacial acetic acid. The system produced a response of 14 mV between $10^{-6}$ and $10^{-7}$ M (35 and 3.5 μg $dm^{-3}$) and thus seems very promising.

Chloride-selective electrodes based on silver chloride are generally robust and well-behaved which makes them useful in many industrial control applications. Pommez and Stachenko (62) report the use of a chloride electrode for in-line control of one stage of the sugar refining process. Raw sugar crystals are washed to remove films of residual mother syrup; chloride concentrates in this outer film and hence the amount of chloride present can be used as a measure of the amount of washing required. The chloride electrode is immersed directly in the raw sugar, which is at a concentration of 60° Brix and at 70°C, and the signal from it is fed directly to a small computer which controls the washing cycle. The chloride measurement is not very accurate, as small variations in the composition of raw washed sugar liquors can affect the activity coefficient of the chloride substantially, and hence upset the calibration (also done at 70°C in 60° Brix standards). However the accuracy is sufficient for the purpose and, although it was shown that greater accuracy can be achieved with a potentiometric titration, this takes too long for process control.

Methods for controlling the chloride content of milk, which is important for the detection of mastitic or abnormal milk, have been described by Sweetsur (63) and by De Clercq, Mertens and Massart (64). Both sets of workers found that direct measurements in milk were impossible by normal procedures because of the effects of fat and protein on both the ion-selective electrode and reference electrode. Sweetsur used a 1:10 sample: buffer ratio in order to try to overcome the difficulties but obtained results approximately 40 percent high which needed correction by a regression equation. Results then correlated well with the British Standard Method. De Clercq et al. suggested that high results are partly due to casein interference and partly due to ionic strength effects. After testing several reagents for reducing the interference they recommend either the Folin-Wu reagent (sodium tungstate in sulphuric acid) or zinc acetate plus potassium ferrocyanide. The relative standard deviation for the chloride determination using either of these reagents lay between 1 and 2 percent, and the accuracy was adequate in comparison with the Volhard method.

Quality control assays of a range of chlorinated pharmaceuticals may be made using the chloride electrode (65); according to the accuracy and precision required either a direct method or a potentiometric titration is used. Many pharmaceuticals, such a tetracycline hydrochloride, contain the chloride in dissociated form and liberate this chloride simply by dissolution in distilled water; in these cases the only sample pretreatment required is the addition of an aliquot of strong sodium nitrate solution to keep the ionic strength constant. Undissociated chloride is determined after Schoniger combustion of the sample.

Studies have been carried out to determine the best technique for the analysis of discrete samples of chloride in paper machine waters (66). These waters contain a wide range of ions, including those of various metals such as aluminium, calcium, magnesium and small amounts of copper, together with pulp and wood extracts. In many machines the water runs in a closed loop, resulting in anaerobic conditions in which bacterial activity causes production of hydrogen sulphide. Direct measurement of chloride in these waters containing hydrogen sulphide is not possible because of the strong interference from sulphide ions (unless a pretreatment reagent to destroy or remove sulphide is used) but the chloride may be analysed reproducibly by potentiometric titration. For samples free of sulphide, a known addition method may be sufficiently reliable but direct measurement is still not feasible because of the variable interference from the sample constituents.

An apparatus for the continuous determination of hydrogen chloride in gases and aerosols has been described by Lee (67). A sample scrubbing tower is used with a water flow in at the top of 1.1 $cm^3\ min^{-1}$ and a sample gas flow in near the bottom of 100 $cm^3\ min^{-1}$. The response range of the system was, using gaseous concentrations, 20 to 6000 ppm HCl and for a concentration change from 0 to 1100 ppm, 99 percent of the final equilibrium reading was reached in 90s. The system uses a calomel reference electrode connected to the flow cell with a salt bridge filled with 1.0M potassium fluoride.

A silver/sulphide ion-selective electrode has been used as working electrode in a relatively rapid potentiometric titration method for the control of chloride in salted meat products such as canned sausage, corned beef and salami (68). In order to avoid interference from components of the meat samples such as histidine and cysteine which are known to form strong silver complexes, all samples are macerated in a reagent consisting of 1M nitric acid plus 0.01M potassium dichromate, which modifies these compounds to form other, non-interfering compounds. Thus, the strong acid causes protonation of the imidazole group in histidine and the oxidant converts cysteine to cystine. The high concentration of the acid also serves to keep constant the sample ionic strength. Meat samples are first minced and then 10g macerated with 50 $cm^3$ reagent plus 0.2 $cm^3$ detergent for not less than 15s; the macerated sample is titrated with silver nitrate solution to a pre-determined potential. This method was compared with the British Standard Method and an ashing method, both of which took at least $3\frac{1}{2}$ hours per sample as against 10 to 12 minutes for the potentiometric titration. The method was applicable to a wider range of meats than the British Standard Method, gave good correlation with both other methods and was generally more precise, with relative standard deviations typically in the range 0.4 to 0.9 percent.

An interesting paper by Vandeputte et al. (69) includes a description of a laboratory continuous flow system using a combination chloride electrode; the system is tested out on the analysis of mineral water samples. The authors comment on the relatively low sampling rates achievable with most reported systems based on ion-selective electrodes (typically 10 to 30 samples per hour) in contrast to the rates regularly achieved with systems having colorimetric sensors (typically 60 or more samples per hour). This is particularly surprising in view of the results of some authors who have measured electrode response times of the order of tens of milliseconds. This discrepancy is partly due to poor cell design which does not allow the sample solution to impinge on the electrode surface in such a way as to replace the solution layers closest to the solid membrane phase as quickly as possible, but is also due to difficulties with streaming potentials at high flow rates (6). The use of a combination chloride electrode made it necessary to have only one flow cell in place of two, and by earthing both the input and output metal tubes all streaming potentials were eliminated. The results showed a notable improvement upon those previously obtained. Tests were done on the system operating at 50 samples per hour and slight carry-over was detected for base-lines of both 1 mg $dm^{-3}$ and 8 mg $dm^{-3}$ although not so much as to make the system unusable. Finally, a series of seven mineral water samples, containing chloride in the range 2 to 350 mg $dm^{-3}$, were analysed. The accuracy was good in comparison with a manual method and the precision also acceptable; the relative standard deviation for determinations of a sample containing 3.7 mg $dm^{-3}$ chloride was 1.1 percent.

A final example of the use of the chloride electrode is in the method for determining chloride in sulphuric acid copper plating baths (70). Samples are treated with an acetate buffer (the strength of which varies according to the concentration of acid in the type of bath concerned) and measured directly. Chloride in the range 30 to 250 mg $dm^{-3}$ was determined with equal precision by either the direct calibration graph method or by a standard addition procedure.

## 4.8 Copper

Studies of copper ion-selective electrodes, with membranes made of mixtures of copper(II) sulphide and silver sulphide, have shown that when used in samples containing copper ions at concentrations below $10^{-6}$M the response of the electrode is strongly influenced by sample pH and dissolved oxygen (71); the membrane slowly dissolves to liberate copper ions into acidic samples and also, more slowly, into neutral samples. When analysing discrete samples this will cause electrode drift and errors due to sample contamination. This probably accounts for the slowly rising potentials in dilute solutions recorded by Blaedel and Dinwiddie (72). Electrodes based on copper selenide are said to be hardly more successful for submicromolar measurements as interference from redox effects can cause irreproducible responses (71). The results obtained are substantially improved in flowing systems as shown in a further paper by Blaedel and Dinwiddie (73), in which they report Nernstian response right down to $10^{-8}$M, without the help of metal-ion buffers, and a further 23mV response between $10^{-8}$M and $10^{-9}$M (the sub-Nernstian response is possibly due to traces of copper in the water used for making the standards). However the response times are too long for practical purposes, being in excess of 1 hour at $10^{-7}$M and 3 days at $10^{-9}$M. Thus, the use of copper electrodes must be ruled out for trace metal analysis.

Copper concentrations in tap water and natural waters have been measured by Smith and Manahan (74) with a standard addition technique. They developed a pretreatment reagent for addition to samples in a volume ratio of 1:1; the buffer consisted of 0.05M sodium acetate, 0.05M acetic acid, 0.02M sodium fluoride and $2 \times 10^{-3}$M formaldehyde. The acetate compounds served to fix the sample pH and ionic strength and also complexed a fixed proportion of the copper, in effect forming a copper(II)-acetate metal buffer and reducing loss of copper by absorption onto the walls of the sample container. The fluoride ions were added to mask ferric ions and the formaldehyde produced a reducing medium (to minimise membrane attack by oxygen as mentioned above). Copper at a concentration of 9 μg $dm^{-3}$ was spiked into natural water samples containing from 3 to 47 μg $dm^{-3}$ copper; the average recovery was 102.9 percent and the standard deviation 7.5 percent. Copper impurities in the pretreatment reagent, about 1-2 μg $dm^{-3}$, were found to limit the method.

A version of the same reagent has been used by Fung and Fung (75) in their method for the determination of copper in palm oil. This assay is particularly important in the production of this oil, as trace amounts of copper (even at levels of 30 μg $kg^{-1}$) catalyse the oxidation of the product, generating aldehydes, ketones and other compounds which give an off-flavour. The copper concentrations found in quality control vary from 10 μg $kg^{-1}$ up to 5 mg $kg^{-1}$; there is also iron(III) present in the range 0.7 to 30 mg $kg^{-1}$. Samples of the palm oil were ashed and the ash digested in nitric acid. The acid is boiled off and the residue dissolved in a solution containing dilute perchloric acid and sodium perchlorate; the resultant solution is treated with a buffer very similar to that used by Smith and Manahan which has been described above. The copper ion concentration was determined by direct potentiometry (in a darkened container to prevent errors due to light effects on the electrode membrane) and for comparison, by atomic absorption spectrophotometry: a standard addition procedure was also tested but no difference between the results of this and the direct method was measured. For a range of twelve palm oil samples containing copper in the range 0.3 to 0.6 mg $kg^{-1}$ no statistically significant difference was found between the potentiometric and photometric methods; it was also shown that at low copper concentrations the potentiometric method gave a standard deviation of about half that of the other method.

Other applications of the copper electrode which may be briefly mentioned involve its use in the measurement of determinands other than copper. Przyborowski (76) has used the copper electrode in the analysis by potentiometric titration of pharmaceuticals, such as methizol and methylthiouracil, which contain mercapto groups. Solutions of methizol are buffered to pH 5.6 with an acetate buffer and titrated with copper sulphate solution; in the case of methylthiouracil, the samples were dissolved in dilute ammonium hydroxide and buffered to pH 11.1 with a borate buffer before titration. End-point breaks in excess of 100 mV were obtained in both analyses.

An electrode method for use in the food industry for the routine determination of reducing sugars has been described (77). Samples of honey, marmalade, jam or fruit juice are treated with a copper sulphate plus sodium carbonate reagent and boiled for 10 minutes. The resultant suspension is filtered and the filtrate analysed for copper by a straightforward standard addition procedure; the decrease in the copper concentration is taken as a measure of the concentration of reducing sugar. Results correlated well with the AOAC method and also showed good precision.

In several of the papers cited and in others, the authors note the special precautions required to keep the electrode in good working order. For example, Papastathopoulos et al. (77) polished the electrode with fine diamond paste and treated the surface with silicone oil every day. This problem of the surface condition of the membrane has been particularly studied by Johansson and Edstrom (78). It is clear that gradual surface deterioration limits the potential use of the electrode in many types of sample and especially in the large majority of potential applications to continuous monitoring, even for quite high copper concentrations.

## 4.9 Cyanide

Cyanide may be measured either by use of the iodide electrode, often referred to as the cyanide electrode when used in cyanide determinations, or by use of the silver/sulphide electrode. The response of the iodide electrode to cyanide depends on the dissolution by cyanide in samples of the silver iodide in the electrode membrane with consequent liberation of iodide ions; the electrode responds to the concentration of these iodide ions, which is directly proportional to the cyanide concentration. A result of this dissolution mechanism is that if the electrode is used for the analysis of strong cyanide solutions the membrane dissolves rapidly and the life of the electrode is unacceptably short; in practice an upper concentration limit of $10^{-3}$M cyanide may be recommended. The limit of Nernstian response found by different authors varies, but a fair average of the various results is probably about 2 x $10^{-6}$M (0.05 mg $dm^{-3}$), with a limit of detection one decade lower. The electrode suffers interference, in respect of the cyanide response, from iodide, sulphide and, to a lesser extent bromide; selectivity coefficients are quoted elsewhere (1). The electrode responds only to cyanide ions, and not to HCN, hence samples must be made alkaline to pH 11-12 to make sure all cyanide is in the ionic form; at higher pH, hydroxyl ions interfere.

Alternatively, cyanide may be measured by use of the 'indicator technique' suggested by Frant et al. (79) samples are treated with a small quantity of a solution of dicyanoargentate(I) ions, $Ag(CN)_2^-$, and the resultant free concentration of silver ions measured with the silver/sulphide electrode. The method has an upper concentration limit of $10^{-2}$M, resulting from the onset of formation of $Ag(CN)_3^-$. The lower limit is set by the background concentration of cyanide ions in the indicator solution as there must be a slight excess in the solution for it to be stable; depending on the various concentrations involved, this may make the limit of detection of the technique about 5 x $10^{-7}$M (0.01 mg $dm^{-3}$) or lower. Samples again need to be made alkaline to pH 11-12. The technique gives a response of 118 mV per decadic change of cyanide concentration. The authors recommend that the sensing technique is used in conjunction with a Gran's plot procedure.

The relative merits of the two methods of cyanide measurement have been examined by Clysters et al. (80) who concluded that the method based on the silver/sulphide electrode was preferable because of the larger signal produced by changes in cyanide concentration (and hence the better precision of measurement) and the better selectivity. Moreover the silver/sulphide electrode lasted longer and was capable of giving a lower limit of detection reliably. The response of the cyanide electrode is also inconveniently dependant on sample flow rate (81). It is notable that most recent methods published for cyanide determination have been based on the silver/sulphide electrode.

Hofton (82) described a continuous monitor for the determination of cyanide in coke oven and blast furnace effluents and also in river water. The monitor was designed for measurements down to and below the 0.05 mg $dm^{-3}$ discharge limit set by many water authorities; the actual limit of detection achieved by the system was 0.01 mg $dm^{-3}$ and the response was linear up to 1 mg $dm^{-3}$. The measurement was based on direct reading with the silver/sulphide electrode, as the Gran's plot procedure used by Frant et al. (79) is difficult to automate and was thought to be unsuitable and unnecessary for an industrial monitor. The sample stream is pumped into the monitor at a rate of 5 $cm^3$ $min^{-1}$ and mixed with 0.5 $cm^3$ $min^{-1}$ 2M NaOH and 0.1 $cm^3$ $min^{-1}$ reagent (2.5 x $10^{-3}$M Ag $(CN)_2^-$); the treated sample is pumped through the flow cell which is thermostatted at 30$^\circ$C. Every 12 hours the system is recalibrated by means of a 0.05 mg $dm^{-3}$ standard which is passed through the system in place of the sample. The response time of the monitor was approximately 10 minutes for virtual equilibrium to be reached after change of sample concentration. The results produced by the monitor agreed well with those from a manual potentiometric method.

Another continuous monitor for cyanide, developed by Durst, (25) works by different means in that the cyanide is separated from the sample before measurement. A diagram of the monitor is shown in Fig. 9. The sample is acidified with 2M sulphuric acid, in order to convert the free cyanide present to hydrogen cyanide, and pumped through a dialyser in which the acidified sample is separated from an alkaline solution of the indicator recommended by Frant et al. (79) by a hydrophobic, microporous membrane. Under the conditions used the HCN is almost completely stripped from the sample stream into the indicator stream, and by having a slower flow rate for the indicator than for the sample stream, the cyanide is concentrated, giving a more sensitive measurement than would otherwise be possible. Although the sample stream is pumped in the conventional manner, the indicator solution is fed through by gravity from a constant head tank so that effects due to pump pulsations are avoided. Samples are analysed at the rate of 30 per hour with accuracy and precision better than 2 percent over the concentration range 30 to 400 μg $dm^{-3}$; the limit of detection of the system, taken as that concentration of cyanide which produced a signal 5 mV above that from a blank, was found to be 0.5 μg $dm^{-3}$. Durst states that the monitor could be further developed by addition of a UV irradiator to decompose heavy metal cyanide complexes, so that the monitor could be used for total cyanide measurements.

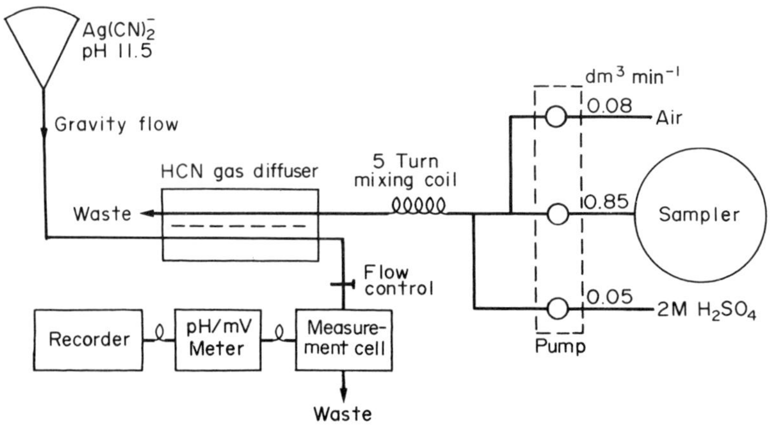


Fig. 9. Diagram of a free cyanide monitor (from Ref. 25, by courtesy of Marcel Dekker, Inc.).

A method for measuring cyanide in well-waters, lakes, streams and effluents using the cyanide electrode has been worked out so that the response of the electrode to interferents is allowed for (83). An aliquot of the sample is acidified to pH 2-3 and the apparent cyanide content measured by a standard addition of iodide. A second aliquot is adjusted to pH 11.5 with a sodium phosphate/sodium hydroxide buffer and free cyanide determined, again by a standard addition procedure. If required, total cyanide may also be measured after UV irradiation of a third aliquot for 2 minutes. The results of the cyanide measurements are corrected for interference by subtracting the result from the acidic sample. Sulphide is found to interfere strongly by forming a precipitate of silver sulphide over the membrane, which can only be removed by polishing: samples containing sulphide are therefore treated with a solution of a bismuth salt to precipitate the interferent as bismuth sulphide. Samples having cyanide concentrations in the range 1 to 500 $\mu g\ dm^{-3}$ were analysed with relative standard deviations varying from 20 percent at 1 $\mu g\ dm^{-3}$, 9 percent at 2 $\mu g\ dm^{-3}$ to 3 percent at 100 $\mu g\ dm^{-3}$. The manual determinations required approximately 10 minutes per sample.

At the opposite end of the concentration scale, Lapatnick (84) has determined the high concentrations of cyanide in silver-plating baths, using a cyanide electrode. The samples need considerable dilution to bring them down into the measurement range of the electrode, consequently 0.1 $cm^3$ sample is added to 100 $cm^3$ 0.25M sodium hydroxide and the cyanide determined by a standard addition procedure.

### 4.10 Fluoride

The measurement of fluoride with the fluoride electrode is very straightforward for a wide variety of samples. The electrode gives a linear response to fluoride ion activity in solutions of concentration ranging from saturated down to about $2 \times 10^{-6}$M (0.04 mg $dm^{-3}$) and has a limit of detection at about $10^{-7}$M. The only interferent is hydroxide (selectivity coefficient $k^{pot}_{F,OH} \triangleq 1$) although some carboxylate ions, such as citrate, slow the electrode response below $10^{-6}$M (85). Thus provided that the fluoride in the sample to be analysed can be converted to uncomplexed ionic form in an aqueous solution, it can readily be measured. Most of the detail in many of the published methods is concerned with procedures for sample dissolution rather than, as with many other potentiometric methods, removing electrode interferents. Potentiometric methods for fluoride determination are generally much faster, more convenient and selective than alternative methods (e.g., Ref. 86).

4.10.1 Ionic strength adjustors and other reagents. Several reagents have been recommended for use in conjunction with the fluoride electrode; the function of these is to keep the ionic strength constant, fix the pH to approximately 5-6 (in which range both hydroxyl ion interference and formation of undissociated HF are virtually nil) and complex any heavy metals which would otherwise complex fluoride. The ionic strength adjustment may be achieved with a high concentration of potassium chloride; this is preferred to sodium chloride as appreciable concentrations of undissociated $NaF^o$ form in fluoride samples treated with strong sodium solutions (87). The buffering component of the pretreatment reagent is usually an acetate or citrate buffer (citrate also acts as a complexing agent) and the most suitable complexing agent is, in most cases, DCTA; this complexing agent will deal satisfactorily with most common interferents such as calcium, magnesium and iron(III) ions. If, however, an appreciable concentration ( >0.5 mg $dm^{-3}$) of aluminium is present, a more satisfactory complexing agent to use is said to be 1M Tiron (4,5 - dihydroxybenzene - 1,3 - disulphonate), in conjunction with 0.2M sulphosalicylic acid and 0.5M EDTA or DCTA at pH 6.0 - 6.5 (85). However, for the large majority of applications when the aluminium concentration is low a buffer consisting 1M KCl, 1M acetic acid (adjusted to pH 5.8 with concentrated KOH) plus 1 g $dm^{-3}$ DCTA is suitable, although up to 5 g $dm^{-3}$ DCTA may be necessary in waters containing high concentrations. This buffer is added to samples at a volume ratio of 1:1, although 1:5 may be equally successful and more economical for treatment of dilute waters (87,88,89). Many minor variants on this recipe have been proposed. Care must be taken in measurements at low fluoride concentrations to use reagents with minimal contamination from fluoride.

4.10.2 Liquid samples. Several systems have been described for the automatic or semi-automatic analysis of fluoride in water samples. Laboratory continuous flow systems, based on Technicon apparatus, have been developed by Erdmann (90) and by Philbert et al. (38), both for water quality monitoring purposes. Erdmann used a pretreatment reagent spiked with 0.2 mg $dm^{-3}$ fluoride as this improved the reproducibility of measurement at the 1.5 mg $dm^{-3}$ level. He also heated the treated sample stream to 55°C before measurement, but it is not clear why this was necessary. The system was calibrated over 0 to 2 mg $dm^{-3}$ and samples were analysed at the rate of 30 per hour. Results on natural water samples agreed well with the results obtained from a manual electrode method (88). Philbert et al. (38) used an unspiked reagent but had instead a 0.1 mg $dm^{-3}$ $F^-$ wash solution in the sampler. The system was usually calibrated over the range 0.02 to 1.00 mg $dm^{-3}$, although alternative, higher top limits could be used. No significant carry-over was apparent when the system operated at 40 samples per hour. For a full-scale range of 2.00 mg $dm^{-3}$, replicate analyses of 0.02 mg $dm^{-3}$ standards gave a relative standard deviation of 6.5 percent. Lingerak et al. (16,91) have used a computer-controlled flow system for determining the quite high concentrations of fluoride (and also chloride and ammonia) in rainwater in a heavily industrialised part of the Netherlands and also for the analysis of fluoride in samples of uranium tetrafluoride. The computer controls a Gran's plot standard addition procedure and also automatically calibrates the electrode. Accuracy was satisfactory down to 0.05 mg $dm^{-3}$, approximately the lowest concentration of fluoride in the rainwater samples tested. The measurements in $UF_4$ could be made both accurately and precisely (for five replicate analyses a mean of 24.3±0.17 percent F was found: the expected value was 24.2 percent F), and could be performed at the rate of 20 to 60 per hour, depending on the requirement for accuracy and precision.

The performance of the fluoride electrode in flow systems has been investigated in detail with a view to optimising the system design (6,19). The results are largely applicable also to other types of ion-selective electrode. Methods for reducing streaming potentials, already mentioned, were investigated and also the importance of a fast sample velocity across the membrane, with adequate stirring of the samples in the measurement cell were emphasised.

An industrial monitor for the continuous monitoring of fluoride in potable water supplies was developed by Collis and Diggens (92). This was primarily intended for use in conjunction with fluoridation schemes and can be linked for this purpose to the controller of a fluoride dosing pump. The calibration range of the monitor may be either 0.01 - 0.15 mg $dm^{-3}$ for bore-hole water analysis or 0.1 to 1.5 mg $dm^{-3}$ for analysis of other waters and for the fluoridation control. The electrode is automatically standardised every 12 hours. The use of a continuous analyser for monitoring fluoride and chloride in subterranean waters has been reported by Berthier (93). The system required only occasional service and the electrode life was greater than one year.

Fluoride has been determined in several types of strong acid. Sample preparation usually consists of a major dilution followed by addition of a standard pretreatment reagent of the type described above. Croomes and McNutt (94) measured the fluoride content of red fuming nitric acid. This acid to which approximately 0.5M HF is added as an inhibitor is an extremely powerful oxidant used in conjunction with the liquid propellant in missiles. For the analysis 1 g sample is diluted in 100 $cm^3$ 0.1M acetate buffer and the fluoride determined directly with the electrode.

A simple method has been developed by Entwistle et al. (95) for determining undissociated hydrofluoric acid, HF, and also nitric acid in the pickling bath liquors used for metal treatment. The hydrofluoric acid is measured by diluting 1 $cm^3$ of the liquor in 50 $cm^3$ 1M $KNO_3$ saturated with quinhydrone and reading the potential of a fluoride electrode in the solution against a quinhydrone pH-sensing electrode (ordinary pH glass electrodes cannot be used in these samples as they dissolve in the HF). The cell potential varies in a Nernstian manner with the concentration of HF and the cell is standardised with HF standards in nitric acid. Liquors containing 5 - 40 g $dm^{-3}$ HF, 50 - 250 g $dm^{-3}$ nitric acid and 0 - 20 g $dm^{-3}$ fluoride complexants (which are said not to affect the measurement) were analysed for HF with a random error of 1 percent. Since the quinhydrone electrode responds to iron(III) ions (a probable component of used liquors) as well as hydrogen ions and because quinhydrone may be oxidised by nitric acid, Johansson (96) suggested that an alternative hydrogen ion-selective electrode, such as his permaplex membrane electrode, would give better results.

Schick (97) has described a Gran's plot procedure, using the special Gran's plot paper, for determining fluoride in a range of liquid household products containing fluoride, such as cleaners, rust removers, stain remover, metal cleaner and fluxes. The fluoride contents ranged from about 23 - 30 percent for rust remover and etching solution down to about 1 percent for cleaners and stain remover. Sufficient of the sample is diluted, if necessary, with water to very approximately $10^{-3}$M, and 50 $cm^3$ of this solution is treated with 50 $cm^3$ pretreatment reagent (an ammonium acetate/citrate buffer at pH 6 was used, but others would probably be equally effective): the treated sample is titrated with ten 1 $cm^3$ aliquots of 1 g $dm^{-3}$ $F^-$ standard and the results analysed using the Gran's plot paper. The overall relative standard deviation was 2 percent and the range of recovery of 1 percent $F^-$ spikes was 98 - 104 percent. These results do not seem sufficiently good to justify use of the cumbersome Gran's plot method for it is likely that a single standard addition method would have served equally well.

4.10.3 Solid samples. Fluoride has also been determined in a wide variety of solid samples. Samples which do not liberate the fluoride to be measured as fluoride ions when added to water, must be pre-treated by procedures that vary considerably from sample to sample. Thus, digestion in strong acid or alkalis, various fusions and pyrohydrolysis have all been used for inorganic samples, such as various ores, soils, metals, glasses, etc. The same procedures may also be applicable to organic samples as are also ashing and hydrogenation. A few examples are given below although the principle from the electroanalytical point of view is the same in each, namely, conversion of the fluorine in the sample to fluoride ions in solution, followed by addition of one of the standard fluoride pretreatment reagents and measurement by direct or standard addition potentiometry.

The determination of the halogens fluorine and iodine, which are important contaminants in selenium, is very important in the production of refined selenium (98). Samples of selenium are melted with sodium hydroxide and the resultant mixture boiled with water for 30 - 45 minutes to extract the sodium salt of the halide. The solution is neutralised with sulphuric acid, the precipitated selenium filtered off, and the resultant solution analysed for fluoride or iodide by the standard addition method. For selenium samples known to be free from alloys, silver or doping compounds, the halide may more simply and safely be extracted by leaching the ground sample with 10M sodium hydroxide for 2 hours. The limit of detection was found to be 0.05 ppm for iodine and 0.1 ppm for fluorine.

Palmer(99) reports the determination of fluoride in a range of ores such as cryolite, fluorspar and sodium fluorosilicate. Cryolite was dissolved by boiling in sodium hydroxide solution, whereas fluorspar was dissolved in a hydrochloric acid solution of aluminium chloride; the aluminium salt prevents volatilisation of the fluoride from the acid solution by retaining it in solution in complex form. Aluminium interference could be removed, as discussed previously, with sulphosalicylate and calcium interference with EDTA. Errors in the methods were less than 2 percent with relative standard deviations of 1 percent. A similar buffer was used by Boniface and Jenkins (100) who analysed steelworks materials such as slags, fluorspar and iron ore, after fusion with a boric acid-sodium carbonate mixture. The method was found to be free from interference by boron. The relative standard deviation from 20 determinations of a standard basic slag was 1.7 percent. The method was suitable for samples containing down to 0.02 percent fluoride.

Various minerals and rock samples have also been analysed by Troll et al. (101), who fused samples with a mixture of sodium and potassium carbonates. The fluoride was determined after solution of the melt by a standard addition technique. Analysis of standards gave accuracies of 0.3 to 0.7 percent and relative standard deviations less then 4 percent. The pretreatment reagent was used as the bridge solution in the double junction reference electrode in order to avoid problems with liquid junction potentials. Fluoride in silicate minerals has been measured after pyrohydrolysis (102) (heating of a ground mixture of the sample with lead dioxide and copper in superheated steam) and collection of the $H_2SiF_6$ reagent before analysis. The precision of determination on a range of samples was excellent but the accuracy remains open to question (103).

Wilson and Marczewski (104, 105) have described a method for estimating total fluorine in petroleum in which the fluorine present is bound to carbon. The C-F bonds are broken by reaction with sodium biphenyl in non-aqueous conditions under nitrogen but, if wet, the samples must be dried with a molecular sieve before analysis. After allowing 15 minutes for the reaction to reach completion, a small quantity of a cationic detergent is added to the mixture to prevent formation of emulsions, and the sodium fluoride extracted with three portions of water. The fluoride was determined, after reagent addition, by a standard addition method. Determinations of fluoride at concentrations down to 10 ppb were possible. It was found to be necessary to run reagent blanks.

In a collaborative study on the determination of fluoride in feeds (106), the fluoride is extracted from the samples by simple dissolution in 1M hydrochloric acid. Fluoride has also been measured in mechanically deboned meat to determine the apparent increase in fluoride concentration over that in meat deboned by hand (107). Samples were defatted by extraction with petroleum ether, mixed with pretreatment reagent, homogenised and left on a steam bath for 1 hour. After breaking up the lumps and diluting the liquid, the sample was filtered and the fluoride determined in the filtrate by a direct, calibration curve method. The method was considerably faster and simpler than alternative, non-potentiometric methods. Recoveries of 10 ppm fluoride spikes tended to be low, ranging from 86 to 99 percent.

4.10.4 Gaseous samples. Mascini (108) has described an automatic monitor for monitoring hydrogen fluoride in the atmosphere. Small amounts of hydrogen fluoride are discharged into the air from factories producing aluminium, glass, fertilizers, ceramics and steel; because of the high toxicity it is desirable or, in some cases essential to monitor the concentration of the gas. This is, however, generally so low (0.1 - 15 $\mu g\ m^{-3}$) that collection of the gas for an hour is necessary before the fluoride is sufficiently concentrated for measurement. A diagram of the collection and measurement apparatus is given in Fig. 10. Sample air is pumped through a dust separator and a large filter (not shown) impregnated with citric acid, and heated to 80$^\circ$C; this filter retains particulate fluoride (e.g., NaF or $AlF_3$ dust) but releases fluoride adsorbed on particulates. The air then passes down the central 1 meter spiral, the inside walls of which are coated with sodium carbonate to absorb the fluoride. By operation of the valves

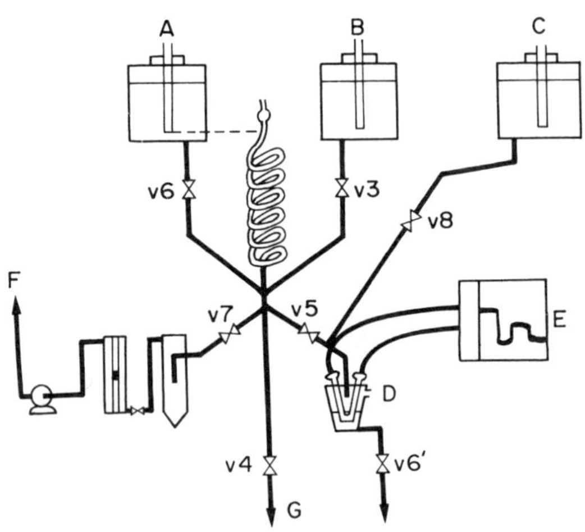


Fig. 10. Diagram of a fluoride-in-air monitor. A. Sodium citrate reservoir. B. Sodium carbonate reservoir. C. Sodium fluoride reservoir. D. Analysis cell. E. Potentiometric recorder. F. Gas exit. G. Waste. (from ref. 108, by courtesy of Elsevier Scientific Publishing Co.)

every hour the air flow is stopped and 5 $cm^3$ citrate solution flows into the spiral; the fluoride dissolves in the citrate solution and, after further valve operation, flows into the cell for measurement with the fluoride electrode. The coating of sodium carbonate is then renewed by solution flowing from the other reservoir, and the cell refilled with a standard fluoride solution to standardise the electrodes. Air flowing through the system at 1.5 $m^3 h^{-1}$, gives a fluoride concentration in the final solution of $10^{-6}$M for an air concentration (averaged over the hour) of 0.1 μg $m^{-3}$. This constituted the limit of detection of the monitor. The absorption efficiency of the spiral remained at 100 percent for fluoride concentrations up to 10 μg $m^{-3}$ and consequently the response of the monitor was linear in this range, at higher concentrations the efficiency dropped sharply, limiting the application of the monitor.

## 4.11 Iodide

In principle it is quite feasible to use the iodide-selective electrode for analysing iodide in industrial samples but there are few reports of such applications. The electrode gives a Nernstian response down to $10^{-6}$M or slightly lower, although measurements on samples containing less than $10^{-5}$M iodide are experimentally difficult because of aerial oxidation of the iodide; samples containing only traces of iodide should be deaerated by bubbling nitrogen through them and treated with ascorbic acid in order to get stable and accurate readings (109). In concentrations of iodide higher than about $10^{-1}$M, the membrane begins to dissolve through complex formation and the electrode response becomes irreproducible. The iodide electrode suffers interference from sulphide (even at low concentrations), cyanide and high concentrations of bromide.

Iodide has been determined in several pharmaceutical compounds after Schöniger combustion (110). In order to reduce the iodate which is formed during the combustion, the residue is dissolved in sodium hydroxide solution and treated with Devarda's alloy. The resultant solution is diluted 1:1 with 2M potassium nitrate and the iodide determined by direct potentiometry.

Traces of iodine in refined selenium have been measured potentiometrically (98) as described in the section on fluoride applications.

## 4.12 Nitrate

Use of the nitrate electrode has greatly simplified the determination of nitrate in very many types of samples. Alternative spectrophotometric methods, such as those based on 2, 4- or 2, 6- xylenol, phenoldisulphonic acid or chromotropic acid, all have their supporters but none stands out as sufficiently accurate or reproducible to gain wide acceptance (111). Moreover, all these methods are unsuitable for use with coloured or turbid samples without time-consuming pretreatment. Many also use unpleasant or dangerous reagents, which are undesirable for routine methods, and have an inconveniently long colour development period. Direct UV spectrophotometric measurement is an attractive alternative for relatively pure waters, but is limited in application by interference from organic material.

Recent advances have resulted in nitrate electrodes with limits of Nernstian response approximately one decade lower (e.g., Refs. 112, 113) than those used previously, although virtually all published results refer to the earlier types of electrode. The limit of Nernstian response of nitrate electrodes is now approximately $10^{-5}$M and the limit of detection $10^{-6}$M or slightly lower; these values vary slightly from one manufacturer's electrode to another. There has, unfortunately, been no concomitant improvement in electrode selectivity. Leaving aside species such as the perhalate and iodide ions, which are unlikely to be present in samples, the most serious interferents which may have to be removed by sample pretreatment, are chloride, nitrite and bicarbonate ions. High concentrations of anionic surfactants also interfere and, because they are more difficult to remove, may prevent the electrode being used in waters heavily polluted with surfactants. The sample pH may lie within the range 3 to 11, although the lifetime of the membrane will be greatest if the range is narrowed to 4 to 6.

In the water industry, the importance of nitrate measurement is increasing as more and more water supplies are becoming polluted with it. This nitrate comes from many sources of which agricultural fertilisers and the oxidation of ammonia in sewage works are particularly important, whilst few processes remove nitrate from the water. Consequently, in some areas nitrate levels in potable waters are coming close to the maximum acceptable concentration fixed by the World Health Organisation, and continuous monitoring, using the nitrate electrode, has become routine.

Many authors have reported the determination of nitrate in various waters using straightforward procedures. Samples are typically buffered to about pH 4 to 5 with a phosphate buffer and a little EDTA is added to prevent precipitation of hardness or heavy metals. If high concentrations of bicarbonate are present, a pH of 4 is necessary to convert all the interferent to dissolved carbon dioxide. If chloride is interfering, discrete samples are treated with silver sulphate solution to precipitate silver chloride; while continuous streams of sample are passed through a silver-loaded ion-exchange column. Nitrite may be removed by addition of sulphamic acid. Milham et al. (114) used a pretreatment reagent consisting of 0.01M aluminium sulphate, 0.01M silver sulphate, 0.02M boric acid and 0.02M sulphamic acid, adjusted to pH 3.0 with sulphuric acid; this reagent was added to samples at a volume ratio of 1:1. The aluminium salt is present to complex anions of organic acids which might interfere. Replicate measurements on a river water sample containing 4 mg $dm^{-3}$ nitrate nitrogen had a relative standard deviation of 1.6 percent, and the results agreed closely with those from a Devarda's alloy reduction method. The authors also measured the nitrate content of several plants and soils. Meredith (36) has described Dutch laboratories which continuously monitor nitrate in river water.

Samples containing ionic interferents which cannot readily be removed may be analysed by reduction of the nitrate with Devarda's alloy (115) and measurement of the ammonia produced with the ammonia probe, which is far more selective than the nitrate electrode. The procedure is firstly to make the solution alkaline by addition of sodium hydroxide solution. If the sample naturally contains a large excess of ammonia over nitrate ions, this ammonia may now be removed by bubbling nitrogen through the alkaline sample at 80-90$^{\circ}$C; otherwise the correction for smaller amounts of ammonia may be made by measurement with the probe before and after reduction. Samples containing nitrite are treated with sulphamic acid. The sample is then reduced with Devarda's alloy either by addition of the alloy to a sample aliquot or by pumping the sample through a column packed with plastic-bonded Devarda alloy beads. For samples containing 1 to 50 mg $dm^{-3}$ nitrate analysed in a laboratory continuous flow system working on this principle, with the flow rates and volumes adjusted so that samples take 1 minute to pass through the column, a 100 percent reduction yield and minimal carry-over has been achieved with a freshly-prepared column. The reduction efficiency of the column drops to 80 percent after 5 days, after which it needs to be replaced. Samples were analysed at the rate of 20 per hour with a relative standard deviation of typically 2 to 3 percent. The manual method had a limit of detection of about 0.05 mg $dm^{-3}$ nitrate, although the calibration curve was non-linear below 0.6 mg $dm^{-3}$. This method was tested on the analyses of sewage from a coke factory, a slaughterhouse and a brewery and shown to give results very similar to those from a colorimetric method.

Burman and Johansson (116) have developed a 'sample addition' method for determining nitrate in pickling bath liquors. These liquors consist of mixtures of nitric and hydrofluoric acids and also contain, as they are used, increasing concentrations of iron(III) ion. It is important to control the nitrate concentration in order to conserve the acid and to maintain the quality of the finish on the dipped articles. In this method, the potential of a nitrate electrode is read in 50 $cm^3$ reagent containing 5 x $10^{-3}$M nitrate, 0.5M disodium EDTA and 0.25M ammonia, all adjusted to pH 8.3. A 0.5 $cm^3$ sample aliquot is added to the reagent and the potential read again, and from the two potentials the nitrate concentration in the sample is calculated. Measurements on a synthetic sample consisting of 1M nitric acid, 0.6M hydrofluoric acid and 0.2M iron(III) chloride gave an error in nitrate concentration of 4 percent and a relative standard deviation of 1 percent, both of which are quite adequate for control purposes.

A method has been described (117) for the measurement of free nitrate in the plutonium oxide suspensions used in nuclear fuel preparation. The suspensions consist of the plutonium oxide in nitric acid; part of the nitrate is absorbed by the solid and the remainder remains free in solution. The free nitrate may be measured directly with a nitrate electrode, in conjunction with a double junction reference electrode. The method was also used to study the variation with pH of the ratio of absorbed to free nitrate.

The nitrate electrode has been used in the food industry for the measurement and control of nitrate in a range of foodstuffs (118,119). Here, the nitrate content of baby food is particularly important because of the susceptibility of babies to nitrate poisoning. In both the papers cited, the authors use a manual method for treating discrete samples. Interference from chloride and organic acids are removed by adding 2-3 g silver- and aluminium-loaded resins to samples. Meat samples are defatted before measurement. No attempt was made to fix the pH or ionic strength of samples or to match the standards to samples in any way; consequently the standard addition method is likely to give better results. The accuracy and precision, which were barely adequate, would probably have been improved if a buffer such as that of Milham et al. (114) had been used. The analysis of foods has also been reported by Mergey and Bonnoit (120).

An apparatus incorporating a nitrate electrode has been developed by Forney and McCoy (121) for monitoring atmospheric particulate nitrate. An aerosol impaction device is used to collect the sample and dissolve the particulate nitrate in a continuously flowing stream of $10^{-2}$M silver fluoride solution. This solution precipitates any halide or sulphide in the sample which would otherwise interfere with the electrode response, whilst also providing a constant level of fluoride ions which enables a fluoride electrode to be used as reference electrode (making a cell without liquid junction). A special flow cell was designed in the shape of an inverted cone which permitted adequate stirring without bubble formation over long periods despite the high working temperature of 38°C, at which the cell was thermostatted. The flow rate through the cell was 6-10 $cm^3$ $min^{-1}$. The apparatus gives answers comparable to those from the more laborious standard techniques.

## 4.13 Nitrite

The nitrogen oxide sensing probe has been used for the analysis of nitrite. The probe has a Nernstian response over the range $10^{-2}$M to 2 x $10^{-6}$M nitrite. Samples must be acidified to pH$<$2 in order to convert all the nitrite to dissolved nitrogen oxide. Despite the large apparent concentration range, the application of the probe is limited in practice by interference from other gases in the sample. Thus, dissolved carbon dioxide prevents the use of the probe for the analysis of nitrite in most natural and potable waters unless this interferent can be removed. Another possible application, the analysis of nitrite in processed cooked meat such as salami, is made difficult by the need to virtually completely defat the sample (as fat clogs the probe membrane) without altering the nitrite content. It is, however, particularly desirable to be able to use a potentiometric method for nitrite as the popular spectrophotometric method (the Griess method) uses a carcinogenic reagent which is difficult to replace satisfactorily.

Sherken has reported a method for determining nitrite in smoked fish (122), such as chub. It is necessary for food processors to control the nitrite in this and other foods as a result of legislation limiting nitrite contents because of the risk of formation of carcinogens. The fish sample is macerated in 100 $cm^3$ water and then heated on a steam bath for 2 hours to eliminate natural reductants. Strong copper sulphate is added to the cooked solution to precipitate protein and the solution is filtered. An aliquot of the resultant filtrate is acidified with a bisulphate buffer and the dissolved nitrogen oxide concentration measured by a Gran's plot procedure using Gran's plot paper. Recoveries of added nitrite were excellent and, indeed, better than with the current standard method.

A range of waters and effluents spiked with nitrite and analysed gave results that agreed well with those from the Griess-Ilosvay method (123). As the samples all contained less than 0.05 mg $dm^{-3}$ $NO_2^-$-N originally the probe was not sufficiently sensitive for their analysis. Several soil samples were analysed satisfactorily, although carbon dioxide interference made measurements on calcareous soils impracticable.

## 4.14 Nitrogen

The measurements of ammonia, nitrate and nitrite with ion-selective electrodes have all been separately described in previous sections, but it is also sometimes desirable to measure the sum of all three (total inorganic nitrogen) or simply total nitrogen (total inorganic plus organic nitrogen). Because the ammonia probe has generally superior performance to the other nitrogen species-sensing electrodes, and because methods are available for converting all the nitrogen species to ammonia, it is customary to convert the nitrogen to ammonia rather than other nitrogen species for measurement.

The system devised by Mertens et al. (115) described in Section 4.12, may equally well be used for measuring total inorganic nitrogen, if the ammonia and nitrite removal stages are omitted. These authors found that Devarda's alloy was the preferred reductant and gave superior results to aluminium powder. This choice of reductant was confirmed by Hansen et al. (124), who also made comparisons with several other reductants; they used the method for analysing river water untreated waste water and biologically treated waste water.

Consequently, it is surprising to note that Vita (125) preferred aluminium as a reductant after comparison with Devarda's alloy. This may be because he chose to distil the ammonia from the alkaline mixture of Devarda's alloy and sample, thus introducing an extra stage with several opportunities for introduction of errors, rather than measure directly in the alkaline mixture. Vita used the method for determining bound nitrogen, an undesirable impurity, in uranium hexafluoride: the nitrogen is present in the form of nitrosyl and nitryl hexafluorouranates and heptafluorouranates. The sample is first hydrolysed in cold acidic potassium permanganate solution in order that all the various nitrogen oxides, formed as hydrolysis products, are converted to nitrate and that none is lost through volatilisation. The resultant solution is treated with sodium fluoride, if the fluoride concentration is low, concentrated hydrochloric acid and aluminium powder to reduce the nitrate. After dissolution of the aluminium, the sample is made alkaline and the ammonia determined with an ammonia probe by a standard addition method. Because of the high total concentration of dissolved solids in the sample, as presented to the probe, the probe internal filling solution was modified to 0.01M ammonium chloride plus 1M potassium chloride to prevent osmotic drift.

The determination of total Kjeldahl nitrogen, using an ammonia probe for the analysis of ammonia directly in the digest, is now well-established; this procedure, by making the distillation step unnecessary, removes a major and time-consuming source of error. Several authors have used the method for the analysis of a range of materials (126 - 130). Buckee (126) has described a laboratory continuous flow apparatus, suitable for use in the brewing industry, for the analysis of wort, beer, malt and barley sample digests at the rate of 60 per hour. Although the digests are diluted before measurement, it was necessary to modify the probe's internal filling solution to prevent osmotic drift. Generally, the matching of the filling solution to the sample may most simply be done by the addition of sodium sulphate to the filling solution. The probe and mixing coil were immersed in a thermostatted water bath, held at 28.0 $\pm$ 0.1$^\circ$C; this helped to achieve the very high reproducibility which was recorded (relative standard deviations for ten replicate measurements on pure acetanilide and glycine were 0.4 percent and 0.1 percent respectively, and for measurements on wort and beer samples always better than 0.4 percent). When the Kjeldahl catalyst contains mercury or copper this must be removed by the addition, in the alkaline pretreatment reagent, of iodide (128) or EDTA (127) to prevent formation of metal-ammine complexes which would cause a low answer. Other samples analysed by this method include meat (127,129), cereals (129), soil (128) and freshwaters (130)(the latter in the concentration range 0.1 - 1.0 mg $dm^{-3}$ N).

Methods for determining nitrogen in steel have been described in two papers (131, 132). The nitrogen is present both as dissolved atomic nitrogen and as nitrides of the various metals forming the steel; the different nitrides have different stabilities. The nitrogen present as dissolved atomic nitrogen and that combined in the less stable nitrides (such as iron and manganese nitrides) is known as mobile or hydrogen extractable nitrogen. The ratio of mobile to total nitrogen (determined by Kjeldahl analysis) can be altered by heat treatment and affects grain growth in the steel. Hence, it is important to determine this ratio in order to control the heat treatment. For the measurement of the mobile nitrogen (131), steel millings are heated in a stream of hydrogen at 500$^\circ$C for 2 hours and the ammonia formed is collected in a triethanolamine buffer at pH 6.8, spiked with 5 mg $dm^{-3}$ $NH_4^+$, in a specially designed absorption cell. The ammonium ion concentration in the collection solution is monitored with an ammonium-selective glass electrode and mercury/mercury(I) sulphate/ magnesium sulphate reference electrode. Mobile nitrogen concentrations in the range 0.0005 - 0.01 percent could be determined with standard deviations of 0.0001 - 0.0003 percent. The precision was superior to that obtained with an indophenol blue finish but the method was unsuitable for steels containing vanadium, niobium or chromium.

The measurement of mobile nitrogen has been taken a stage further by Hirst and Maucione (132) who have developed an apparatus which permits continuous monitoring of the nitrogen evolved by silicon-iron samples, containing various trace metals, whilst the temperature is varied. The apparatus is shown in Fig. 11. Samples are gradually heated up to 1250°C in the stream of hydrogen. After passing over the sample the gas stream containing the ammonia produced is cooled and then impinges on the end of an ammonia probe. A plot is produced of nitrogen evolution against temperature, the shape of which is found to be characteristic of the composition and conditioning of the samples. The apparatus is calibrated by passing through gas mixtures of known ammonia content.

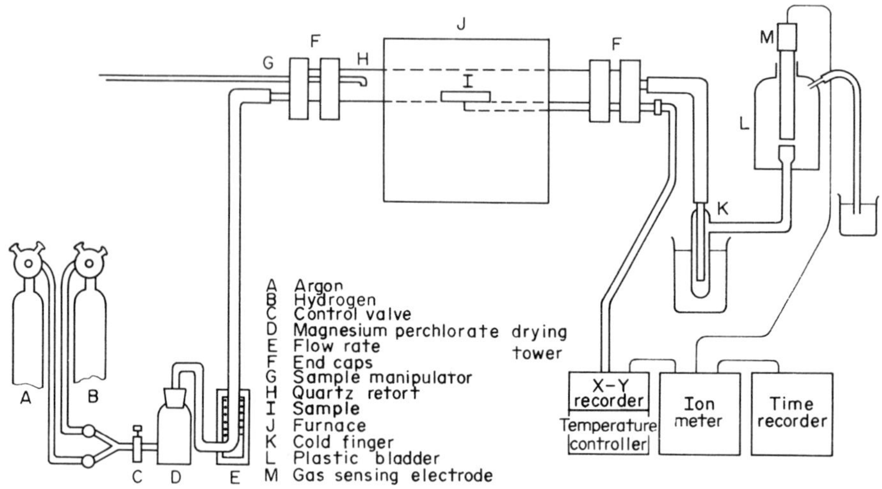


Fig. 11. Diagram of apparatus for determination of nitrogen in silicon-iron (from Ref. 132, by courtesy of the American Chemical Society).

## 4.15 Silver

The silver/sulphide electrode may be used for determining silver over a very wide concentration range, from saturated solutions down to $10^{-7}$M, and the sample pH may lie between 2 and about 9, but very few industrial applications have been reported. Measurements on discrete samples below $10^{-5}$M, although certainly feasible (133), are made more difficult by slow response times (134) and absorption of the sample onto the walls of the sample container (135). The only major interferent is mercury(II) ion. The electrode cannot be used for sulphide measurements immediately after silver measurements unless the surface coating is removed from the membrane.

Lapatnick (83) describes a method for determining silver in silver-plating bath liquors containing cyanide. A 0.1 $cm^3$ sample aliquot is diluted with 100 $cm^3$ water and the silver content of the resultant solution measured by a standard addition procedure in which an aliquot of a standard $NaAg(CN)_2$ solution is added to the solution.

## 4.16 Sodium

Sodium-selective glass electrodes have been commercially available for longer than any other ion-selective electrode (except the pH electrode); because of this, and the large response range and relatively high selectivity of the electrode, very many applications for the electrode have been found of which a few will be cited as examples.

The principal interferents are hydrogen ions and silver ions; potassium ions interfere to a lesser extent (typically $k^{pot}_{Na,K} \simeq 10^{-3}$). Consequently, as silver ions are only rarely a sample constituent, sample pretreatment is usually restricted to pH adjustment to ensure pH > pNa+3. Samples containing a large excess of potassium ions over sodium ions cannot usually be satisfactorily analysed. It is, however, unwise to generalise too much on procedures with sodium electrodes as very many of their behaviour characteristics (e.g., response range, selectivity, response time, temperature coefficient and durability) vary considerably from one make of electrode to another. Several comparisons of commercial electrodes have been published (136 - 140).

The most widespread application of the sodium electrode is in the continuous monitoring of the very low concentrations of sodium in the high-purity water in power stations. Sodium is usually monitored at the output of cation exchange and mixed ion exchange columns, in order to detect breakthrough, and in the boiler feed lines to ensure water purity and minimise corrosion. Measurements may also be taken to detect leakage of sodium into the water system. Sodium concentrations in these sample streams may fall well below 1 μg $dm^{-3}$ (pNa = 7.3) so that the different makes of industrial monitor designed for these applications are calibrated down to 0.1 μg $dm^{-3}$, ($\simeq 5 \times 10^{-9}$M), a very low level for an electrode to measure in pure waters.

In order to obtain useful readings at such a low concentration samples must be made alkaline (pH ⩾ 11) without contaminating the sample with sodium from a buffer; this is achieved by adding either ammonia, as a gas or entrained in air, or a volatile amine. To some extent the choice of system will depend on the selectivity of the electrode being used to monitor hydrogen ions and ammonia. Where the electrode and the concentration of ammonia to be measured permit, it is preferable to use ammonia as a buffer because of the extra cost and hazard of the volatile amine. However, because the electrodes are less sensitive to the amines than to ammonia the amine allows a lower limit of detection. In a comparison of several amines, Eckfeldt and Proctor (141) found di-isopropylamine to permit readings at the lowest level.

The electrode is housed in a plastic flow cell with the reference electrode down-stream to prevent interference from the bridge solution. It has been shown (5) that a lower limit of detection is reached if the cell shape is such that the sample velocity over the electrode bulb is high, and also that it is advantageous to keep the sample temperature low (142) (although this can lead to excessive response times). Goodfellow et al. (140) deduced from similar results and others that the solubility of the glass membrane itself, that is, the rate at which the glass dissolved with liberation of monovalent cations, limited the electrode response to sodium concentrations below 0.1 μg $dm^{-3}$.

The performance of complete industrial monitors (143) and of laboratory monitors (140,144) have been described. The interior of a typical industrial monitor is shown in Fig. 12. In the higher concentration ranges there is good agreement between the potentiometric and flame photometric results.

Sodium electrodes are used for measuring the high sodium concentrations in the black, green and white pulping liquors in paper mills (145): sulphide and hydroxide are also measured. The measurements are used for process control and to reduce waste. The strongly alkaline samples are diluted with 1M ammonium carbonate in the volume ratio 1 part sample : 20 parts buffer to fix the ionic strength and pH and reduce the sodium ion concentration to a reasonable level. Direct potentiometry is sufficiently accurate and precise for the purpose of process control and, for the green and white liquors, gave results equivalent to

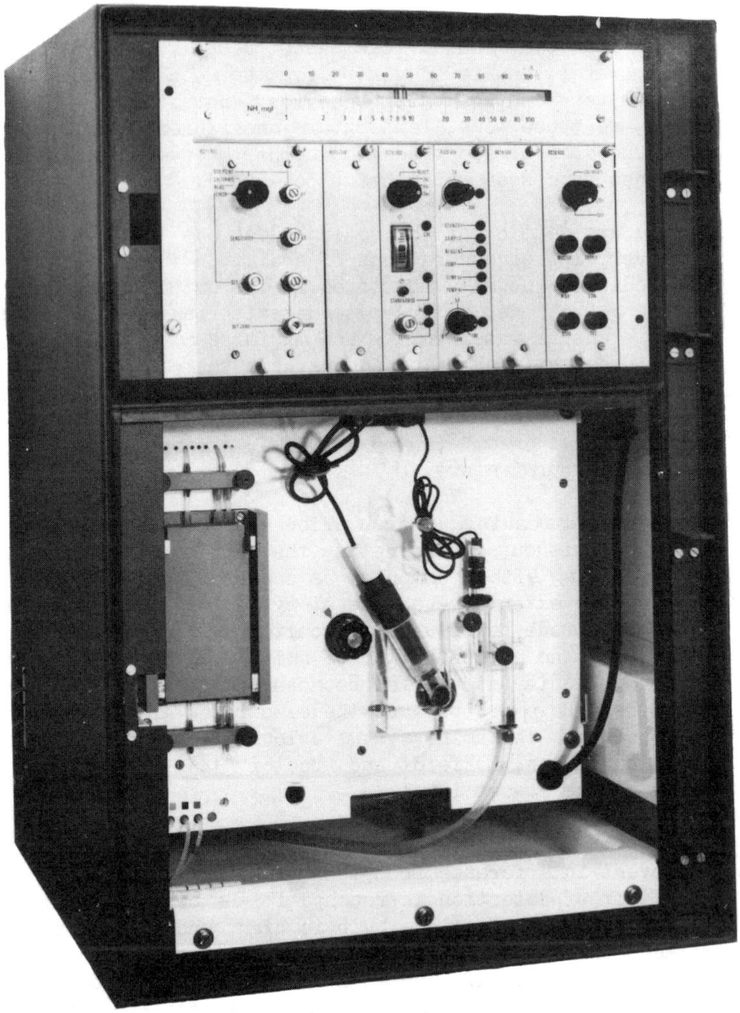

Fig. 12. The interior of an industrial sodium monitor (by courtesy of Electronic Instruments Ltd.).

those from flame photometric and zinc uranyl acetate methods; for the black liquors the results appeared to be better than with the other methods. Particulate matter, even pulp slurry, did not interfere. The method has been automated by Bergner (139) who analysed discrete samples in the range 1-5M $Na^+$ to an accuracy of about 1 percent and with relative standard deviations of less than 1 percent. The system was standardised with 4M sodium hydroxide an aliquot of which was interspersed between every five samples. Bergner also treated several commercial sodium electrodes to find which performed best in the application and two brands were ruled out for instability or too much interference from ammonium ions.

The sodium content of batches of raw material, containing soda ash, for glass-making have been measured by Knupp (52). Samples were digested in hydrochloric acid and adjusted to pH 9 with ammonium hydroxide; the sodium was measured using a calibration graph procedure. Results were in fair agreement with those from determinations with zinc uranyl acetate.

Sodium has been measured in a variety of foodstuffs. Halliday and Wood (146) determined salt in bacon by inserting the sodium electrode and reference electrode directly into macerated bacon samples without any type of chemical pretreatment. The results agreed well with the Volhard method. An even simpler method, which was satisfactory for most types of bacon, was to punch two holes in a roll of bacon with a cork borer and insert the two electrodes; the readings were taken directly. Secor et al. (147) describe a method for measuring sodium in brine-frozen vegetables and sodium hydroxide-treated rice straw, used for cattle feed. The sodium is extracted by soaking or macerating the sample in 0.05M ammonium carbonate and an aliquot of the resultant solution analysed directly. An ashing method was also tested, but found to offer no advantages for these samples. The measurements agreed well with those by AAS and the need to filter samples was avoided. The results of a collaborative study on the determination of sodium in dietetic foods, vegetables in particular, have been presented by McNerney (148). Samples were preserved with formaldehyde and homogenised in a blender and a sample aliquot subsequently treated with triethanolamine buffer. Sodium was determined in the resultant solution by a Gran's plot method. The coefficient of variation of the results from the six laboratories was 4.6 percent for samples of beet containing 680 ppm $Na^+$ but increased sharply for samples containing less than 100 ppm $Na^+$ up to 32.5 percent for samples of corn with about 10 ppm $Na^+$; the results were however quite satisfactory for the purpose and in good agreement with those of the standard AAS method.

A soda loss monitor has been constructed by Tretter (149) to monitor flue gases from recovery furnaces and lime kilns in paper mills. The monitor can detect changes in the firing rates of the furnaces and failure or partial failure of the flue gas scrubber. A sample of the gas stream is scrubbed in the monitor with sufficient solution to give a sodium concentration, in normal running conditions, of 6-8 mg $Na^+$ $dm^{-3}$. The time for the monitor to show the first response to a change in the soda content of the flue gas is 5 minutes, and 15 minutes are required for an equilibrium reading.

## 4.17 Sulphide/sulphur

The silver/sulphide electrode has been used for measuring sulphide and sulphur in a variety of liquid and solid samples. The concentration range of the electrode is from saturated solutions down to $<10^{-7}$M (in solutions unbuffered with respect to sulphide), although measurements in solutions below $10^{-5}$M are difficult because of their susceptibility to aerial oxidation. It is also necessary to buffer solutions to a very alkaline pH, at least 13, to convert as much as possible of the $H_2S$ and $HS^-$ species to $S^{2-}$, the species sensed by the electrode. In order to obtain this pH and minimise sample oxidation, samples are treated, at a volume ratio of 1:1 with a buffer consisting of 2M sodium hydroxide, 2M sodium salicylate and 0.4M ascorbic acid (150). Several authors have found that this solution deteriorates inconveniently rapidly, necessitating almost daily preparation; the life of the solution may be prolonged by preparation and storage under nitrogen (151). This has been successfully used by Korhonen and Lumme (152) for the determination of sulphide down to $10^{-5}$M in paper machine water. The sulphide may rise from the anaerobic conditions that may develop in parts of the water circuit and which allow sulphate reducing bacteria to generate sulphide; this in turn destroys the passivation layer on stainless steel surfaces and leads to corrosion. The calibration graph method gave good accuracy over the whole measurement range. For the analysis of effluents from anaerobic digestion plants (153) treating a trade waste, the high buffering power of the sample made it necessary to resort to a 10M potassium hydroxide plus 0.1M ascorbic acid buffer (stored under nitrogen). Measurements were satisfactory down to $10^{-4}$M but Baumann (154) points out that heavy metal impurities in the sodium salicylate may limit the method.

Gulens and Labbate (155) describe a continuous monitor for analysing sulphide in effluent from a heavy water plant. The effluent from the first stage of the plant is saturated with hydrogen sulphide; this gas is removed in a steam stripper. The monitor continuously checks that the concentration of sulphide has been reduced to ca. 0.1 mg $dm^{-3}$, showing satisfactory operation of the stripper, as the water is discharged to the environment. The sample stream (5 $cm^3$ $min^{-1}$) is mixed at a ratio of 1:1 with a stream of deaerated 2M sodium hydroxide: no reductant is required as the sample is oxygen-free and the measurement cell sealed from the air. Because the water is being discharged at the rate of $31 \times 10^3$ $dm^3$ $min^{-1}$, the response time of the monitor needs to be fast; in practice, virtual equilibrium was achieved in 2 - 3 minutes. It was shown that the electrode gave a Nernstian response in the monitor down to $2 \times 10^{-7}$M $S^{2-}$.

Measurements of sulphur in steel, petroleum products and aerosols have been made by Clysters and Adams (156). Inorganic sulphur compounds, including sulphate, are reduced to sulphide with a strong reductant; a mixture of hydriodic acid, sodium hypophosphite and acetic acid was found to be suitable (except for samples containing high concentrations of nitrate, which had to be removed). Elementary sulphur and sulphur in some organic compounds is reduced to sulphide with Raney nickel; alternatively, for other organic samples, a Schöniger flask ashing method may be used to convert the sulphur to sulphate, with subsequent conversion of the sulphate to sulphide. When using either the acid reductant or Raney nickel, the sulphide is evolved from the reduction medium as hydrogen sulphide gas and absorbed in sodium hydroxide solution. The sulphide in this solution was measured most accurately by potentiometric titration with lead nitrate solution. The methods allow the sequential determination of different forms of sulphur in a sample. A sample of standard Bessemer steel was analysed by the acid reduction method, which gave results close to the known sulphur content and with a relative standard deviation of 0.8 percent. Light distillate fuel was determined by the Raney nickel method, also with good results, but poor recoveries were obtained for sulphur in crude oil.

Low concentrations of sulphur in metallic chromium and low-carbon ferrochromium have been measured by Bozon and Bozon (157). Samples are dissolved in a mixture of phosphoric and hydrochloric acids and the hydrogen sulphide evolved carried in a stream of nitrogen into an acetate buffer containing excess silver ion. The residual excess, after absorption is complete, is estimated by addition of increments of standard silver solution in a Gran's plot procedure. Average errors amounted to 2.01 percent for estimations of sulphur in chromium and 4.07 percent for sulphur in ferrochromium.

Sulphide may be determined in the black liquor used for pulping in the paper industry. Direct readings may be made with the alkaline buffer already described (150). Alternatively, as titrations with silver nitrate are impossible due to reduction of the silver, a sample addition method may be used. Here, a sample aliquot is added to a cadmium nitrate solution in an acetate buffer and the cadmium ion activity measured before and afterwards with a cadmium electrode. The sulphide electrode has also been used for the study of the migration of the various components of the polysulphide liquor into wood chip sections (158).

For the measurement of sulphur in coal, Schoniger flask combustion gives very low results (156) but combustion in a bomb gives satisfactory conversion to sulphate (159). The combustion residue is extracted with water and neutralised to phenolphthalein; 50 $cm^3$ 1,4-dioxane is added and the sulphate titrated potentiometrically with lead perchlorate solution using a lead electrode as indicator electrode. The lead electrode needed polishing with emery paper after every ten or fifteen titrations. Ten replicate determinations on a sample found to contain 0.79 percent sulphur had a standard deviation of 0.027 percent. Titrations of mixtures of sulphide and sulphate with lead perchlorate (156) in 1,4-dioxane gave low sulphate results but use of a water/methanol mixture as titration medium gave better results.

Concentrations of hydrogen sulphide in air in the range 50 to 1000 ppb have been measured (160). The sample gas is passed through 10 $cm^3$ 1M sodium hydroxide, 0.1M ascorbic acid in a glass impinger at the rate of 200 $cm^3\ min^{-1}$ for 20 minutes. The sulphide was measured by potentiometric titration with $6 \times 10^{-6}$M cadmium sulphate solution using a sulphide electrode as indicator electrode. The method gave good accuracy and linearity over the quoted range. No interference was detected from nitrogen dioxide or sulphur dioxide at concentrations similar to that of hydrogen sulphide or from a large excess of ozone.

Several sulphur-containing organic compounds, such as thiourea (161) and thioacetamide (162), may be measured by potentiometric titration with silver nitrate solution but thiourea may also be measured directly with the sulphide electrode.

## 4.18 Sulphur dioxide

A sulphur dioxide probe may be used for the determination of sulphur dioxide (as well as the ionic forms, sulphite and bisulphite) over the approximate concentration range $5 \times 10^{-2}$M down to $10^{-5}$M; these values vary between the different commercial products as do other performance characteristics (1). The only important interferents are strong acetic acid and hydrogen sulphide. The most common application is in the food industry for the control of the sulphur dioxide added to many types of food and beverages as a preservative. In many cases the food manufacturer is anxious to have a rapid method for ensuring that close to the maximum concentration of sulphur dioxide permitted by law is present in the product when it leaves the factory so that the shelf life is as long as possible.

Both "free" and "total" sulphur dioxide may be measured with the probe (163). "Free" sulphur dioxide is measured in liquid samples by acidifying to pH $\leqslant$ 0.7, with sulphuric or perchloric acid, in order to convert nearly all of the sulphite and bisulphite species to dissolved sulphur dioxide. The solution is then analysed by a calibration graph method. A bisulphate/sulphuric acid buffer has also been suggested; in this case a fixed but smaller proportion of the sulphite species are present as sulphur dioxide. This is satisfactory for samples with relatively low buffering power. Speed of measurement is essential to minimise loss of sulphur dioxide from the sample by volatilisation. Solid samples are macerated and extracted with water. "Free" sulphur dioxide has been successfully measured in sucrose, sausage seasoning, fruit juices, fruit squashes and fruit purées. "Total" sulphur dioxide, which is at least theoretically what is controlled by law, is determined by alkaline pretreatment of the sample (pH $\geqslant$ 12 for a short period), in order to decompose aldehyde-bisulphite addition compounds, and subsequent acidification and measurement as above. Such a procedure has proved satisfactory for beer, wine, glucose syrups, sausages, beefburgers, fruit squashes and fruit concentrates.

Two papers have appeared specifically on the determination of sulphur dioxide in wine (164,165). Binder et al. (164) measured both "free" and "total" sulphur dioxide. They added 1 volume of 1M sodium hydroxide to every two volumes of wine, to make the wine alkaline, and stirred for 15 minutes in order to release the 'bound' sulphur dioxide. For acidification they used the bisulphate/ sulphuric acid buffer. In view of the disagreements between the results from the several methods with which the probe results were compared (a serious problem in developing methods for sulphur dioxide), the probe appeared to give very fair answers in tests with several types of wine. The authors also used the probe for the analysis of horseradish sauce and gelatine. Hansen et al. (165) report measurements of "free" sulphur dioxide in several wines: a buffer consisting 20 percent v/v phosphoric acid in aqueous 20 percent ethanol was added 1:1 to samples.

Fuzzi (166) reports the measurement of free sulphur dioxide in the spent liquor from 'cooking' wood by the acid sulphite process and in waste water from paper mills. The spent liquor samples, containing from 2 to 5 g $SO_2$ $dm^{-3}$, were diluted a hundredfold before measurement. This dilution caused some problems concerned with perturbation of the equilibrium between 'free' and 'bound', resulting in a small overestimation of the sulphur dioxide. Analysis of pure acid samples did not show this effect. Satisfactory agreement is obtained between the potentiometric results and results from iodometric titrations, after allowing for the consumption of iodine by other reductants in the samples.

Sulphur in petroleum (167) may be determined after conversion of the sulphur to sulphur dioxide. A sample is burnt in a lamp sulphur apparatus and the sulphur dioxide produced is drawn in a stream of air into an absorbing solution consisting of tetrachloromercurate in a phosphate buffer at pH 6.9, stabilised with EDTA and glycerol. After collection of the sulphur dioxide, sulphamic acid is added to the solution to destroy nitrite and the solution is acidified with the bisulphate/ sulphuric acid buffer; this buffer lowers the pH to 1.2 leaving about 20 percent of the sulphur dioxide as bisulphite. The sulphur dioxide may be determined by a known addition procedure. Results from two samples of jet fuel agreed with those from the standard barium sulphate gravimetric method to within the known range of accuracy of that method. The potentiometric method was suitable for analysing fuels containing down to below 0.005 percent sulphur.

ACKNOWLEDGEMENT

The author thanks Electronic Instruments Ltd. for permission to publish this review.

REFERENCES

1. P.L. Bailey, 'Analysis with Ion-selective Electrodes, Heyden, London, 1976.
2. J.W. Ross, J.H. Riseman and J.A. Krueger, Pure Appl. Chem., 36, 473 (1973).
3. P.L. Bailey and M. Riley, Analyst, 100, 145 (1975).
4. G.K. Buckee, J. Inst. Brewing, 80, 291 (1974).
5. E.L. Eckfeldt and W.E. Proctor, Anal. Chem., 47, 2307 (1975).
6. P. Van den Winkel, J. Mertens and D.L. Massart, Anal. Chem., 46, 1765 (1974)
7. A.Hulanicki and M. Trojanowicz, Analytica Chim. Acta, 87, 411 (1976).
8. C. Rushton, Kent Tech. Review, No. 21, 9 (1978).
9. D. Midgley and K. Torrance, 'Potentiometric Water Analysis', Wiley, Chichester, 1978.
10. L. Snyder, J. Levine, R. Stoy and A. Conetta, Anal. Chem., 48, 942A (1976).
11. D. Betteridge, Anal. Chem., 50, 832A (1978).

12. W.D. Basson, Lab. Practice, 26, 541 (1977).
13. J.R. Potts in 'Advances in Automated Analysis - Technicon International Congress 1976', Vol. 2, Mediad Inc., Tarrytown, N.Y., 1977, p. 38.
14. I. Sekerka and J.F. Lechner, Anal. Letters, 7, 463 (1974).
15. J.J. Zipper, B. Fleet and S.P. Perone, Anal. Chem., 46, 2111 (1974).
16. W.A. Lingerak, F. Bakker and J. Slanina, 'Proceedings of the Conference on Ion-selective Electrodes - Budapest 1977' ed. E. Pungor, Akadémiai Kiadó, Budapest, 1978, p. 453.
17. P.D. Gaarenstroom, J.C. English, S.P. Perone and J.W. Bixier, Anal. Chem., 50,811 (1978).
18. I. Sekerka and J. Lechner, Intern. J. Environ. Anal. Chem., 2, 313 (1973).
19. J. Mertens, P. Van den Winkel and D.L. Massart, Anal. Chem., 48, 272 (1976)
20. R. Wawro and G.A. Rechnitz, Anal. Chem., 46, 806 (1974).
21. K. Tóth, I. Gavallér and E. Pungor, Analytica Chim. Acta, 57,131 (1971).
22. K. Tóth and E. Pungor, Analytica Chim. Acta, 64, 417 (1973).
23. E.H. Hansen, A.K. Ghose and J. Růžička, Analyst, 102, 705 (1977).
24. P.L. Bailey, J. Wilson, S. Karpel and M. Riley, as Ref. 16, p. 201.
25. R.A. Durst, Anal. Letters, 10, 961 (1977).
26. A. Homola and R.O. James, Anal. Chem., 48, 776 (1976).
27. Nj. Radić, Analyst, 101, 657 (1976).
28. G.I. Goodfellow and H.M. Webber, Analyst, 97, 95 (1972).
29. J. Mertens, P. Van den Winkel and D.L. Massart, Bull. Soc. Chim. Belg., 83, 19 (1974).
30. P.L. Bailey and M. Riley, Analyst, 102, 213 (1977).
31. R. Dewolfs, G. Broddin, H. Clysters and H. Deelstra, Z. Anal. Chem., 275, 337 (1975).
32. N. Shibata, Analytica Chim. Acta, 83, 371 (1976).
33. D. Midgley and K. Torrance, Analyst, 97, 626 (1972).
34. D. Midgley and K. Torrance, Analyst, 98, 217 (1973).
35. W.H. Evans and B.F. Partridge, Analyst, 99, 367 (1974).
36. M.J. Beckett and A.L. Wilson, Water Research, 8, 333 (1974).
37. D. Meredith, Kent Tech. Review, No. 22, 3 (1978).
38. F.J. Philbert, M.N. Smith and O. El Kei, as Ref. 13, p. 43.
39. T.C. Woodis and J.M. Cummings, J. AOAC, 56, 373 (1973).
40. C. Benz and J.P. Simko, Soap, Cosmet., Chem. Spec., 53, 40 (1977).
41. T. Jacobsen, Medd. Norsk Farm. Selskap, 37, 145 (1975).
42. F. Drawert and T. Nitsche, Brauwissenschaft, 29, 299 (1976).
43. R.M. Carlson and J.L. Paul, Anal. Chem., 40, 1292 (1968).
44. H.E. Wilde, Anal. Chem., 45, 1526 (1973).
45. Y-S. Su, T.S. Magliocca, K.F. Sugawara, W.R. Strzegowski and J.P. Williams, Analytica Chim. Acta, 98, 115 (1978).
46. H.B.S. Conacher and A.D. McKenzie, J. AOAC, 60, 918 (1977).
47. J.E. Graf, T.E. Vaughn, W.H. Kip, J. AOAC, 59, 53 (1976).
48. A. Hulanicki and M. Trojanowicz, Analytica Chim. Acta, 68, 155 (1974).
49. A. Hulanicki and M. Trojanowicz, Chem. Anal. (Warsaw) 18, 235 (1973).
50. M. Mascini, Analytica Chim. Acta, 56, 316 (1971).
51. T.P. Hadjiioannou and D.S. Papastathopoulos, Talanta, 17, 399 (1970).
52. T.F. Christiansen, J.E. Busch and S.C. Krogh, Anal. Chem., 48, 1051 (1976).
53. R.C. Knupp, Ceram Bull., 49, 773 (1970).
54. U. Fiedler, E.H. Hansen and J. Růžička, Analytica Chim. Acta, 75, 423 (1975).
55. T. Almgren, Analytica Chim. Acta, 73, 420 (1974).
56. D. Midgley, Analyst, 100, 386 (1975).
57. K. Tomlinson and K. Torrance, Analyst, 102, 1 (1977).
58. J.W. Ross, J.H. Riseman and M.S. Frant, U.S. Patent No. 3,563,874 (16th February 1971)

59. J.F. Lechner and I. Sekerka, J. Electroanal. Chem., 57, 317 (1974).
60. I. Sekerka, J.F. Lechner and R. Wales, Water Research, 9, 663 (1975).
61. G.B. Marshall and D. Midgley, Analyst, 103, 438 (1978).
62. P. Pommez and S. Stachenko, Report ARS 72-90, U.S. Dept. of Agric. Res. Serv., 1971, p. 82.
63. A.W.M. Sweetsur, Analyst, 99, 690 (1974).
64. H. De Clercq, J. Mertens and D.L. Massart, J. Agr. Food Chem., 22, 153 (1974).
65. Y.M. Dessouky, K. Tóth and E. Pungor, Analyst, 95, 1027 (1970).
66. J. Korhonen and P.O. Lumme, Paperi ja Puu - Papper o. Trä., 59, 558 (1977).
67. T.G. Lee, Anal. Chem., 41, 391 (1969).
68. M. Kapel and J.C. Fry, Analyst, 99, 608 (1974).
69. M. Vandeputte, L. Dryon and D.L. Massart, Analytica Chim. Acta, 91, 113 (1977).
70. G. Czerwenka, R. Koch and E. Scheubeck, Z. Anal. Chem., 279, 269 (1976).
71. D. Midgley, Analytica Chim. Acta, 87, 7 (1976).
72. W.J. Blaedel and D.E. Dinwiddie, Anal. Chem., 46, 873 (1974).
73. W.J. Blaedel and D.E. Dinwiddie, Anal. Chem., 47, 1070 (1975).
74. M.J. Smith and S.E. Manahan, Anal. Chem., 45, 836 (1973).
75. Y.S. Fung and K.W. Fung, Analyst, 103, 149 (1978).
76. L.W. Przyborowski, as Ref. 16, p. 519.
77. D.S. Papastathopoulos, D.P. Nikolelis and T.P. Hadjiioannou, Analyst, 102, 852 (1977).
78. G. Johansson and K. Edström, Talanta, 19, 1623 (1972).
79. M.S. Frant, J.W. Ross and J.H. Riseman, Anal. Chem., 44, 2227 (1972).
80. H. Clysters, F. Adams and F. Verbeek, Analytica Chim. Acta, 83, 27 (1976).
81. B. Fleet and H. von Storp, Anal. Chem., 43, 1575 (1971).
82. M.E. Hofton, Environ. Sci. Technol., 10, 277 (1976).
83. I. Sekerka and J.F. Lechner, Water Research, 10, 479 (1976).
84. L.N. Lapatnick, Analytica Chim. Acta, 72, 430 (1974).
85. J. Vesely, D. Weiss and K. Stulik, 'Analysis with Ion-selective Electrodes,' Ellis Horwood Ltd., Chichester, 1978, p. 125.
86. S.F. Deane, M.A. Leonard, V. McKee and G. Svehla, Analyst, 103, 1134 (1978)
87. J. Bagg, Anal. Chem., 48, 1811 (1976).
88. J.E. Harwood, Water Research, 3, 273 (1969).
89. I. Sekerka and J.F. Lechner, Talanta, 20, 1167 (1973).
90. D.E. Erdmann, Environ, Sci. Technol., 9, 252 (1975).
91. J. Slanina, F. Bakker, C. Lautenbag, W.A. Lingerak and T. Sier, Mikrochim. Acta, 519 (1978).
92. D.E. Collis and A.A. Diggens, Water Treat. Exam., 18, 192 (1969).
93. P. Berthier, Analusis, 2, 722 (1974).
94. E.F. Croomes and R.C. McNutt, Analyst, 93, 729 (1968).
95. J.R. Entwistle, C.J. Weedon and T.J. Hayes, Chem. Ind. (London) 433 (1973).
96. G. Johansson, Analytica Chim. Acta, 77, 283 (1975).
97. A.L. Schick, J. AOAC, 56, 798 (1973).
98. U. Westerland-Helmerson, Anal. Chem., 43, 1120 (1971).
99. T.A. Palmer, Talanta, 19, 1141 (1972).
100. H. J. Boniface and R.H. Jenkins, Analyst, 102, 739 (1977).
101. G. Troll, A. Farzaneh and K. Cammann, Chem. Geol., 20, 295 (1977).
102. A. Farzaneh and G. Troll, Geochem, J., 11, 177 (1977).
103. G. Troll and A. Farzaneh, Geostandards Newsletter, 2, 43 (1978).
104. J.N. Wilson and C.Z. Marczewski, Anal. Chem., 45, 2409 (1973).
105. J.N. Wilson and C.Z. Marczewski, Anal. Chem., 50, 1584 (1978).
106. L. Torma, J. AOAC, 58, 477 (1975).
107. T. Dolan, L. Legette, J. McNeal and A.J. Malanoski, J. AOAC, 61, 982 (1978)
108. M. Mascini, Analytica Chim. Acta, 85, 287 (1976).

109. J. Kontyoannokos, G.J. Moody and J.D.R. Thomas, Analytica Chim. Acta, 85, 47 (1976).
110. M. Vandeputte, L. Dryon, L. De Hertogh and D.L. Massart, Proc. Analyt. Div. Soc., 15, 80 (1978).
111. D. Myles, Appendix A in First Report May 1973 - January 1977, Standing Committee of Analysts to Review Standard Methods for Quality Control of the Water Cycle, Standard Technical Committee report no. 7, Department of the Environment/National Water Council; London 1977.
112. H.J. Nielsen and E.H. Hansen, Analytica Chim. Acta, 85, 1 (1976).
113. J.A. Wright and P.L. Bailey, as Ref. 16, p. 603.
114. P.J. Milham, A.S. Awad, R.E. Paull and J.H. Bull, Analyst, 95, 751 (1970).
115. J. Mertens, P. Van den Winkel and D.L. Massart, Anal. Chem., 47, 522 (1975)
116. J.O. Burman and G. Johansson, Analytica Chim. Acta, 80, 215 (1975).
117. K.P. Louwrier and J.R. de Rijk, Z. Anal. Chem., 270, 203 (1974).
118. S.L. Pfeiffer and J. Smith, J. AOAC, 58, 915 (1975).
119. M.A. Leidtke and C.E. Meloan, J. Agric. Food Chem., 24, 410 (1976).
120. C. Mergey and J.M. Bonnoit, Analusis, 6, 164 (1978).
121. L.J. Forney and J.F. McCoy, Analyst, 100, 157 (1975).
122. S. Sherken, J. AOAC, 59, 971 (1976).
123. M.A. Tabatabai, Comm. Soil Sci. Plant Anal., 5, 569 (1974).
124. E.H. Hansen, J. Růžička and N.R. Larsen, Analytica Chim. Acta, 79, 1 (1975)
125. O.A. Vita, Analytica Chim. Acta, 81, 45 (1976).
126. G.K. Buckee, J. Inst. Brew., 80, 291 (1974).
127. P.M. Todd, J. Sci. Food Agric., 24, 488 (1973).
128. J.M. Bremner and M.A. Tabatabai, Comm. Soil Sci. Plant Anal., 3, 159 (1972)
129. A.R. Deschreider and R. Meaux, Analusis, 2, 442 (1973).
130. R.J. Stevens, Water Research, 10, 171 (1976).
131. J.B. Headridge and G.D. Long, Analyst, 101, 103 (1976).
132. R.G. Hirst and C.M. Maucione, Anal. Chem., 50, 1046 (1978).
133. D.C. Muller, P.W. West and R.H. Muller, Anal. Chem., 41, 2038 (1969).
134. D.J. Crombie, G.J. Moody and J.D.R. Thomas, Analytica Chim. Acta., 80, 1 (1975).
135. R.A. Durst and B.T. Duhart, Anal. Chem., 42, 1002 (1970).
136. M.F. Wilson, E. Haikala and P. Kivalo, Analytica Chim. Acta, 74, 395, 411 (1975).
137. S. Phang and B.J. Steel, Anal. Chem., 44, 2230 (1972).
138. Gy. Bencz, A. Horvath and L. Szeredai, Meres Autom., 23, 92 (1975).
139. K. Bergner, Analytica Chim. Acta, 87, 1 (1976).
140. G.I. Goodfellow, D. Midgley and H.M. Webber, Analyst, 101, 848 (1976).
141. E.L. Eckfeldt and W.E. Proctor, Anal. Chem., 43, 332 (1971).
142. E. Eckfeldt, Anal. Chem., 47, 2309 (1975).
143. A.A. Diggens, K. Parker and H.M. Webber, Analyst, 97, 198 (1972).
144. H.M. Webber and A.L. Wilson, Analyst, 94, 209 (1969).
145. B.L. Lenz and J.R. Mold, Tappi, 54, 2051 (1971).
146. J.H. Halliday and F.W. Wood, Analyst, 91, 802 (1966).
147. G.E. Secor, G.M. Mc Donald and R.M. Mc Cready, J. AOAC, 59, 761 (1976).
148. F.G. Mc Nerney, J. AOAC, 59, 1131 (1976).
149. V.J. Tretter, Tappi, 52, 2324 (1969).
150. M.S. Frant and J.W. Ross, Tappi, 53, 1753 (1970).
151. E.L. Donaldson and D.C. Mc Mollan, Anal. Letters, A11, 39 (1978).
152. J. Korhonen and P.O. Lumme, Paperi ja Puu - Papper o. Trä, 60, 5 (1978).
153. F. Mosey and D.A. Jogo, Technical Report TR53, Water Research Centre, 1977.
154. E.W. Baumann, Anal. Chem., 46, 1345 (1974).
155. J. Gulens and B. Labbate, AECL-5542, Atomic Energy of Canada Ltd., Ontario, 1976.
156. H. Clysters and F. Adams, Analytica Chim. Acta, 92, 251 (1977).
157. H. Bozon and S. Bozon, Analusis, 6, 243 (1978).

158. W.C. Harris, E.P. Crowell and D.H. McMahon, Tappi, 57, 82 (1974).
159. J.E. Hicks, J.E. Fleenor and H.R. Smith, Analytica Chim. Acta, 68, 480 (1974).
160. D.L. Ehman, Anal. Chem., 48, 918 (1976).
161. M.K. Pápay, K. Tóth and E. Pungor, Analytica Chim. Acta, 56, 291 (1971).
162. M.K. Pápay, K. Tóth, V. Izvekov and E. Pungor, Analytica Chim. Acta, 64, 409 (1973).
163. P.L. Bailey, J. Sci. Food Agr., 26, 558 (1975).
164. A. Binder, S. Ebel, M. Kaal and T. Thron, Deutsche Lebensmittel-Rundschau, 71, 246 (1975).
165. E.H. Hansen, H.B. Filho and J. Růžička, Analytica Chim. Acta, 71, 225 (1974).
166. S. Fuzzi, Industria della Carta, 15, 69 (1977).
167. J.A. Krueger, Anal. Chem., 46, 1338 (1974).

# THE AUTHOR

Dr. P. L. Bailey

Peter Bailey studied for his Ph.D. under Professor E. Bishop at the University of Exeter, graduating in 1972. After a short spell as Staff Demonstrator at Exeter, he worked for a year on ion-selective electrodes in Professor Pungor's laboratories at the Technical University, Budapest on a Royal Society exchange fellowship. Upon return to the United Kingdom, he joined E.I.L. and was engaged in research and development work on various gas sensing membrane probes and ion-selective electrodes. He is now Factory General Manager of the E.I.L. factory at Richmond, at which the Company's range of electrodes and many of their meters are made.

*Ion-Selective Electrode Rev.* Vol.1, pp.139-143.
Pergamon Press Ltd. 1979. Printed in Great Britain

INTERNATIONAL UNION OF
PURE AND APPLIED CHEMISTRY

ANALYTICAL CHEMISTRY DIVISION
COMMISSION ON ANALYTICAL NOMENCLATURE

# RECOMMENDATIONS FOR PUBLISHING MANUSCRIPTS ON ION-SELECTIVE ELECTRODES*

Prepared for Publication by
G. G. GUILBAULT

## INTRODUCTION

Papers on ion-selective electrodes are concerned with fundamental aspects, developments, appraisal and applications. While it may be difficult to lay down guidelines for publishing manuscripts on fundamental aspects, prospective readers of papers on ion-selective electrode developments, appraisal and applications can more easily reach intelligent decisions on advantages and limitations if attention is given to set specification. Of prime importance is the objective that all papers should aim at consistency in preferred usages. In this respect special attention is directed to the report of The Analytical Nomenclature Commission of the International Union of Pure and Applied Chemistry on " Recommendations for Terms and Symbols in the Field of Ion-Selective Electrodes. " The definitions of most frequently used terms taken from this report are given below, and are followed by summaries of the essential points to be included in papers on new electrode developments and applications respectively. This present report was prepared by a committee consisting of G. G. Guilbault ( Chairman ), R. A. Durst, M. S. Frant, H. Freiser, E. H. Hansen, T. S. Light, G. J. Moody, E. Pungor, G. Rechnitz, N. M. Rice, T. J. Rohm, J. Ruzicka, W. Simon and J. D. R. Thomas. It has been circulated in first, second and third draft forms and was discussed in a joint meeting of IDCNS, Commission I. 3, V. 3, and V. 5 in Warsaw. This final report is the result of proposed modifications made during this meeting, and as a result of comments made by various scientists.

### ION-SELECTIVE ELECTRODE NOMENCLATURE AND DEFINITIONS

#### A. DEFINITIONS OF MOST FREQUENTLY USED TERMS

1. Calibration Curve is a plot of the potential ( emf ) of a given ion-selective electrode cell assembly ( ion-selective electrode combined with an identified reference electrode ) versus the logarithm of the ionic activity ( or concentration ) of a given species. For uniformity, it is recommended that the potential be plotted on the ordinate ( vertical axis ) with the more positive potentials at the top of the graph and that $pa_A$ ( -log activity of the species measured, A ) or $pc_A$ plotted on the abscissa ( horizontal axis ) with increasing activity to the right.

*Reprinted from *IUPAC Information Bulletin*, No. 1, pp. 69-74 (1978), with permission from the International Union of Pure and Applied Chemistry.

2. Limit of Detection. A calibration curve ordinarily has the following shape.

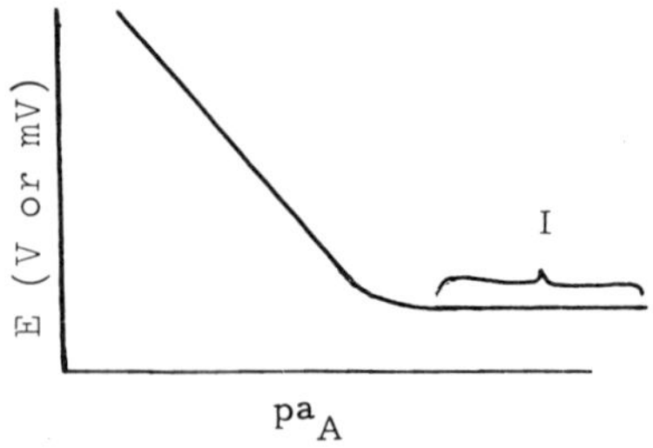


By analogy with definitions adopted in other fields, the limit of detection should be defined as that concentration for which, under the specified conditions, the potential E deviates from the average potential in region I by some arbitrary multiple of the standard error of a single measurement of the potential in region I.

In the present state of the art, and for the sake of practical convenience, a simpler ( and more convenient ) definition is recommended at this time. The practical limit of detection may be taken as the activity ( or concentration ) of A at the point of intersection of the extrapolated illustration:

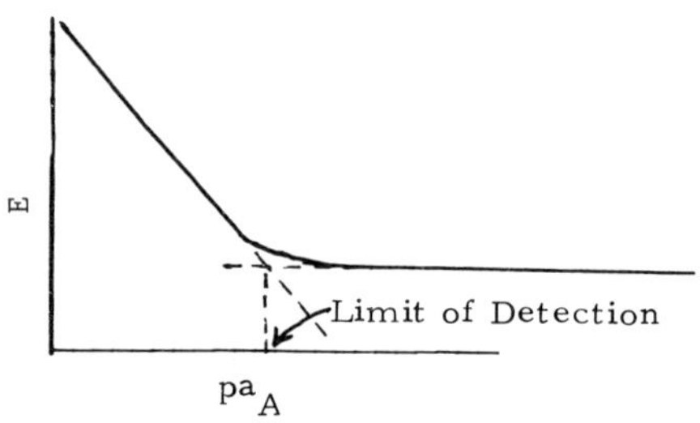


Since many factors affect the detection limit, the experimental conditions used should be reported, i. e. , composition of the solution, the history and preconditioning of the electrode, stirring rate, etc.

3. Drift is the slow non-random change with time in the potential ( emf ) of an ion-selective electrode cell assembly maintained in a solution of constant composition and temperature.

4. Interfering Substance is any species, other than the ion being measured, whose presence in the sample solution affects the measured emf of a cell.

Interfering Substances fall into two classes: Interferences with the electrode response and those with the analytical procedure. Examples of the first class would be those substances which give a response similar to the ion being measured and whose presence generally results in an apparent increase in the activity ( or concentration ) of the ion to be determined ( e. g. , $Na^+$ for the $Ca^{++}$ electrode ), those species which interact with the membrane so as to change its chemical composition ( i. e. , organic solvents for the liquid or polyvinylchloride ( PVC ) membrane electrodes ), or electrolytes present at a high concentration giving rise to appreciable liquid-junction potentials. The second class of interfering substances are those which interact with the ion being measured so as to decrease its activity or apparent concentration, but where the electrode continues to report the true activity ( e. g. , $CN^-$ present in the measurement of $Ag^+$ ).

5. The modified Nernst Equation for Ion-Selective Electrodes and Definition of $k^{pot}_{A,B}$.

$$E = \text{constant} + \frac{2303\,R\,T}{z_A\,F}\,\text{Log}\,[a_A + k^{pot}_{A,B}(a_B)^{z_A/z_B} + k^{pot}_{A,C}(a_C)^{z_A/z_C}]$$

$E$ is the experimentally observed potential of an I. S. E.

$R$ is the gas constant ( and is equal to 8. 31441 J $K^{-1}$ $mol^{-1}$ ).

$T$ is the thermodynamic temperature ( in K ).

$F$ is the Faraday constant ( and is equal to 9. 6485 x $10^4$ C $mol^{-1}$ ).

$a_A$ is the activity of the ion, A ( for concentrations measured in moles $l^{-1}$ or molality ).

$a_B$ and $a_C$ are the activities of the interfering ions, B or C ( for concentrations measured in moles $l^{-1}$ or molality ).

$k^{pot}_{A,B}$ is the potentiometric selectivity coefficient.

$z_A$ is an integer with sign and magnitude corresponding to the charge of the principal ion, A.

$z_B$ and $z_C$ are integers with sign and magnitude corresponding to the charge of an interfering ion, B or C.

The " constant " term includes the standard potential of the indicator electrode, $E^o_{ISE}$, the reference electrode potential, $E_{Ref}$, and the junction potential, $E_j$ ( all in millivolts ). This equation can only be rigorously derived when $z_A = z_B = z_C$.

6. Nernstian Response. An ion-selective electrode is said to have Nernstian response over a given range of activity ( or concentration ) for which a plot of the potential of such an electrode in conjunction with a reference electrode vs. the logarithm of the ionic activity of a given species ( $a_A$ ) is linear with a slope of ( approximately ) 2303 $RT/z_A F$ ( 59. 16/$z_A$ mV at 25° C ).

7. Practical Response Time. The length of time that elapses between the instant at which an ion-selective electrode and a reference electrode are brought into contact with a sample solution ( or at which the concentration of the ion of interest in a solution in contact with an ion-selective electrode and a reference electrode is changed ) and the first instant when the potential of the cell has reached 90 % of the final value. The experimental conditions used should be stated, i. e. , the stirring rate, the composition of solution of which the response time is measured, the composition of the solution to which the electrode was exposed prior to this measurement, the history and preconditioning of the electrode, and the temperature.

8. Potentiometric Selectivity Coefficient, $k^{pot}_{A,B}$ defines the ability of an ion-selective electrode to distinguish between different ions in the same solution. It is not identical to the similar term used in separation processes. The selectivity coefficient should preferably be evaluated by measuring the response of an ion selective electrode in solutions of the primary ion, A, and interfering ion, B ( fixed interference method ). Alternatively, the separate solution method could be used to calculate the selectivity coefficient. This method is less desirable because it does not approximate as well the conditions under which the electrodes are used. It should only be used when the electrode exhibits a Nernstian response. The activity of the primary ion A and the interference B at which $k^{pot}_{A,B}$ is determined should always be specified since the value of $k^{pot}_{A,B}$ is defined by the modified Nernst equation. The smaller value of $k^{pot}_{A,B}$, the greater the electrode's preference for the principal ion, A.

9. Fixed Interference Method. The emf of a cell comprising an ion-selective electrode and a reference electrode is measured with solutions of constant level of interference, $\underline{a}_B$, and varying activity of the primary ion, $\underline{a}_A$. The potential values obtained are plotted versus the logarithm of the activity of the primary ion. The intersection of the extrapolation of the linear portions of this curve ( i. e., where the two terms under the log sign in the expanded Nernst equation have an equal contribution ) will indicate the values of $\underline{a}_A$ which are to be used to calculate $\underline{k}^{pot}_{A,B}$ from:

$$\underline{k}^{pot}_{A,B} = (\underline{a}_A / \underline{a}_B)^{\underline{z}_A / \underline{z}_B}$$

10. Separate Solution Method. The emf of a cell comprising an ion-selective electrode and a reference electrode is measured with each of two separate solutions, one containing the ion, A, at the activity $\underline{a}_A$ ( but no B ), the other containing the ion B, at the same activity $\underline{a}_B = \underline{a}_A$ ( but no A ). If the measured values are $\underline{E}_1$ and $\underline{E}_2$, respectively, the value of $\underline{k}^{pot}_{A,B}$ may be calculated from the following equation if the Nernstian equation holds for the indicator ion:

$$\log \underline{k}^{pot}_{A,B} = \frac{\underline{E}_2 - \underline{E}_1}{2303\, \underline{RT} / \underline{z}_A \underline{F}} + \left(1 - \frac{\underline{z}_A}{\underline{z}_B}\right) \log \underline{a}_A$$

This method is not recommended except in those cases where the mixed solution method is inconvenient or unfeasible.

## B. CLASSIFICATION OF ION-SELECTIVE ELECTRODES

Electrodes should be classified using the system described in Pure and Applied Chem. 48, 127 ( 1976 ).

## C. PAPERS ON NEW OR MODIFIED ELECTRODES

Such papers should include the following essential information :

( a ) Constructional details, including the membrane and conditions of its preparation, inner filling solution, internal reference electrode and contacts. Pretreatment should be specified as should the cell assembly - including the reference electrode employed. For commercial electrodes it is adequate to quote the manufacturer's name and model designation.

( b ) Calibration range, including slope, detection limit and influence of pH, the latter expressed as a potential/pH diagram for at least two levels of activity of the measured ion.

( c ) Stability, including details of storage conditions and drift and information on its susceptibility to attack by chemical agents such as acids, bases, or complexing agents.

( d ) Static or dynamic response times with specific information on how these are obtained with respect to solution conditions, stirring rate, pretreatment, etc. Any change in response time with electrode age should be noted.

( e ) Interfering substances ought to be classified according to type with selectivity for the primary ion over other counter-ions being computed by the Fixed Interference Method at specified interference levels. The selectivity coefficients should be measured on a newly made electrode and indication given how the coefficients change during electrode lifetime.

( f ) Any limitations of the electrode should be clearly stated along with details of operational lifetimes and rejuvenating treatment ( which should be evaluated in terms ( b ) and ( e ) ).

( g ) Temperature coefficient.

## D. PAPERS ON APPLICATIONS FOR DIRECT CONCENTRATION/ACTIVITY MEASUREMENTS

These papers can with advantage refer to papers describing characteristics of the electrode employed and which would fall into category C above. Attention ought also to be given to:

( a ) Details of the cell assembly employed and especially of the reference electrode.

( b ) Interferences and their possible elimination.

( c ) Details of experimental procedure, including ionic-strength adjustment and calibrating standards.

( d ) A comparison of activity/concentration values obtained with those obtained by traditional or alternative established procedures ( if they exist ) with appropriate statistical tests.

( e ) Recovery tests.

( f ) The electrometer or potentiometer used in the emf measurements, as well as its sensitivity, should be specified by manufacturer and model designation and input impedance.

## E. PAPERS ON TITRATION APPLICATIONS

In addition to a mention of papers describing electrode characteristics and attention to items listed in D, special attention must be given to :

( a ) Method used for correct end-point location.

( b ) Possible interferences.

( c ) The possibility of electrode damage during usage.

## F. PAPERS ON ON-LINE PROCESS APPLICATIONS AND AUTOMATIC METHODS

Reports on applications to on-line continuous monitoring systems ought to include reference to associated work on electrode characteristics and preliminary experimentation on direct concentration/activity measurement. This, and especially measurements on intermittent samples, will help to characterize possible errors due to the additional parameters of the flowing system. Specific mention should be made of:

( a ) Dynamic response-time behavior of the cell.

( b ) Full details of cell design including the reference electrode and the supplementary addition of reagents.

( c ) Definition of calibration procedure ( i. e. , on-line, continuous flow or batch, etc. ) and frequency of calibration requirements.

*Ion-Selective Electrode Rev.*, Vol.1, pp.145-186. 0191-5371/79/1201-0145 $05.00/0


# MICRO ION-SELECTIVE ELECTRODES FOR INTRACELLULAR IONS

H. Mack Brown and J. D. Owen

Department of Physiology, University of Utah, Salt Lake City, Utah, U.S.A.

CONTENTS

KEYWORDS: Micro ion-selective electrodes; ion-selective electrodes for intracellular ions; intracellular ions with ion-selective electrodes

## 1. INTRODUCTION

Ion-selective microelectrodes (hereafter referred to as an electrode) and their application to measurement of intracellular ionic activities in biological preparations have been reviewed previously by Walker and Brown to the mid-1976 literature [153]. That information will be summarized here, as well as extending the review to papers published to approximately mid-1978. We apologize to anyone who was inadvertently excluded. A computer-based keyword search was used to obtain some of the material for the present review which is considered from the standpoint of general principles and of the various types of ion-selective microelectrodes, namely, calcium, pH, sodium, potassium and chloride.

## 2. GENERAL PRINCIPLES

### 2.1 Calculating the intracellular activity of an ion

In order to determine the intracellular activity of an ion with an ion-selective electrode, the transmembrane potential, $\Delta E$, of the ion must be found where,

$$\Delta E = E_{in} - E_{out} \tag{1}$$

The outside potential for the ion is given by the Nikolskii equation [114].

$$E_{out} = E^{o} + \frac{RT}{nZ_{A}F} \ln \left(a_{A}^{out} + \sum_{B} k_{A,B}^{pot} (a_{B}^{out})^{Z_A/Z_B}\right) \quad (2)$$

where A represents the primary ion or the ion of interest, B, the interferent ion, $k^{pot}$, the selectivity coefficient and $E^{o}$ the reference potential. The other terms are the same as those found in the well-known Nernst equation describing the primary ion potential in the absence of an interferent ion. The inside potential for the ion is represented by,

$$E_{in} = E^{o} + \frac{RT}{nZ_{A}F} \ln \left(a_{A}^{in} + \sum_{B} k_{A,B}^{pot} (a_{B}^{in})^{Z_A/Z_B}\right) + E_{m} \quad (3)$$

where $E_m$ is the membrane potential as measured with a conventional potassium chloride filled microelectrode. Substituting eqs. 2 and 3 into eq. 1 and solving for the intracellular activity of the primary ion, yields

$$a_{A}^{in} = \left(a_{A}^{out} + {}_{B}\, k_{A,B}^{pot} (a_{B}^{out})^{Z_A/Z_B}\right) \exp\left[(E_{in} - E_{out} - E_{m}) \frac{nZ_{A}F}{RT}\right] - \sum_{B} k_{A,B}^{pot} (a_{B}^{in})^{Z_A/Z_B} . \quad (4)$$

## 2.2 Plotting the intracellular activity of an ion

At low primary ion activity, a typical calibration curve gradually bends away from linearity. Yet this region can still be used in determining low intracellular activity of the primary ion ($a_A^{in}$). Calibration solutions must be prepared to mimic the intracellular ionic activities of the major interferent ions for this procedure. The transmembrane potential ($E_A - E^{o} - E_m$) is read off the $E_{out}$ ordinate of the calibration curve and $a_A^{in}$ is read off the $a_1^{out}$ abscissa of the calibration curve. An example of this is shown in Fig. 1, curve 3, which shows the calibration curve for a $Ca^{2+}$-selective electrode in the presence of 200 mM KCl to provide the potassium interferent ion which closely resembles the amount of potassium in giant neurons for sea animals.

## 2.3 Calibrating an electrode

The calibration procedure which most closely resembles the situation in practical use of the electrode in biological situations is referred to as the mixed solution method. Here the primary ion activity is varied while the background level of interferent ion remains constant (see Fig. 1). Values for the selectivity coefficient, $k_{A,B}^{pot}$, are found by substituting values for $k_{A,B}^{pot}$ until a best fit is found for $E_{out}$ vs $a_A^{out}$. The higher the values for $k_{A,B}^{pot}$, the less selective is the electrode for the primary ion. Prior to and after an experiment, an electrode should be calibrated to obtain an indice of the function throughout the experiment. A check of the pre- and post-parameters includes the following questions. Did the slope remain constant? Did the selectivity coefficients change? Is the time response the same?

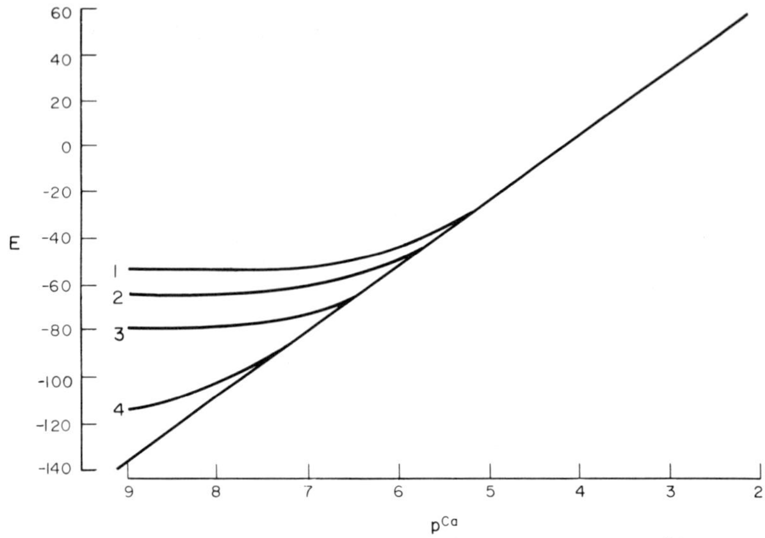


Fig. 1. Calibration curve for a 1.5 μm tip diameter t-HDOPP $Ca^{2+}$ electrode [36]. All solutions buffered with Tris to 7.25. The first three $Ca^{2+}$ solutions are not $Ca^{2+}$ buffered, the middle three solutions are $Ca^{2+}$ buffered with NTA (nitrilotriacetic acid) and the last three solutions are buffered with EDTA (ethylenediaminetetraacetic acid). See Owen and Brown [117] for a detailed list of the total and ionized ions in the solution. 1. $[K^+]$ = 200 mM and $[Mg^{2+}]$ = 3 mM. 2. $[K^+]$ = 200 mM and $[Mg^{2+}]$ = 1 mM. 3. $[K^+]$ = 200 mM. 4. $[K^+]$ = 50 mM. Values of $k^{pot}_{Ca,K}$ = $2 \times 10^{-6}$ and $k^{pot}_{Ca,Mg}$ = $2.5 \times 10^{-4}$ yield the best fit of the data (see eq. 2).

## 3. TYPES OF ION-SELECTIVE ELECTRODES

### 3.1 Calcium

3.1.1 Technical developments. The section on $Ca^{2+}$ electrodes is the smallest in terms of application, yet it contains the largest variety of different types of electrodes. This represents the difficulties associated with measuring intracellular $Ca^{2+}$ and the many attempts to overcome those difficulties.

Růžička [129] demonstrated the superiority of a straight-chained aryl phosphate $Ca^{2+}$ exchanger over the standard Orion alkyl phosphate $Ca^{2+}$ exchanger macro-electrode. In 1976, Christofferson [44] used the straight-chained compound to prepare a 20 μm tip-diameter miniature electrode. In the same year (1976), Brown et al. [36] incorporated a highly branched aryl phosphate and PVC in the tip of a 1 μm microelectrode. This compound was originally thought to be less selective towards $Ca^{2+}$ than the unbranched compound, but it was found to behave surprisingly similar to the straight-chained compound in an electrode on the micron scale. Brown et al. [36] demonstrated that functional microelectrodes could only be constructed when the percentage of PVC was decreased as the tip diameter of the electrode was decreased. Instead of using about 28% PVC, as used routinely for constructing PVC macroelectrode [129,44] they used 18% PVC in 1.5 μm tip-diameter electrodes and 10% PVC in 1 μm electrodes.

In 1976, Oehme et al. [115] reported the use of a neutral carrier compound in a microelectrode which was reported to be sensitive for $Ca^{2+}$. Ammann et al. [10] have recently compared selectivity coefficients of the Orion macroelectrode, the Růžička et al. [129] macroelectrode and the microelectrode developed by Brown et al. [36] with the performance of the carrier macroelectrode. It should be emphasized that there is only one microelectrode in this comparison. Table 1 summarizes these results plus additional results obtained by others more recently.

Ammann et al. [10] reported only a small loss of selectivity in microelectrodes compared to their macroelectrodes, but a detailed report [115] of the microelectrode performance suggests a considerably diminished loss of selectivity. The Ammann et al. ETH 1001 macroelectrode selectivity coefficients for $K^+$ and $Mg^{2+}$ will be considered separately.

3.1.1.1 Potassium interference. In two earlier publications [8,9], the $pk^{pot}_{Ca,K}$ (negative log of the selectivity coefficient) for the ETH 1001 microelectrode was reported to be similar to the $pk^{pot}_{Ca,K}$ for the Orion electrode: 3.8 vs 4.0 (see Table 1). In 1978 [10] values of 5.22 and 6.2 for $pk^{pot}_{Ca,K}$ were given. The 5.22 value was determined in non-buffered $Ca^{2+}$ solutions. This means the $Ca^{2+}$ in the solutions was probably not much lower than $10^{-5}$ or $10^{-6}$ M. This follows since, even though $10^{-8}$ M $Ca^{2+}$ containing water is feasible [117], there is μM $Ca^{2+}$ contamination in 200 mM $K^+$ solutions, i.e. 0.001% calcium contamination is routinely found in reagent grade potassium chloride. This is an undesirable procedure for assessing the effect of $K^+$ on a $Ca^{2+}$ electrode in the sub-micromolar region. The 6.2 value was "estimated from Na/K discrimination and experience in PVC systems". Indirect results usually are not considered accurate when they disagree with direct measurements. "Experience . . ." is somewhat subjective and difficult to assess. Heinemann et al. [64] have reported a $pk^{pot}_{Ca,K}$ value of approximately 5 for a somewhat larger electrode. We also found a $pk^{pot}_{Ca,K}$ of 5 for the ETH 1001 electrode in calcium calibration solutions with 200 mM KCl as the background interferent (refer to Table 1 and Fig. 2).

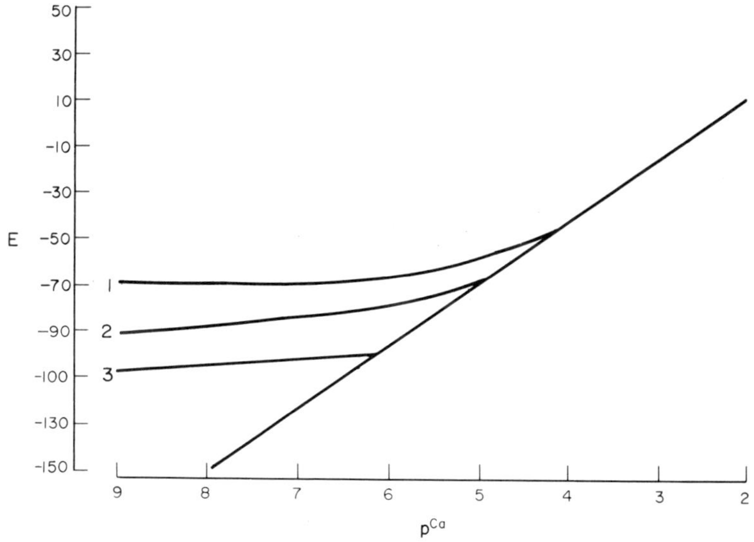


Fig. 2. Calibration curve for a ETH 1001 1.5 μm tip diameter $Ca^{2+}$ electrode using the solutions described in Fig. 1, except solution 4. Approximate values were obtained for the selectivity coefficients since the experimental data did not fit eq. 2 (see Fig. 1). Values of $k^{pot}_{Ca,K} = 1 \times 10^{-5}$ and $k^{pot}_{Ca,Mg} = 1 \times 10^{-3}$ were obtained.

3.1.1.2 Magnesium interference. Twice Ammann et al. [8,9] reported a value of 4.5 for $pk^{pot}_{Ca,Mg}$ (see Table 1). In 1978 [10] a value of 5.1 was obtained in non-buffered $Ca^{2+}$ solutions. As with $K^+$, this means obtaining a selectivity coefficient at a $Ca^{2+}$ activity which is much greater than in actual intracellular experiments, so it should probably not be considered accurate in that context. Heinemann et al. [64] have reported a $pk^{pot}_{Ca,Mg}$ value of about 4. The $pk^{pot}_{Ca,Mg}$ value

TABLE 1. COMPARISON OF SELECTIVITY COEFFICIENTS AMONG VARIOUS $Ca^{2+}$ ELECTRODES

| Exchanger | Type of electrode | Negative log of the selectivity coefficients ($pk^{pot}_{A,B}$) $Na^{+}$ | $K^{+}$ | $Mg^{2+}$ | $H^{+}$ |
|---|---|---|---|---|---|
| HDDP | Macro (Orion[a]) | 3.5 | 4.00 | 1.85 | -8.00 |
| ETH 1001 | Macro (Ammann *et al.*[b]) | 3.4<br>3.5[c]<br>5.0[d](I)<br>6.1[d](II)<br>5.0[e] | 3.8<br>3.8[c]<br>5.22[d](II)<br>6.2 [d](III)<br>5.0[e] | 4.5<br>4.5[c]<br>5.1[e](I)<br>4.0[e] | 4.5[c]<br>0.05[d](IV) |
| HDOPP | Macro (Růžička *et al.* [a]) | 5.2<br>4.70 to 5.20[d](I) | 5.7 | 3.6 | -4.2 |
| ETH 1001 | Micro (Amman *et al.* [b]) | 3.1 | 2.8<br>5.0[f] | 3.3<br>3.0[f] | -5.0[g] |
| t-HDOPP | Micro (Brown *et al.* [h]) | 5.2<br>5.62 to 6.22[e](I) | 5.7<br>5.57[e](I)<br>6.0[i] | 3.3 to 3.6[i] | -3.2[j]<br>-5.0[d](IV) |

TABLE 1 LEGEND

(a) Růžička *et al.* [129].
(b) Oehme *et al.* [115] their Fig. 4.
(c) Ammann *et al.* [9].
(d) Ammann *et al.* [10] Table I. (I) Fixed interference method: not $Ca^{2+}$-buffered. (II) Fixed interference method: $Ca^{2+}$-buffered. (III) Estimated from Na/K discrimination and experience gained with PVC-based systems. (IV) Fixed primary ion method (vary pH).
(e) Heinemann *et al.* [64] for 2 to 3 µm electrodes.
(f) This work with a 1.5 µm tip diameter electrode, see Fig. 2.
(g) Oehme *et al.* [115] their Fig. 3.
(h) Brown *et al.* [36], using a corrected constant.
(i) Owen and Brown [117] and this work with 1.5 µm tip diameter electrode, see Fig. 1.
(j) Owen and Brown, unpublished results.

HDDP = di-decyl phosphoric acid
HDOPP = di(n-octylphenyl)phosphoric acid
t-HDOPP = di[4-(1,1,3,3-tetramethylbutyl)phenyl]phosphoric acid
ETH 1001 = neutral carrier sensor for calcium ions

we obtained was approximately 3 (see Table 1 and Fig. 2). In our experience, the carrier electrode behavior does not fit the Nikolskii equation in complex solutions containing high levels of interferents.

It is obvious that selectivity coefficients can vary considerably for a variety of circumstances. It is important that comparisons are made on the same type of electrode configuration, i.e. macro or micro. It is also important that selectivity coefficients be obtained according to standard procedures which are detailed enough so that others can either reproduce them or criticize them. In the Ammann et al. [10] article, "Design and use of Ca microelectrodes", they stated that although the loss of selectivity was seen in microelectrodes, it was probably "due to imperfect silanization and/or variation in the methodology of selectivity determination. This statement stresses the point of electrode figuration. They also reported ". . . on careful silanization comparable selectivities can be found for corresponding macro and microelectrodes as shown by Heinemann et al. (1977)". However, as seen in Table 1, Heinemann [64] used 2-3 μm-tipped extracellular electrodes and the $k^{pot}_{A,B}$'s they obtained for $Mg^{2+}$ and $K^+$ fall below the best values for the macroelectrode. Brown et al. [36] have found that 2-3 μm tipped electrodes behave similarly to macroelectrodes but they are too large to have much utility in intracellular work. They also do not predict workability in a microelectrode (approx. 1 μm). Therefore, when evaluating microelectrode selectivities, it is important to consider under that conditions they were obtained, i.e. were the values obtained directly or extrapolated from macroelectrode data.

It is obvious from Table 1 that a wide range of $k^{pot}_{A,B}$ values for the various $Ca^{2+}$ electrodes have been reported, especially the ETH 1001 electrode. Our values for the ETH 1001 1.5 μm tipped electrode, which could possibly be considered a "large microelectrode" since it is greater than 1 μm, yields selectivity values similar to those reported by Heinemann et al. [64] for their ETH 1001 2-3 μm electrode.

### 3.1.2 Applications.

3.1.2.1 Intracellular studies. Owen, et al. [119] reported $a^i_{Ca} = 7 \times 10^{-7}$ M in Aplysia giant cells in early 1977 using the t-HDOPP sensor in a $Ca^{2+}$ microelectrode. $a^i_{Ca}$ is the intracellular calcium activity. Their value for $Ca^{2+}$ compares with Meech's [104] estimate of $3 \times 10^{-6}$ M for $Ca^{2+}$ using the technique of pressure-injecting Ca/EGTA to produce prolonged action potentials. Similar pressure injecting experiments have been reported [25] to abolish the light response of the giant cell, which were interpreted to indicate a level of ionized calcium of between $10^{-6}$ and $10^{-7}$ M. Stinnarke and Tauc [139] estimated that the $Ca^{2+}$ equilibrium potential [$E_{Ca}$] was +165 mV from the decline of aequorin light emission at large positive membrane potentials. The $Ca^{2+}$ calculated from their $E_{Ca}$ is approximately $10^{-8}$M. Our estimate of $E_{Ca}$ from our $Ca^{2+}$ electrode measured $a^i_{Ca}$ is almost 120 mV. Late in 1977, Christofferson and Simonsen [44,45] using the reduced PVC percentage procedure for microelectrodes reported by Brown et al. [36] obtained $a^i_{Ca} = 4.5 \times 10^{-7}$ M in snail giant neurons. A $Ca_i$ value of 3 to $8 \times 10^{-8}$M was reported [105] for Helix neurons, determined from voltage-clamp experiments in the presence and absence of extracellular $Ca^{2+}$. A value of about $10^{-7}$ M $Ca^{2+}$ was reported by Ashley et al. [14] for Balanus nubilus muscle fibers using the ETH 1001 carrier in a 20 μm $Ca^{2+}$ electrode. During contraction produced by caffeine or KCl, $a^i_{Ca}$ rose to approximately $10^{-5}$ M.

3.1.2.2 Extracellular studies. Bosher and Warren [23] measured $a_{Ca}$ in rat cochlear endolymph, an extracellular fluid, with a microelectrode (tip ~ 2 μm) containing the ETH 1001. They obtained a value of $3 \times 10^{-5}$ M. During anoxia, they found $a_{Ca}$ dramatically increased. Changes in extracellular $Ca^{2+}$ were measured by Nicholson et al. [113] in rat cerebellum with double-barrelled micropipettes, one side containing ETH 1001. During repetitive stimulation of the parallel fiber purkinje cell cerebellar circuit, extracellular $Ca^{2+}$ fell to about 80% of the base line level ($a_{Ca}$). During the spreading depression of Leão, extracellular $Ca^{2+}$ fell to about 10% of the baseline. Similar large decreases occurred during terminal anoxia.

A summary of the applications of the $Ca^{2+}$ microelectrode, as well as for all the following ion selective electrodes is found in Table 2.

### 3.2 pH

#### 3.2.1 Technical developments

3.2.1.1 Glass microelectrodes (sealed tip). By far, the most successful type of pH microelectrode has been developed from Corning 0150 glass. Several different types have been developed each having some unique advantage. The miniaturization of pH sensitive electrodes can be traced to Caldwell's initial work in 1954 [38]. Caldwell successfully made and inserted a rather large 100 micrometer (μm) glass assembly intracellularly into squid axon to measure intracellular pH ($pH_i$). A "drop-through" electrode was later developed by Hinke in 1959 [66]. This development was dependent upon a successful glass-to-glass seal between an external insulating pipette and a small piece of 0150 glass dropped through the tip of the insulating glass. This seal met the rather stringent insulation criterion necessary to reduce the potential leakage shunt around the high impedance pH glass at the tip.

Recessed tip microelectrodes with glass-to-glass seals were first designed for kidney micropuncture [83] but these were too large (~10 μm) for neurons. In 1974, Thomas [44] recessed the sealed pH glass within a 1 μm insulating pipette forming a microwell at the tip of the microelectrode. A glass-to-glass seal was also involved in the construction of this electrode but sharper electrodes were possible than with the Hinke-type. Both models possessed the general features of excellent selectivity over other intracellular competing cations especially $K^+$. They possessed excellent structural properties and relatively fine tips. Details of electrode construction from glass pipettes have been reviewed by Hinke [68], Khuri [81] and Thomas [144].

Several other models of the glass microelectrode have also appeared. Notably Carter and Rector [42,43] have applied a glaze to insulate the pH glass except at the tip. An open-tipped pH glass microelectrode is the result and has been claimed to function adequately under some conditions [93]. A double-barrelled microelectrode constructed with a film of pH glass welded to the tip of theta-tubing has recently been described by Levy and Coles [98]. Antimony microelectrodes have also been used for pH measurements [16,103]. But they have been found to be subject to interference from changes in oxygen tension. Caldwell [38] attempted to measure $pH_i$ in large muscle fibers of the crab with a 15 μm tungsten electrode. However, results from this electrode were not in concert with subsequent measurements obtained using glass microelectrodes. Table 3 summarizes results from a number of different studies with glass pH microelectrodes.

3.2.1.2 Glass microelectrodes (open-tipped). Some comment is required on studies utilizing open-tip pH sensitive microelectrodes. In general, these produced results that are at variance with those presented in Table 3. An open-tipped pH microelectrode was used for investigating rat skeletal muscle fiber [42,43]. These experiments are technically difficult in that single muscle fibers are penetrated with double-barrelled or triple-barrelled microelectrodes. The construction of these electrodes is quite unusual. The tips are very small (0.05 μm) and are constructed by first applying a glaze to the capillary tubes and then pulling the electrode with special electrode-puller parameters. The electrode tip is left unsealed. The internal filling solution

TABLE 2. $Ca^{2+}$ MICROELECTRODE STUDIES

| Electrode Type | Preparation | $a_x^i$/M | $E_x$/mV | Factors examined | Reference |
|---|---|---|---|---|---|
| A. Nerve cells | | | | | |
| $Ca^{2+}$-t-HDOPP | <u>Aplysia</u> giant cell (<u>Aplysia californica</u>) | $7.7 \times 10^{-7}$ | +120 | Resting level | 118,119 |
| $Ca^{2+}$-HDOPP | Snail giant cell (<u>Helix pomatia</u>) | $4.5 \times 10^{-7}$ | +106 | Resting level | 45 |
| B. Skeletal Muscle | | | | | |
| $Ca^{2+}$ - ETH 1001 | Barnacle muscle fiber (<u>Balanus nubilus</u>) | $10^{-7}$ ($10^{-5}$) | +147 | Resting level (Contraction) | 14 |

TABLE 3. pH MICROELECTRODE STUDIES

| Electrode type | Preparation | $pH_i$ | $\beta$[a] | Factors examined | Reference |
|---|---|---|---|---|---|
| A. Axon | | | | | |
| Glass electrode assembly, 100 μm | Squid (*Loligo forbesi*) | 6.6-7.0 | -- | DNP[b], Donnan equilibrium | 37,38,40,41 |
| Glass electrode assembly, 100 μm | Squid (*Loligo pealii*) | 7.3-7.5 | -- | Injury, hot sea water, ethylene-dichloride, $pH_o$, membrane potential | 138 |
| Glass drop-through 10-30 μm tip | Squid (*Loligo pealii*) | 7.3 | -- | $CO_2/HCO_3$ saline; $pH_i$, "alkaline creep" and rebound with $CO_2$ and $NH_4Cl$ | 20,21,22 |
| Unspecified | Squid (*Loligo pealii*) | | | Cyanide, DNP, proton pump | 51 |
| B. Nerve cells | | | | | |
| 1 μm varnish insulated | Snail (*Helix pomatia*) (*Planorbis carneus*) | 7.0-7.7 | -- | Seasonal variations | 137 |
| Recessed tip | Snail (*Helix aspersia*) | 7.4<br>25 | 12.3<br>48<br>25 | HCl and $HCO_3^-$ iontophoresis, $CO_2/HCO_3^-$ saline, $[K]_o$-free, $CO_2$ saline, $N_2$, azide, DNP, proton pump, creep and rebound in $CO_2/HCO_3$ saline, diamox[c], carbonic anhydrase | 143,144,145,146 |
| C. Skeletal muscle | | | | | |
| 100 μm glass | Crab (*Carcinus maenus*) | 6.9 | -- | Injury | 39 |
| 100 μm glass | Crab (*Maia squinado*) | 7.0 | -- | Membrane potential, $[K]_o$, $[pH]_o$, buffers, $CO_2$ | 41 |
| 1 μm shellac-insulated drop-through | Frog | 7.0 | -- | DNP, membrane potential | 88,89,136 |
| Drop-through | Barnacle (*Balanus nubilus*) | 7.6<br>7.1 | 7.1 | Heat contracture, $a^i_{Na}$, $pH_i$ with DMO[d], $CO_2$, $NH_4^+$, membrane potential | 70,73 |
| $HCO_3$ liquid ion exchanger | Frog, rat | 7.0<br>7.4 | 7.4 | Respiratory acidosis | 82,84 |

| Electrode type | Preparation | $pH_i$ | β[a] | Factors examined | Reference |
|---|---|---|---|---|---|
| D. Cardiac muscle | | | | | |
| Open-tipped glass | 7 μm rat atrial fibers | 6.9 | -- | Careful evaluation of electrode | 93,94 |
| Recessed tip | Sheep purkinje fibers | 7.2 | 35 | Reference electrode and pH electrode not in same cell. Cell coupling, | 56,57 |
| | rat, guinea pig and ferret ventricular fibers | | 70 | HEPES and Tris buffer, $CO_2/HCO_3$ solution, $[K]_o$, SITS[e] | |
| E. Photoreceptors | | | | | |
| Recessed tip | Barnacle photoreceptor (<u>Balanus eburneus</u>) | 7.2-7.4 | 20-50 | $CO_2$, $CO_2/HCO_3$, membrane potential, creep and rebound in $CO_2/HCO_3$, light-induced changes in $pH_i$, $pH_i$ induced changes in K conductance | 32,33,34 |
| Double barrel | Limulus ventral photoreceptor (<u>Limulus polyphemus</u>) | 7.0 | -- | $pH_i$ in dark | 98 |

[a] Buffer capacity (mM/l·pH)

[b] 2,4-dinitrophenol

[c] acetazolamide

[d] 2,4-dinitrophenol

[e] 4-acetamido-4'-isothiocyanatostilbene-2,2'-disulfonic acid

consists of distilled water, rather than more conventional acid or buffer solutions. In the double-barrel configuration, the second barrel is used for membrane potential measurement. Reported pH responses of these electrodes has ranged from 30 to 50 mV per pH unit.

In general, the findings from these studies are at variance with most others conducted with the more conventional pH microelectrodes:
(a) $H^+$ distribution across the membrane is governed by a Donnan equilibrium,
(b) $pH_i$ is approximately 6.0 and
(c) $pH_i$ varies as the resting membrane potential is varied by passing an extrinsic current across the cell membrane.
These authors have concluded that different results are obtained with different techniques due to compartmentalization of pH in the cytoplasm. Other investigators using the same type of electrode have obtained results similar to those reported by Carter et al. [43]. However, other pH methods such as DMO used in conjunction with the electrodes provided values closer to those found with other types of pH electrodes.

Paillard [121] has attempted to reconcile results from the Carter type electrode and the more conventional ones. These can be summarized as follows:
(a) $pH_i$ was measured in rat and crab muscle fibers.
(b) Differences in $pH_i$ were dependent on the depth within the muscle of the fibers penetrated with the microelectrode.
(c) Superficial fibers were more acid ($pH_i$ 6.1) than deep fibers.
(d) It was concluded that the lower pH in the superficial fibers was due to inadequate microelectrode insulation and that the latter value is probably the most accurate one.

It is noteworthy that the recessed tip pH microelectrode developed by Aickin and Thomas has provided values of about 7.2 for both superficial and deep fibers in the same preparation [1]. They found more acid values immediately after impalement with the microelectrodes than after a period for stabilization of $pH_i$. This can be attributed to some cell damage due to penetration of the electrodes. Thomas also found that $pH_i$ was little affected by changes in the membrane potential of the cell and prior results on this preparation were attributed to inadequate insulation of the pH glass. It should be noted at this juncture that Hinke and Menard [73] have recently found that the membrane potential may affect $pH_i$. They used a conventional type of pH electrode. Thus, some aspects of the controversy remain unsettled.

Another attempt has been made recently to fabricate and use open-tipped uninsulated pH glass microelectrodes. They were constructed from Corning 0150 glass [116] to measure $pH_i$ in rat epitheleal cells. The performance of the pH electrodes was compared with conventional 3 M KCl-filled microelectrodes. The pH electrodes were filled 0.5 M KCl. The former had resistances of 15-40 MΩ, whereas the latter had resistances of 100-2,000 MΩ. The tip potentials or the potential measured between a given electrode and a reference electrode produced positive slopes for both the pH electrode and convential KCl electrode of about 35 mV per decade of KCl solution. However, as pH was varied in the presence of background KCl (0.1 M), the conventional KCl electrode showed no pH response, whereas the open-tipped pH glass electrodes yielded slopes of between 24 and 32 mV per pH unit. It is noteworthy that the pH response flattened considerably beyond pH 8. This is not unexpected since the potassium interference is a possibility in dilute solutions of protons. Those pH electrodes with highest resistance produced the best pH response which indicates they had better sealing at the tip. The pH electrodes with maximum resistance had a slope of 46 mV per pH unit. With appropriate corrections, it was possible to obtain a $pH_i$ value of 7.0 in duodenal cells. These cells are small (15 μm by 50 μm) which augment the difficulties. The $pH_i$ of 7.0 is comparable to that obtained in other preparations.

3.2.1.3 Double-barrel glass microelectrode. A unique double barrel pH microelectrode was developed for $pH_i$ measurements in small cells [98]. This electrode consists of a double barrelled theta-glass electrode which has been beveled at the tip. A bubble is blown from pH sensitive glass and with the aid of a microforge a small piece of this glass is sealed to the tip of the theta glass. Since one side of the electrode is to be used for membrane potential recording, a high voltage AC current is passed across the membrane on this side of the electrode which ruptures the pH glass. The pH electrode response is 56-58 mV per pH unit with a measured resistance of 1.5 x $10^{11}$ Ω. Intracellular $pH_i$ of *Limulus* ventral photoreceptors was measured and the reported value was approximately 7.0 in the dark-adapted state.

3.2.1.4 Antimony microelectrode. Antimony microelectrodes obtained commercially from Transidyne General (Ann Harbor, Mich.) have been used to measure the $pH_i$ of the slime mould *Physarum polycephalum* [61]. The measurements were compared with a fluorescence monitoring technique for $pH_i$, 4-methylesculetin. The authors state that 75% of the electrodes changed calibration by more than ± 10 mV by the end of an experiment. It is not clear whether this is a change in reference potential at the same pH or if it represents a change in slope. If it is the former, this is not an uncommon problem with most types of pH microelectrodes especially after cell penetration. They obtained evidence from both the microelectrode technique and the fluorescence technique of an alteration in $pH_i$ that was associated with the mitotic cycle of the slime mould. One cycle represented approximately eight hours and when the $pH_i$ reached its most alkaline state (6.6), mitotic division took place. It is noteworthy that newly fertilized sea urchin eggs become alkaline after fertilization (see section 3.1.2.6).

Antimony electrodes have also been utilized to study $pH_i$ in squid axon [16]. Conventional glass microelectrodes were pulled and coated with antimony and then insulated to within 1-2 μm of the tip with a coat of epoxy paint. The electrodes showed different pH responses in different ranges of pH. In the main area of interest (pH 7 to pH 8) the electrode slope was only 43 mV per pH unit, but improved to 53 mV per pH unit between pH 4 and 7. These authors report a $pH_i$ of about 7.0 in good agreement with early studies on squid axons with glass electrodes. More alkaline values (approx. 7.3) were obtained in later studies (Table 3). The antimony electrode provided results in substantial agreement with other studies indicating that $pH_i$ is unaffected by changes of membrane potential. However, it was found that $pH_i$ was altered when $pH_i$ was changed with a phosphate buffer. Caldwell ([41], Table 3) and Spyropoulos [138] found no comparable changes under similar conditions utilizing glass electrodes. This suggests that this finding may be due to some sort of electrode artifact.

3.2.1.5 Bicarbonate microelectrode. The $HCO_3^-$ liquid ion exchanger developed by Corning has been utilized in one side of a double-barrelled electrode assembly to measure $pH_i$ in frog and rat skeletal muscle fiber [84]. The selectivity of this exchanger against $Cl^-$ is quite good (0.02) and the electrode is insensitive to $OH^-$ and $H^+$ in the pH range 4 to 8. However, it is sensitive to $pCO_2$. If suitable precautions are taken to maintain $pCO_2$ at a constant value, the electrode has good utility. These authors calculated a $pH_i$ value based on the distribution of $HCO_3^-$ inside and outside frog muscle fibers of 7.04 and 7.14 in rat muscle fibers. Intracellular $HCO_3^-$ values *in situ* were 4.4 mM and 12.6 mM in frog and rat fibers, respectively.

*In vivo* intracellular bicarbonate in rat skeletal muscle has recently been studied under different metabolic states [82]. The $pH_i$ and $E_m$ was measured during
(a) experimental alkalosis induced by drinking 75 mM of sodium bicarbonate,
(b) experimental acidosis produced by drinking 75 mM $NH_4Cl$,
(c) a high and low potassium diet,
(d) a low sodium diet
(e) a low chloride diet.

Under normal control conditions a $pH_i$ of 7.03 was obtained in mammalian Ringer equilibrated with 5% $CO_2$. As expected, bicarbonate is not in electrochemical equilibrium. Either bicarbonate is accumulated or hydrogen is extruded from the muscle cell. The buffer value under normal conditions was approximately 16 meq/liter per pH unit. Experimental alkalosis was associated with an increased $pH_i$ and some decrease in the membrane potential. It was usually found that a metabolic acidosis or alkalosis produced intra- and extracellular pH changes in the same direction. However, the intracellular changes were always greater than the extracellular changes indicating that these muscle cells have less buffering capacity than blood. By contrast, $K^+$ deficiency produced a small blood alkalosis (0.05 pH unit) but a large intracellular acidosis (0.25 pH unit).

3.2.2 Applications. Generally speaking, very small pH sensitive microelectrodes have found their main application in the study of intracellular activities in living cells, especially nerve cells. Initial studies in this regard were centered around the question of normal levels of the $pH_i$. It is now established with a great deal of certainty that most nerve cells have $pH_i$ values of between 7.1 and 7.3. These values are generally more acid than the extracellular environment (generally $pH_o \sim 7.5$-$7.8$). In the face of an electrical gradient of -60 to -90 mV across the cell membrane, the $pH_i$ still remains on the basic side of electrochemical equilibrium:

$$pH_i = \frac{E_m F}{2.303\ RT} + pH_o \tag{5}$$

For example, if a Donnan distribution of $H^+$ ions prevailed, the resting $pH_i$ of most nerve cells with membrane potentials ranging from -60 to -90 mV would range from 6.6 to approximately 6. This observation has forced investigation into the problem of the mechanisms of intracellular pH regulation.

3.2.2.1 pH indicator and pH electrode studies. Very few detailed studies have been conducted comparing the indicator methods with glass pH microelectrodes. In one study [63], the indicator (DMO, morpholine and nicotine) and electrode method were in poor agreement. The open-tipped pH microelectrode provided much lower values of $pH_i$ than the indicators (6 vs 7). A subsequent study [73] produced good agreement provided that DMO data were extrapolated to 0-time. $pH_i$ was determined with fibers equilibrated in

(a) normal Ringer solution,
(b) $CO_2$ Ringer, and
(c) $NH_4^+$ Ringer.

After a long period of equilibration (3 hrs), the final $pH_i$ in each of these solutions was 7.1, 6.5 and 7.1, respectively. The behavior of $pH_i$ in these solutions could be fitted by a model involving passive movement of protons and a proton pump.

In a related study [21], very good agreement was found between measurement of $pH_i$ in Balanus muscle fibers with pH electrodes and with the indicator method. Agreement was within 0.01 pH unit. However, using the method of distribution of methylamine there was a difference of 0.11 pH unit, methylamine producing the higher values. The same authors found very good agreement between the indicator method and pH electrode methods in squid axons ($pH_i = 7.35$). They account for the difference between the methods (generally they give greater credence to the electrode) by the following sources of error:

(a) the $pK_a$ of methylamine might be different inside than outside the cell,
(b) the membrane is conceivably permeable to the marker, and
(c) there could be intracellular compartmentalization of pH which could influence readings of the marker and the pH microelectrode.

3.2.2.2 Axon. Reference to Table 3 summarizes electrode type, the preparation and major findings of some earlier studies on nerve axons. This preparation was the first and easiest one to investigate because of its size. Both large electrodes (100 μm) and small ones (1-10 μm) have been utilized to investigate a variety of pH effects in squid axon. The results summarized in Table 3 are for glass microelectrodes. The use of antimony electrodes was described earlier.

3.2.2.3 Nerve cells. Table 3 summarizes the earlier investigations. The more recent studies focus on pH regulation and are described subsequently under that heading in section 3.2.2.8.

3.2.2.4 Skeletal muscle. Aickin and Thomas [2,3,4,5] have recently studied intracellular pH of mouse soleus muscle fibers. Different stable values of $pH_i$ are obtained in different concentrations of external $CO_2$. In this respect, mouse soleus behaves like cardiac fiber [57]. Protons are not distributed in a Donnan equilibrium in this preparation. The $pH_i$ is dependent on temperature being on the average 7.07 at 37°C and 7.23 at 28°C. Removal of $CO_2$ and maintaining $pH_o$ at a constant value depolarizes the membrane by approximately 10 mV and produces an intracellular alkalosis of about 0.3 pH unit. The carbonic anhydrase inhibitor acetazolamide has no effect on $pH_i$ changes consequent to changes of $CO_2$, indicating little carbonic anhydrase in this preparation. Various buffer values are obtained in this preparation depending upon the experimental paradigm used. Changes of external $CO_2$ produced buffer values ranging between 42 and 50 meq/liter per pH unit. Removal of ammonium sulfate, on the other hand, produced buffer values between 60 and 80. The latter paradigm, that is addition and removal of ammonium sulfate, was used to good advantage by these investigators to study the mechanism of pH regulation in this preparation [4,5]. They observed that the time course of pH recovery following application of ammonium sulfate was not altered in potassium-free, ouabain or reduced $Ca^{2+}$ (10 to 1 mM). However, they did observe that sodium-free solutions and amiloride inhibited recovery almost totally. It should be noted that Johnson et al. [74] observed the same amiloride effect in sea urchin eggs. Aickin and Thomas invoke a similar method of pH regulation, namely Na-H exchange as proposed by Johnson et al. They also obtained some evidence for bicarbonate chloride exchange mechanism in that SITS also had a small effect on the pH recovery following application of ammonium sulfate.

Intracellular pH has been recently studied in frog skeletal muscle with recessed tip microelectrodes [17]. Some intracellular acidification was observed in $CO_2$-$HCO_3$ Ringer solution compared to a HEPES-$O_2$ Ringer with the same $pH_o$. Under these conditions, $pH_i$ increases from 6.9 to 7.0. The buffer value was 35 based on $CO_2$ removal. This preparation exhibited one unusual feature in that application of $CO_2$-$HCO_3$ produced no "alkaline $pH_i$ creep". In most preparations examined in this respect, application of $CO_2$-$HCO_3$ saline produces a slow recovery of $pH_i$ towards the initial value.

3.2.2.5 Photoreceptors. The large Balanus photoreceptor has been studied with the recessed tip type microelectrode [32,33,34,35] and a resting $pH_i$ of 7.3 was reported for dark adapted Balanus photoreceptors. Immediately upon penetration of the cells with a KCl recording electrode and the pH electrode, the cell is more acid ($pH_i \simeq 7.0$). It was observed that the cell acidifies slightly upon illumination (maximum change approx. 0.3 pH unit) [33]. Although this $pH_i$ change is relatively small, it represents a very large increase in protons when the buffer value of the cell is considered. The buffer value in normal saline (no $CO_2$) is approximately 20. Thus, a 0.3 decrease in $pH_i$ during illumination represents approximately 100 protons released per each photon of light absorbed by the photopigment.

Application of $CO_2$-$HCO_3$ saline produced an intracellular acidification followed by rebound alkalinization upon return to normal saline. The amplitude of the photoreceptor response to illumination closely followed the $pH_i$ of the cell becoming less at more acid pH and larger (as much as 50%) at the more alkaline pH. Thus, it appears that the conductance change to a fixed amount of light is altered significantly by small changes of $pH_i$ at the internal membrane. The $\Delta pH_i$ in 5% $CO_2$-$HCO_3$ Ringer at constant $pH_o$ is comparable to the $\Delta pH_i$ observed with a saturating flash of light. Interbarrel iontophoresis of bicarbonate or acid mimics the effect of $CO_2$-$HCO_3$ saline on the receptor potential. The light-induced current-voltage relation in $CO_2$ saline show a 50% reduction with no alteration of the reversal potential. Thus, a decrease of $pH_i$ does not appear to alter the selectivity of the light sensitive channel.

Internal acidification by $CO_2$ solution decreased the resting membrane potential and increased membrane resistance. It was demonstrated that the $CO_2$ produces a conductance decrease to $K^+$ ions in the "dark channel". It was suggested that the liberation of protons might be linked to $Na^+$ entry during illumination and that acidification at the internal membrane may modify the amplitude of the receptor potential. This is a potential mechanism for reducing sensitivity under continual illumination. The $pH_i$ of Limulus ventral photoreceptors, measured with a new double-barrelled pH electrode, is reportedly 7.0 [98].

3.2.2.6 Fertilization. Following the lead obtained from measuring the pH of homogenates of sea urchin eggs after fertilization [99,74], Shen and Steinhardt [132] have obtained direct evidence for an increase in intracellular pH of sea urchin eggs after fertilization. They obtained other results that are at variance with those based on cell homogenates. In the earlier reports, an initial acidification (pH 6.5 to 6.4) preceded the final cell alkalinization to pH 6.7. One minute was required for the former and 2 min for the latter. With the recessed tip microelectrode no evidence was found for an initial acidification preceding the increase in intracellular pH. Furthermore, the intracellular pH was maintained for up to half an hour at the same level. A gradual decrease in $pH_i$ had been indicated in the homogenate studies. It had been observed that ammonium chloride will produce the same effects as insemination [74]. Shen and Steinhardt observed the same effect using intracellular pH electrodes. These authors obtained evidence to show that given sufficient time in different ammonium chloride solutions that the same increase in pH is obtained before a steady state value is obtained. The $pH_i$ in ammonium chloride or procaine is usually higher with ammonium chloride induction than it is in normally fertilized eggs (7.65 vs 7.20).

3.2.2.7 Miscellaneous. The $pH_i$ of sunflower root cells has been measured with pH sensitive microelectrodes [24]. The membrane potential recorded from these cells was approximately -35 mV. However, $pH_i$ increased from cell to cell as the microelectrode was advanced from the epidermis to the protoxylem.

3.2.2.8 pH regulation. Considerable recent progress has been made on the method of intracellular pH regulation with the aid of pH sensitive microelectrodes of the drop-through or recessed tip design. The most elaborate of these experiments has been conducted by Thomas [146,147,148] on a 200 μm snail neuron in which microelectrodes are inserted to record membrane potential and $H^+$ ion activity. In addition, electrodes are inserted into the cell for electrophoretic injection of acid without affecting the potential drop across the membrane. The recovery from acid load produced by ionophoretic injection of $H^+$ was observed under different conditions. $Na^+$ injections did not affect pH recovery. pH recovery was blocked when the preparation was in the presence of DIDS (4,4'-diisothiocyanostilbene-2,2'-disulfonic acid); DIDS was without effect on $Na_i$ recovery [148] after $Na^+$ injection. $pH_i$ recovery was reduced by decreasing external bicarbonate or internal $Cl^-$ (sulfate Ringer). Reduction of $Cl_o$ was without effect during $pH_i$ recovery from acid load produced by $CO_2$. These studies provide evidence that Cl-$HCO_3$ exchange is involved in pH regulation as is some H-Na. Thomas suggests that the energy for these mechanisms is provided by the downhill entry of sodium.

Similar types of studies have been performed with much larger cylindrical preparations namely, the giant squid axon and the giant Balanus muscle fiber

[19,20,21,22,125]. These preparations have the additional advantage that internal dialysis can be performed to control the internal environment of the cell [125]. Boron and DeWeer [21] have studied $pH_i$ recovery with the Hinke-type drop-through pH electrode in squid axon. They utilized the technique of acid load produced by exposing the axon to ammonium chloride. This produces the mirror image effect of acid load produced by $CO_2-HCO_3$, namely an initial rapid alkalinization and a creep back to the pre-load $pH_i$ level. Upon removal of ammonium chloride, the $pH_i$ decreases rapidly to a value below the control level and then slowly returns to control. It was observed that $pH_i$ recovery following ammonium chloride application is augmented in $CO_2-HCO_3$ Ringer. The recovery is blocked by cyanide and DNP (reversibly). Data with $CO_2$ indicated a proton (bicarbonate) pump or active hydrogen sequestration is involved in pH regulation.

Internal dialysis of squid giant axon [125] has produced evidence for anion exchange involved in pH regulation. In this case, acid load was produced by internal dialysis of a low pH solution and $pH_i$ recovery was followed with a Hinke-type pH electrode. As in snail, recovery was augmented by bicarbonate solutions and inhibited if the internal chloride was reduced by dialyzing with a $Cl^-$-free solution. pH recovery was also inhibited by SITS (4-acetamido-4'-isothiocyanato-stilbene-2,2'-disulfonic acid). The involvement of ATP was studied since the dialysis medium can be rendered ATP-free. In an ATP-free dialysis solution, recovery of $pH_i$ due to acid load was inhibited. A good quantitative relation was shown between $HCO_3$-stimulated acid extrusion and $HCO_3$-stimulated chloride efflux. Thus, as in snail, evidence was obtained for bicarbonate chloride exchange but in this case the pump is driven by an ATP energy source. The mechanism acts essentially like an hydrochloric acid pump. In contrast to snail, the sodium gradient is not invoked to account for action of the pump. Equivalent studies on snail neuron have not been conducted.

Evidence for chloride bicarbonate exchange has also been obtained in Aplysia neuron [128]. In this case, $a^i_{Cl}$ was found to decline in the presence of bicarbonate/$CO_2$ Ringer solution when an acid load was produced by application of ammonium chloride saline. This change in $a^i_{Cl}$ was blocked by SITS.

The relation between intracellular $Ca^{2+}$ injection and pH changes have been examined recently by Meech and Thomas [106]. Pressure injection of calcium chloride into the cell interior resulted in a small but reliable cell acidification (0.2 to 0.3 pH units). On some occasions, the calcium injection was sufficiently small that the usual indice of calcium injection, a membrane hyperpolarization, were not observed. However, a small decline in $pH_i$ could be detected. Calcium injection did not produce a reliable change in $a^i_{Na}$ as monitored with recessed tip $Na^+$ microelectrodes. From known estimates of the buffer value of this cell, the meq of protons per $Ca^{2+}$ injection could be calculated. It was estimated that the protons released would be in the millimolar range. $CO_2$/bicarbonate saline ($pH_o$ 7.25) decreased the $pH_i$ produced by calcium injection which is in keeping with the observation that the intracellular buffering power is augmented in $CO_2$/bicarbonate salines. The acidification, per se, does not appear to be responsible for the membrane hyperpolarization during calcium chloride injection since hydrochloric acid injection does not induce a hyperpolarization. Pressure injection of magnesium chloride was without effect. Cobalt chloride exerted an effect similar to $Ca^{2+}$ injection in the sense that it decreased $pH_i$, however, the recovery phase overshot the resting level in the alkaline direction. Injection of lanthanum and calcium chloride enhanced $pH_i$ recovery, but in this case the membrane hyperpolarization outlasted the pH change. This is the opposite of what is observed during calcium chloride injection. If intracellular acid changes are tightly coupled to the internal $Ca^{2+}$ level, it would appear that the time course of $pH_i$ changes and membrane potential changes would be similar. However, in this paper it was shown that the two do not behave similarly most of the time. Consequently, the hyperpolarization may be linked to $Ca^{2+}$ in some more complex fashion.

## 3.3 Sodium

3.3.1 Technical Developments. $Na_i$ has been investigated in a large variety of nerve cells with Na microelectrodes made primarily from Corning NAS 11-18 Na sensitive glass [54]. Microelectrodes suitable for intracellular recording were initially the drop-through type developed by Hinke in 1959 [66]. A recessed tip model was developed later by Thomas [142]. Both electrodes require a glass-to-glass seal between the insulating micro-pipette and the Na-sensitive tip. The Hinke design has an inherently better response time, but the Thomas model has the virtue of a sharper tip and is thus more suitable for smaller cells (<200 µm dia).

Two carrier-based microelectrodes have been more recently described. Although one of the carriers was specifically designed for a Na electrode [109], it does not possess sufficient discrimination against $K^+$ ($K_{Na,K}$ ~ 2 x $10^{-1}$) to be of much use for intracellular work where $K^+$ is 1-2 orders higher than $Na^+$. Furthermore, the utility of the compound in a microelectrode has not been described. The most promising electrode of the carrier type is based on the naturally occurring compound monensin [90]. This microelectrode incorporates monensin and a nitro-benzene solvent in the tip of a 1 µm siliconized micropipette. The electrode is siliconized with Dow Corning 1107 silane, then filled with the monensin mixture. The $k^{pot}_{A,B}$, where A equals the primary ion and B, the interferent, are 0.07 for $Na^+$ and $H^+$, 0.01 for $Ca^{2+}$ and $NH_4^+$ and 0.14 for $Mg^{2+}$. Thus, the electrode shows good promise for work on intracellular sodium activities although it would have some diadvantages compared to the glass electrode because of generally lower selectivities. For example, the glass microelectrode has a reported $k^{pot}_{Na,K}$ of $10^{-3}$ [29]. The electrode has the virtue, especially when compared to the recessed-tip type of $Na^+$ glass microelectrode that the time response is much better (2 s to ten-fold change in $Na^+$) compared to 30 s. The monensin electrode was used to study spreading depression in the catfish cerebellum. It was known from past investigations that spreading depression is associated with large decreases in the external activities of $Cl^-$ and large increases in the external activity of $K^+$. In the present study, it was found that $C^o_{Na}$ fell from 149 mM to 57 mM within 6-7 min after spreading depression was initiated by microinjection of a concentrated $K^+$ solution on the cerebellum.

There has been a recent review of findings with sodium (and $K^+$) electrodes in epithelial transport with emphasis on subcellular compartmentalization [46]. This review deals with techniques that have received considerable attention in the past in cellular neurophysiology. One report [79] documents the damaging effect of ion sensitive electrodes on small cells (liver cells).

For the most part, the more current sodium microelectrode studies have been conducted with recessed tip type of microelectrodes [50,55,122,158]. Table 4 summarizes most of the older findings utilizing this type of electrode. The more current findings are discussed below.

### 3.3.2 Applications

3.3.2.1 Neurons. The prevalent use of $Li^+$ as a therapeutic agent for human depression has prompted the investigation of the relation between intracellular $Na^+$ and $Li^+$ on membrane properties [149,122]. An iontophoretic injection of $Li^+$, which was calculated to raise intracellular $Li^+$ to 22 mM per liter, produced a membrane hyperpolarization and an increase in membrane conductance in snail brain neurones. Raising the external $Li^+$ concentration in the bathing medium produced the same effect and was found with a $Na^+$ microelectrode to be associated with a reduction of $C^i_{Na}$ from 5.5 mM to 3.5 mM within 15-20 min. Ouabain, a $Na^+$-$K^+$ pump inhibitor had the opposite effect, that is, it increased $C^i_{Na}$ from 4 to 30 mM within 90 min. One possible effect for the $Li^+$ initiated membrane hyperpolarization would be that $Li^+$ in some way affects an accumulation of $Ca^{2+}$ resulting in a membrane hyperpolarization (increased $K^+$

TABLE 4. $Na_i$ MICROELECTRODE STUDIES

| Electrode type | Preparation | $a^i_{Na}$/mM | $E_{Na}$/mV | $\gamma_{Na}$ | Factors examined | Reference |
|---|---|---|---|---|---|---|
| A. Axon and neuron | | | | | | |
| Drop-through type with 11-18 glass | Squid (*Loligo forbesi*) | 27 | 68 | -- | Spike height and $E_{Na}$ | 67 |
| | | 37 | 67 | -- | Resting potential | |
| | | | | | Artificial saline and sea-$H_2O$ | |
| | | 42 | | | Extruded axoplasm | |
| | | 85-90 [a] | | | $C^i_{Na}$. Osmotic $Na^+$ | |
| | | | | | Repetitive stimulation | |
| Inverted tip 11-18 glass | Snail (*Helix aspersa*) | 6.3 [b] | -- | -- | Na-K pump current | 141 |
| Recessed tip 11-18 glass | | 3.5 [b] | -- | -- | $\Delta a^i_{Na}$ ouabain | 142 |
| | | 2.5 | | | K-free solution, membrane potential, Na-free, injection of $Na^+$ | |
| Shellac insulated 11-18 glass | Snail (*Helix pomatia*) | 13 | -- | -- | Species differences | 89 |
| | (*Planorbis corneus*) | 8 | -- | -- | | |
| Recessed tip 11-18 glass | Aplysia (*Aplysia californica*) | 40.3 | 56.8 | -- | Ouabain | 53 |
| B. Receptors | | | | | | |
| Recessed tip | Barnacle (*Balanus eburneus*) | 10-50 | +64 | -- | Illumination, reversal potential, wavelength, K-free solution, $Ca^{2+}$, Na-K exchange pump | 29,31,35 |

| Electrode type | Preparation | $a^i_{Na}$/mM | $E_{Na}$/mV | $\gamma_{Na}$ | Factors examined | Reference |
|---|---|---|---|---|---|---|
| C. Muscle | | | | | | |
| Drop-through 11-18 glass | Crab (*Carcinus maenus*) | 13.5 | -- | 0.3 | $Na_i$ activity coefficient | 66 |
| Drop through 11-18 glass | Barnacle (*Balanus nubilus*) | 9 | -- | 0.1-0.3 | Heat contracture, solvent $H_2O$, $Na^+$ flux | 7,59,60,69, 70,72 |
| Shellac-insulated 11-18 glass | Frog | 7-10 | -- | 0.2-0.56 | $Na_i$ activity coefficient | 89,97,135 |
| Drop-through with paraffin seal | Frog | | -- | 0.25-0.3 | Cold, K-free low $Ca^{2+}$ $Na_i$ compartmentalization | 95 |
| NAS 27-04 glass | Rabbit papillary muscle | 5.7 | -- | 0.18 | $Na_i$ compartmentalization | 96 |
| NAS 27-05 glass | Frog | 6.5 | -- | 0.33 | K-free | 11 |

[a] Concentration of total Na in axoplasm

[b] Electrode calibrated in terms of concentration

conductance). The experiments conducted in the present paper did not support this mechanism.

3.3.2.2 Skeletal muscle. White and Hinke [158] have recently studied the $Na^+$-$K^+$ exchange pump with radiosodium movement and changes in $a^i_{Na}$ (the intracellular Na activity, $C_{Na} \times \gamma_{Na}$) with a recessed-tip type microelectrode in frog sartorius muscle fibers. It was found that the steady-state $a^i_{Na}$ was not obtained with these electrodes from 1 to 3 min after initial penetration. The measured $a^i_{Na}$ with the glass microelectrode was reported to be between 10 and 14 mM. Total Na measured with atomic absorption spectroscopy was 40 mM, which includes a 22% contribution of extracellular $Na^+$. The empirical $\gamma_{Na}$ based on these measurements was reported to be 0.5. $Na_{22}$ efflux in $Na^+$-free solution (Tris, $Li^+$ or Choline) required three rate constants to account for the data obtained. The two fastest ones were considered to be associated with movement of extracellular $Na^+$. The reduction in $a^i_{Na}$ consequent to a replacement of extracellular $Na^+$ with a $Na^+$ substitute could be accounted for with two rate constants (0.15 and 0.006 $min^{-1}$). The two time constants suggest two paths for loss of intracellular sodium. Alternatively, the data could be fitted with a cubic equation which had been suggested from other experiments to be the result of a hypothetical carrier molecule linked to three $Na^+$. If the latter is taken literally, it would suggest but a single pathway for $Na^+$ extrusion. Reduction of external $Na^+$ in the presence of $10^{-4}$ ouabain did not affect the rate constants obtained but did reduce A in equation $Na^i = Ae^{-kt}$. This reduction of A might be interpreted as a reduction in the pump activity. That is, instead of approximately two thirds of the $Na^+$ leaving with the fast rate constant, only 33% leaves by this route in the presence of ouabain and the remainder leaves by the slower route. The ouabain data could also be fitted by a power function of six. Since modification of a carrier by ouabain was not an attractive alternative to the loss of $Na^+$ involving two separate pathways, the carrier model seems unlikely.

A ouabain poisoned fiber in frog Ringers solution produces a slow rate constant similar to slow efflux in 0-$Na^+$ solution. This suggests the existence of a slow leakage pathway independent of pump action. An increase of $a^i_{Na}$ in 0-$K^+$ solution was also investigated in this study. Although the increase of $a^i_{Na}$ was arrested in 0-$K^+$, it was generally not reversed by return to normal Ringer solution.

3.3.2.3 Cardiac muscle. Recessed-tip $Na^+$ glass microelectrodes have been recently applied to intracellular investigations of $Na^+$ activity in sheep heart Purkinje fibers [50,55]. These investigators obtained a $C^i_{Na}$ of 9.7 mM in standard Tyrodes solution with a $C^o_{Na}$ of 140 mM. Assuming a $\gamma_{Na}$ of 0.75, this yields an $a^i_{Na}$ of 7.3 mM. A 10-fold reduction in $C^o_{Na}$ (140 mM to 14 mM) produced a concomitant reduction of $C^i_{Na}$ from 10 to 3 mM. Tris or lithium substitution of $Na^+$ produced the same effect. The same rate of $Na^+$ loss was experienced in ouabain and $K^+$-free solutions as it was in reduced $Na^+$, indicating that the $Na^+$ pump was not involved in this phenomenon. However, reduction of $Ca^{2+}$ also produced the same rate of $Na^+$ loss as the reduction of external $Na^+$. In the presence of strophanthidin, it was found that only $Ca^{2+}$ amongst $Mg^{2+}$ and $Mn^{2+}$ produced the drop in $C^i_{Na}$. The $Na^+$-$K^+$ pump should be blocked with this procedure which suggests the existence of $Na^+$-$Ca^{2+}$ exchange mechanism independent of a $Na^+$-$K^+$ exchange mechanism.

### 3.4 Potassium

These electrodes generally consist of silanized micropipettes with the tips containing Corning's #477317 $K^+$ liquid exchanger.

3.4.1 Technical developments. The importance of electrode configuration in making $K^+$ electrodes was discussed by Vyskocil and Kriz [161]. It was found that the greater the angle set by the walls, the more advantageous was the shape for filling the microelectrode. Also discussed are procedures for preparing double-barrelled electrodes and side-pore type of electrode for measuring extracellular $K^+$ during muscle contraction. Walker [150] reported the problems for silanizing micropipettes for making $K^+$ electrodes.

A method of making fine double-barrelled $K^+$ electrodes was described by Coles and Tsacopoulos [47]. Basically, micropipettes were pulled from borosilli-cate theta capillary and beveled. One barrel was silanized by blowing dichloro-dimethylsilane through it. Intracellular measurements of $K^+$ were made in the retina of the drone with these electrodes. In dark adapted cells a $K^+$ activity of 89 mM was reported. During light the $K^+$ activity decreased by 14 $\pm$ 9 mM.

Quaternary amine iontophoresis has been characterized with $K^+$ electrodes [52]. By altering the composition of the reference solution trace, ideal respon-ses would be illicited, e.g. acetylcholine assay electrodes contained a reference solution of 1.5 M KCl with 1.2 M acetylcholine $Cl^-$ and those for carbamylcholine measurements contained 1.4 M KCl with 1 M carbamycholine $Cl^-$.

Comparisons were made between a macro $K^+$ electrode containing valinomycin in silicone rubber and a conventional micro Corning exchanger $K^+$ electrode's ability to measure $K^+$ in the cortex of an unanesthetized rat [49]. The macro $K^+$ electrode was more durable, long-lived and had a greater sensitivity but since it had a longer time response and a larger field of measurement, it was felt that the macro was best suited to measurements of phenomena such as spread-ing depression.

3.4.2 Applications.

Table 5 summarizes the potassium ion microelectrode studies.

3.4.2.1 Neurons. $a_K^i$ in Aplysia giant neurons, where $a_K^i$ is the intracellular $K^+$ activity, was found by Russell [124] to decrease upon treatment with the antibiotic nystatin. The nystatin channel was suggested to be monovalent selective, since the loss of internal $K^+$ was reduced when external $Na^+$ was re-placed with $Ca^{2+}$ or $Mg^{2+}$. The greatly reduced loss of $K^+$ when $Ca^{2+}$ or $Mg^{2+}$ were the only external cations suggests that neither $Ca^{2+}$ nor $Mg^{2+}$ were able to readily move through the nystatin channel.

Lux [101] used double-barrelled larger diameter (2-4 μm tip) electrodes, with $K^+$ exchanger on one side and 0.1 M NaCl on the reference side to measure extracellular $K^+$ activity around snail neurons. It was suggested that the long-term suppression of net outward current by a prior conditioning pulse consists of two components. One component is a true time- and voltage-dependent partial inactivation of the delayed potassium system. The second component contributing to depression of delayed outward current during subsequent pulses is an inward current which partially short circuits the potassium current.

Lux [100] has also used $K^+$ electrodes to measure fast extracellular $K^+$ changes during spike activity or voltage clamp pulses of exposed neurons. A com-parison between transferred charge during outward current and the amount of extra $K^+$ outside the cell indicated a deficit in net outward current.

$K^+$ electrodes were used by Heyer and Lux [65] in bursting pace-maker neurons of the snail Helix pomatia to define the deficit in net outward charge transfer under voltage clamp compared with the charge carried by the $K^+$ effect.

$a_K^i$ was determined in absorptive cells lining the villi of isolated, stripped proximal segments of Amphiuma small intestine. With single-barrelled electrodes, $a_K^i$ = 41.6 $\pm$ 1.5 mM and with double-barrelled electrodes, $a_K^i$ = 38.5 $\pm$ 2.4 mM [157].

$K^+$ activity in the epithelial cells of toad bladder was reported as 41.2 $\pm$ 0.5 mM by Kimura et al. [86]. It was also reported that vasopressin decreased $a_K^i$ with an elevation in $E_m$, whereas prostaglandin increased the $a_K^i$ with elevation in the $E_m$.

TABLE 5. $K^+$ MICROELECTRODE STUDIES

| Electrode type | Preparation | $a_x^i$/mV | $E_x$/mV | Factors examined | Reference |
|---|---|---|---|---|---|
| A. Axon and nerve cells | | | | | |
| Corning 27-8 in glass | Squid (*Loligo forbesi*) | 208 | -76 | Resting level | 67 |
| $K^+$ Corning exchanger | Crayfish giant axon (*Procambarus clarkii*) | 210 | -76 | Resting level | 48 |
| $K^+$ Corning exchanger | Aplysia giant cell (*Aplysia californica*) | 165 | -76.9 | Resting level, cooling, ouabain | 126 |
| $K^+$ Corning exchanger | Aplysia giant cell (*Aplysia californica*) | 197 | -83 | Resting level, ouabain light response | 26,30 |
| $K^+$ Corning exchanger | Aplysia giant cell (*Aplysia californica*) | 197 | -83 | Resting level, ouabain | 53,27 |
| $K^+$ Corning exchanger | Aplysia $L_1$-$L_6$ cells (*Aplysia californica*) | 142 | -76 | Resting level | 91 |
| B. Skeletal muscle | | | | | |
| $K^+$ NAS 27-04 glass | Frog | 95.5 | -100 | Resting level, zero-$K^+$ | 95 |
| $K^+$ NAS 27-04 glass | Frog | 97 | -102 | Resting level | 97 |
| $K^+$ NAS 27-05 glass | Frog | 89.9 | -- | K-free | 11 |
| $K^+$ NAS 27-04 glass | Frog | 82.6 | -- | Resting level | 86 |
| $K^+$ NAS 27-04 glass | Frog | 96 | -100 | Resting level | 89 |
| $K^+$ Corning exchanger | Rat | 123 | -86 | Resting level and various $[K]_o$ changes | 77 |

| Electrode Type | Preparation | $a_x^i$/mV | $E_x$/mV | Factors examined | Reference |
|---|---|---|---|---|---|
| $K^+$ Corning exchanger | Frog | 105 | -89 | Resting level, zero-$Cl^-$ high $K^+$ | 85 |
| $K^+$ Corning exchanger | Frog sartorius muscle | 91.2 | -100.3 | Resting level, cold zero $K^+$ | 76 |
| $K^+$ Corning exchanger | Frog striated muscle | 96 | -- | Resting level | 135 |
| $K^+$ Corning exchanger | Barnacle (*Balanus nubilus*) | 111 | -90 | Varying $[K]_o$, constant $[Cl]_o$ | 71 |
| $K^+$ Corning exchanger | Barnacle (*Balanus nubilus*) | 93 | -- | Resting level | 69,71,72 |
| Corning NAS 27-8 in glass | Lobster (*H. vulgaris*) | 84 | -- | Resting | 66 |
| C. Heart muscle | | | | | |
| $K^+$ Corning exchanger | Rabbit atrial muscle | 86 | -85 | Resting level and effect of pH | 133 |
| $K^+$ Corning exchanger | Frog heart ventricle muscle | 86 | -97 | Resting level | 155 |
| $K^+$ Corning exchanger | Beating canine cardiac Purkinje fiber | 310 (112) | -101 | Resting level (ouabain) | 107,108 |
| $K^+$ selective glass | Rabbit papillary muscle | 83 | -79 | Resting level 96 | |
| D. Receptor | | | | | |
| $K^+$ Corning exchanger | Barnacle photoreceptor (*Balanus eburneus*) | 120 | -77 | Resting level $[Ca]_o$ on $a_K^i$ | 29,35 |

| Electrode type | Preparation | $a_x^i$/mV | $E_x$/mV | Factors examined | Reference |
|---|---|---|---|---|---|
| E. Epithelia and others | | | | | |
| $K^+$ Corning exchanger | *Amphiuma* small intestine cell | 41.6 | -80 | Resting level | 157 |
| $K^+$ Corning exchanger | Toad bladder epithelial cells | 41.2 | -73 | Resting level | 86 |
| $K^+$ Corning exchanger | Frog lens | 83.6 (52.4) | 8.2 | Resting level (ouabain) | 123 |
| $K^+$ Corning exchanger | *Calliphora* salivary gland cytoplasm | 143 | -50 | Resting level | 62 |
| $K^+$ Corning exchanger | *Calliphora* salivary gland cytoplasm | 133<br>139 | -66 | Resting level | 15 |
| $K^+$ Corning exchanger | Bullfrog renal proximal tubule cell | 60 | 0 | Resting level | 58 |

3.4.2.2 Skeletal muscle. $K^+$ electrodes were used by Kernan and MacDermott [77] to measure $K^+$ in rat skeletal muscles immersed in normal, hypotonic and high-$K^+$ Krebs fluid. The values reported for $a_K^i$ were 194, 86 and 290.1 mM, respectively.

3.4.2.3 Cardiac muscle. Miura et al. [107] measured intracellular $K^+$ activity in beating canine cardiac purkinje fibers and reported it to be 130.0 ± 2.3 mM. $a_K^i$ was unaffected by changes in $a_K^o$, although the maximum diastolic potential and the resting transmembrane potential was altered by changes in $a_K^o$. Ouabain (2 x $10^{-7}$M) for 30 min decreased $a_K^i$ to 112.2 mM.

$a_K^o$ (where o denotes outside cell layers and not the outside bathing medium) and $a_K^i$ in isolated superfused rabbit atrial muscle were measured in steady-state conditions as well as during decreased pH [133]. As pH fell, $a_K^o$ increased and $a_K^i$ decreased.

3.4.2.4 Brain. Changes in the extracellular $K^+$ activity in the cerebral cortex of lightly anesthetized unrestrained rats were determined with a head-mounted microdrive system by Korytova [87]. Nociceptive stimuli (pinching of the tail) elicited EEG arousal lasting 47 s with a concomitant increase in external $K^+$ from 3.0 mM to 3.31 ± 0.04 mM.

Lux [102] observed steady state levels, transient rises and decreases of intracortical $K^+$ activity in cats. Pentobarbital and diphenylhydantoin had a decreasing effect on stimulation-induced $K^+$ increase.

When gray matter in cat brains was excited by afferent input, or by direct electrical stimulation, Somjen et al. [134] found that an increase in extracellular $K^+$ concentration was linearly correlated with a sustained shift in electrical potential and a response of oxidative metabolism.

3.4.2.5 Receptors. Intracellular $K^+$ in Balanus photoreceptors was found to be 120 mM [29]. This yielded an $E_K$ of -77 mV. Also, $E_m$ was reported to be generally more negative than $E_K$ as $a_K^o$ was increased.

$a_K^o$ for frog retinas, conditioned for spreading depression by exposure on one side of the retina to $Cl^-$ free Ringers, was observed by Mori et al. [110] to increase and the increase was maximal in the inner plexiform layer and subsiding towards the distal retina.

The $K^+$ permeability of the outer capsule of the frog isolated muscle spindle was studied using $K^+$-selective microelectrodes [6]. It was found that the outer capsule of the frog muscle spindle acts as a diffusion barrier for $K^+$ since the changes in $K^+$ concentration in the intracapsular space were seven times slower than in the external solution.

3.4.2.6 Miscellaneous. $K^+$ electrodes were used to measure the intracellular level of $K^+$ in the salivary glands of the blowfly Calliphora by Berridge and Schlue [15] as well as Gupta et al. [62]. During stimulation with 5-hydroxy-tryptamine, the $K^+$ concentration increases from a resting level of 133.2 to 139.0 mM.

Fujimoto and Kubota [58] used double-barrelled $K^+$ selective microelectrodes to measure $K^+$ in renal proximal tubule of the bullfrog. The $K^+$ electrochemical profile in the bullfrog proximal tubule was reported as 2.9 ± 0.2 mM (tubular fluid), 59.7 ± 4.5 mM (cell) and 2.5 ± 0.2 mM (plasma).

$a_K^i$ in minute sites located in different compartments of plant cells was described by Verkhouskaya et al. [160].

## 3.5 Chloride

The majority of studies have used electrodes prepared from the Corning exchanger #477315 [151]. Two papers have reported on Ag/AgCl electrodes. One involves a Ag depositing technique and the other involves Ag/AgCl wire sealed in a glass insulating micropipette.

3.5.1 Technical developments. The first AgCl technique involves filling the tip of the glass microelectrode with a solution of ammoniacal silver nitrate [75]. The tip is then immersed in a formaldehyde solution resulting in a Ag metal plug in the tip of the microelectrode. Often blobs are formed at the end and these electrodes must be rejected. These electrodes yield Nernstian slopes in pure KCl solutions, but unfortunately few data were obtained on the anion selectivity of the electrode. It was observed that there was a 5-10 mV shift of the electrode potential in the presence of approximately 100 mM acetate, but no estimation of the junction potential change was given so this potential change is difficult to evaluate.

Another development has been the Ag wire electrode inserted into the protective sleeve of a glass micropipette. Nield and Thomas [111] attempted to make a chlorosilane barrier between the reference and sample solutions at the tip of the electrode. The barrier was found to become leaky and it also moved up and down the electrode shaft forming a variable dead-space in the tip of the electrode. Saunders and Brown [131] modified this procedure by using a wax seal at the tip of the electrode which overcame most of the objectionable features of the chlorosilane barrier. A detailed analysis of the performance of this electrode and a comparison of the selectivities of the Ag/AgCl electrode with the liquid ion exchanger electrode has been presented [131]. Both types of electrode are significantly more selective to $I^-$ and $Br^-$ than they are to $Cl^-$. However, the AgCl electrode is not sensitive to common anions used for extracellular substitution of $Cl^-$ such as acetate, propionate or methanesulfonate whereas the liquid ion exchanger responds almost as well to proprionate as it does to $Cl^-$. The liquid ion exchanger electrode does show good selectivity against aspartate and glutamate ($k^{pot}_{A,B}$ is 0.017 and 0.077 respectively).

Another Ag/AgCl electrode involving a wax seal was recently described by Armstrong *et al*. [12]. The electrode was of the "drop-through" type; the novelty of the electrode was in vacuum deposited silver on a sealed microelectrode tip (approx. 0.5 μm diameter). This tip protruded beyond an insulating pipette by 2-5 μm and was sealed to this pipette by paraffin. The slopes obtained were 55 mV per 10-fold change in $a_{Cl}$ and like the Ag wire electrode there was no observed interference from $HCO_3^-$ or $H_2PO_4^-$. $I^-$ and $Br^-$ were reported to produce "erratic" responses as might be anticipated from considerations discussed by using the wire electrodes [111,131].

3.5.2 Applications.

Table 6 summarizes chloride ion microelectrode studies.

3.5.2.1 Neurons. Russell [124] reported that pretreatment of 50 mM $NH_4^+$ solution followed by a 10 mM $HCO_3^-$/0.4 $CO_2$ solution produces a decline in $a^i_{Cl}$ in *Aplysia* neurons, where $a^i_{Cl}$ is the intracellular chloride activity. It was stated that intracellular pH regulation involves a $Cl^-/HCO_3^-$ exchange process. The effect of nystatin on the permeability of chloride in *Aplysia* giant neuron was discussed earlier by Russell *et al*. [127].

Thomas [148] has discussed the role of $Cl^-$ in the regulation of intracellular pH in snail neurons. Reducing external $Cl^-$ did not inhibit $pH_i$ recovery, whereas exposure of the cell to sulfate Ringer reduced $pH_i$ recovery from $CO_2$ application.

3.5.2.2 Muscles. $a^i_{Cl}$ of quiescent sheep cardiac Purkinje fibers was reported to be 13.8 mM ± 0.74 by Vaughan-Jones [159]. A passive $Cl^-$ distribution would indicate an $a^i_{Cl}$ of 5.7 mM.

Walker and Ladle [154] reported $a^i_{Cl}$ in frog ventricle as 17.6 ± 0.57 mM and it was concluded that $Cl^-$ was actively transported into these cells.

$a^i_{Cl}$ in frog skeletal muscle was reported to be 2.8 mM by Bolton and Vaughan-Jones [18], or about 0.5 mM higher than would be expected due to passive distribution of $Cl^-$. This small difference could be due to interference from intracellular

TABLE 6. $Cl^-$ MICROELECTRODE STUDIES

| Electrode type | Preparation | $a_x^i$/mM | $E_x$/mV | Factors examined | Reference |
|---|---|---|---|---|---|
| A. Axon | | | | | |
| $Cl^-$ Ag/AgCl | Squid (*Loligo*) | 121 | -27 | Resting level | 80 |
| $Cl^-$ Ag/AgCl | Crayfish (*Procambarus clarkii*) | 252 | -53 | Resting level | 140 |
| $Cl^-$ Corning exchanger | Crayfish (*Procambarus clarkii*) | 134 | -57 | Resting level | 48 |
| B. Neuron | | | | | |
| $Cl^-$ Ag/AgCl | Snail (*Helix aspersa*) | 9-11<br>28 | -59 (H cells)<br>-39 (D cells) | Resting level | 75 |
| $Cl^-$ Ag/AgCl | Snail (*Helix aspersa*) | 8.3<br>8.3 | -46 (H cells)<br>-43 (D cells) | Resting level | 112 |
| $Cl^-$ Ag/AgCl | Aplysia giant cell (*Aplysia californica*) | 152 (10) (Considering a correction for the presence of about 1 mM $Br^-$ or $\ll$1 mM $I^-$ in the cell) | | Resting level | 120 |
| $Cl^-$ Corning exchanger | Aplysia (*Aplysia californica*) | 28<br>42 | -64 (H cells)<br>-53 (D cells) | Resting level<br>pH | 28, 152, 153 |
| $Cl^-$ Corning exchanger | Aplysia (*Aplysia californica*) | 37 | -56 | Resting level, cooling | 126 |
| $Cl^-$ Corning exchanger | Aplysia (*Aplysia californica*) | 35.6 | -55.8 | Resting level, ouabain | 53 |
| $Cl^-$ Corning exchanger | Aplysia (*Aplysia californica*) | 38<br>(10) (Corrected for the presence of about 1 mM $Br^-$ or $\ll$1 mM $I^-$ in the cell) | | Resting level | 120 |

| Electrode Type | Preparation | $a_x^i$/mM | $E_x$/mV | Factors examined | Reference |
|---|---|---|---|---|---|
| $Cl^-$ Corning exchanger | Aplysia $L_1$-$L_6$ cells (Aplysia californica) | 34 | -57 | Resting level | 91 |
| $Cl^-$ Corning exchanger | Aplysia medial pleural cells | -- | -58 | Resting level | 13 |
| C. Skeletal muscle | | | | | |
| Ag/AgCl | Balanus nubilus striated muscle fiber | 29.5 | -71.5 | Resting level $[K]_o$ $[Cl]_o$ constant solution | 71 |
| | (Summer, 1966) | 28.8 | -63.5 | Resting level | 60 |
| | (Spring, 1967) | 24.4 | -70.9 | | |
| Ag/AgCl | Frog sartorius muscle fiber | -- | -80.3 | Resting level | 78 |
| $Cl^-$ Corning exchanger | Frog skeletal muscle | 2.8 | -- | Resting level | 18 |
| D. Heart cells | | | | | |
| $Cl^-$ Corning exchanger | Sheep cardiac Purkinje fibers | 14 | -52 | Resting level | 159 |
| $Cl^-$ Corning exchanger | Frog heart ventricle muscle | 18 | -43 | Resting level | 92 |
| E. Receptors | | | | | |
| $Cl^-$ Corning exchanger | Barnacle (Balanus eburneus) | 65 | -42 | Resting level | 29 |
| $Cl^-$ Corning exchanger | Barnacle (Balanus eburneus) | $59.3 \pm 10$ | $-44.1 \pm 4.5$ | Resting level | 130, 131 |
| Ag/AgCl | Barnacle (Balanus eburneus) | 70 | -39 | Resting level | 130, 131 |

| Electrode type | Preparation | $a_x^i$/mM | $E_x$/mV | Factors examined | Reference |
|---|---|---|---|---|---|
| F. Miscellaneous | | | | | |
| $Cl^-$ Corning exchanger | Bullfrog renal proximal tubule cell | 9.2 | 0 | Resting level | 58 |
| $Cl^-$ Corning exchanger | *Amphiuma* small intestine absorptive cells | 20 | -41 | Resting level | 156 |

bicarbonate.

When the chloride permeability ($P_{Cl}$) in frog skeletal muscle was reduced by acid, $E_{Cl}$ became more positive than $E_m$, which indicated an accumulation of internal chloride [17]. When $P_{Cl}$ was increased by returning to a more alkaline solution, $E_m$ was more positive than $E_{Cl}$. The results suggest the presence of a small, active movement of chloride, which is normally masked by larger passive movements of chloride when $P_{Cl}$ is large.

3.5.2.3 Photoreceptors. Saunders and Brown [131] used a technique of intracellular replacement of chloride in *Balanus* photoreceptors. They used standard liquid ion exchanger electrodes as well as a newly developed Ag/AgCl electrode to monitor changes in $a^i_{Cl}$.

Brown [29] measured an $a^i_{Cl}$ of 65 mM in *Balanus* photoreceptors, giving an $E_{Cl}$ of -42 mV.

3.5.2.4 Miscellaneous. The $Cl^-$ electrochemical profile of the bullfrog proximal tubule was reported by Fujimoto and Kubota [58]. The values for $a_{Cl}$ given were 77.8 ± 2.3 (tubular fluid), 9.2 ± 1.3 (cell and 70.7 ± 2.5 mM (plasma).

$a^i_{Cl}$ in the secretory canaliculi and the lumen of *Calliphora* salivary glands was measured by Gupta *et al.* [62]. Stimulation with $10^{-8}$ M 5-hydroxy-tryptamine caused a fall in $a_{Cl}$ in the canaliculi fluid and an increase in $a_{Cl}$ in the liminal fluid.

Double-barrelled chloride-selective microelectrodes were used to determine $a^i_{Cl}$ in the absorptive cells of isolated, stripped proximal and distal segments of *Amphiuma* small intestine [156]. $Cl^-$ was passively distributed across the mucosal membranes in fall and winter animals ($a^i_{Cl}$ = 20 mM), but summer animals actively accumulated $Cl^-$ in both proximal and distal segments.

## 4. REFERENCES


1. C.C. Aicken and R.C. Thomas, Micro-electrode measurements of the internal pH of crab muscle fibers, *J. Physiol.* 252, 803 (1975).

2. C.C. Aickin and R.C. Thomas, Intracellular pH of mouse soleus muscle, *J. Physiol.* 260, 25P (1976).

3. C.C. Aicken and R.C. Thomas, The effect of external Na and amiloride on $pH_i$ recovery in mouse soleus muscle, *J. Physiol.* 269, 80P (1977).

4. C.C. Aickin and R.C. Thomas, An investigation of the ionic mechanism of intracellular pH regulation in mouse soleus muscle fibres, *J. Physiol.* 273, 295 (1977).

5. C.C. Aicken and R.C. Thomas, Microelectrode measurement of the intracellular pH and buffering power of mouse soleus muscle fibres, *J. Physiol.* 267, 791 (1977).

6. N.P. Alekseev, Study on the permeability of the outer capsule of the frog muscle spindle to potassium ions using ion-selective microelectrodes, *Chem. Abs.* 85, 18 181 (1976).

7. R.C. Allen and J.A.M. Hinke, Sodium compartmentalization in single muscle fibers of the giant barnacle, *Can. J. Physiol. Pharmac.* 48, 139 (1970).

8. D. Ammann, R. Bissig, Z. Cimerman, U. Fiedler, M. Guggi, W.E. Morf, M. Oehme, H. Osswald, E. Pretsch and W. Simon, Synthetic neutral carriers for cations, In: Ion and Enzyme Electrodes in Biology and Medicine, Ed: M. Kessler,

L.C. Clark, D.W. Lubbers, I.A. Silver, W. Simon, University Park Press, Baltimore. (1976) p. 22.

9. D. Ammann, M. Guggi, E. Pretsch and W. Simon, Improved calcium ion-selective electrode based on a neutral carrier, Anal. Lett. 8, 709 (1975).

10. D. Ammann, P.C. Meier and W. Simon, Design and use of $Ca^{2+}$ selective micro-electrodes, In: Detection and Measurement of Free Calcium Ions in Cells, Ed: C.C. Ashley and A.K. Campbell. Elsevier/North-Holland (1979) in press.

11. W.M. Armstrong and C.O. Lee, Sodium and potassium activities in normal and "sodium rich" frog skeletal muscle, Science 171, 413 (1971).

12. W.M. Armstrong, W. Wojtkowski and W.R. Bixenman, A new solid-state micro-electrode for measuring intracellular chloride activities, Biochim. Biophys. Acta 465, 165 (1977).

13. P. Ascher, D. Kunze and T.O. Neild, Chloride distribution in Aplysia neur-ones, J. Physiol. 256, 441 (1976).

14. C.C. Ashley, T.J. Rink and R.Y. Tsien, Changes in free Ca during muscle contraction, measured with an intracellular Ca-selective electrode, J. Physiol. 280, 27P (1978).

15. M.J. Berridge and W.R. Schlue, Ion-selective electrode studies on the effects of 5-hydroxytryptamine on the intracellular level of potassium in an insect salivary gland, J. Exp. Biol. 72, 203 (1978).

16. H.I. Bicher and S. Ohki, Intracellular pH electrode experiments on the giant squid axon, Biochim. Biophys. Acta 255, 900 (1972).

17. T.B. Bolton and R.D. Vaughan-Jones, Continuous direct measurement of intra-cellular chloride and pH in frog skeletal muscle, J. Physiol, 270, 801 (1977).

18. T.B. Bolton and R.D. Vaughan-Jones, The use of electrodes containing liquid ion-sensitive resin to measure the intracellular chloride activity of skeletal muscle, J. Physiol. 260, 6P (1976).

19. W.F. Boron. Intracellular pH transients in giant barnacle muscle fibers. Am. J. Physiol. 233, C61 (1977).

20. W.F. Boron and P. DeWeer, Active proton transport stimulated by $CO_2/HCO_3$ blocked by cyanide, Nature 259, 240 (1976).

21. W.F. Boron and P. DeWeer, Intracellular pH transients in squid giant axons caused by $CO_2$, $NH_3$ and metabolic inhibitors, J. Gen. Physiol. 67, 91 (1976).

22. W.F. Boron and P. DeWeer, Studies on the intracellular pH of squid giant axons, Biophys. J. 15, 42a (1975).

23. S.K. Bosher and R.L. Warren, Very low calcium content of cochlear endolymph, an extracellular fluid, Nature 273, 377 (1978).

24. D.J.F. Bowling, Measurement of intracellular pH in roots using a proton sensitive microelectrode, Memb. Trans. Plants, (Int. Workshop) 386 (1974).

25. A.M. Brown, P.S. Baur and F.J. Tuley, Phototransduction in Aplysia neurons: calcium release from pigmented granules is essential, Science 188, 157 (1975).

26. A.M. Brown and H.M. Brown, Light response of a giant Aplysia neuron, J. Gen. Physiol. 62, 239 (1973).

27. A.M. Brown, J.M. Russell and D.C. Eaton, Resting ionic permeabilities in a neuron, In: Ion and Enzyme Electrodes in Biology and Medicine, Ed: M. Kessler, L.C. Clark, D.W. Lubbers, I.A. Silver, W. Simon. University Park Press; Baltimore (1976) p. 283.

28. A.M. Brown, J.L. Walker and R.B. Sutton, Increased chloride conductance as the proximate cause of hydrogen ion effects in Aplysia neurons, J. Gen. Physiol. 56, 559 (1970).

29. H.M. Brown, Intracellular $Na^+$, $K^+$ and $Cl^-$ activities in Balanus photoreceptors, J. Gen. Physiol. 68, 281 (1976).

30. H.M. Brown and A.M. Brown, Ionic basis of the photoresponse of Aplysia giant neuron: $K^+$ permeability increase, Science 178, 755 (1972).

31. H.M. Brown and M.C. Cornwall, Ionic mechanism of a quasi-stable depolarization in barnacle photoreceptor following red light, J. Physiol. 248, 579 (1975).

32. H.M. Brown and R.W. Meech, Effects of pH and $CO_2$ on large barnacle photoreceptors, Biophys. J. 15, 276a (1975).

33. H.M. Brown and R.W. Meech, Intracellular pH and light adaptation in barnacle photoreceptors, J. Physiol. 263, 218P (1976).

34. H.M. Brown, R.W. Meech and R.C. Thomas, pH changes induced by light in large Balanus photoreceptors, Biophys. J. 16, 33a (1976).

35. H.M. Brown and D. Ottoson, Dual role for $K^+$ in Balanus photoreceptor: antagonist of $Ca^{2+}$ and suppression of light induced current, J. Physiol. 257, 355 (1976).

36. H.M. Brown, J.P. Pemberton and J.D. Owen, A calcium-sensitive microelectrode suitable for intracellular measurement of calcium(II) activity, Anal. Chim. Acta 85, 261 (1976).

37. P.C. Caldwell, Measurement of intracellular pH by means of a tungsten microelectrode, J. Physiol. 120, 31P (1953).

38. P.C. Caldwell, A glass microelectrode suitable for the measurement of the intracellular pH of large cells, J. Physiol. 124, 1P (1954).

39. P.C. Caldwell, An investigation of the intracellular pH of crab muscle fibres by means of micro-glass and micro-tungsten electrodes, J. Physiol. 126, 169 (1954).

40. P.C. Caldwell, The sensitivity to pH of the inhibitory effects of dinitrophenol on squid axons, Biochem. J. 67, 1P (1957).

41. P.C. Caldwell, Studies on the internal pH of large muscle and nerve fibres, J. Physiol. 142, 22 (1958).

42. N.W. Carter. Direct measurement of intracellular pH with glass microelectrodes, Clin. Res. 9, 177 (1961).

43. N.W. Carter, F.C. Rector, D.S. Campion and D.W. Seldin, Measurement of intracellular pH of skeletal muscle with pH-sensitive glass microelectrodes, J. Clin. Invest. 46, 920 (1967).

44. G.R.J. Christoffersen and E.S. Johansen, Microdesign for a calcium-sensitive electrode, Anal. Chim. Acta 81, 191 (1976).

45. G.R.J. Christoffersen and L. Simonsen, $Ca^{2+}$ sensitive microelectrode: intracellular steady state measurement in nerve cell, Acta physiol. Scand. 101, 492 (1977).

46. M.M. Civan, Intracellular activities of sodium and potassium, Am. J. Physiol. 234, F261 (1978).

47. J.A. Coles and M. Tsacopoulos, A method of making fine double-barrelled potassium-sensitive microelectrodes for intracellular recording, J. Physiol. 270, 12P (1977).

48. M.C. Cornwall, D.F. Peterson, D.L. Kunze, J.L. Walker and A.M. Brown, Intracellular potassium and chloride activities measured with liquid ion exchanger microelectrodes, Brain Res. 23, 433 (1970).

49. W. Crowe, A. Mayevsky and L. Mela, Application of a solid membrane ion-selective electrode to in vivo measurements, Am. J. Physiol. 233, C56 (1977).

50. J.W. Deitmer and D. Ellis, Effects of divalent cations on the intracellular sodium ion concentration of sheep heart Purkinje fibres, J. Physiol. 271, 16P (1977).

51. P. DeWeer and D. Geduldig, Effect of cyanide on the resting membrane potential of squid giant axon, Biophys. J. 15, 42a (1975).

52. V.E. Dionne, Characterization of drug iontophoresis with a fast microassay technique, Biophys. J. 16, 705 (1976).

53. D.C. Eaton, J.M. Russell and A.M. Brown, Ionic permeabilities of an Aplysia giant neuron, J. Memb. Biol. 21, 353 (1975).

54. G. Eisenman, D.O. Rudin and J.U. Casby, Glass electrodes for measuring sodium ion, Science 126, 831 (1957).

55. D. Ellis, The intracellular sodium ion concentration of sheep heart Purkinje fibres and its relationship to external sodium, J. Physiol. 266, 74P (1977).

56. D. Ellis and R.C. Thomas, Direct measurement of the intracellular pH of mammalian cardiac muscle, J. Physiol. 262, 755 (1976).

57. D. Ellis and R.C. Thomas, Microelectrode measurements of the intracellular pH of mammalian heart cells, Nature 262, 224 (1976).

58. M. Fujimoto and T. Kubota, Physicochemical properties of a liquid ion exchanger microelectrode and its application to biological fluids, Jap. J. Physiol. 26, 631 (1976).

59. D.C. Gayton, R.D. Allen and J.A.M. Hinke, The intracellular concentration and activity of sodium in giant barnacle muscle fibers, J. Gen. Physiol. 54, 433 (1969).

60. D.C. Gayton and J.A.M. Hinke, The location of chloride in single striated muscle fibers of the giant barnacle, Can. J. Physiol. Pharamac. 46, 213 (1968).

61. D.F. Gerson and A.C. Burton, The relation of cycling of intracellular pH to mitosis in the acellular slime mould Physarum polycephalum, J. Cell. Physiol. 91, 297 (1976).

62. B.L. Gupta, M.J. Berridge, T.A. Hall and R.B. Moreton, Electron microprobe and ion-selective microelectrode studies of fluid secretion in the salivary glands of Calliphora, J. Exp. Biol. 72, 261 (1978).

63. S.F. Hannan and P.M. Wiggins, Intracellular pH of frog sartorius muscle. Biochim. Biophys. Acta 428, 205 (1976).

64. U. Heinemann, H.D. Lux and M.J. Gutnick, Extracellular free calcium and potassium during paroxysmal activity in the cerebral cortex of the cat, Exp. Brain. Res. 27, 237 (1977).

65. C.B. Heyer and H.D. Lux, Properties of a facilitating calcium current in pace-maker neurones of the snail, Helix pomatia, J. Physiol. 262, 319 (1976).

66. J.A.M. Hinke, Glass microelectrodes for measuring intracellular activities of sodium and potassium, Nature 184, 1257 (1959).

67. J.A.M. Hinke, The measurement of sodium and potassium activities in the squid axon by means of cation-selective glass microelectrodes, J. Physiol. 156, 314 (1961).

68. J.A.M. Hinke, Cation-selective microelectrodes for intracellular use, In: Glass Electrodes for Hydrogen and Other Cations. Ed: G. Eisenman. Dekker, New York (1967) p. 464.

69. J.A.M. Hinke, Solvent water for electrolytes in the muscle fiber of the giant barnacle, J. Gen. Physiol. 56, 521 (1970).

70. J.A.M. Hinke, J.P. Caille and D.C. Gayton, Distribution and state of monovalent ions in skeletal muscle based on ion electrode, isotope and diffusion analysis, Ann. N.Y. Acad. Sci. 204, 274 (1973).

71. J.A.M. Hinke and D.C. Gayton, Transmembrane $K^+$ and $Cl^-$ activity gradients for the muscle fiber of the giant barnacle, Can. J. Physiol. Pharmac. 49, 312 (1971).

72. J.A.M. Hinke and S.G.A. McLaughlin, Release of bound sodium in single muscle fibers, Can. J. Physiol. Pharmac. 45, 655 (1967).

73. J.A.M. Hinke and M.R. Menard, Intracellular pH of single crustacean muscle fibres by the DMO and electrode methods during acid and alkaline conditions, J. Physiol. 262, 533 (1976).

74. J.D. Johnson, D. Epel and M. Paul, Intracellular pH and activation of sea urchin eggs after fertilization, Nature 262, 661 (1976).

75. G.A. Kerkut and R.W. Meech, Microelectrode determination of the intracellular chloride concentration in nerve cells, Life Sci. 5, 453 (1966).

76. R.P. Kernan and M. MacDermott, Changes in potassium activity within frog sartorius muscle fibres during sodium enrichment in potassium-free Ringer fluid, J. Physiol. 249, 25P (1975).

77. Kernan, R.P. and M. MacDermott, Intracellular potassium concentrations and extracellular spaces in rat skeletal muscles immersed in normal, hypotonic and high-K modified Krebs fluid, determined by potassium-selective microelectrodes, J. Physiol. 263, 158P (1976).

78. R.P. Kernan, M. MacDermott and W. Westphal, Measurement of chloride activity within frog sartorius muscle fibres by means of chloride sensitive microelectrodes, J. Physiol. 241, 60P (1974).

79. M. Kessler, J. Hoper, D. Schafer and R. Strehlau, Measurements of extracellular and of interstitial cation ativity (pK, pNa, pCa) with ion-selective electrodes, Bibl. Anat. 15, 237 (1977).

80. R.D. Keynes, Chloride in the squid giant axon, J. Physiol. 169, 690 (1963).

81. R.N. Khuri, Glass microelectrodes and their use in biological systems. In: Glass Electrodes for Hydrogen and Other Cations, Ed: G. Eisenman. Dekker, New York (1967) p. 478.

82. R.N. Khuri, S.K. Agulian and K.K. Bogharian, Intracellular bicarbonate of skeletal muscle under different metabolic states, Am. J. Physiol. 230, 228 (1976).

83. R.N. Khuri, S.K. Agulian and R.I. Harik, Internal capillary glass microelectrodes with a glass seal for pH, sodium and potassium, Pflüegers Arch. 301, 182 (1968).

84. R.N. Khuri, K.K. Bogharian and S.K. Agulian, Intracellular bicarbonate in single skeletal muscle fibres, Pflüegers Arch. 349, 285 (1974).

85. R.N. Khuri, J.J. Hajjar and S.K. Agulian, Measurement of intracellular potassium with liquid ion-exchanger microelectrodes, J. Appl. Physiol. 32, 419 (1972).

86. G. Kimura, S. Urakabe, S. Yuasa, S. Miki, Y. Takamitsu, Y. Orita and H. Abe, Potassium activity and plasma membrane potentials in epithelial cells of toad bladder, Am. J. Physiol. 232, F196 (1977).

87. H. Korytova, Arousal-induced increase of cortical ($K^+$) in unrestrained rats, Experienta 33, 242 (1977).

88. P.G. Kostyuk and A.Z. Sorokina, On the mechanism of hydrogen ion distribution between cell protoplasm and the medium, In: Membrane Transport and Metabolism, Ed: A. Kleinzeller and A. Kotyk. Academic Press, New York, (1961) p. 192.

89. P.G. Kostyuk, Z.A. Sorokina and Y.D. Kholodova, Measurement of activity of hydrogen, potassium and sodium ions in striated muscle fibers and nerve cells, In: Glass Microelectrodes, Ed: M. Lavallee, O.F. Schanne and N.C. Hebert. Wiley, New York (1969) p. 322.

90. R.P. Kraig and C. Nicholson, Sodium liquid ion exchanger microelectrode used to measure large extracellular sodium transients, Science 194, 725 (1976).

91. D.L. Kunze and A.M. Brown, Internal potassium and chloride activities and the effects of acetylcholine on identifiable Aplysia neurons, Nature New Biol. 229, 229 (1971).

92. R.O. Ladle and J.L. Walker, Intracellular chloride activity in frog heart, J. Physiol. 251, 549 (1975).

93. M. Lavallee, Intracellular pH of rat atrial muscle fibers measured by glass micropipette electrodes, Circ. Res. 15, 185 (1964).

94. M. Lavallee and G. Szabo, Cation sensitive glass micro-electrodes for intracellular measurements, Digest 6th Intern. Conf. Med. Electr. Biol. Eng. Tokyo. 568 (1965).

95. C.O. Lee and W.M. Armstrong, State and distribution of potassium and sodium ions in frog skeletal muscle, J. Memb. Biol. 15, 331 (1974).

96. C.O. Lee and H.A. Fozzard, Activities of potassium and sodium ion in rabbit heart muscle, J. Gen. Physiol. 65, 695 (1975).

97. A.A. Lev, Determination of activity and activity coefficients of potassium and sodium ions in frog muscle fibres, Nature 201, 1132 (1964).

98. S. Levy and J.A. Coles, Intracellular pH of Limulus ventral photoreceptor measured with with a double-barrelled pH microelectrode, Experienta 33, 553 (1977).

99. A. Lopo and V.C. Vacquier, The rise and fall of intracellular pH of sea urchin eggs after fertilization, Nature 269, 591 (1977).

100. H.D. Lux, Change of potassium activity associated with membrane current flow, Fed. Proc. 35, 1248 (1976).

101. H.D. Lux, Simultaneous measurements of extracellular potassium-ion activity and membrane currents in snail neurones, In: Ion and Enzyme Electrodes in Biology and Medicine. Ed: M. Kessler, L.C. Clark, D.W. Lubbers, I.A. Silver and W. Simon, University Park Press, Baltimore (1976) p. 302.

102. H.D. Lux, Fast recording ion specific microelectrodes: their use in pharmacological studies in the CNS, Neuropharmac. 13, 509 (1974).

103. G. Malnic and F.L. Vieira, The antimony microelectrode in kidney micropuncture, Yale J. Biol. Med. 45, 356 (1972).

104. R.W. Meech, Prolonged action potentials in Aplysia neurones injected with EGTA, Comp. Biochem. Physiol. 48A, 397 (1974).

105. R.W. Meech and N.B. Standen, Potassium activation in Helix aspersa neurones under voltage clamp: a component mediated by calcium influx, J. Physiol. 249, 211 (1975).

106. R.W. Meech and R.C. Thomas, The effect of calcium injection on the intracellular sodium and pH of snail neurones, J. Physiol. 265, 867 (1977).

107. D.S. Miura, B.F. Hoffman and M.R. Rosen, The effect of extracellular potassium on the intracellular potassium ion activity and transmembrane potentials of beating canine cardiac purkinje fibers, J. Gen. Physiol. 69, 463, (1977).

108. D.S. Miura and M.R. Rosen, The effects of ouabain on the transmembrane potentials and intracellular potassium activity of canine cardiac purkinje fibers, Circ. Res. 42, 333 (1978).

109. D. Morf, D. Ammann, E. Pretsch and W. Simon, Carrier antibiotics and model compounds as components of selective ion-sensitive electrodes, Pure Appl. Chem. 36, 421 (1973).

110. S. Mori, W.H. Miller and T. Tomita, Microelectrode study of spreading depression (SD) in frog retina--Muller cell activity and $(K^+)$ during SD, Jap. J. Physiol. 26, 219 (1976).

111. T.O. Neild and R.C. Thomas, New design for a chloride-sensitive microelectrode, J. Physiol. 231, 7P (1973).

112. T.O. Neild and R.C. Thomas, Intracellular chloride activity and the effects of acetylcholine in snail neurones, J. Physiol. 242, 453 (1974).

113. C. Nicholson, G. Ten Bruggencate, R. Steinberg and H. Stockle, Calcium modulation in brain extracellular micro-environment demonstrated with ion-selective micropipette, Proc. Natl. Acad. Sci. USA 74, 1287 (1977).

114. B.P. Nikolskii, The theory of the glass electrode, Zh. Fiziol. Khim. 10, 495 (1937).

115. M. Oehme, M. Kessler and W. Simon, Neutral carrier $Ca^{2+}$ microelectrode, Chimia 30, 204 (1976).

116. Y. Okada and A. Inouye, pH-sensitive glass microelectrodes and intracellular pH measurements, Biophys. Struct. Mech. 2, 21 (1976).

117. J.D. Owen and H.M. Brown, Comparison between $Ca^{2+}$ sensitive microelectrode and arsenazo III, In: Detection and Measurement of Free Calcium Ions in Cells. Ed: C.C. Ashley and A.K. Campbell. Elsevier/North-Holland, Amsterdam (1979) in press.

118. J.D. Owen, H.M. Brown and J.P. Pemberton, $Ca^{2+}$ in the Aplysia giant cell and the Balanus eburneus muscle fiber, Biophys. J. 16, 34a (1976).

119. J.D. Owen, H.M. Brown and J.P. Pemberton, Neurophysiological applications of a calcium-selective microelectrode, Anal. Chim. Acta 90, 241 (1977).

120. J.D. Owen, H.M. Brown and J.H. Saunders, Chloride activity in the giant cell of Aplysia, Biophys. J. 15, 45a (1975).

121. M. Paillard, Direct intracellular pH measurement in rat and crab muscle, J. Physiol. 223, 297 (1972).

122. L.D. Partridge and R.C. Thomas, The effects of lithium and sodium on the potassium conductance of snail neurones, J. Physiol. 254, 551 (1976).

123. C.A. Paterson, M.C. Neville, R.M. Jenkins and D.K. Nordstrom, Intracellular potassium activity in frog lens determined using ion specific liquid ion-exchanger filled microelectrodes, Exp. Eye Res. 19, 43 (1974).

124. J.M. Russell, Effects of ammonium and bicarbonate-$CO_2$ on intracellular chloride levels in Aplysia neurons, Biophys. J. 22, 131 (1978).

125. J.M. Russell and W.F. Boron, Role of chloride transport in regulation of intracellular pH, Nature 264, 73 (1976).

126. J.M. Russell and A.M. Brown, Active transport of potassium and chloride in an identifiable molluscan neuron, Science 175, 1475 (1972).

127. J.M. Russell, D.C. Eaton and M.S. Brodwick, Effects of nystatin on the Aplysia giant neuron, Biophys. J. 16, 75a (1976).

128. J.M. Russell, D.C. Eaton and M.S. Brodwick, Effects of nystatin on membrane conductance and internal ion activities in Aplysia neurons, J. Memb. Biol. 37, 137 (1977).

129. J. Růžička, E.H. Hansen and J.C. Tjell, Selectrode--The universal ion selective electrode. Part IV. The calcium(II) selectrode employing a new ion exchanger in a nonporous membrane and a solid-state reference system, Anal. Chim. Acta 67, 155 (1973).

130. J.H. Saunders and H.M. Brown, Determination of intracellular chloride in Balanus eburneus photoreceptor, Biophys. J. 15, 323a (1975).

131. J.H. Saunders and H.M. Brown, Liquid and solid-state $Cl^-$-sensitive microelectrodes. Characteristics and application to intracellular $Cl^-$ activity in Balanus photoreceptor, J. Gen. Physiol. 70, 507 (1977).

132. S.S. Shen and R.A. Steinhardt, Direct measurement of intracellular pH during metabolic derepression of the sea urchin egg, Nature 272, 253 (1978).

133. R.B. Skinner and D.L. Kunze, Changes in extracellular potassium activity in response to decreased pH in rabbit atrial muscle, Circ. Res. 39, 678 (1976).

134. G.G. Somjen, M. Rosenthal, G. Cordingley, J. LaManna and E. Lothman, Potassium neuroglia and oxidative metabolism in central gray matter, Fed. Proc. 35, 1266 (1976).

135. Z.A. Sorokina, The activity of potassium and sodium ions in the protoplasm of striated muscle fibers in the frog, Eksperim. Biol. Med. 12, 17 (1964).

136. Z.A. Sorokina, Relation between intra and extrafiber pH, resting potential and potassium ion concentration in frog muscle, Tsitologiya 3, 48 (1961).

137. Z.A. Sorokina, Extra- and intracellular H-ion activity determination in mollusc nerve ganglia, Zh. Evolyutsionnoi Biokhim. Fiziol. 1, 343 (1965).

138. C.S. Spyropoulos, Cytoplasmic pH of nerve fibers, J. Neurochem. 5, 185 (1960).

139. J. Sinnakre and L. Tauc, Calcium influx in active Aplysia neurons detected by injected aequorin, Nature New Biol. 242, 113 (1973).

140. A. Strickholm and B.G. Wallin, Intracellular chloride activity of crayfish giant axons, Nature 208, 790 (1965).

141. R.C. Thomas, Membrane current and intracellular sodium changes in a snail neurone during extrusion of injection sodium, J. Physiol. 201, 495 (1969).

142. R.C. Thomas, Intracellular sodium activity and the sodium pump in snail neurones, J. Physiol. 220, 55 (1972).

143. R.C. Thomas, The effect of bicarbonate on the intracellular buffering power of snail neurones, J. Physiol. 239, 103P (1974).

144. R.C. Thomas, Intracellular pH of snail neurones measured with a new pH-sensitive glass microelectrode, J. Physiol. 238, 159 (1974).

145. R.C. Thomas, The effects of $CO_2$ and bicarbonate on the intracellular pH of snail neurones, In: Ion and Enzyme Electrodes in Biology and Medicine. Ed: M. Kessler, L.C. Clark, D.W. Lubbers, I.A. Silver and W. Simon, University Park Press, Baltimore (1976) p.288.

146. R.C. Thomas, The effect of carbon dioxide on the intracellular pH and buffering power of snail neurons, J. Physiol. 255, 715 (1976).

147. R.C. Thomas, Ionic mechanism of the $H^+$ pump in a snail neurone, Nature 262, 54 (1976).

148. R.C. Thomas, The role of bicarbonate, chloride and sodium ions in the regulation of intracellular pH in snail neurones, J. Physiol. 273, 317 (1977).

149. R.C. Thomas, W. Simon and M. Oehme, Lithium accumulation by snail neurones measured by a new $Li^+$-sensitive microelectrode, Nature 258, 754 (1975).

150. J.L. Walker, Ion-selective liquid ion exchanger microelectrodes, In: Ion and Enzyme Electrodes in Biology and Medicine. Ed: M. Kessler, L.C. Clark, D.W. Lubbers, I.A. Silver and W. Simon, University Park Press, Baltimore (1976) p. 116.

151. J.L. Walker, Ion specific ion exchanger microelectrodes, Anal. Chem. 43, 89A (1971).

152. J.L. Walker and A.M. Brown, Unified account of the variable effect of carbon dioxide on nerve cells, Science 167, 1502 (1970).

153. J.L. Walker and H.M. Brown, Intracellular ionic activity measurements in nerve and muscle, Physiol. Rev. 57, 729 (1977).

154. J.L. Walker and R.O. Ladle, Intracellular chloride activity in heart muscle, In: Ion and Enzyme Electrodes in Biology and Medicine. Ed: M. Kessler, L.C. Clark, D.W. Lubbers, I.A. Silver and W. Simon, University Park Press, Baltimore (1976). p. 351.

155. J.L. Walker and R.O. Ladle, Frog heart intracellular potassium activities measured with potassium microelectrodes, Am. J. Physiol. 225, 263 (1973).

156. J.F. White, Activity of chloride in absorptive cells of Amphiuma small intestine, Am. J. Physiol. 232, E553 (1977).

157. J.F. White, Intracellular potassium activities in Amphiuma small intestine, Am. J. Physiol. 231, 1214 (1976).

158. J.F. White and J.A.M. Hinke, Use of the sodium microelectrode to define sodium efflux and the behavior of the sodium pump in the frog sartorius, In: Ion and Enzyme Electrodes in Biology and Medicine. Ed: M. Kessler, L.C. Clark, D.W. Lubbers, I.A. Silver and W. Simon, University Park Press,

Baltimore (1976) p. 355.

159. R.D. Vaughan-Jones, Intracellular chloride activity of quiescent cardiac Purkinje fibres, J. Physiol. 272, 32P (1977).

160. M.L. Verkhovskaya, E.S. Imasheva, G.A. Kurella and L.G. Yaglova, Potassium-specific intracellular microelectrodes based on a liquid ion exchanger and their use on turgid cells, Chem. Abs. 85, 222 (1976).

161. F. Vyskocil and N. Kriz, Modifications of single and double-barrel potassium specific microelectrodes for physiological experiments, Pflügers Arch. ges. Physiol. 337, 265 (1972).

# THE AUTHORS

H. Mack Brown, Ph.D.

Jeffrey D. Owen, Ph.D.

H. Mack "Bud" Brown graduated from the University of Utah with a B.S. in Zoology and Chemistry in 1958. He graduated with a Ph.D. in 1964 having completed clinical work in Psychology and a thesis on chemical changes associated with behavior in invertebrates. He minored in Photomolecular Biology. The thesis won the Flannagan award in 1965.

He was a post-doctoral fellow in Neurology and Physiology in 1965-1966 and an NIH Special Fellow (Neurophysiology) at Scripps Institute of Oceanography, La Jolla, CA, from 1967 to 1969. He was Assistant Research Professor of Physiology at UCLA, Los Angeles, CA, in 1969-1970. In 1970, he joined the Physiology Department at the University of Utah, where he became Professor in 1976. He has been a visiting Research Scientist to the Department of Physiology at the Karolinska Institute on two separate occasions (1971 and 1976).

His research interests have focused on the visual transduction process in invertebrate photoreceptors and development of techniques to aid understanding the process. The latter accounts for his interest in ion-selective micrcelectrodes for intracellular ion analysis.

J.D. Owen is an Assistant Professor of Physiology at the University of Utah Medical Center. He graduated from the University of Utah in 1971 with a Ph.D. in Physical Chemistry. His current research interests include the use of calcium ion selective microelectrodes to study the effect of intracellular calcium ions on invertebrate photoreceptor's response toward light.

*Ion-Selective Electrode Rev.*, Vol.1, pp.187-206.
Pergamon Press Ltd. Printed in Great Britain.

# THE APPLICATION OF ION-SELECTIVE ELECTRODES IN DENTAL HEALTH AND MINERALISED TISSUE PROGRAMMES

G. J. Moody and J. D. R. Thomas
Chemistry Department, U.W.I.S.T., Cardiff CF1 3NU, Wales

CONTENTS



KEYWORDS: Fluoride ion-selective electrodes; calcium-ion selective electrodes; X-ray photoelectron spectroscopy; argon ion-etching.

## 1. INTRODUCTION

In 1938 Dean observed that the percentage of caries-free children residing in areas receiving domestic water with a high fluoride content was greater than those living in areas supplied with water containing less fluoride [1]. Since that time the significance of fluoride in dental health programmes has been widely studied.

A common method for fluoride assay in saliva, plaque, enamel and bone samples involved a diffusion procedure to separate the fluoride from interferences prior to a colorimetric assay [2]. Such methods are tedious and "faith in one's method seems to be as important as scientific rectitude" [3].

The commercial availability of the fluoride ion-selective electrode [4] and simple techniques for its use [5] revolutionised the whole field of fluoride analysis [6] and this is crucial in the dental research outlined in this review. The utility of the Orion 94-09 fluoride ion-selective electrode for certain dental studies was soon established with saliva [7], plaque [8] and teeth [9] and rapidly extended to many interesting programmes [10-183]. Since 1966 many other ion-selective electrodes have been designed and used for a wide variety of analyses [6]. Except for more limited applications of the calcium [183-190] and iodide [191] ion-selective electrodes only the fluoride electrode has found extensive use. Thus, it has especially facilitated research on the role of fluoride in plasma and saliva, the relation between fluoride in tooth enamel and caries and the deposition of fluoride from ingested and topical agents on a more accurate and lavish scale than hitherto possible.

## 2. SALIVA

Until recently the uncertainty of actual fluoride levels in saliva and its relation to ingested fluoride reflected the lack of sufficiently sensitive analytical techniques. In 1968 Grøn, McGann and Brudevold clearly established the effectiveness of the fluoride electrode for the rapid assay of fluoride in relatively small volumes of parotid duct saliva (5 $cm^3$) and the agreement with chemical assays demonstrated its reliability [7]. The fluoride levels of <0.01 to 0.05 ppm in fasting parotid saliva were 5 to 10-fold below previous literature values for whole saliva. This was related to the incorporation of fluoride into the whole saliva from the plaque which itself is rich in fluoride. Indeed, after centrifugation whole saliva levels approached these very low resting values of parotid and sub-maxillary saliva [147]. Fasting levels of 0.02 to 0.05 ppm determined by an alternative gas chromatography technique has amply confirmed these findings [192]. Complexation of fluoride in saliva was insignificant while similar levels were found in samples collected simultaneously from parotid and submaxillary glands [7]. No differences were evident in the salivary elimination of fluoride in children with and without kidney disfunction [65].

The low fasting parotid levels in humans are very sensitive to ingested fluoride and rise rapidly within 5 to 15 minutes of swallowing aqueous sodium fluoride in gelatin capsules, maximise after 30 to 60 minutes and gradually decrease to the original levels within 2 to 6 hours depending on the dose of fluoride. Typical maxima for parotid salivary fluoride were 0.06, 0.2 and 0.3 ppm after single 1, 5 and 10 mg doses respectively [7]. Shannon and Edmonds also recorded the remarkable sensitivity of parotid gland activity to exogenous doses of fluoride [85].

Following the administration of sodium fluoride with aluminium chloride in gelatin capsules the rates of parotid fluoride secretion and peak levels are both depressed and even more so when using sodium monofluorophosphate in place of sodium fluoride [86]. In neither case was aluminium detected in the saliva. It was concluded that the hydrolysis of the $PO_3F^{2-}$ anion was necessary prior to absorption of the fluoride ion product in the gastrointestinal tract while fluoroaluminate (III) complexes are less readily absorbed. Aluminium in beverages and foods may therefore influence the pre-eruptive deposition of fluoride in enamel [86]. However, Forsman and Ericsson consider that there is little difference between fluoride absorption from ingested fluoride of $PO_3F^{2-}$ owing to rapid hydrolysis by gut phosphatases [82]. Whole saliva also showed a broad range of enzymatic activity towards monofluorophosphate - as measured by fluoride release - with two major peaks at ~pH 7.4 and 8.3. On centrifugation 90 percent of the monofluorophosphate hydrolytic activity of whole saliva was retained in the sediment [119]. Pearce and More reported whole saliva to be inactive towards monofluorophosphate in various dentrifices possibly due to the anti-enzymatic behaviour of the associated detergents [76]. The amount of fluoride that could be released during the (brief) few minutes that a monofluorophosphate-based toothpaste is in the oral cavity would be insignificant when compared with free fluoride ion already present as an impurity [76].

Investigations have been undertaken with healthy male dental students on the effect of the rate of gland function on parotid fluoride levels during 20 minute exogenous stimulation with various sweets (Table I) [41]. Two striking observations were reported. First, the mean fluoride in unstimulated parotid fluid was significantly greater than in (any) of the faster flowing stimulated parotid fluid runs. Secondly, fluoride did not change in response to significant increases in the stimulated flow rates [41].

The flow rate independence of the fluoride profiles are distinctly different

TABLE I. The effect of flow rate on fluoride concentration in parotid duct fluid [41]

| Exogenous Stimulant | Parotid fluid variables | | | |
|---|---|---|---|---|
| | Flow rate/$cm^3 min^{-1}$ | | Fluoride/ppm | |
| | Mean[a] | sd | Mean | sd |
| None | 0.06 | 0.018 | 0.018 | 0.003 |
| Peppermint | 0.307 | 0.211 | 0.013 | 0.003 |
| Cherry | 0.464 | 0.211 | 0.013 | 0.003 |
| Grape | 0.776 | 0.166 | 0.012 | 0.002 |
| Cherry | 0.506 | 0.209 | 0.012 | 0.003 |
| Peppermint | 0.386 | 0.205 | 0.012 | 0.002 |

a 20 samples (4 subjects, 5 days)

compared with parotid fluid chloride which increases with stimulated fluid rates [41]. This suggests that fluoride is carried passively in bulk flow with the fluid of secretion and this is supported by the fact that fluoride levels in saliva and plasma are of the same order [80].

The interesting relation between fluoride in saliva and plasma has been studied with 6 male students after a 12 hour fast [124,125]. While the fluoride in fasting saliva and plasma are closely similar, considerable differences arise following a single oral dose of 3, 6 or 10 mg of sodium fluoride. The rapid rise-slow fall profile observed for saliva lies closely parallel to, but lower than, the corresponding plasma fluoride levels sampled during 12 hours after dosing. The saliva-plasma fluoride ratio for the 3 subjects dosed with 3 mg tablets was very stable, $0.63 \pm 0.01$ and it is suggested that saliva fluoride may prove a substitute for blood sampling in pharmacokinetic studies of fluoride [125]. Fluoride secreted by the parotid glands seems to originate from exchangeable fluoride in plasma and peak fluoride in stimulated saliva coincided with peak plasma values [125].

Irrespective of the role of fluoride in saliva Ekstrand has emphasised that any elevation following oral administration is of but short duration [125]. It is thus doubtful whether intake of conventional fluoride tablets is a suitable, or indeed desirable, way of achieving elevated steady-state salivary levels of fluoride.

Most studies normally utilise analates of 1 to 5 $cm^3$. A simple modification of the Orion 94-09 fluoride electrode as a "hanging-drop" model permitted the assay of fluoride in 5 $mm^3$ of parotid saliva samples at $0.013 \pm 0.0008$ ppm compared with $0.012 \pm 0.0003$ ppm using 2 $cm^3$ samples [18]. The use of 6 fluoride electrodes and a switch enabled 40 to 50 samples to be analysed per day [18]. Even more analates can be processed with just one electrode and the Orion 901 microprocessor [193].

The total calcium levels in parotid saliva are elevated in patients with cystic fibrosis, uraemia, Conn's syndrome and glycoside intoxication but little is known about the ionised calcium fraction and which is of interest regarding complexed fluoride [190]. Thus, parotid saliva samples were collected anaerobically from 10 healthy, fasting males following subcutaneous pilocarpine injection. At saliva flow rates of 0.1 to 0.6 $cm^3 min^{-1}$, ionised calcium rose from 4.53 to 7.05 ppm and then remained reasonably constant from 0.6 to 2.4 $cm^3 min^{-1}$. A similar, but higher, profile was observed for total calcium. The $Ca^{2+}$:Total Ca ratio of 0.54 $\pm$ 0.035 did not change significantly within this flow rate range [190]. It is interesting that the total calcium level in submaxillary duct fluid is about twice that in the parotid fluid [194].

The concentration of iodide and thiocyanate in whole saliva collected from 27 smokers and 92 non-smokers has been evaluated using a Beckman iodide selective electrode and a colorimetric method respectively. The concentration of thiocyanate was significantly higher in the saliva of smokers than non-smokers and also increased with the smoking habit while the reverse held true for iodide (Table II) (191). Within either group of smokers or non-smokers the saliva of females contained significantly less thiocyanate than males but more iodide [191].

TABLE II. The concentration of iodide and thiocyanate in the supernatant fluid of whole mouth saliva [191]

| Cigarettes /day | Number in test group | Test group | Mean levels/ppm Iodide | Thiocyanate |
|---|---|---|---|---|
| 0 → 5 | 8 | Smokers | 1.26 ± 0.3 | 95.0 ± 50.0 |
| 6→10 | 5 | Smokers | 1.39 ± 0.74 | 163.0 ± 69.9 |
| > 10 | 14 | Smokers | 0.97 ± 0.14 | 194.0 ± 61.7 |
| 0 | 92 | Non-smokers | 1.8 ± 1.1 | 72.0 ± 38.4 |

The highest mean ratio of SCN:I in any group was fortunately < 30 and interference from thiocyanate interaction with the silver iodide in the sensor membrane was therefore unlikely.

## 3. PLAQUE

Plaque comprises mainly bacteria, e.g., *S.mutans* and *S.sanguis*, bacterial products and salivary substances. Experimental and natural plaques have similar inorganic compositions which implies a non-enamel origin. Owing to its variable inorganic content, e.g. calcium and fluoride, Grøn, Yao and Spinelli decided to examine plaque formation under controlled conditions [8]. Thus, plaque material was allowed to form either on natural human dentitions or on human or shark enamel blocks positioned by an oral acrylic device. Samples were taken from the buccal surfaces of teeth of 5 adults (who had refrained from tooth brushing) after 1, 2 and 4 days, and treated with 0.5M perchloric acid (1 $cm^3$) before assay with a fluoride electrode. The average content of 25.2 ppm (dry weight) for the buccal plaque was about 10 → 20 times less than previously recorded but still 1000 → 2000 times higher than fasting saliva [8].

The plaque grown on shark enamel, which is essentially fluoroapatite, did not differ significantly from that formed on the intra-oral human enamel block and which itself had a low fluoride content, namely, 400 → 700 ppm. However, human enamel blocks pretreated with titanium tetrafluoride then acidulated phosphate fluoride (APF) increased the fluoride in the plaque to 113.5 ppm (dry weight) after 4 days [8].

Still higher levels have been found in plaque collected from 55 children aged 11 → 13 years who had participated for 4 years in a programme of bi-monthly mouth rinsings with 0.2 per cent sodium fluoride [44]. As with enamel, the bulk of fluoride taken up by plaque is rapidly lost within a few days of the last mouthing rinsing event [44]. Thus, frequent and regular supplies of fluoride are desirable to maintain any anti-caries momentum in plaque [48] where fluoride inhibits acid production and thus helps to protect erupted teeth from decay [84].

Topical application of teeth with an APF gel caused a persistent 45 → 75 per cent drop of *S.mutans* in the occlusal samples of plaque but had no effect on the corresponding approximal plaque [104]. This pattern was still evident 3 months after the last APF gel treatment. While the nature of this effect is uncertain, topical APF offers great promise as an anti-microbial agent for certain plaque bacteria. However, it must be noted that the APF gel had no effect on the incidence of *S.Sanguis* in either type of plaque sample except at one week after treatment where there was a significant rise in approximal samples [104]. The fluoride which reacts slowly with the plaque is considered to be bound by low molecular mass cations and not by proteins, carbohydrates or the Streptococci [49], but later *in vitro* work has shown that fluoride is bound to cell walls [195].

The extracellular aqueous phase of plaque represents the environment of the underlying tooth enamel surface and is considered to be the interface site of the interactions between enamel and that environment. Centrifugation of plaque samples at 5000 g provides this free aqueous phase called plaque fluid [130,186, 196]. Control fluid collected in this manner from 24 hour plaque samples of 97 healthy dental students showed a mean fluoride value of 0.038 ppm compared with 36.7 ppm for the residual dry plaque (Table III).

Each subject then rinsed once with 380 ppm sodium fluoride (10 $cm^3$) for 1 minute when the fluid-dry plaque was assayed over 3 hours (Table III). After 1 minute the plaque fluoride increased over 200-fold compared with only a 11-fold rise in plaque residue [130]. The elevation of fluoride in plaque after sodium fluoride rinsing suggests that the solid phase contributes to the maintenance of fluoride in plaque fluid [130].

TABLE III Fluoride content (ppm) of plaque fluid and dried plaque residue before and at several times after one sodium fluoride rinse[130]

| System | | Plaque fluid | | Plaque residue | |
|---|---|---|---|---|---|
| | | Mean | sd | Mean | sd |
| Control | | 0.038 | 0.013 | 36.7 | 21.5 |
| After 380 ppm rinse: | | | | | |
| 1 | min | 93.7 | 15.2 | 414 | 32 |
| 15 | min | 6.8 | 2.3 | - | - |
| 30 | min | 0.76 | 0.17 | 104 | 7 |
| 1 | h | 0.53 | 0.12 | 76 | 10 |
| 2 | h | 0.15 | 0.06 | 80 | 28 |
| 3 | h | 0.04 | 0.013 | 26 | 18 |

Buffering capacity studies with 24 hour-old plaque samples using pH glass electrodes suggested the presence of at least 3 buffer components in the fluid. The substantial buffering capacity is a major factor in limiting acid production during bacterial growth [196].

Fluoride in some beverages, especially tea, provides an important source of plaque fluoride [197]. Dietary fluorine provided by camellia or tea leaves was cariostatic in the case of Wistar rats and Gershon-Cohen and McClendon [198] posed the question "why are healthy (human) teeth not prevalent in tea drinking countries?". The fluoride complexation reactions in tea infusates, e.g., $\beta_1 AlF^{2+} = 10^{6.13}$, could be a contributory factor since aluminium and iron in one commercial tea averaged 2000 and 2800 ppm respectively [199]. Moreover, only about 20 per cent of the fluoride in 5 common brands of teas can be detected in their brews using a fluoride electrode [200].

## 4. ENAMEL

In general, solid samples must be dissolved before assay with an ion-selective electrode. Enamel is no exception and quantitative information on fluoride levels in intact and extracted teeth at various stages of growth and treatment is fundamental in dental health studies. Thus, McGann etched three or more layers, each about 10 µm thick from intact tooth enamel with 0.5M perchloric acid for just 30 seconds. Over 50 assays per day were effected with 2 or 3 fluoride ion-selective electrodes coupled to one meter and a recorder [23].

A biopsy technique involving polishing off about 0.2 mg of enamel sample from a single, intact human tooth was an important development in dental research. This was achieved in <1 minute with a rotary midget felt cone impregnated with silicon carbide abrasive and coated with glycerine slurry to trap the enamel particles before transfer to a plastic assay test tube containing 0.5M perchloric acid. First biopsies for fluoride in the outer 1 → 2 µm of enamel of intact human teeth were similar but whose mean concentration for a group of 200 military personnel varied from 430 → 2540 ppm [9], a value of 1000 ppm being considered the minimum required for effective caries prevention [35]. However, second biopsies on the same tooth the next day or even after 5 weeks showed much less fluoride, e.g., the fluoride in a left central tooth of a 20 year old fell from 1500 → 600 ppm over 2 days. Most teeth except those located posteriorly could be biopsied in this fashion [9].

A similar biopsy study of 33 Danish children revealed 2000 → 5800 ppm fluoride in the first 0.5 µm layer and 1500 → 4800 ppm in the next one of similar thickness [50]. This work confirmed the previous well-known fluoride gradient reported for extracted teeth [9]. The gradient decreases dramatically with depth from the enamel surface towards the amelodentinal junction. It arises largely through post-eruptive surface uptake of fluoride from oral fluids superimposed on a baseline value related to the quantity of fluoride ingested during tooth development.

Minor variations in the thickness of the outermost enamel layers can thus profoundly affect assays unless close attention is given to applying even abrasive pressure and to avoid extraneous sources of contamination. Biopsy techniques can therefore be difficult since the actual sampling can modify teeth surfaces and two identical surfaces do not exist (50). Any biopsy technique must control the depth and total enamel removed since a thick layer will be invariably lower in fluoride than a thin layer from the same enamel site.

A major criticism of the above biopsy technique is that the pressure, area and depths of enamel regions examined are not carefully standardised and an electrically driven abrasive device is considered to effect greater control for quantitative removal of surface enamels [19,52,53]. The procedures resemble dental prophylaxis and biopsies have been conducted on hundreds of patients without apparent discomfort. No more enamel is lost than in pumice treatments and it also gives a highly polished surface while chair-time is brief, only about 2 minutes. Several very thin layers of *in vivo* enamel can be successively removed from individual teeth under a variety of oral environments. Hence, long-term programmes involving interaction between enamel and fluoride (or, of course, other ions) can very conveniently be studied [50].

Other interesting biopsy studies concern the fluoride in the right maxillary central incisors of several hundred young men and women residing in 6 locations with varying fluoride levels in potable water supplies. The mean fluoride at comparable enamel depths ranging from 1.7 $\pm$ 0.4 µm to 2.2 $\pm$ 0.7 µm was generally about 10 per cent higher in the male subjects [109]. The mean surface enamel concentrations of a group of Marine Corps recruits suffering from dental fluorosis was 1550 $\pm$ 868 ppm compared with 841 $\pm$ 395 ppm for a control group [26]. A total of 224 biopsies taken from the buccal surfaces of maxillary teeth of 18 adult caries-free mediaeval skulls [129] revealed fluoride levels similar to a group of present day 14-year old children [143].

## 4.1 Dental caries studies

Dental caries is commonly treated by restoration which in 1972 cost £72 million in the United Kingdom alone. The effectiveness of fluoride in reducing caries after incorporation into the tooth enamel structure is well-established. Thus, alternative and more asthetic dental procedures frequently include topical application of aqueous fluoride or gels, particularly in non-fluoridated water regions. Here, the fluoride ion-selective electrode has played a considerable role in studies on the control of dental caries especially regarding the most effective fluoride agency and the method of its application to teeth enamel. Numerous investigations have shown an elevation of fluoride in enamel, albeit temporary, after exposure to various fluorides, including sodium fluoride.

The fluoride in the enamel of 63 boys and girls before, immediately after, and 1 week after topical application of ammonium fluoride, was considerably higher than with acidulated phosphate fluoride but 3 months later was not significantly different. No differences were evident between the sexes at any time [53]. However, Arends and Schutof found quite the opposite effect 24 hours after topical application of the same 2 compounds to 140 incisors and premolars - not only on the outer teeth surfaces but even at 4 different, lower enamel depths [77].

The most common topical agents include sodium fluoride and acidulated sodium fluoride phosphate. More recently, other preparations which also enhance fluoride uptake by *in vivo* and *in vitro* tooth enamels include ammonium fluoride [53,77]; sodium monofluorophosphate [94,120]; titanium tetrafluoride [30]; tin(II) fluoride [4,16,17,35,55,93]; sodium hexafluorostannate(IV) [120]; ammonium hexafluorosilicate(IV) [29] and cetylaminehydrofluoride [120].

The fluoride uptake by human enamel in freshly extracted premolars from 6 such topical agents during 5 minutes showed a significant but widely variable increase (Table IV). At 3 µm from the enamel surface a significantly higher uptake occurred with APF and cetylaminehydrofluoride while the latter compound also induced especially deep penetration of fluoride [120]. Considerable quantities of tin as well as fluoride have also been found in the outer layers of extracted permanent bovine incisors following topical application of 8 per cent tin(II) fluoride [93].

### 4.1.1 Protective anti-caries coatings.

Topical prophylaxis is usually effected by repeated mouth rinsing, teeth painting or brushing with dentrifices. Extremely high levels of fluoride can be achieved in enamel by such treatments. Thus, the mean level in intact teeth of young children of 2912 ppm (mean depth 2.8 µm) rose

TABLE IV. The fluoride uptake in the buccal surfaces of premolars at various depths from the enamel surfaces of newly erupted premolars[120]

| Topical fluoride solution | Fluoride at various depths from surface/ppm ± se[a] | | | | |
|---|---|---|---|---|---|
| | 3 μm | 6 μm | 10 μm | 15 μm | 20 μm |
| NaF | 2073 ± 229 | 810 ± 108 | 485 ± 67 | 247 ± 37 | 178 ± 30 |
| NaF-$H_3PO_4$ | 2655 ± 172 | 1123 ± 85 | 564 ± 82 | 230 ± 47 | 174 ± 44 |
| $Na_2PO_3F$ | 424 ± 103 | 174 ± 97 | 115 ± 73 | 74 ± 45 | - |
| $SnF_2$ | 376 ± 89 | 133 ± 62 | 99 ± 31 | 37 ± 31 | - |
| $Na_2SnF_6$ | 1816 ± 149 | 953 ± 108 | 430 ± 84 | 177 ± 57 | 108 ± 41 |
| Amine fluoride | 3516 ± 508 | 3241 ± 396 | 2834 ± 441 | 2164 ± 555 | - |

a The fluoride found in control areas of teeth has been subtracted from fluoride in experimental zones

to 26 189 ppm (mean depth 3.6 μm) immediately following a 1 minute pretreatment with 0.05M phosphoric acid then ammonium fluoride for 3 minutes [53]. In many cases such topical treatments provide 10 → 20 times the 1000 ppm necessary to afford minimum caries protection.

The fluoride concentration to which teeth are exposed varies enormously. Thus, erupted teeth are constantly exposed to fluoride, but of exceedingly low levels, (0.01 → 0.05 ppm) in fasting saliva, compared with say, 5000 → 10 000 ppm in topical solutions but of course for very brief periods - minutes Unfortunately, the acquired fluoride eventually falls below the anticaries protection level. This loss is thermodynamically inevitable [62]. Hence, following APF treatment, the mean fluoride of 12 243 ± 1100 (sd) ppm in extracted teeth had fallen to 2184 ± 400 (sd) ppm after washing for 24 hours in running tap water [77]. Periodic topical treatment must be maintained to derive long term benefit.

Protective fluoride-based coatings, if possible of high adhesive strength, which provide a long term slow release of fluoride, constitute a novel approach to solving this important dental health problem. Except for a synthetic hydroxy-apatite [97], most research has been conducted with extracted teeth and fluorine containing glass ionomer [97]; silicate [97] and carboxylate cements [20]; zinc phosphate [21]; polyurethane [64,77]; varnishes [137]; copal resins [137] and polyphosphonates [108].

*In vitro* studies with albino rat [64] and human [77] teeth have established a reduction in acid solubility of enamel [64] and a retention of about 2000 ppm fluoride in the outer 8 μm layer of human enamel [77]. In the *in vivo* programme a single application of lacquer releasing fluoride effectively reduced caries lesions of the molar teeth of albino rats fed on higher sugar diets inoculated with a cariogenic bacterium [64].

4.1.2 Studies with animals. While studies have established that sound bovine enamel will not necessarily predict the fluoride uptake of human enamel [27], a variety of interesting investigations have involved lions [12], pigs [126-128], monkeys [33], hamsters [36,162], and especially rats [34,37,38,64,79,102,117,128].

The corrosive influence of acidic beverages on tooth enamel depends on the frequency of their consumption. Thus, rats and hamsters were provided with various drinks for 4 weeks and then killed. Mandibular and maxillary molars were extracted and exposed to 0.5M hydrochloric acid for 30 minutes to remove about 100 μm of enamel for fluoride assay (Table V). Cola and citrus drinks fortified with fluoride reduced erosion presumably owing to the increased uptake of fluoride.

The mean fluoride levels were generally greater in the mandibular molars, probably due to their more lengthy exposure to oral fluids [38].

TABLE V. The fluoride content of molar tooth surfaces of rats and hamsters [34,36,38]

| Drinking fluid for animal group | No. of animals | Mean fluoride content of enamel with sd/ppm | | Comments |
|---|---|---|---|---|
| | | Maxillary molars | Mandibular molars | |
| Tap water control(~0.5)[a] | 10 | 90 ± 22(0)[b] | 88 ± 29(0)[b] | Degree of caries in all molars of hamsters drinking fluoride-supplemented juice was notably lower than hamsters on natural juice.[36] |
| Pure grapefruit juice (~0.25) | 9 | 102 ± 54(43±9) | 131 ± 83(42±15) | |
| Fluoridated juice (1.9) | 10 | 114 ± 65(22±3) | 142 ± 65(21± 9) | |
| Tap water control (0.5) | 6 | 14.6 ± 3.6 | 15.5 ± 6.6 | No increase in fluoride of molar teeth of rats drinking water up to 2 ppm of fluoride compared with the acid beverage supplemented to the same content especially in the mandibulars.[38] |
| Tap water (2) | 9 | 14.7 ± 5.2 | 21.5 ± 23.9 | |
| Pure grapefruit juice (~0.2) | 10 | 20.3 ± 9.5 | 30.3 ± 10.7 | |
| Fluoridated juice (2) | 8 | 96.6 ± 53.8 | 115.9 ± 76.6 | |
| Distilled water | 9 | 43.8 ± 12.3(0) | 41.3 ± 29.0(0) | Increasing fluoride supplementation to cola diet decreased erosion in rats. A significant difference noted between mean erosion scores of maxillary molars of rats drinking pure cola and those on the highest fluoride cola. [34]. |
| Distilled water (1.7) | 10 | 30.6 ± 20.8(0) | 37.6 ± 18.4(0) | |
| Cola (0.2) | 10 | 62.9 ± 51.2 (50.5 ± 18.7) | 91.8 ± 75.6 (127 ± 48.4) | |
| Cola (5.20) | 10 | 74.1 ± 35.5 (27.1 ± 25.2) | 107.0 ± 97.1 (88.9 ± 36.8) | |

a Parenthesised values are ppm fluoride.

b Parenthesised values represent erosion scores assessed with a microscope by scoring the loss of buccal and lingual surfaces of molars.

Mellberg and Larson determined the fluoride in the first and second layers of first, second and third maxillary molars of rat groups fed on either a non-fluoridated diet or on one fortified to 20 ppm fluoride [28]. First molars contained the least and third molars the most fluoride even with the non-fluoride diet group which seems to correlate with the period of systemic exposure to fluoride. Again, the outermost enamel layer contained more fluoride than the second layer but even the highest mean value of 376 ± 36 (se) ppm for the third molars was still low compared with the 600 → 750 ppm range for a roughly comparable layer in human deciduous teeth from caries-prone low water fluoride regions [28].

Fluoride administered over eight weeks by gastric intubation to rats aged 21 → 22 days is interesting since these animals cannot vomit and presumably up to that age their un-erupted teeth had experienced only systematic exposure to fluoride [102].

At the start of the eight week experiment, the first and second molars were at near complete eruption and thus their exposure was restricted to the post-eruptive period. Animals in the group drinking fluoridated water with consequent direct and systemic exposure showed greater fluoride uptake and caries inhibition in the first and second molars than the intubated rat group. Rat third molars, however, erupt later at about 35 days of age and thus in this work experienced some 14 days of pre-eruptive fluoride exposure [102]. Rat molars rapidly accumulate fluoride from dilute solutions after eruption. The fact that the third molars contained more fluoride than the first or second molars after the total of eight weeks implies that the 14 day interval before eruption is highly critical for fluoride uptake by rat enamel and both pre- and post-eruptive exposure to fluoride is necessary for maximum uptake [102].

4.1.3 Mechanism of anti-caries effect. The anti-caries mechanism of fluoride ions is not fully understood [201]. Apatite is a major component of mature enamel and the application of topical fluoride agencies initiates the dissolution of a very thin section of enamel.

The kinetic interfacial reactions of fluoride and hydrogen ions have been studied in a closed system at the surface of specially framed enamels using pH glass and fluoride electrodes. Rates increased with increased enamel surface, pH and pF as well as circulation flow rates [16,17].

One mechanistic proposal invokes surface interaction between the fluoride ions and the bulk hydroxy-apatite phase in enamel to create a concentration gradient of a less soluble, i.e., more caries resistant phase comprising fluoroapatite:

$$Ca_5(PO_4)_3OH(s) + H_3O^+ + F^- \rightleftharpoons Ca_5(PO_4)_3F(s) + 2H_2O; \quad \Delta H° = -96 \text{ kJ mol}^{-1}$$

Infrared and X-ray studies [60] show that calcium fluoride is also formed:

$$2Ca_5(PO_4)_3OH(s) + 2HF \rightleftharpoons 3Ca_3(PO_4)_2(s) + CaF_2(s) + 2H_2O; \quad \Delta H° = -12 \text{ kJ mol}^{-1}$$

The fluoride levels found in the outer 1 → 2 μm of human enamel (*in vivo* and *in vitro*) within 1 → 2 hours of topical fluoride treatment never exceeded 35 000 → 40 000 ppm [62]. The theoretical value for fluoroapatite is 38 000 ppm compared with 490 000 ppm for calcium fluoride. Stearns thus proposed the major anti-caries species to be fluoroapatite [62]. The calcium fluoride detected is considered to be easily sloughed from such enamel surfaces [62]. In any event, the fluoride levels fall rapidly from 38 000 ± 6000 ppm to 13 000 ± 200 ppm and 8100 ± 800 ppm within 7 and 14 days of post-topical treatment with APF gels [61]. These conclusions [62] need re-examination in the light of recent studies on enamels pretreated in the same way with APF gels [202,203].

The inability to distinguish the surface composition from bulk material in such enamels could produce a distorted analysis of the surface species. However, X-ray photoelectron spectroscopy (ESCA) coupled with argon ion-etching can be effectively employed to evaluate concentration gradients not only for fluoride but also carbon, calcium and phosphorus in enamel layers <1 nm thick [202,203]. This technique revealed a heavy outer organic coating in all untreated human enamels comprising carboxyl-, carbonyl-, carbanol- and hydrocarbon-type carbon atoms about 8 nm thick [203]. Complete removal of this coating by first brushing then argon ion-etching revealed an inorganic phase with a Ca:P ratio of 1.4:1 which remained constant for successive etchings. This accords with the known Ca:P ratio of

1.6:1 for bulk enamel.

Extracted teeth were then brushed and treated with APF. Quite different ESCA profiles were evident. Thus, the calcium concentration at the surface or in the interior enamel was unaffected by the fluoridation whereas the surface phosphorus was substantially depleted and largely replaced by fluorine, the Ca:F ratio being 1:2. Calcium fluoride, not fluoroapatite, is thus the major product formed following acid fluoridation of human enamel. Below the surface, fluorine continually declined and reached ~80 per cent of the initial value at 400 nm[203]. The initial slight fall in phosphorus to about 50 nm was quickly followed by a steady increase and at ~300 nm reached the value required for unfluoridated enamel. This pattern reflects the steady transition from calcium fluoride on the surface to the interior of unreacted hydroxy apatite.

The ESCA profiles were considerably different using tin(II) fluoride. The surface comprised mainly a tin-oxygen compound but no fluorine [202]. This again contrasts with previous reports of fluoride assays in the outer layers of teeth treated with tin(II) fluoride [93]. Removing a 15 nm layer with an argon ion beam revealed moderate concentrations of fluoride as well as tin, but thereafter both elements steadily decreased with depth. In contrast, calcium and phosphorus which were low at the surface steadily increased with depth [202].

Thus, ESCA in association with argon etching is an elegant technique to examine surface and sub-surface compositions of dental enamels, but unfortunately only for *in vitro* purposes. Auger spectroscopy has also been used to examine fluoride release patterns from polyphosphonate coatings applied to the surface of human enamels [108].

4.1.4 Dentrifices, mouth rinses and tablets. The fluoride in dentrifices is generally present as sodium fluoride, or monofluorophosphate or tin(II) fluoride [45,73,82,88,89]. Fluoride in the aqueous extracts or slurries of toothpastes can be analysed in <10 minutes with an ion-selective electrode. The results closely matched those obtained by a diffusion thorium(IV) nitrate colorimetric method but which required 45 minutes per sample [88].

Apart from small initial "losses" of uncomplexed fluoride in toothpastes containing fluoride or monofluorophosphate, caused by absorption on calcium carbonate polishing particles, no major shelf depletion arose for at least a year [89]. However, another study indicated considerable losses of active ionic fluorine after just 4 months of shelf life [73]. Calcium-based polishing agents in 5 commercial and 4 experimental toothpastes prevent the absorption of a considerable fraction of fluoride and monofluorophosphate anions from the gut of rats and human adults [82].

The quantity of fluoride in solutions expectorated following oral rinsing is significantly less than in the original solution. Birkeland has proposed a screening test for dentrifices based on correlation between fluoride in such expectorates and enamel uptake [45]. Parkins has also measured the fluoride in expectorates collected from patients chewing or rinsing with three different preparations [68]. The greatest retention was realised with sodium fluoride tablets (41 per cent) compared with APF tablets (20 per cent) and APF rinses (7 per cent) [68] . A very modest uptake of fluoride from such mouth rinses can significantly reduce caries. Thus, an increase of 200 → 300 ppm fluoride in the outermost enamel layer was associated with about 30 per cent reduction in caries [204].

It has been proposed that monofluorophosphate has no anti-caries role in current dentrifices and its use in preventitive dentistry needs to be reassessed[76].

This view is supported by more recent work although considerable differences were evident in the two monofluorophosphate toothpastes owing to their different abrasives [121].

The caries-reducing effect achieved by daily use of fluoride containing dentrifices does not seem impressive [121,201] whereas the alternative topical application procedure results in a much higher fluoride concentration in the enamel [120].

Water-fluoridation programmes for caries prevention cannot be used in rural and mountainous regions [92]. The role of fluoridated domestic salt has been examined using fluoride electrodes as an alternative or comple-ment to fluoride in potable water [10,92,135]. Thus, a packeted salt with 250 ppm fluoride content is estimated to be equivalent in anti-caries capacity to that of fluoridated waters [92].

## 5. BONE

Bone samples are readily analysed after ashing and acid dissolution. Singer and Armstrong obtained good agreement for fluoride in rat bones using the fluoride ion-selective electrode and a diffusion-colorimetric method [24]. An autopsy survey of ribs and vertebrae has confirmed that mature bones can acquire fluoride to an extent depending on the local water fluoride. One group of autopsies came from Michigan with water supplies fluoridated for 20 years at 1 ppm and the second group from the New York area with 0.1 ppm fluoride. The incidence and severity of intervertebral osteoarthritis and osteoporosis seemed unaffected at least over this ten-fold range of fluoride [25]. It has also been established that the fasting plasma fluoride concentration is in equilibrium with bone fluoride and both increase with age in a human population ranging from 17 → 82 years old [149].

## 6. CONCLUSION

The fluoride ion-selective electrode is cheap, robust and simple to use and provides results very quickly. In addition, new biopsy techniques provide more realistic samples of *in vivo* and *in vitro* enamel. As a result it has been possible to undertake long-term studies of the relationship between fluoride in enamel and its susceptibility to caries as well as the deposition of fluoride from dietary and topically applied fluoride in many forms on a massive scale.

## 7. REFERENCES


1. H.T.Dean, *Public Health Report, Washington*, 53, 1443 (1938).

2. H.W.Wharton, *Analyt.Chem.*, 34, 1296 (1962).

3. A.G.MacDonald, *Adv.Analyt.Chem.and Instr.*, 4, 103 (1965).

4. M.S.Frant and J.W.Ross, *Science*, 154, 1553 (1966).

5. M.S.Frant and J.W.Ross, *Analyt.Chem.*, 40, 1169 (1968).

6. G.J.Moody and J.D.R.Thomas, In: *Ion-Selective Electrodes in Analytical Chemistry*, Ed. Henry Freiser, Plenum Press, New York and London, Volume 1 (1978). Page 336.

7. P.Grøn, H.G.McCann and F.Brudevold, *Archs Oral Biol.*, 13, 203 (1968).

8. P.Grøn, K.Yao and M.Spinelli, *J.Dent.Res.*, 48, 799 (1969).

9. F.Brudevold, H.G.McCann and P.Grøn, *Archs Oral Biol.*, 13, 877 (1968).

10. Y.Ericsson, *Acta Odontol.Scand.*, 29, 43 (1971).

11. Y.Ericsson, *Caries Res.*, 7, 56 (1973).

12. C.Van Peteghem and A.Heyndrickx, *J.Pharm.Belg.*, 29, 63 (1974).

13. R S.Levine, *J.Dent.Res.*, 52, 1249 (1973).

14. F.W.Barnes and R.Runcie, *J.Clin.Path.*, 21, 668 (1968).

15. H.Luoma, S.K.J.Helminen, H.Ranta, I.Rytömaa and J.H.Meurman, *Scand.J.Clin. Lab.Investig.*, 32, 217 (1973).

16. B.G.Jeansonne and F.F.Feagin, *J.Dent.Res.*, 51, 767 (1972).

17. F.F.Feagin, B.G.Jeansonne and A.McCaghren, *J.Dent.Res.*, 51, 1457 (1972).

18. P.Ventakeswarlu, *Clin.Chim.Acta*, 59, 277 (1975).

19. S.H.Y.Wei, *J.Dent.Res.*, 52, 1268 (1973).

20. T.J.Bitner and S.H.Y.Wei, *J.Dent.Res.*, 52, 157 (1973).

21. S.H.Y.Wei and D.L.Sierk, *J.Amer.Dent.Assn.*, 83, 621 (1971).

22. I.Gedalia, P.Goldhaber and L.Golieb, *J.Dent.Res.*, 49, 1553 (1970).

23. H.G.McCann, *Arch oral Biol.*, 13, 475 (1968).

24. L.Singer and W.D.Armstrong, *Analyt.Chem.*, 40, 613 (1968).

25. A.M.Pochomis and F.D.Griffiths, *Amer.Ind.Hyg.Assn.J.*, 32, 557 (1971).

26. H.J.Keene, F.D.Grossman, E.D.Peterson, J.R.Mellberg and C.R.Nicholson, *Caries Res.*, 9, 247 (1975).

27. J.R.Mellberg and K.L.Loertscher., *J.Dent.Res.*, 52, 447 (1973).

28. J.R.Mellberg and R.H.Larson, *J.Dent.Res.*, 50, 600 (1971).

29. J.R.Mellberg and K.L.Loertscher, *Archs oral Biol.*, 17, 1107 (1972).

30. S.A.Mundorff, M.F.Little and B.G.Bibby, *J.Dent.Res.*, 51, 1567 (1972).

31 R.Aasenden, F.Brudevold and B.Richardson, *Archs oral Biol.*, 13, 625 (1968).

32. J.Granbart, I.Gedalia and P.Pisanti, *J.Dent.Res.*, 51, 1677 (1972).

33. J.Ehrlich, N.Hochman, I.Gedalia and M.Tal, *J.Dent.Res.*, 54, 897 (1975).

34. E.Shabat, J.Anaise, V.Westreich and I.Gedalia, *J.Dent.Res.*, 54, 426 (1975).

35. D.Kochavi, I.Gedalia and J.Anaise, *J.Dent.Res.*, 54, 304 (1975).

36. I.Gedalia, .Anaise, V.Westreich and A.Fuks, *J.Dent.Res.*, 54, 496 (1975).

37. V.Westreich and I.Gedalia, *J.Dent.Res.*, 52, 1508 (1972).

38. L.Lehman, I.Gedalia and V.Westreich, *Ann.Dent.*, 33, 2 (1974).

39. L.B.Schulman, I.Gedalia and R.M.Feingold, *J.Dent.Res.*, 52, 1314 (1973).

40. R.P.Feller and I.L.Shannon, *J.Dent.Res.*, 54, 1213 (1975).

41. I.L.Shannon, R.P.Suddick and E.J.Edmonds, *Caries Res.*, 7, 1 (1973).

42. J.M.Birkeland, *Caries Res.*, 7, 11 (1973).

43. J.M.Birkeland, *Caries Res.*, 7, 39 (1973).

44. J.M.Birkeland, L.Jorkjend and F.R.von der Fehr, *Caries Res.*, 5, 169 (1971).

45. J.M.Birkeland, *Caries Res.*, 5, 193 (1971).

46. J.M.Birkeland, *Caries Res.*, 4, 243 (1970).

47. J.M.Birkeland and P.Lökken, *Caries Res.*, 6, 325 (1972).

48. J.M.Birkeland, *Caries Res.*, 9, 110 (1975).

49. J.M.Birkeland and G.Rölla, *Arch. oral Biol.*, 9, 110 (1972).

50. M.J.Larsen, M.Kold and F.R.von der Fehr, *Caries Res.*, 6, 193 (1972).

51. T.M.Athanassouli, D.S.Papastatthopoulos and T.P.Hadjiioannou, *Caries Res.*, 7, 368 (1973).

52. S.H.Y.Wei and J.S.Wefel, *J.Dent.Res.*, 54, 201 (1975).

53. S.H.Y.Wei and E.M.Schultz, *Caries Res.*, 9, 50 (1975).

54. M.Gonsalez, B.G.Jeansonne and F.F.Feagin, *J.Dent.Res.*, 52, 261 (1973).

55. B.G.Jeansonne and F.F.Feagin, *J.Dent.Res.*, 52, 348 (1973).

56. M.Gonsalez, *J.Dent.Res.*, 50, 1056 (1971).

57. L.C.Chow and W.E.Brown, *J.Dent.Res.*, 52, 1220 (1973).

58. L.C.Chow and W.E.Brown, *J.Dent.Res.*, 54, 1089 (1975).

59. R.S.Levine, *J.Dent.Res.*, 51, 1025 (1972).

60. R.I.Stearns and A.F.Berndt, *J.Dent.Res.*, 52, 1253 (1973).

61. R.I.Stearns, *J.Dent.Res.*, 52, 30 (1973).

62. R.J.Stearns, *J.Dent.Res.*, 49, 1444 (1970).

63. R.I.Stearns, *J.Dent.Res.*, 50, 1575, (1971).

64. H.Lee, D.E.Ocumpaugh and M.L.Swartz, *J.Dent.Res.*, 51, 183 (1972).

65. L.Capozzi, F.Marci, P.L.Negri and A.Pepe, *Caries Res.*, 6, 68 (1972).

66. A.C.Ramsey, E.J.Duff, L.Paterson and J.L.Stuart, *Caries Res.*, 7, 231 (1973).

67. J.A.Weatherell, C.Robinson and A.S.Hallsworth, *Caries Res.*, 6, 312 (1972).

68. F.M.Parkins, *J.Dent.Res.*, 51, 1346 (1972).

69. L.Singer, B.A.Jarvey, P.Venkateswarlu and W.D.Armstrong, *J.Dent.Res.*, 49, 455 (1970).

70. R.S.Riggins, F.Zeman and D.Moon, *Calcif.Tissue Res.*, 14, 283 (1973).

71. P.Hotz, H.R.Mühleman and A.Schait, *Helv.Odontol.Acta*, 14, 26 (1970).

72. G.Charlton, B.Blaining and R.G.Schamschula, *Archs oral Biol.*, 19, 139 (1974).

73. E.J.Bushee, D.K.Grisson and D.R.Smith, *J.Dent.Children*, 38, 279 (1971).

74. R.Aasenden, E.C.Moreno and F.Brudevold, *Archs oral Biol.*, 18, 1403 (1973).

75. P.Grøn, F.Brudevold and R.Aasenden, *Caries Res.*, 5, 202 (1971).

76. E.I.F.Pearce and R.D.More, *Caries Res.*, 9, 459 (1975).

77. J.Arends and J.Schuthof, *Caries Res.*, 9, 363 (1975).

78. D.S.Magrill, *Caries Res.*, 9, 45 (1975).

79. H.Luoma, L.Turtola, Y.Collan, J.Meurman and S.Helminen, *Caries Res.*, 6, 183 (1972).

80. Y.Ericsson, *Caries Res*, 3, 159 (1969).

81. R.Aasenden, E.C.Moreno and F.Brudevold, *Archs oral Biol.*, 17, 355 (1972).

82. B.Forsman and Y.Ericsson, *Comm.Dent.Oral Epidemiol.*, 1, 115 (1973)

83. E.J.Duff and J.L.Stuart, *Analytica Chim.Acta*, 52, 155 (1970).

84. W.M.Edgar, G.N.Jenkins and A.Tatevossian, *Brit.Dental J.*, 128, 129 (1970).

85. I.L.Shannon and E.J.Edmonds, *Archs oral Biol.*, 17, 1303 (1972).

86. F.Brudevold, Y.Bakhos and P.Grøn, *Archs oral Biol.*, 18, 699 (1973).

87. E.J.Duff and J.L.Stuart, *Analytica Chim.Acta*, 57, 233 (1971).

88. N.Shane and D.Miele, *J.Pharm.Sci.*, 57, 1260 (1968).

89. B.Noren and C.Harse, *J.Soc.Cosmet.Chem.*, 25, 3 (1974).

90. A.F.Berndt and R.I.Stearns, *Analytica Chim.Acta*, 74, 446 (1975).

91. C.A.Ostrom, T.Koulorides, F.Hickman and P.Phantumvantit, *J.Dent.Res*, 56, 213 (1977).

92. H.J.Wespi and W.Bürgi, *Caries Res.*, 5, 89 (1971).

93. M.S.Putt, J.F.Beltz and J.C.Muhler, *J.Dent.Res.*, 57, 772 (1978).

94. E.D.Eans, *Caries Res.*, 10, 59 (1976).

95. E.D.Eanes and J.L.Mayer, *J.Dent.Res.*, 57, 617 (1978).

96. S.Duke and G.C.Forward, *Caries Res.*, 12, 12 (1978).

97. L.Forsten, *Scand.J.Dent.Res.*, 85, 503 (1977).

98. V.Caslavska, E.C.Moreno and F.Brudevold, *Archs oral Biol.*, 20, 333 (1975).

99. P.F.DePaula, F.Brudevold, R.Aasenden, E.C.Moreno, H.Englander, Y.Bakhos, F.Bookstein and J.Warram, *Archs oral Biol.*, 20, 859 (1975).

100. J.R.Mellberg, C.R.Nicholson and F.E.Law, *J.Dent.Res.*, 51, 551 (1972).

101. J.R.Mellberg, C.R.Nicholson, L.W.Ripa and J.R.Barenie, *J.Dent.Res.*, 55 976 (1976).

102. R.H.Larson, J.R.Mellberg and R.Senning, *Archs oral Biol.*, 22 437 (1977).

103. J.R.Mellberg and C.R.Nicholson, *Caries Res.*, 8, 148 (1974).

104. W.J.Loesche, S.A.Syed, R.J.Murray and J.R.Mellberg, *Caries Res.*, 9, 139 (1975).

105. J.R.Mellberg and R.H.Larson, *Caries Res.*, 12, 137 (1978).

106. F.Brudevold, R.Aasenden, B.N.Srinivisan and Y.Bakhos, *J.Dent.Res.*, 56, 1165 (1977).

107. B.D.Goodman and H.W.Kaufman, *J.Dent.Res.*, 56, 1201 (1977).

108. M.Anbar, E.P.Farley, D.D.Denson and K.R.Maloney, *J.Dent.Res.*, 58, 1134 (1979).

109. R.Aasenden, *Archs oral Biol.*, 19, 697 (1974).

110. S.Kashket and V.M.Rodriguez, *Archs oral Biol.*, 21, 459 (1976).

111. S.Kashket and F.J.Bunick, *Archs oral Biol.*, 23, 993 (1978).

112. M.J.Larsen, F.R.van der Fehr and J.M.Birkeland, *Archs oral Biol.*, 21, 723 (1976).

113. D.Clarkson and J.S.Wefel, *J.Dent.Res.*, 58, 600 (1979).

114. B.H.Clarkson and L.Silverstone, *J.Int.Ass.Dent.Child.*, 5, 27 (1974).

115. B.H.Clarkson, *J.Dent.Res.*, 57, 784 (1978).

116. B.H.Clarkson, *J.Int.Ass.Dent.Child*, 12, 9 (1976).

117. D.H.Reteif, J.M.Navia and H.Lopez, *Archs oral Biol.*, 22 207 (1977).

118. G.N.Jenkins and W.M.Edgar, *Caries Res.*, 11, (Suppl.1), 226 (1977).

119. E.I.F.Pearce and G.N.Jenkins. *Archs oral Biol.*, 22, 405 (1977).

120. E.Kirkegaard, *Caries Res.*, 11, 16 (1977).

121. E.Kirkegaard, *Caries Res.*, 11, 24 (1977).

122. P.Grøn, *Caries Res.*, 11, (Suppl.1), 172 (1977).

123. J.Ekstrand, *Caries Res.*, 12, 123 (1978).

124. J.Ekstrand, *Calcif.Tiss.Res.*, 23, 225 (1977).

125. J.Ekstrand, *Scand.J.Dent.Res.*, 85, 16 (1977).

126. R.L.Speirs, *Archs oral Biol.*, 20, 877 (1975).

127. R.L.Speirs, *Archs oral Biol.*, 23, 1013 (1978).

128. R.L.Speirs, *Archs oral Biol.*, 23, 1019 (1978).

129. A.Richards, M.J.Larsen and O.Fejerskov, *Archs oral Biol.*, 24, 83 (1979).

130. A.Tatevossian, *Archs oral Biol.*, 23, 893 (1978).

131. H.Edenholm, G.Johnson, G.Koch and L.Petersson, *Swedish Dent.J.*, 1, 59 (1977).

132. H.Luoma and H.Tuompo, *Archs oral Biol.*, 20, 749 (1975).

133. A.Rahkamo and H.Tuompo, *Proc.Finn.dent.Soc.*, 71, 71 (1975).

134. P.Pohto and E.Alanen, *Proc.Finn.dent.Soc.*, 72, 27 (1976).

135. E.Alanen and P.Pohto, *Proc.Finn.dent.Soc.*, 73, 225 (1977).

136. K.Helgeland and J.Leirskar, *Scand.J.Dent.Res*,., 84, 37 (1976).

137. H.Nordbö and H.M.Eriksen, *Scand.J.Dent.Res.*, 84, 386 (1976).

138. E.L.Lakomaa and I.R.Rytömma, *Scand.J.Dent.Res.*, 85, 89 (1977).

139. L.O.Turtola, *Scand.J.Dent.Res.*, 85, 373 (1977).

140. L.O.Turtola, *Scand.J.Dent.Res.*, 85, 380 (1977).

141. L.O.Turtola, *Scand.J.Dent.Res.*, 85, 535 (1977).

142. C.M.Patterson, B.J.Kruger and T.J.Daley, *Archs oral Biol.*, 22, 419 (1977).

143. A.Richards, M.J.Larsen, O.Fejerskov and A.Thylstrup, *Archs oral Biol.*, 22, 425 (1977).

144. A.Helle and K.Haavikko, *Proc.Finn.dent.Soc.*, 73, 76 (1977).

145. A.Helle and K.Haavikko, *Proc.Finn.dent.Soc.*, 73, 144 (1977).

146. H.Hanhijärivi, *Proc.Finn.dent.Soc.*, 70, Suppl III (1974).

147. K.Yao and P.Grøn, *Caries Res.*, 4, 321 (1970).

148. G.Ahrens, *Caries Res.*, 10, 85 (1976).

149. F.M.Parkins, N.Tinanoff, M.Moutinlo, M.B.Anstey and M.H.Waziri, *Calcif.Tiss.Res.*, 16, 335 (1974).

150. H.M.Messer, W.D.Armstrong and L.Singer, *Calcif.Tiss.Res.*, 13, 217 (1973).

151. E.Weiss, I Gedalia and Y.Zilberman, *J.Dent.Res.*, 56, 1345 (1977).

152. I.L.Shannon and E.J.Edmonds, *J.Dent.Res.*,56, 1521 (1977).

153. I.L.Shannon, R.P.Feller and H.H.Chauncey, *J.Dent.Res.*, 55, 506 (1976).

154. I.L.Shannon, *Comm.Dent.Oral Epidemiol.*,2, 214 (1974).

155. J.Zadok, I.Gedalia, J.Weinman and L.Daphni, *J.Dent.Res.*, 55, 310 (1976).

156. M.Goldstein and I.Gedalia, *J.Dent.Res.*, 55, 912 (1976).

157. J.I.Bischoff, E.H.M.van der Merwe, D.H.Retief, F.H.Barbakow and P.E.Cleaton-Jones, *J.Dent.Res.*, 55, 37 (1976).

158. G.K.Stookey and D.B.Stahlman, *J.Dent.Res.*, 55, 333 (1976).

159. S.H.Y.Wei, D.M.Soboroff and J.S.Wefel, *J.Dent.Res.*, 55, 426 (1976).

160. M.L.Swartz, R.W.Phillips, R.D.Norman, S.Elliason, B.F.Rhodes and H.E.Clark, *J.Dent.Res.*, 55, 757 (1976).

161. A.A.Hussain, J.H.Kraal and H.Wahner, *J.Dent.Res.*, 57, 872 (1978).

162. A.Nathaniel, I.Gedalia, J.Anaise and L.Daphni, *J.Dent.Res.*, 57, 534 (1978).

163. Y.Ericsson, *Caries Res.*, 11 (Suppl.1), 2 (1977).

164. I.Gedalia, D.Almog and S.Yariv, *Caries Res.*, 11, 287 (1977).

165. P.Phantumvanit, F.F.Feagin and T.Koulourides, *Caries Res.*, 11, 52 (1977).

166. J.M.Tenlate and J.Arends, *Caries Res.*, 11, 277 (1977).

167. J.A.Weatherell, G.Naylor and A.S.Hallsworth, *Caries Res.*, 11, 231 (1977).

168. J.A.Weatherell, D.Deutsch, C.Robinson and A.S.Hallsworth, *Caries Res.*, 11, (Suppl.1), 85 (1977).

169. G.Rölla, *Caries Res.*, 11, (Suppl.1), 243 (1977).

170. I.L.Shannon, *Caries Res.*, 11, (Suppl.1), 206 (1977).

171. C.Robinson, P.Fuchs, D.Deutsch and J.A.Weatherell, *Caries Res.*, 12, 1 (1978).

172. J.M.Birkeland and G.Charlton, *Caries Res.*, 10, 72 (1976).

173. W.L.Jongebloed, I.Molenaar and J.Arends, *Caries Res.*, 9, 388 (1975).

174. B.Angmar-Monsson, Y.Ericsson and O.Ekberg, *Calcif.Tiss.Res.*, 22, 77 (1976).

175. J.M.Birkeland and L.Tronstad, *Scand.J.Dent.Res.*, 84, 200 (1976).

176. L.Hellström and Y.Ericsson, *Scand.J.Dent.Res.*, 84, 255 (1976).

177. G.Rölla and W.H.Bowen, *Scand.J.Dent.Res.*, 85, 149 (1977).

178. M.J.Larson, S.J.Jensen and A.Thorsen, *Scand.J.Dent.Res.*, 85, 327 (1977).

179. P.F.de Paola, P.Soparkar, S.Foley, F.Bookstein and Y.Bakhos, *Comm.Dent. Oral Epidemiol.*, 5, 7 (1977).

180. T.M.Athanassouli and D.S.Papastathopoulis, *Comm.Dent.Oral Epidemiol.*, 6, 185 (1978).

181. C.Bruun and H.Givskov, *Comm.Dent.Oral Epidemiol.*, 6, 27 (1978).

182. P.C.Spector and M.E.J.Curzon, *Caries Res.*, 13, 227 (1979).

183. A.Tatevossian and C.T.Gould, *Archs oral Biol.*, 21, 319 (1976).

184. O.Wolf, I.Gedalia, J.Reisstein, J.Goldman and H.Steiglitz, *J.Dent.Res.*, 52, 467 (1973).

185. A.Tatevossian and G.N.Jenkins, *Caries Res.*, 8, 317 (1974).

186. A.Tatevossian and C.T.Gould, *Archs oral Biol.*, 21, 313 (1976).

187. G.L.Gardner and G.H.Nancollas, *J.Dent.Res.*, 55, 342 (1976).

188. Ö.Häsvold and S.Dahn, *Calcif.Tiss.Res.*, 22, 425 (1977).

189. G.Rölla, S.Dana Hsu and W.H.Bowen, *Caries Res.*, 11, 308 (1977).

190. H.Maier, M.T.Coroneo, G.Antonczyk and A.Heidland, *Archs oral Biol.*, 24, 225 (1979).

191. J.Tenovuo and K.K.Mäkinen, *J.Dent.Res.*, 55, 661 (1976).

192. C.Bruun and H.Givskov, *Scand.J.Dent.Res.*, 87, 1 (1979).

193. G.J.Moody and J.D.R.Thomas, *Lab.Practice*, 28, 125 (1979).

194. S.Wotman, I.D.Mandel, J.Mercadante and C.R.Denning, *Archs oral Biol.*, 16, 663 (1971).

195. S.Kashket and V.M.Rodriguez, *J.Dent.Res.*, 54, 201 (1975).

196. A.Tatevossian, *Caries Res.*, 11, 216 (1977).

197. J.C.Tamacus, A.C.Ramsay and J.L.Hardwick, *Brit.div. I.A.D.R. Abstract No.179* (1974).

198. J.Gershon-Cohen and J.F.McClendon, *J.Albert Einstein Med.Center*, 5, 153 (1957).

199. B.Vickery and M.L.Vickery, *Analyst*, 101, 445 (1976).

200. G.J.Moody, B.Ong, K.Quinlan and J.D.R.Thomas, to be published.

201. R.S.Levine, *Brit.dent.J.*, 9, 140 (1976).

202. N.L.Craig and D.M.Hercules, *Analyt.Letters*, 8, 831 (1975).

203. D.M.Hercules and N.L.Craig, *J.Dent.Res.*, 55, 829 (1976).

204. R.Aasenden, P.F.de Paola and F.Brudevold, *Archs oral Biol.*, 17, 1705 (1972).

*Ion-Selective Electrode Rev.*, Vol.1, 207-250.
Pergamon Press Ltd. Printed in Great Britain.

# APPLICATION OF ION-SELECTIVE ELECTRODES IN WATER ANALYSIS

Adam Hulanicki and Marek Trojanowicz
Department of Chemistry, University of Warsaw, Warsaw, Poland

CONTENTS

KEYWORDS: Ion-selective electrodes for water analysis; gas sensors for water analysis; on-line electrode monitors for water analysis; water analysis by ion-selective electrodes and gas sensors.

## 1. INTRODUCTION

Many analytical techniques are used in water analysis, but just four techniques are normally used for inorganic species. These are atomic and molecular absorption spectroscopy, voltammetry and potentiometry. Atomic absorption spectroscopy deals with metal determination only, which is also the main area of application of voltammetry. Several voltammetric procedures have been developed for other species, including organic compounds, but in pollution and water quality studies the principal interest lies in the determination of trace amounts of cadmium, lead, zinc and copper by stripping voltammetry. The largest range of determined compounds is by direct molecular absorption spectroscopy, or indirectly through chemical reactions designed to produce absorbing species. However, supplementary reactions have the disadvantages of more complicated procedures, of not being easily automated, and of not having sufficiently low detection limits.

Potentiometric procedures are normally of intermediate detectability, usually estimated as $10^{-5}$ - $10^{-6}$ M. This is comparable to molecular absorption spectroscopy, but worse than atomic absorption spectroscopy and inverse voltammetry. There are, however, several features which make potentiometry very attractive as an analytical method in water analysis.

The response time of the electrode, depending on type of electrode, concentration range and geometry of the system, varies from seconds to several minutes, and this is consistent with the requirements of many analytical real time *in-situ* monitors. Only in extreme cases, usually at very low concentrations, is the response time of several tens of minutes.

An important characteristic of potentiometric measurements is the logarithmic relationship between potential and concentration (or more correctly, activity). This makes possible the use of ion-selective electrodes over several orders of magnitude of concentration range, usually with constant precision. In practical determinations precision is influenced by other non-instrumental factors such as the presence of interfering species and complexants, adsorption, chemical contamination, and so on.

Potentiometry is readily applied in laboratories for routine determinations, because its instrumentation is simplier than for many other techniques, and hence is relatively inexpensive. Also, potentiometric measurements can be used in field conditions with portable instruments and this is of considerable importance in environmental monitoring of water quality, especially when storage and transportation of samples may be sources of errors in analytical determinations.

A virtue of potentiometric measurements is connected with the non-disturbance of the chemical equilibria in which the sensed species participate. This enables the determination of a given form apart from the total amount of this substance. This speciation has been widely used in physicochemical studies of ionic equilibria[1], but it is also of importance in practical analytical measurements in aquatic media. For example, ion-selective electrodes permit the determination of free fluoride in the presence of aluminium, calcium or magnesium species[2], but they may be used also for evaluating the total fluoride after demasking the bound fluoride with citrate or DCTA. A similar procedure may be applied for the determination of total and free cyanide.

The property of potentiometric measurements of sensing just ionic species must be taken into account especially when metal ions have to be determined. Metal ions occur in natural water in various forms[3-7] : namely, as solvated ions, as dissolved inorganic complexes, mainly with hydroxyl and carbonate ions, as organic complexes with a variety of more-or-less defined ligands such as humic and fulvic acids[8,9], and as adsorbed or precipitated modes on organic or inorganic particulate matter. It is obvious therefore, that the absolute and relative degrees of complexation of metals depend significantly on the origin, and hence of the composition , of water, particularly its electrolyte and organic matter content.

Ion-selective electrodes may be used principally for determining free ions, but the ionic balance is often disturbed by the addition of reagents in order to achieve satisfactory measuring conditions. Thus, purposeful addition of some complexing agent with a stronger binding capacity than that of ligands present in natural waters provides a constant complexation medium, but when properly calibrated this enables determination of the total content of a given metal[10].

The response of potentiometric sensors depends primarily on the activity, not concentration, of analytes. This offers an unique possibility of measuring activities which are the direct driving factors of chemical and biochemical processes, and are responsible for the equilibrium of aquatic media[11]. It must, however, be realized that for the measurements of activities the following three conditions must be fulfilled:

(1) The sample which is investigated must be measured without any pretreatment; i.e.,it must not be diluted, concentrated or mixed with reagents or their solutions.
(2) The electrode, at least in the range of interest, must responde only to the ion whose activity is being determined. If such an electrode is not "ideally" specific it is useless in activity measurements because addition of masking agents for overcoming interference jeopardizes the first condition.
(3) The electrode response should be theoretical, or at least perfectly defined, on the basis of activity standards[12,13].

The need for obeying these three conditions is applicable to special cases, rather than as a general rule in water analysis. This is because in potentiometric measurements the analyst normally looks for concentration data, which are generally obtained by providing in samples and standards a constant, usually high, ionic strength level.

## 2. SCOPE OF ION-SELECTIVE ELECTRODES IN WATER ANALYSIS

The successful application of ion-selective electrodes to water analysis is without any doubt connected with the fact that the sample matrix creates a natural medium for potentiometric measurements. In general, the samples usually need only minor and simple modification or pretreatment in order to achieve the appropriate analytical conditions. However, it should not be forgotton that even water analysis involves different types of samples, which may differ considerably in their composition. Thus, concentrations of dissolved substances in atmospheric water (rain, snow, etc.) are quite different from those of sea water, in spite of the fact that both may be seriously polluted. Ground water, especially mineral water, may contain quite different species, and in different concentrations, from other types of water. Finally, some surface waters in extreme cases may be considered to be akin to effluents or waste water.

There are many criteria for defining the purity of water; these are usually established by various public services. They differ in grade depending on the intended use of water, e.g., for domestic, agricultural, industrial, recreation or other purposes. The defining parameters are physical, chemical and physiological in nature. In special cases, the water supplier or user may need special parameters of water quality, but in the majority of cases those listed in Table 1 are adequate for water control and analysis. In this Table, the italicized items represent those for which potentiometric methods may be employed. However, this needs qualifying, for some species may occur in various types of water at very different levels , which may or may not be within the range of ion-selective electrodes. Despite the commercial availability of electrodes for some species , the devices have never been used in water analysis because the permissible levels of the species concerned are much lower than the detection limits of the potentiometric sensors. This is the case for lead, zinc, cadmium and mercury.

For comparison of the real possibilities of electrodes for determination of low concentration levels with the range which may be of interest for an analyst , some data have been collected in Table 2. These data must be treated cautiously, because the detection limit may vary according to various conventions, working conditions, chemical background due to contamination, adsorption , redox and other processes, and most obviously from the type of the sensor, including its preparation and preconditioning. Also , the detection limit can be quite different in static and flow measurement conditions. All these factors can influence the presented data by one order of magnitude or more. Similarly, the permissible level data are just guidelines because they are taken from different standards and regulations. Nevertheless, their comparisons with detection limits throw some light on the strengths and weakness of potentiometry.

Additional data to the comparison are given in Table 3 for the most abundant elements occurring in sea water. However, the data concerning speciation may vary with total salinity, temperature, pressure and locality of sampling.

It was mentioned above that the sample in water analysis is in nearly final form for measurements. It must also be remembered that the procedures reviewed below may be often applied with success to other types of samples - namely, solid, liquid or gaseous - after their dissolution or absorption in aqueous media. This is especially important in relation to environmental studies, where the analysis of water samples by itself does not provide sufficient information about the circulation of pollutants or their sources. Thus, the methods of water analysis may relatively easily be extended to other media, such as soil extracts, biological samples or air particulates, after adequate sample preparation.

TABLE 1. Water Parameters Determinable by Chemical Analysis. (the italicized items can be evaluated by potentiometric measurements)

| | |
|---|---|
| General characteristics: | dissolved oxygen<br>biochemical oxygen demand (BOD)<br>chemical oxygen demand (COD)<br>*total carbon*<br>*hardness*<br>*alkalinity, acidity* and *pH*<br>salinity, electrical conductance |
| Natural water components and nutrients: | *calcium, magnesium*<br>*sodium, potassium*<br>aluminum, iron<br>*chloride, bromide, iodide*<br>nitrogen - *total, ammonia, nitrate, organic*<br>phosphorous - phosphate, polyphosphate, organic |
| Other and toxic components: | *boric acid*<br>*fluoride*<br>*sulphide*, sulphite<br>beryllium, zinc, cadmium, mercury, *copper*<br>chromium, manganese, molybdenum, silver<br>bismuth, lead, selenium, arsenic<br>detergents<br>pesticides, insecticides<br>grease and oil<br>phenols |

With regard to the specialized literature on water analysis, the previous reviews[14,23] are mainly devoted to measuring techniques and the specific characteristics of potentiometry, rather than to general descriptions of developments in this area. The recently published monograph, "Potentiometric Water Analysis"[24], deals in detail with general principles and techniques of potentiometry, and presents for each component very careful technical instructions of determination. This review is directed to the details of the general developments, and for convenience is taken mainly from the standpoint of the determination of individual species.

## 3. THE DETERMINATION OF INDIVIDUAL SPECIES

The various species are considered according to the periodic table, taken from left to right. However, measurements connected with hydrogen ion concentration - i.e. acidity, alkalinity and pH - are purposely omitted, because, although they are obviously of primary significance, they centre on an area requiring a special and different treatment.

TABLE 2. Detection Limits for Potentiometric Measurements with Ion-Selective Electrodes

| | Determination limit (mg $dm^{-3}$) | Permissible level for domestic use (mg $dm^{-3}$) |
|---|---|---|
| Ammonia | 0.02 | 0.5 |
| Boron ($BF_4^-$-electrode) | 0.03 | 1.0 |
| Bromide | 0.4 | 1.0* |
| Carbon dioxide | 4.0 | satd.* |
| Cadmium | 0.1 | 0.01 |
| Calcium | 0.4 | 106-500* |
| Chloride (based on Hg salts) | 0.01 | 250 |
| (based on Ag salts) | 1.0 | 250 |
| Chlorine | 0.05 | 1.0 |
| Copper | 0.003 | 1.0 |
| Cyanide | 0.02 | 0.05 |
| Fluoride | 0.02 | 1.0 |
| Iodide | 0.03 | 0.01 |
| Lead | 0.2 | 0.05 |
| Nitrate | 0.4 | 10 |
| Perchlorate | 0.2 | ? |
| Potassium | 0.4 | 10* |
| Silver | 0.01 | 0.05 |
| Sodium | 0.001 | 100* |
| Sulphide | 0.03 | 0.05 |
| Sulphur dioxide | 0.2 | ? |
| Thiocyanate | 0.3 | 1.0 |
| Water hardness | 0.005 | 0-200* |

*For species that are not harmful, the most commonly encountered range or limit is given.

TABLE 3. Chemical Speciation and Average Concentration of more Abundant Elements in Sea Water

| Element | Main species | Concentration (mg $dm^{-3}$) |
|---|---|---|
| Chlorine | $Cl^-$ | 19 000 |
| Sodium | $Na^+$ | 11 000 |
| Magnesium | $Mg^{2+}$ | 1 350 |
| Sulphur | $SO_4^{2-}$, $NaSO_4^-$, $MgSO_4$ | 900 |
| Calcium | $Ca^{2+}$ | 400 |
| Potassium | $K^+$ | 400 |
| Nitrogen | $NO_3^-$, $NO_2^-$, $NH_4^+$ | 150 |
| Strontium | $Sr^{2+}$ | 80 |
| Bromine | $Br^-$ | 67 |
| Carbon | $HCO_3^-$, $CO_3^{2-}$, $CO_2$ | 28 |
| Boron | $B(OH)_3$, $B(OH)_4^-$ | 4.5 |
| Silicon | $H_4SiO_4$ | 2.0 |
| Fluorine | $F^-$, $MgF^+$ | 1.3 |
| Phosphorus | $HPO_4^{2-}$, $PO_4^{3-}$, $H_2PO_4^-$ | 0.06 |
| Iodine | $IO_3^-$, $I^-$ | 0.06 |

## 3.1. Sodium

Sodium is a common component of most natural waters, ranging from less than 1 mg $dm^{-3}$, to 11 g $dm^{-3}$ which is the average sodium content in sea water. The most serious demand for sodium determination is connected with power station water, where sodium content needs to be below 10 μg $dm^{-3}$. The continuous control of sodium level is necessary at this low level because of the danger of corrosion and formation of sediments.

The greatest problem in the determination of low contents of sodium are connected with the possibility of contamination from glassware, distilled water and reagents. Therefore, the preparation of water containing below 2 μg $dm^{-3}$ sodium is difficult, and reagents used for pH adjustment should also be of special purity. Fewer difficulties may arise in flow systems, as long as they are constructed from plastics or stainless steel. Eckfeldt and Proctor[25] found that for higher flow rates (100 $cm^3$ $min^{-1}$), sodium is not washed out of glass, as happens for smaller flow rates when the sodium concentration in the flowing water is smaller than 1 μg $dm^{-3}$.

The common electrode used for sodium determination is the glass electrode manufactured by about a dozen makers. The selectivity coefficient $k^{pot}_{Na,K}$ is usually smaller than $10^{-2}$, while $k^{pot}_{Na,NH_4}$ is smaller than $10^{-3}$. The neutral carrier electrode (Phillips IS 561 Na) has a somewhat worse selectivity towards potassium ($5 \times 10^{-1}$) and ammonium ($2 \times 10^{-1}$), which is restrictive in some applications.

Relatively less attention has been paid to stationary or static measurements. Reynolds[26], determining low sodium levels (0.07 - 1.4 mg $dm^{-3}$) in glacier water, found the precision equal to 3 %. The samples were made alkaline with triethanolamine. Tanzen and Solovieva[27] used stationary measurements for sodium determination in wastes (>40 mg $dm^{-3}$), and found an error of up to ± 30 % when the excesses of interfering ions were not greater than 2 for potassium, 1 for iron(II), and 100 for calcium. Rather large amounts of water softeners were not deleterious to the determination, and the acidic samples were even neutralized with solid calcium oxides.

An automatic system[28] permits 20 measurements per hour for water samples (1.4 - 50 mg $dm^{-3}$ sodium) or wastes (∿200 mg $dm^{-3}$ sodium). The samples were mixed (4:1) with 0.04 M ammonium chloride containing 1 % triethanolamine. There was good agreement with flame photometry, where the latter determination is more rapid, but more difficult for continuous work, and more expensive.

Several workers have found the significant influence of sodium contamination in flow systems. A relatively high (1 mg $dm^{-3}$) limit of determination was found by Van der Winkel *et al.*[29], who used an EDTA-ammonia buffer for proper pH adjustment and elimination of precipitation of alkaline earth metal salts. High electrolyte concentration was found advantageous in order to eliminate meter oscillations. Other authors[30,31] determined low concentrations (0.25 μg $dm^{-3}$ - 25 mg $dm^{-3}$) by adjusting the samples to pH 11 with gaseous air-ammonia mixture. In flow systems with an E.I.L. sodium electrode, disturbances from potassium originating from the reference electrode may be minimized by proper construction of the measuring unit, but below 50 μg $dm^{-3}$ it took at least 30 min to obtain steady potential readings, which is inferior compared to flame photometry, but the electrode results were considered more precise.

Several amines other than triethanolamine have been tested as alkalizing reagents. Octadecylamine was found inconvenient because of possible adsorption at the electrode surface, but morpholine was suitable[32].

Eckfeldt and Proctor[33], who used a Leeds and Northrup electrode, found that at low sodium levels ammonia may cause erroneous results, and used dimethylamine or di-isopropylamine instead. A small increase of pressure of a inert gas in the reference electrode was advantageous in order to decrease the liquid junction potential. At 1 μg $dm^{-3}$ sodium it seems useful to carry out the measurements at lower temperatures, when the ion-product of water and hence the effect of hydrogen ions is decreased[34].

### 3.2. Potassium

The potassium content in natural waters seldom exceeds 20 mg $dm^{-3}$, except for sea water. The calibration curve is linear down to ∿ 1 mg $dm^{-3}$ for the neutral carrier sensor electrode based on valinomycin, and somewhat further for the glass electrode. The former needs more care when used, and its lifetime is shorter, being commonly less than 1 yr, compared with up to a few years for the glass electrode, when properly maintained. The main advantage of the neutral electrode is its significantly better selectivity for potassium over sodium ($k^{pot}_{K,Na}$, $10^{-3}$ - $10^{-5}$), which for glass electrodes is never expressed by selectivity coefficients smaller than $k^{pot}_{K,Na} = 5\times10^{-2}$ (Beckman 39137). This is a most important aspect of the use of potassium electrodes, because in most samples the sodium content exceeds that of potassium.

The contamination by potassium from non-sample sources is less severe than for sodium, but nevertheless direct use of calomel electrodes or salt bridges containing potassium salts should be avoided. Because of relatively small differences in sensitivity for potassium and ammonium, the presence of the latter should be avoided, and hence the aliphatic amines are used for pH adjustment. When working with the neutral carrier electrode , however, the pH can be brought to above 10 with sodium hydroxide, although it must not be added in excessive amounts.

Reynolds[26] used the neutral carrier electrode for potassium determination in surface water without any chemical treatment in the range 0.12 - 4 mg $dm^{-3}$. The precision was quoted as ±3 %, but parameters such as mixing rate and cell geometry influence the results and selectivity deteriorates with electrode lifetime. 1.2 - 7.3 mg $dm^{-3}$ of potassium has been determined[28] in an automatic system after addition of 0.04 M sodium chloride at pH 12. Agreement to better them 0.1 mg $dm^{-3}$ was obtained with flame photometry, in determinations at 20 samples per hour. For potassium in wastes in the range 50 - 75 mg $dm^{-3}$ the differences ranged up to 0.8 mg $dm^{-3}$.

A PVC membrane electrode containing valinomycin has been used for the determination of up to 400 mg $dm^{-3}$ of potassium in sea water[35]. The potassium standards were prepared in a background of 0.47 M sodium chloride. Anfält and Jagner[36] have analysed sea water by the standard addition procedure, with Gran's linearization. These authors also found that the selectivity of the electrode deteriorates in the course of its work.

### 3.3. Magnesium

Magnesium is rarely determined as such in waters. Instead, its content is found as a difference between total calcium and magnesium content (corresponding to water hardness), and calcium content, determined separately. The liquid ion-exchanger electrode, called divalent metal electrode, is commonly used in the calcium plus magnesium measurements. Calcium is selectively complexed with EGTA and magnesium may then be determined in the sea water[37]. The restricted selectivity of the electrode needed careful choice of the neutralizing agent if the samples were acidic. For this purpose , 0.5 M diethanolamine was used, and led to a linear calibration curve for magnesium in the range 0.2 mg $dm^{-3}$ - 2 g $dm^{-3}$. In the presence of EGTA, the harmful effect of polyvalent metals is fully suppressed[38]. The potentiometric results were found to be in good agreement with those of atomic absorption spectroscopy and compleximetric titration.

### 3.4. Calcium

Calcium is present in all natural waters, but only a fraction remains uncomplexed and free as $Ca^{2+}$ aquo-ions. Thompson and Ross[39] showed that ∿ 85 % of calcium in sea water occurs as the aquo-ions, the remainder being in the form of sulphate or carbonate complexes. Qualitatively similar situations occur in other types of water. It is quoted that when the sulphate concentration exceeds 5 x $10^{-4}$ M (∿ 50 mg $dm^{-3}$), there is an error of not less than 10 %, as also occurs for more than 300 mg $dm^{-3}$ of hydrogen carbonate[40]. In laboratory practice it must be always taken into account that the carbon dioxide from the air may seriously disturb the results of direct determination.

Frequent occurrence of magnesium and sodium in waters requires relatively good selectivity for calcium over these cations. All practical calcium electrodes are of the liquid membrane or of the plastics (PVC) membrane type[41-43]. They may contain a liquid ion-exchanger or a neutral carrier. The former are usually based on salts of organophosphoric acids; initially they contained calcium di(n-decyl) phosphate[44], but the somewhat larger radical in di(n-octylphenyl)phosphate, used by Růžička *et al.*[45] in the Selectrode or in PVC membranes[46-48] improves the selectivity. Also, the neutral carrier electrode[49] commercially available from Philips, and now form Orion Research Inc., shows a response not much disturbed by magnesium and sodium (Table 4).

TABLE 4. Selectivity Coefficients of Calcium Ion-Selective Electrodes

| | Ross[44] | Růžička *et al.*[45] | Ammann *et al.*[49] |
|---|---|---|---|
| $k^{pot}_{Ca,Mg}$ | $(1.4 \div 5) \times 10^{-2}$ | $2.5 \times 10^{-4}$ | $4 \times 10^{-5}$ |
| $k^{pot}_{Ca,Na}$ | $1 \times 10^{-4}$ | $6 \times 10^{-6}$ | $3 \times 10^{-4}$ |
| $k^{pot}_{Ca,K}$ | $7 \times 10^{-5}$ | $2 \times 10^{-6}$ | $2 \times 10^{-4}$ |

In the pH range 7 - 11, the Orion calcium electrode response is pH-independent. The improved phosphate exchanger extends this range slightly in the acid direction, and the neutral carrier electrode is not influenced by pH, not even at pH 3.

For direct measurements of calcium in water, all the above-mentioned interferences should be eliminated. For measurements in tap water (70 - 175 mg $dm^{-3}$) a Constant Complexation Buffer has been developed[50]. This contains 0.4 M potassium nitrate, 0.02 M sodium iminodiacetate, 0.04 M acetylacetone, and 0.02 M ammonia-ammonium chloride buffer. The buffer, when mixed in 1:1 volume ratio with water samples, provides constant complexation of calcium by iminodiacetate , which overshadows the influence of weaker complexing agents present in water. Acetylacetone masks magnesium, but is necessary only when the Orion model 92 - 20 calcium electrode is used. With the newer electrodes, mentioned above, the concentration of magnesium usually occurring in water does not interfere. Such conditions enable measurements with an error less than 4 % in comparison to compleximetric titrations. The same buffer has been used in flow conditions[51], and the results for 30 - 100 mg $dm^{-3}$ of calcium were in good agreement with those obtained in static measurements.

Another solution was used by Hansen, Růžička and Ghose[52] for determining calcium in water and waste water in the 70 - 144 mg $dm^{-3}$ range; namely, one made up of $10^{-3}$ M borax, $10^{-3}$ M sodium hydroxide and 0.14 M sodium chloride. However, some negative errors occurred, and were attributed to the presence of carbonates. In measurements performed in a Mobile Laboratory using a 2.9 M ammonia-ammonium chloride buffer. Thurnau[53] pointed out the necessity of a proper control of ionic strength.

A somewhat longer but more accurate procedure for calcium determination still used is titration, with a polyaminopolycarboxylic acid as a ligand, and an ion-selective electrode as an indicator. Whitfield *et al.*[34] titrated up to 0.4 % calcium in sea water, using a tetra-ammonium salt of EGTA. The end point was determined using the Gran method, with a precision of about 0.5 %. By using the Gran and Fortuin methods 80 mg $dm^{-3}$ of calcium was found in a tap water sample[58] with the tetramethylammonium salt of EGTA as buffer. The result was within 1 % of those obtained by other titration techniques. The calcium PVC electrode has also been used by Jagner and Østergaard-Jensen[56] for the titrimetric determination of calcium in sea water. For 30 samples, the relative standard deviation was ∿ 0.8 %, which is superior to photometric titrations, but the determination time was longer.

In the above titrations, the presence of interferents, mainly sodium, may seriously diminish the potential break at the end point in comparison to that which can be calculated from compleximetric titrations. To avoid this, other electrodes have been used in the presence of small amounts of indicator ions, in order to trace the course of the titration potentiometrically. In this way, Mascini[57] used a copper-sensitive electrode for calcium titration in borate buffer, in the presence of a small amount of the copper complex with EDTA. Similarly, a cadmium electrode has been used for titration in a 0.1 M ammonium buffer[58] containing 2 x $10^{-4}$ M CdEDTA. It is interesting that such procedures enable determinations of low concentrations of calcium, i.e. down to 2 mg $dm^{-3}$. When the presence of magnesium is expected or suspected, EDTA should be replaced by EGTA , for this reacts more selectively with calcium.

Proper regard must be paid to the choice of masking agents, because of their effect on the conditional stability constants of calcium and magnesium complexes. Christiansen, Busch and Krogh[59] have mentioned that with EDTA titrant, and using a calcium-sensitive Selectrode, calcium may be determined in the presence of magnesium. When the solution contains 3.4-dihydroxybenzoic acid or acetylacetone, two separate potential breaks are observed, although the magnesium content is not larger than 4 % of that of calcium. The reported relative standard deviation of calcium in drinking water was better than 0.5 %.

Calcium may be also determined by gradient titration with EGTA as a titrant[51]. Precision of such procedure is worse than for other titrations, but alkali metals which give rise to errors in stationary measurements based on a calibration curve do not interfere.

### 3.5. Water hardness

Water hardness is conventionally determined as the calcium plus magnesium content. It is obvious that methods used for determination of individual concentrations of the metals may be applied for the evaluation of the separate contribution of calcium and magnesium. The most commonly used EDTA titrations with visual indicators give the total content of both metals by accepting a reasoned assumption that other EDTA-reacting ions are present in only negligible amounts. In such titrations, the visual indicator may be replaced by a potentiometric end point, using a calcium ion selective electrode[54,55] or a copper ion-selective electrode[57], which acts through the equilibrium between copper and calcium (or magnesium) chelates. EDTA is the complexant that is most widely used as a titrant[57], but may be replaced by DCTA[54,58]. As in the case of determinations of calcium with a liquid membrane electrode, a large excess of sodium may either decrease the end point break or even cause negative errors. Therefore, replacement of sodium ions by ammonium or protonated amine ions in added reagents can be advantageous.

Direct potentiometric measurements of water hardness are possible with the application of the so-called divalent electrode[26,60]. Like the calcium electrode, it is based on organophosphoric acid, but depends on a different solvent (e.g. decanol exhibits equal sensitivity towards calcium and magnesium). In practice, the value of $k^{pot}_{Ca,Mg}$ does however, deviate from unity, either at different levels of both metals, or after prolonged use of the electrode. This can be the source of discrepancies between direct measurements and titrimetry.

Relatively good accuracy (±3 %) was mentioned by Reynolds[26] for very soft waters (1-7 mg $dm^{-3}$ calcium), following measurements in solutions containing added sodium hydrogen carbonate. Thus, it was possible to keep pH constant and add the complexing species in constant and controlled amounts. Olivier and Mannion[62], who analysed samples containing up to 30 mg $dm^{-3}$ Ca, used the standard addition technique, or relied on a calibration curve when they used a flow technique. Positive errors were observed in the latter measurements when the flow rate was too large, for the exchanger was washed out of the membrane phase. The control of hardness in demineralized water (1-16 mg $dm^{-3}$ calcium) has indicated that good analytical results were obtained when the pH was above 6, and the sodium content was controlled at a constant level.

In general, the effect of anions in complexing calcium or magnesium has not been investigated in detail. However, Sekerka and Lechner[62] indicate that the presence of carbonate makes direct measurements impossible, and therefore used the standard addition and dilution technique in an automatic system. After the first measurement with the sample, 1 $cm^3$ of calcium solution was added, measured for the second time, diluted in a 1:1 ratio, and once more measured. Apart from data processing, the whole system is automatically controlled. A precision of better than ±2 % was quoted for the calcium carbonate range from 2 to 100 mg $dm^{-3}$, and the results compared with atomic absorption spectroscopy are accurate to ±5 %. The emf. in the first measurement makes possible the determination of the non-carbonate hardness[63].

Extensive studies on the performance of the hardness electrode were made by Thurnau[53], by testing it in a Mobile Water Quality Laboratory. For accurate results, the ionic strength should be controlled carefully. Because for more dilute solutions the hardness electrode response increases, the constant ionic background was kept at 2.1 M.

It must, however, be mentioned that despite the theoretically favourable conditions for direct measurements of water hardness, the divalent (calcium plus magnesium) ion-selective electrode is not commonly used in routine determinations.

### 3.6. Copper

At the earlier stage of electrode development, a liquid-state copper ion-selective electrode was used, but at present only solid-state electrodes are of importance. The membranes consist of mixed crystals of copper(II) sulphide and silver sulphide, or single crystals of copper(I) sulphide or selenide. Both types have been used in measurements of copper(II) concentrations, but some differences in their linear range, sensitivity to interferences, and surface passivation have been observed. It is not obvious how far these differences can be ascribed to differences of chemical composition or to structural modifications of chalcogenides prepared in various conditions[64].

The most serious interferences for the copper electrode when applied to natural water are iron(III) and chloride. The former causes either a membrane oxidation reaction, or an exchange reaction based on sulphides. The latter, in the silver-salt-containing electrodes, may be explained[65,66] by formation of silver chloride (or chloride complexes), according to the reaction:

$$Ag_2S + 2\bar{n}Cl^- + Cu^{2+} \longrightarrow CuS + 2AgCl_{\bar{n}}.$$

In measurements of copper concentration in natural water, it must be taken into account that besides free (hydrated) copper ions, other soluble complexed copper species are present in varying proportions. Also, they remain in equilibrium with particulate organic and inorganic matter, which also contain chemically or physically bound copper. The ion-selective electrodes are well suited for evaluation of equilibrium with these species and of the pH-dependent reactions which, with carbonate, may lead to copper removal from the solution phase[4,7], or to formation of humic and fulvic acids complexes[9].

The relatively complex systems and the rather low level of copper in natural water, in spite of favourable characteristics, make the practical determination of copper difficult. Therefore, extreme precautions are advisible, particularly concerning the purity of reagents and solutions used, especially in view of the reports[65,67,68] anomalous electrode response following exposure to chloride in the presence of copper(II).

Smith and Manahan[69] used the Orion model 94-29A copper electrode for determining 3.3 - 47 mg $dm^{-3}$ copper in natural and tap water. The proper medium for such determinations was provided by adding Complexing Antioxidant Buffer (CAB) in 1:1 ratio to the sample. The CAB solution contained 0.05 M acetate buffer, 0.02 M sodium fluoride, and 0.002 M formaldehyde. It complexes iron(III), makes the medium reducing, and the degree of complexation of copper was maintained constant by the acetate ions. Multiple standard addition was used for quantitature, but errors arose because of the slow establishment of steady potential (up to 20 min at 30 μg $dm^{-3}$ copper), and of the effect of light on the electrode membrane. Despite the difficulties, the relative standard deviation was 0.075, and the recovery for 9 μg $dm^{-3}$ was 103 %.

The CAB solution was not recommended for sea-water analysis because of coprecipitation of copper on alkaline earth fluorides[70]. Determination of total copper content is possible when the electrode is calibrated with sea-water standards, and in the absence of organic chelating agents. Such a procedure was successful in the 1 - 20 μg $dm^{-3}$ range, whereas for lower copper concentrations the electrode membrane itself serves as a source of copper ions. It was also found that the presence of chelating agents in coastal water influences the results.

An electrode based on copper(I) sulphide has been applied to the analysis of tap water[71]. The procedure with CAB solution was found unsuitable for low concentrations. This may be attributed to presence of uncomplexed copper(I) ions at the electrode surface. Successful determinations were obtained in solutions containing 0.02 M Tris, 0.1 M potassium nitrate, and 0.02 M potassium fluoride. To increase the rate of iron and manganese complexation, the samples were heated to boiling before making measurements. No detrimental effect of chloride ions was found, and the copper content, which could be semiquantitatively evaluated, was equal to 6 μg $dm^{-3}$, but above 20 μg $dm^{-3}$ the errors did not exceed 25 %.

## 3.7. Boron

The boron content of natural water is usually a few mg $dm^{-3}$ and originates from mineral deposits as well as from industrial and municipal pollution. Potentiometric determination of the boron present in water as boric acid may be carried out after its transformation into the tetrafluoroborate ion, which is sensed by the liquid ion-exchanger electrode of the nitrate type but with the exchanger in the tetrafluoroborate form.

The procedure was proposed by Carlson and Paul[72]. Because of interferences by nitrate and iodide the sample of tap water was first passed through an Amberlite XE-243 ion-exchange column to retain boric acid, which was then transformed into tetrafluoroborate and eluted with sodium hydroxide solution. The method gave results comparable to the curcumin procedure at the 0.66 mg $dm^{-3}$ level and the analytically useful range extends from 0.10 to 20 mg $dm^{-3}$. It should be mentioned that at low concentration levels, the long reaction time of tetrafluoroborate formation limits the scope of the procedure.

## 3.8. Carbon

Determination of carbon content in water, according to chemical pretreatment, can give information concerning total organic carbon and total inorganic carbon. In turn, knowledge of the inorganic carbon content in combination with pH measurements gives data on the distribution of carbonate species in water, but some carbonate, especially in sea water, is also complexed with magnesium, sodium or calcium.

The electrode used in these determinations is the carbon dioxide gas sensor, which has a linear response down to 0.05 mg $dm^{-3}$ carbon dioxide, but which may be inadequate for power feed-water of low carbon dioxide content[73].

Fiedler, Hansen and Růžička[74] used the air-gap electrode in the analysis of natural waters and sewages. The total inorganic carbon content was determined after acidification, the standards being prepared from sodium hydrogen carbonate solutions. Stable potential readings were obtained after a few minutes; however, at low concentrations the time may be much longer. The total carbon content was determined after preliminary oxidation of the sample with alkaline potassium persulphate. When the carbon content was smaller than 2 mM, preconcentration as basic lead-carbonate was found helpful by giving about 100-fold increase in the signal. The quoted relative standard deviation was below 0.03. A similar procedure was used for sea[75] and fresh water[76] containing 75 - 300 mg $dm^{-3}$ of carbonate. In these procedures, sodium hydrogen sulphate was added as an acidifying agent, but dilute sulphuric acid may also be used. To prevent the interferences of hydrogen sulphide, addition of copper(II) sulphate seems to be advantageous.

In general, it must be concluded that the potentiometric carbon dioxide determination is more rapid and simpler than other instrumental methods, especially when these need preliminary separation of carbon dioxide in a separate step.

### 3.9. Cyanide, thiocyanate and cyanate

The determination of cyanide in water is of special importance because of its toxicity. Cyanide usually originates from leakage of industrial waste waters, and is often accompanied by metal ions which form complexes of various degrees of stability and inertness. Direct measurements give information on the amount of free cyanide, which is usually low, and subject to variations due to oxidation and hydrolysis of cyanide, and also due to volatility of hydrogen cyanide when pH is sufficiently below 9. Therefore, the samples must be preserved and measured in an alkaline medium, usually 0.1 M sodium hydroxide.

The solid-state electrodes used for cyanide determination are either iodide-sensitive or silver-sulphide-sensitive electrodes. The action of the former is based on the equilibrium

$$AgI + 2CN^- \rightleftharpoons Ag(CN)_2^- + I^-$$

which causes slow dissolution of the membrane material, and therefore limits the electrode lifetime, especially when the cyanide concentration exceeds $10^{-3}$ M (26 mg $dm^{-3}$)[77,78]. Such levels are practically never found in ordinary waters, but waste waters of high cyanide content should be diluted prior to the measurements.

The second type of electrode used for cyanide monitoring, namely the silver electrode, senses the change of dissociation equilibrium of silver cyanide complex[78,79]. According to reaction stoichiometry, the slope of electrode response is doubled, and may approach 118 mV per decade. This procedure seems to be more suitable for determining a few µg $dm^{-3}$ of cyanide, compared to the former electrode, for which a level of 10 times greater has been mentioned as detection limit[80].

The main interference for both electrodes arises from anions which form insoluble silver compounds of high stability, such as sulphide, iodide, and to a smaller extent bromide and thiocyanate[78]. Concentrations of phosphate, carbonate, chloride and ammonia which are found in natural water are sufficiently low, and do not interfere[79]. The presence of sulphide may be masked by the addition of lead[79] or bismuth nitrates[81]. For the determination of total cyanide, 0.02 M $dm^{-3}$ EDTA may be added as the demasking agent[79], acidified with acetic acid to pH 4, heated for

5 min to 50° C, and made alkaline. The procedure has been found to be successful in the presence of cadmium, zinc, nickel, copper and chromium. Alternatively, hydrogen cyanide may be distilled-off from the sulphuric acid medium to the alkaline absorbing solution[77,80].

In a continuous determination, free cyanide released with 2 M sulphuric acid has been measured after gas dialysis through a polytetrafluoroethylene membrane in a Technicon dialyzer assembly[82] (Fig. 1).

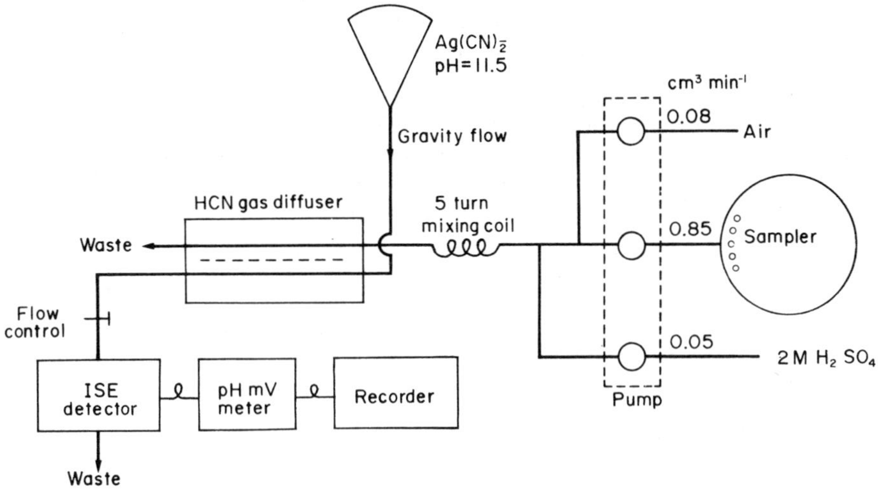


Fig. 1. Flow diagram of cyanide analyser[82]. (By courtesy of Marcel Dekker, Inc.)

Weiss[77] determined cyanide in natural water in the range 0.1 - 5 mg $dm^{-3}$, using the calibration curve procedure. He carried out a differential measurement in the presence of iodide, with and without acidifying the sample.

In a continuous cyanide monitoring with a silver electrode as a sensor, and a sodium glass electrode as reference[78,79], the calibration curve was linear from 0.2 to 250 mg $dm^{-3}$. The curvature at the upper end occurs by the formation of $Ag(CN)_3^{2-}$ complex.

The cyanide electrode has been used by Nota[83] for thiocyanate (0.6 - 6 mg $dm^{-3}$) determination in waters. For this, the sample was oxidized with bromine water in a phosphoric acid medium, the excess of oxidant being bound with an aqueous solution of phenol. The bromine cyanide was decomposed with sulphur dioxide for measurement of the cyanide. The metal ions present in the solution should be removed by ion exchange before the determination. Cyanides are measured simultaneously with thiocyanate, so they should be determined initially, and their amount substracted from the sum $CN^-$ + $SCN^-$, or they should be removed by boiling.

The possibility of cyanate determination in waters[84] depends on the decomposition of the anion by boiling for 30 min with 50 % sulphuric acid, and subsequent measurement of the formed ammonia with an ammonia gas sensor.

3.10. Ammonia

The ammonia and ammonium ion content of waters change significantly as a consequence of contamination and decomposition processes of organic matter. In rivers, the ammonia concentration may be as low as 0.01 mg $dm^{-3}$, or even as high as 2 - 3 mg $dm^{-3}$. Its content in sewage may certainly be higher depending on the type of contamination and treatment processes. However, determination of concentration up to several tens of mg $dm^{-3}$ presents no difficulties for the potentiometric ammonia sensors.

In general, determination of ammonia (or ammonium ions) may be carried out with four types of electrodes: namely, the gas-sensing probe with a hydrophobic membrane, the gas-sensing probe without a membrane (i.e. the air-gap electrode), the ammonium ion-sensitive glass electrode, and the ammonium ion-sensitive neutral carrier electrode. There is no univocal preference for any of the four types, but in particular cases conditions may exist for a more advantageous use of a given electrode.

Precision and accuracy of ammonia determination in waters depend largely on sampling and storage of analytical samples. At elevated temperature (i.e. above 50° C) there is a danger of ammonia losses; hence, the samples should be kept at low temperature and ought to be acidified. It has been reported[85] that adsorption of ammonia on the electrode body may cause low results.

By using extreme precautions the determination of ammonia can be carried out down to 2 μg $dm^{-3}$. On the other hand, at very low ammonia levels there is a danger of contamination from reagents, water, or even the laboratory atmosphere. Distilled and de-ionized water may contain 2 mg $dm^{-3}$,[86] or even as much as 10 μg $dm^{-3}$ [86,87] of ammonia nitrogen. Measurements with gas-sensing probes are carried out in alkaline solutions to ensure completeness of ammonia formation. In order to prevent precipitation of insoluble species in hard waters, as well as complexation of ammonia by transition-metal ions, 0.01 M EDTA was added[86-88] to samples and standards. Formation of precipitates is especially harmful when the surface of the membrane becomes blocked. In such a case, or when significant amounts of surface active agents are present, the air-gap electrodes seem to be most useful.

The gas-sensors[89] generally exhibit a very good selectivity. Serious interferences have been observed only in the presence of volatile amines, such as methylamine, ethylamine[90] or methanolamine[87]. Less volatile bases such as hydrazine, cyclohexylamine or morpholine, have a smaller effect on ammonia determination[91]. Chlorine is also harmful because of the formation of volatile chloramines[86], but this effect may be eliminated by the addition of sodium sulphide or thiosulphate. The alkali-metal ions are practically without adverse effect, but on the other hand they strongly influence the determinations with glass electrodes, and to a smaller extent with neutral carrier electrodes. The selectivity coefficient, $k^{pot}_{NH_4,Na}$, for glass electrodes is of the order of a few tenths, while for the neutral carrier electrode it is of the order of $10^{-3}$.

The gas diffusion processes which are responsible for the action of gas-sensors are strongly affected by temperature changes, hence the measurements ought to be performed in a thermostatted cell. The temperature coefficient of the E.I.L. ammonia electrode is ∿1.5 mV per degree between 16 and 29° C[91]. This is a source of both systematic and accidental error when the whole measuring system is not properly thermostatted.

The temperature effects are all the more important, because the response times of the gas sensor are usually long. Before measurements, washing of the electrode to the "blank level" takes a long time for a stable potential reading[86-88]. For concentrations below 100 μg $dm^{-3}$, the steady potential is reached in 10 - 15 min, but happily the increase of concentration reduces the waiting time. At extremely low levels (i.e. at ∿1.5 μg $dm^{-3}$ [92] the readings should be taken after 1 hr, because of the slow diffusion of ammonia out of the inner part of the sensor. Therefore, it is useful at such low levels to concentrate ammonia by distillation, especially since blockage of the membrane surface by precipitates and adsorbed species tends to extend the measurements to even longer times than those mentioned.

Banwart, Tabatabai and Brenner[90], when determining 1.6 - 3.4 mg $dm^{-3}$ ammonia in natural waters, found the standard deviation to be 0.008 - 0.017 mg $dm^{-3}$, but other workers[86,87] quote a relative standard deviation of 1 - 4 % for their data. Petrochemical waste waters[93] with high ammonia levels (i.e. up to 108 mg $dm^{-3}$), gave standard deviations of the same order of magnitude (2 mg $dm^{-3}$).

Potentiometric measurements in natural water and waste water with gas sensors[94] in the range 0.1 - 0.4 mg $dm^{-3}$ are in accord with the indophenol method, while Gilbert and Clay[92] in their analysis of sea water (0.1 - 4.7 mg $dm^{-3}$) described the reproducibility to be equal to 7 %. Vandevenne and Oudewater[95] also found good precision for ammonia determinations in waters in the 0.2 - 9.6 mg $dm^{-3}$ range, and quote satisfactory agreement with spectrophotometric measurement preceded by distillation.

The ammonia gas sensors are produced by Orion Research Inc. in the U.S.A. or E.I.L. in Britain. A similar construction has been described by Hawker, Midgley and Torrance[96] for the determination of ammonia (0.1 - 1.0 mg $dm^{-3}$) in boiler feed-water and condensed vapour. The air-gap electrode designed by Růžička *et al.*[97] was applied to the analysis of waste-water sample, containing up to 13 mg $dm^{-3}$ ammonia. At that level, stable readings were obtained after 2 - 3 min, and the relative standard deviation was 0.7 - 1.6 %.

The ammonium ion-selective glass electrode has been used by Barica[98,99], to the detection limit of 0.1 mg $dm^{-3}$. At concentrations below 40 mg $dm^{-3}$, the mean error was nearly 5 %, which seems to be comparable with other procedures. The possibility of alkali metal interferences is reduced when the electrolyte content is low, as for boiler feed water. The ammonium ion glass electrode may be especially advantageous under such conditions and Goodfellow and Webber[100] determined 0.01 - 1 mg $dm^{-3}$ ammonia after buffering the sample with triethanolamine. Nevertheless, sodium interferences may still occur, because 100 μg $dm^{-3}$ sodium corresponds to 25 μg $dm^{-3}$ ammonium ion. However, when appropriate corrections were taken into account, the results did not differ by more than ±5 % from those obtained with the indophenol method.

The PVC membrane electrode with a neutral carrier sensor (Philips IS 550-$NH_4$) has been used by Mertens, Van der Winkel and Massart[101] in continuous ammonia determinations in boiler feed-water, in the range of 0.1 - 2 mg $dm^{-3}$. In these measurements, the calibration curve was linear down to 0.4 mg $dm^{-3}$, followed by a curvature attributed to the ammonia impurities in triethanolamine used for buffering. The constant ionic strength was provided by calcium chloride. For feed-water samples, the results of potentiometric and colorimetric determinations did not differ by more than 0.03 mg $dm^{-3}$.

With regard to automatic and continuous determinations, good results have been obtained with electrodes having a hydrophobic membrane. In this respect, Sekerka and Lechner[28] have designed an automatic, computer-controlled system for simultaneous monitoring of sodium, potassium and ammonium in natural waters and sewage. The lowest determined concentration of ammonium ions was 0.1 mg $dm^{-3}$, for which the relative standard deviation did not exceed 10 %. The system enabled the analysis of 20 samples per hour for the three components. In the systems developed for the analysis of high-purity feed-water[102,108] the samples were mixed with one tenth of volume of 1 M sodium hydroxide. When the time interval between calibration and analysis of samples was not greater than 12 h, the relative deviation was below below 10 %.

The monitoring station for ammonia and ammonium ion determination described by

Leschber and Schumann[104] was provided with an automatic calibration system. The simultaneous measurement of pH and ammonium ion concentration makes possible the evaluation of the acid-base equilbrium for ammonia.

### 3.11. Nitrate

Nitrate content is an important parameter of water quality and may originate from industrial and municipal sewage as well as in significant degree in the natural waters from intensively cultivated fields. The nitrate level in water for domestic use is at most a few mg $dm^{-3}$, but quite often it may occur in concentrations of up to 50 mg $dm^{-3}$*. Therefore, much effort has been devoted to the construction and application of the nitrate electrode.

All nitrate electrodes are of the liquid-exchanger type. They are based either on the tris (substituted 1.10-phenathroline) nickel(II) complex or on the tetralkyl-ammonium ion, but often differ in construction. The Orion 92 or 93 series electrodes consist of a hydrophobic membrane saturated with the chelate exchanger, being in contact with an internal reference solution. Other commercial electrodes have the PVC membrane with the tetra-alkylammonium exchanger, and constructions have also been described with a solid internal contact directly to the exchanger[105,106]. The selectivity of both systems is similar. Among the more likely interfering ions are hydrogen carbonate and chloride, for which the selectivity coefficients are of the order of $10^{-2}$. Higher interferences would be expected from nitrate and bromide, but their concentrations are usually so small that they do not disturb nitrate measurements.

The nitrate concentration is influenced by redox processes,[107,108] which on the one hand oxidize other nitrogeneous compounds to nitrate, and on the other hand, especially in anaerobic conditions, decrease the nitrate level by their reduction to nitrite. The nitrite content is usually well below 1 mg $dm^{-3}$, but in doubtful cases its presence should be controlled.

In one of the early papers on the application of the nitrate electrode, Abbot[109] said that the electrode cannot be used below 6 mg $dm^{-3}$ $NO_3^-$ in high salinity waters, but that for the range 6 - 120 mg $dm^{-3}$ nitrate, 0.25 % salinity does not interfere. Sommerfeldt, Milne and Kozut[107] have determined nitrate in surface water without any buffering solutions, and their results were in good agreement with the phenoldisulphonic acid procedure, when the sale content was surprisingly low. On the other hand, Keeney, Byrnes and Jensen[110] found that in similar conditions the results were low by 50 % for 20 mg $dm^{-3}$, but the error was smaller for higher nitrate concentrations. It was stated that acidification with phosphoric acid or precipitation of sulphate and carbonate with barium hydroxide did not influence the results; nevertheless the procedure was considered to be satisfactory for more than 10 mg $dm^{-3}$ nitrate.

The effect of interfering ions in waters and sewage has been investigated in detail by Langmuir and Jacobson[111]. On the basis of estimated selectivity coefficients for chloride and hydrogen carbonate, and the known content of these ions, they proposed corrections for nitrate determination. Results, which agreed within ±1 mg $dm^{-3}$ of those of the brucine method were obtained in the 1.5 - 120 mg $dm^{-3}$ range, but above 50 mg $dm^{-3}$ it was not necessary to take into account the effect of the interfering ions. In spite of these satisfactory results, the correction procedures are not widely used and most effort has been devoted to developing suitable masking media.

---

*The concentrations are given in mg $dm^{-3}$ of nitrate, but these are often reported as nitrate nitrogen, where 1 mg as $NO_3^-$ is ∿ 0.23 mg.

In drinking water (0.1 - 30 mg $dm^{-3}$ $NO_3^-$) and waste waters (25 - 85 mg $dm^{-3}$), good agreement with several comparative procedures has been obtained when the samples and standards were mixed with 0.1 M potassium dihydrogen phosphate, in the ratio 10:1[112]. A more complex mixture was used by Milham *et al.*[113] who for direct measurements of 4 - 80 mg $dm^{-3}$ added a solution containing 0.01 M aluminium sulphate, 0.01 M silver sulphate, 0.02 M sulphamic acid and 0.02 M boric acid. Such a solution should eliminate interferences caused by organic anions, chloride and nitrite. Acidification to pH 3 removes the effect of hydrogen carbonate. The accuracy was not worse than ±3 %. A similar composition was proposed by Weiss[114] who analysed natural waters containing above 0.6 mg $dm^{-3}$ nitrate, and also mentioned that the error in slightly mineralized samples was of the same magnitude. An even stronger version of the buffer of Milham *et al.*[113] was proposed by Sweetsur and Wilson[115], whereby the main ingredients were increased by factor of 2.5, but this was used for the analysis of nitrate in grasses.

Another procedure was proposed by Mertens and Massart[116] who insisted that an excess of sulphate and aluminum is undesirable, but the silver sulphate should be added until chloride is completely precipitated. The sample is acidified with sulphuric acid to pH 3, and then neutralized with barium hydroxide to pH 7. To sample and standards a fixed amount of fluoride was added in order to provide a constant potential of the fluoride electrode used as reference. Such a procedure was found successful for the analysis of mineral waters, containing 0.5 - 5 mg $dm^{-3}$ nitrate; however, the precision and accuracy was lower in comparison with spectrophotometric techniques.

Hulanicki, Lewandowski and Maj[105] used a "home-made" construction of a nitrate-selective electrode, without internal reference solution, for direct measurement of nitrate in tap water. They stated that the main interference is caused by chloride and hydrogen carbonate, and added to 100 $cm^3$ of solution, 2 $cm^3$ of 0.5 M phosphate buffer of pH 2.2 and 10 $cm^3$ of 0.45 % silver sulphate. For 10 mg $dm^{-3}$ nitrate, the errors were below ±5 %. The same electrode and procedure was also applied to surface waters[117], and satisfactory agreement with the phenoldisulphonic acid procedure was found when the nitrate concentration was above 1 mg $dm^{-3}$. Chloride concentrations of above 100 mg $dm^{-3}$ caused positive errors. Similar conditions were used in continuous flow measurements[51] for drinking water. The potential stability over 24 hr was better than ±2 mV. Deviations smaller than 1 mg $dm^{-3}$ appeared for concentrations close to 10 mg $dm^{-3}$, but for the range below 4 mg $dm^{-3}$, systematic positive errors were eliminated by adding a known amount of nitrate to the samples in order to shift the potential measurements to the linear part of the calibration curve. Because hydrogen sulphate has some interference effect on the linear range, it is advantageous to replace silver sulphate by silver phosphate dissolved in acidic phosphate buffer[106].

Flow injection analysis has been adapted for nitrate determinations in a variety of environment samples[118]. The electrode was based on a tetraoctylammonium salt[119], and its linear range extends down to 0.6 mg $dm^{-3}$. The background electrolyte solution used was 0.01 M borax with addition of sodium hydroxide to pH 9.5 in the presence of 0.2 % glycine. Such an alkaline medium was not advantageous for prolonged electrode life, but functioning of the electrode was more reliable. The presence of a high concentration of chloride was also unfavourable, because of a time-dependent effect on the zero line of the electrode.

The above data are a corollary to the statement in an early Orion Newsletter, that at 1 mg $dm^{-3}$ nitrate interfering ions at a level of 100 mg $dm^{-3}$ cause a positive error equal to 92 % for hydrogen carbonate, 70 % for chloride and 6 % for sulphate[120]. Therefore, when large concentrations of these interfering ions are present they should be removed using cation exchangers of silver and hydrogen forms respectively, in series. Because there is no effect by fluoride, potassium fluoride may be used for keeping the ionic strength constant[121].

Ion exchange for removal of chloride and carbonate has also been used by the addition of resin to the sample solutions, before potentiometric measurements[122] of nitrate content in surface water. A different procedure was proposed by Mertens, Van der Winkel and Massart[123], who bound nitrate on the column and eluted it with hydrogen carbonate, which was in turn removed by the addition of acid.

Determination of nitrate in waters has also been carried out after chemical reduction to ammonia, which was detected by an ammonia gas sensor[123]. The high selectivity of this sensor eliminates the possible interferences of anions. The best results were obtained in 0.1 mol $dm^{-3}$ sodium hydroxide solution, using Devarda's alloy formed into beads with polystyrene or PVC. Such a procedure was rather long, because the reduction itself in static measurements takes about 30 min. When the sample contained ammonia in addition to nitrate two measurements became necessary; i.e. one before and one after reduction. When nitrite was present, sulphamic acid was added before the reduction stage. Such a procedure, when applied to the continuous analysis of waters and sewage (Fig. 2), permitted a sample throughput of 20 $h^{-1}$. The calibration curve was linear from 1 - 50 mg $dm^{-3}$, but afterr some days the reduction efficiency dropped below 100 %.

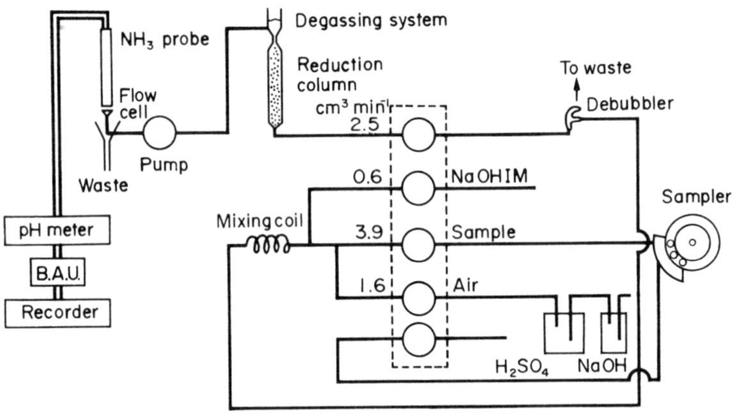


Fig. 2. Flow diagram of a continuous system for the determination of nitrate with an ammonia gas sensor[123]. (By courtesy of the American Chemical Society.)

McKenzie and Young[23] also used an ammonia gas sensor for nitrate determination. Reduction with Devarda's alloy was carried out in hydrochloric acid, in the presence of sodium fluoride. The reduction time took up to 24 hr, and a recovery of 95 % was obtained at the 1 mg $dm^{-3}$ level. For larger amounts of nitrate, the recovery decreased to 75 % but despite this the method was described as satisfactory for stream, river, estuarine and marine survey.

## 3.12. Total nitrogen and total inorganic nitrogen

Total inorganic nitrogen determination is based on the reduction of all inorganic forms of nitrogen to ammonia, which is determined by the ammonia gas sensor. Hansen, Růžička and Larsen[124] used the air-gap electrode for analysis of river water, and untreated and biologically treated waste-water. The reduction with Devarda's alloy was carried out in an alkaline medium in the electrode chamber. The stable potential of the sensor was obtained within 3 min of the end of reduction. The precision was ±3 % for concentrations down to 3 mg $dm^{-3}$ of nitrogen, and the reduction was quantitative. The presence of *L*-aminoacids and urea did not introduce an error larger than the experimental error. However, the total inorganic nitrogen content in river water is usually too small for potentiometric measurement, but this is not the case for waste water.

A similar measuring procedure was used for total nitrogen determination, except that the preliminary treatment of the samples is now by Kjeldahl digestion. The normal Kjeldahl determination is simplified, because there is now no need for the distillation step for final ammonia separation. In the total nitrogen determination there is no danger of interference from volatile amines. Should a mercury catalyst mixture be used for digestion, then in order to avoid complexation of ammonia by mercury(II) an excess of sodium iodide (300 g $dm^{-3}$) should be added[90], because the alkaline EDTA solution used for acid neutralization is insufficient to prevent the mercury(II) based errors. Another source of errors is connected with the osmotic effect which appears when the total concentrations of the solutions on both sides of the hydrophobic membrane differ significantly[91]. This effect may be calculated on the basis of known volumes and concentrations. The digest may be diluted[125], but this unfavourably influences the detection limit. Alternatively, the composition of the electrode-filling solution should be changed by inclusion of potassium sulphate[126].

The time-consuming step-wise analysis procedure has been made semiautomatic by Buckee[126] and Stevens[125] with throughputs of 10 - 60 samples per hour.

## 3.13. Sulphate

The sulphate concentration of natural water ranges from a few mg $dm^{-3}$ to 1000 mg $dm^{-3}$; the upper limit relates to sea water. In river and ground water the sulphate content results from mineral deposits and from industrial contamination; however, even water containing 250 mg $dm^{-3}$ may be used for drinking purposes.

In spite of these relatively high concentrations of sulphate, there is a lack of rapid and sufficiently accurate methods of their determination, and this has promoted searches for sulphate ion-sensitive electrodes. Some papers on this subject have been published[127-131] but in general the results have not been sufficiently encouraging for the commercial production of sulphate ion-selective electrodes. Hence, the lead ion-selective electrode is normally used as a indicator in potentiometric sulphate titrations with lead salts as titrants.

Potentiometric titrations of sulphate with a lead indicator electrode were first proposed by Ross and Frant[132]. They used a 50 % dioxan medium at pH 4, and mentioned that for 0.1 mg $dm^{-3}$ the potential break of ~60 mV was adequate for good results. Nevertheless, the precipitate formation at such low concentration is so slow that in practice this procedure can only be used satisfactorily when the sulphate content is greater than 5 mg $dm^{-3}$. The presence of chloride and nitrate in 50-fold excess, and hydrogen carbonate in 100-fold excess, is permissible, but phosphate interferes[133]. More recently, a titration in 50 % methanol has been recommended[84] in the presence of 0.01 M potassium nitrate and 0.001 M lanthanium nitrate, the latter component should suppress the effect of phosphate. The end point is determined using the Gran method for the points of the titration curve following the equivalent point.

Goerzen and Oster[134] investigated the above method in the analysis of water and soil extracts for sulphate. They were concerned that dioxan often contains peroxides, which deteriorate the electrode surface through passivation. Among other solvents they investigated - namely, methanol, ethanol, propan-1-ol, propan-2-ol, butan-2-ol and acetone - the largest end-point breaks were obtained for methanol and ethanol. The unfavourable decrease of calcium sulphate solubility in the presence of organic solvents was compensated by the addition of 0.6 $cm^3$ of 0.1 M sodium perchlorate and 11 $cm^3$ methanol to a 5 $cm^3$ water sample containing 2 mg sulphate. The samples were titrated at pH 4.4, and good accuracy was obtained, but large concentrations of chloride decrease the potential break.

Titration of sulphate in sea and mineral water (50 - 7500 mg $dm^{-3}$) in 50 % dioxan has been described by Mascini[135], who removed any great excess (10 fold) of chloride and hydrogen carbonate by passing the samples through cation exchangers, respectively, in the silver and hydrogen forms. The detection limit was 10 mg $dm^{-3}$, but because of slow response the method is more practically useful at 100 mg $dm^{-3}$.

Smaller concentrations of sulphate were determined by Hulanicki, Lewandowski and Lewenstam[136], who used commercial electrodes (Orion) and sensors made by them which were based on lead selenide or telluride. In 75 % methanol, a 200-fold excess of nitrate was without effect, but chloride caused a small positive error. Serious interference was observed in the presence of calcium, which co-precipitates with lead sulphate, making the potential response sluggish. Increase of ionic strength eliminates this effect, and 0.5 M or 0.1 M sodium perchlorate was necessary at the 4- or 20-fold excess of calcium, respectively. Below 10 mg $dm^{-3}$ it was recommended that calcium be removed with a hydrogen-form cation exchanger.

Scheide and Durst[137] analysed sea and rain water, and these corresponded to the two extreme concentration levels of sulphates. The determination at low concentrations (3 - 10 mg $dm^{-3}$) gave comparable or better results than gravimetric or turbidimetric determinations, respectively. Below 3 mg $dm^{-3}$ losses of sulphate were observed, probably because of adsorption on the walls of vessels.

A slightly different procedure was used by Reynolds[26], who added 50 $cm^3$ ethanol and 0.1 mM of lead perchlorate to a 25 $cm^3$ sample of natural water, having 1 - 5 mg $dm^{-3}$ sulphate content. The precipitation of lead sulphate was effected at pH 4, and the excess lead was measured with electrodes calibrated with a standard lead solution.

The continuous determination of sulphate in a flow system has been developed by Trojanowicz[138], who recorded the potential difference between two similar lead electrodes immersed in a stream of 2 x $10^{-4}$ M lead perchlorate solution, before and after addition of the sample containing 30 - 90 mg $dm^{-3}$ sulphate (Fig. 3). The lead perchlorate solution was prepared in 75 % methanol containing 0.05 M sodium perchlorate at pH 4. Despite the use of a differential system the passivation of lead electrodes led to loss of sensitivity. However, significant improvement was obtained when both electrodes were cathodically polarized with a current of 0.025 μA. Using $10^{-3}$ M lead perchlorate, the method permitted determination of sulphate in the range from 100 to 400 mg $dm^{-3}$. The results for tap water did not differ by more than 10 % from the titrimetric data. Calcium does not interfere and neither does a 5-fold excess of chloride.

Procedures based on other electrodes have also been described. Shpeizer, Rubinchick and Atamova[139] determined the excess of lead, after precipitation of lead sulphate, by titration with sodium sulphide, using a sulphide ion-selective indicator electrode. Good results were obtained when the sulphate concentration was ∿20 mg $dm^{-3}$.

Jasinski and Trachtenberg[140] used an electrode of chalcogenide glass responding to iron(III)ions for the analysis of sea water, effluents and saline water. When the samples are titrated with barium salts, iron is liberated from its sulphate complexes and alters the electrode potential. Similar interferences to those observed for other precipitation titrations mentioned above were noted. The presence of 0.1 M potassium chloride causes an error of ∿6 %. The calcium sulphate coprecipitation may be allowed for by empirical corrections.

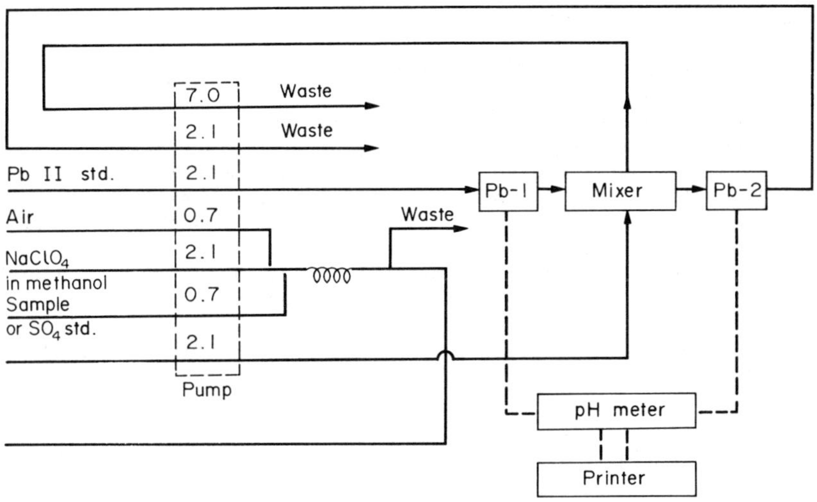


Fig. 3. Flow diagram of a continuous differential sulphate analyser[138].

A quite different type of electrode has been used by Ouzounian and Michard[141]. This is patterned on a liquid exchanger, based on the oxonium salt of p-nonyl(phenoxypoly-ethoxy)ethanol (Triton *N* 101) in *p*-nitroethylbenzene, and is similar to that developed by Jaber, Moody and Thomas[142]. The sensitivity of the electrode towards barium ions was used in the direct titration of sea-water samples containing 2.7 g $dm^{-3}$ sulphate, or of thermal water containing 73 mg $dm^{-3}$, with 0.1 or 0.005 M barium chloride solutions.

## 3.14. Sulphide

The sulphide content of water is usually very low, and any increase indicates pollution, mainly from organic material. Its determination is subject to errors because of oxidation, mainly by dissolved oxygen, by volatilization of hydrogen sulphide from acidic samples[143] and by precipitation with metal ions. Therefore, despite having an electrode, which functions unfettered by other species, the problem of accurate determination of sulphide is not easily carried out. Sulphide ion-selective electrodes are based on solid-state membranes prepared from silver sulphide, or they may be electrodes of the second kind - namely silver/silver-sulphide - but which virtually do not differ from the former type. The detection limit of these electrodes is as low as $10^{-14}$ M and relates to free sulphide ions in a sulphide buffer, i.e. in equilibrium with protonated or complex species. For reliable measurements the total sulphide concentration should not fall much below $10^{-6}$ M (30 μg $dm^{-3}$), which gives the practical analytical detection limit.

The total content of sulphide is determined in strongly alkaline solution, where protonation is significantly suppressed. Otherwise, information about sulphide concentrations may be obtained from direct measurements at lower, but accurately known, pH values. Such a procedure has been used for the computer evaluation of hydrogen sulphide equilibria[144]. This, however, needs careful standardization, because the first protonation constant is not known with sufficient accuracy and depends on temperature and ionic strength.

A similar principle was used by Mor *et al.*[145], who measured sulphide concentration at intermediate pH values, and obtained the final data on the basis of calculations or experimental standardization with solutions prepared from sea water.

The oxidation of sulphide is at least partially prevented when the Sulphide Anti-Oxidant Buffer (SAOB) is added to sample and standard in a 1:4 ratio[146]. This buffer contains in 1 $dm^3$, 80 g sodium hydroxide, 320 g sodium salicylate, and 72 g ascorbic acid. In the concentration range 0.1 - 0.5 mg $dm^{-3}$, the potential readings can be taken after 2 - 3 min, but for higher sulphide levels the electrode response is faster. The lower limit of applicability of this method has been extended to 30 μg $dm^{-3}$ by extracting heavy metals with dithizone[147]. In such cases a modified buffer composition is recommended, and which contains in 1 $dm^3$, 186 g EDTA (disodium salt), 120 g sodium hydroxide and 72 g ascorbic acid. This Alkaline Antioxidant Reagent (AAR)[147] is added in a 1:4 ratio to standards and samples. The calibration curve is linear above 50 mg $dm^{-3}$, but for smaller concentrations the results were also reproducible in the 18 - 180 μg $dm^{-3}$ range, with relative standard deviation being below 0.1.

When the sulphide concentration is below 30 μg $dm^{-3}$ it may be coprecipitated as zinc sulphide on zinc hydroxide after addition of zinc acetate and sodium carbonate[147]. The precipitate is centrifugated, dissolved in the AAR buffer, and the resulting solution used for potential measurements. The procedure gave good recovery and precisions, even for 1.8 μg $dm^{-3}$.

An electrolytically-prepared electrode of the second kind enabled determination of 10 - 250 mg $dm^{-3}$ sulphide in natural water with a linear calibration curve[148]. When the sulphide concentration exceeds 300 mg $dm^{-3}$ it is better to use a titrimetric procedure, with cadium sulphate as titrant. In such determinations, thiosulphate, sulphite and chloride do not interfere, and the results are in good agreement with iodometric titrations.

Gulens and Labbate[149] used the sulphide ion-selective electrode in a continuous flow monitor (Fig. 4). In the 0.01 - 10 mg $dm^{-3}$ range the errors were ±10 % and these were associated with a potential drift. In practice, the solutions should be properly de-aerated and ascorbic acid or hydrazine added to the stream. Hydrazine reductant is, however, less convenient because it reduces silver at the electrode surface. Contamination of samples with oil is another source of errors. Similar conditions have been used for the analysis of heavy water[150].

### 3.15. Sulphite

Sulphite and sulphur dioxide may occur in industrial waste water, or sometimes in other kinds of water, as a pollutant. They are unstable, undergo oxidation, and can be lost even during measurement. Therefore, glycerol is added as a preservative. The gas sensors used for sulphur dioxide determination show Nernstian response down to 2 mg $dm^{-3}$ (E.I.L.)[151] or to 0.15 mg $dm^{-3}$ (Orion Research)[152]. The sensors are generally very selective, and only molecular species which diffuse through a hydrophobic membrane may interfere; nevertheless, some excesses of acetic acid and ammonia are tolerable.

Sulphite has been determined in sea water in the concentration range 0.1 - 60 mg $dm^{-3}$, with an error of 5 %[153].

### 3.16. Fluoride

The fluoride content of waters originating from natural resources, as well as from industrial pollution or water fluoridation is usually of the order of a few mg $dm^{-3}$. Excessive concentration of fluoride is hazardous to health and should therefore be controlled, especially in potable water. Indirect colorimetric methods are, as a rule, complicated and influenced by the presence of other ions[154-157], and this has stimulated the development of potentiometric methods for fluoride. However, a relatively long time is needed for stable potentials at extremely low concentrations of fluoride. Nevertheless, potentiometric fluoride determination has been introduced as a standard procedure for water analysis[158].

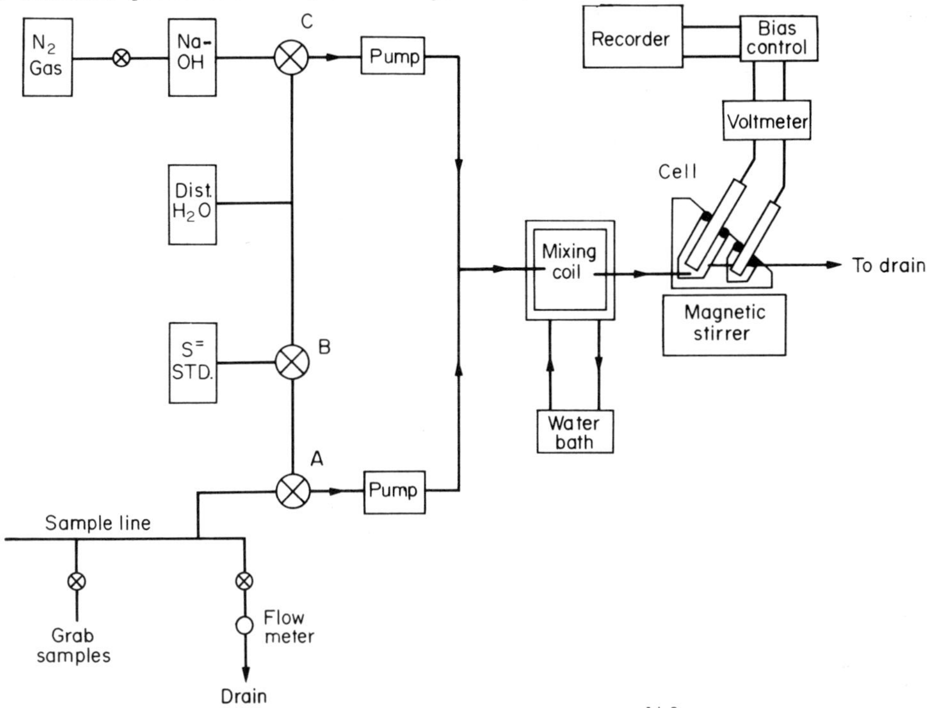


Fig. 4. Flow diagram of a sulphide analyser[149].(By courtesy of the Atomic Energy of Canada Ltd.)

The flouride electrode is one of the most selective ion-sensors; only the presence of large quantities of hydroxyl ions may spoil the results - the selectivity coefficient $k^{pot}_{F,OH}$ is ∿0.1. This is easily eliminated when measurements are carried out in solutions of pH <7, which is satisfactory even at flouride levels $10^{-7}$M.

Fluoride in natural waters is present in the free form, or complexed by metal ions. On the one hand, this makes possible determinations of the contributions of free and complexed forms separately, but, on the other hand, samples require special treatment when total fluoride is to be determined. It is commonly supposed that the main difficulties normally arise from the presence of aluminium or iron(III), which form strong fluoride complexes. Brudevold, Moreno and Bakhas[2] found that in city water containing 0.1 - 0.5 mg $dm^{-3}$, the fluoride existed mainly as aluminium complexes. In sea water, the presence of $AlF^{2+}$ may be neglected, but when samples were spiked with fluoride only half of the fluoride remained uncomplexed. Magnesium is known for binding fluoride in sea water, but probably part of fluoride occurs in the form of collodial suspension[160] and it is these species that are responsible for discrepancies between potentiometric and spectrophotometric methods.

Some discrepancies occur in the presence of large concentrations of sodium, which forms ion pairs more readily with fluoride than does potassium. The net result is a parallel shift in the calibration curve[161].

The detection limit of the fluoride electrode was initially assumed to be $\sim 10^{-6}$ M (20 μg $dm^{-3}$), but Baumann[162] found that the Nernstian response was valid even down to 0.01 μg $dm^{-3}$ in the presence of a fluoride buffer. This has led to the development of fluoride buffers on the basis of thorium compounds[163]. It therefore seems probable that in practical determinations of fluoride, impurities in water and reagents limit determinations in the low concentration range. In this respect, potassium salts are generally less contaminated than are sodium salts[161], but, of course, reagents used as media for measurements may also introduce impurities[164-168]. Therefore, the level of 1 μg $dm^{-3}$ should be considered as a satisfactory practical limit for most purposes.

Many authors have devoted much effort to elucidate the reactions of fluoride demasking and to develop an optimal medium for total fluoride determination. Different opinions have resulted in various media compositions, but which probably arises from different levels of fluoride concentration, and of interfering metal concentrations, together with the presence of other substances, kinetic effects, ionic strength, and the occurrence of particulate matter or colloidal forms of one or more reacting species. Therefore, a demasking mixture of a given composition developed for some type of samples may be not suitable for samples of different origin, even when it seems that the sample composition is similar.

Frant and Ross[164] proposed the use of Total Ionic Strength Adjustment Buffer (TISAB) for measurement of total fluoride. It contained 58 g $dm^{-3}$ sodium chloride, 0.35 g $dm^{-3}$ sodium citrate, 1 M acetate buffer of pH 5, and was mixed in a 1:1 ratio with samples of waters or waste waters having fluoride levels above 0.1 mg $dm^{-3}$. In the analysis of natural and industrial waters good agreement (±0.05 mg $dm^{-3}$) was obtained between ion-selective electrode data and the spectrophotometric SPADNS procedure using a Technicon Autoanalyzer. A similar buffer has also been used for the analysis of sea water[169]. Pakalns and Farrar[170] studied the effect of detergents and found that with the TISAB buffer as little as 40 μg $dm^{-3}$ could be determined in the presence of 1 g $dm^{-3}$ detergents.

The original composition of TISAB has been criticized by several authors. Patterson, Bunton and Crosby[155] consider a solution containing phosphate buffer, EDTA and citrate to be more advantageous. Calcium and other anions do not interfere at the 200 - 500 mg $dm^{-3}$ level, and neither does iron up to 2.5 mg $dm^{-3}$. The study was, however, confined to natural water samples containing from 0.1 to 2 mg $dm^{-3}$ fluoride. Recovery of 0.8 mg $dm^{-3}$ was found accurate to within 0.02 mg $dm^{-3}$. The advantages of potentiometric procedures compared to spectrophotometry have been stated by Crosby, Dennis and Stevens[164], but below 0.3 mg $dm^{-3}$ stable potentials took more than 5 min to be reached.

Harwood[171] investigated the possibility of replacing citrate in TISAB by other chelating ligands such as EDTA or DCTA. According to him, TISAB allows for the presence of 3 g $dm^{-3}$ chloride and sulphate, 10 mg $dm^{-3}$ iron(III) and phosphate, but 0.2 mg $dm^{-3}$ aluminium gives an error of -21 % while 1 mg $dm^{-3}$ of the metal gives as much as -75 %. Replacing citrate by EDTA does not improve the determination, but 1 g $dm^{-3}$ DCTA is very advantageous. For 1 mg $dm^{-3}$ fluoride and 5 mg $dm^{-3}$ aluminium the error was only -10 %. Iron(III) is also more effectively masked, up to 25 mg $dm^{-3}$ by DCTA. In certain samples this concentration level of DCTA is insufficient, and must be increased to 5 g $dm^{-3}$ [172], but for some potable waters it may be decreased to 0.1 g $dm^{-3}$. The commercially-recommended TISAB[173] contains 4 g $dm^{-3}$ DCTA, and is adjusted to pH 5.0 - 5.5 with 5 M sodium hydroxide. Its use has been recommended by other authors for the analysis of fluoridated water[157], and 50 g $dm^{-3}$ sulphate and phosphate, 20 g $dm^{-3}$ chloride, 5 g $dm^{-3}$ free chlorine, 200 mg $dm^{-3}$ iron(III) and 2 mg $dm^{-3}$ aluminium do not introduce an error larger than 10 % for 1 mg $dm^{-3}$ fluoride. This procedure was applied to analysis of city, ground and mineral water[174].

## Applications in Water Analysis

Sekerka and Lechner[166] found linearity of the calibration curve in TISAB only 100 μg $dm^{-3}$ fluoride, but in the absence of any reagents in excess the calibration could be made to 5 μg $dm^{-3}$. Distilled water and sodium chloride were the largest sources of contamination, but sodium hydroxide and acetic acid were found to be practically free from contaminating fluoride. Extension of the linear part to 5 μg $dm^{-3}$ was also possible by using a three times more concentrated buffer, from which sodium chloride was eliminated. Such a solution was used in an automated system, which permitted the determination of 5 - 1000 μg $dm^{-3}$ fluoride with a relative standard deviation of 2 % and good recoveries.

Other authors differ in their conclusions about the efficiency of metal masking, e.g. for fluoride in mineral (0.2 - 8.5 mg $dm^{-3}$)[175] or in fluoridated city water[176]. Here, citrates were claimed to give better results with respect to spectrophotometric determination with zirconium-eriochromcyanine *R* complex as reference. In the presence of larger amounts of aluminium, a solution containing 100 g $dm^{-3}$ sodium chloride, 38 g $dm^{-3}$ citric acid and 258 g $dm^{-3}$ trisodium citrate should be added to samples in a ratio of 1:100[176]. Another buffer may also be used and contains 5 M acetate and 0.5 M citrate adjusted to pH 5[177]. The buffer has been used in a 1:5 ratio.

Measurements based on the standard addition technique gave stable potential readings after 15 min[178]. In sea water with 0.3 - 1.3 mg $dm^{-3}$ of fluoride, recoveries from 98 to 102 % were obtained, but for 0.1 mg $dm^{-3}$, the bias was +30 %. A multiple standard addition technique was used by Anfält and Jagner[159] in sea-water analysis. They did not use a demasking agent or ionic strength adjustor, but only samples acidified with hydrochloric acid to pH 6.6. The results were interpreted graphically or by using the Gran method. A similar procedure was used by Rix, Bond and Smith[168].

In the above procedures, the calibration curve or standard addition techniques were used. The latter is usually recommended in the presence of large amounts of interfering species, as happens in the case of sea water, and gives a measure of the total fluoride. A quite different principle was proposed by Warner and Bressan[167], who on analysing atmospheric water (rain and fog) containing 1 μg $dm^{-3}$, observed the rate of potential change for the electrodes when immersed in the sample solution and in the standard. They showed that for 0.3 - 20 μg $dm^{-3}$ fluoride, as much as 80 - 90 % of the results were accurate to better than 20 %. When the concentration was above 100 μg $dm^{-3}$ the single standard addition method was more efficient.

Potentiometric titration of fluoride in waters, as carried out by Light, Mannion and Fletcher[179], was based on a preliminary concentration of fluoride on Dowex 1X4 ion exchanger, followed by elution with 0.1 M potassium chloride to give a 35-fold increase of concentration. After the addition of ethanol, the sample was titrated at pH 2 with 0.005 M thorium nitrate. For 1 mg $dm^{-3}$, the standard deviation was 0.013 mg $dm^{-3}$, which is more than three times better than direct measurements.

Another procedure[180] is based on a preliminary extraction of iron and aluminium with a chloroform solution of oxine, and was found to be useful for waste waters, sea, city and well water. The results were compared with those obtained after fluoride distillation. Thus, direct determination without separation gave results with a large negative error in relation to those with separation. In the extraction procedure, 10 mg $dm^{-3}$ aluminium and 20 mg $dm^{-3}$ iron do not interfere.

Many successful continuous determinations have beem made by potentiometry. Babcock and Johnson[18] used a monitor with a fluoride electrode for the control of tap water , which when compared with spectrophotometry indicated a -10 % error, attributed to the presence of aluminium and iron. For an average water composition, a numerical correction could be used, which takes an activity coefficient into account. The system needs calibration every two weeks. Diggens and Collis[182] used on-line analysis for raw water (0.01 - 0.15 mg $dm^{-3}$ fluoride) and fluoridated water (0.1 - 1.5 mg $dm^{-3}$ fluoride) and used the results for controlling the fluoride dosing equipment.

Other authors have developed other flow systems for fluoride determination[166,183]. The solution stream used for fluoride measurements may be also combined with the determination of chloride and fluoride in the range 20 µg and $dm^{-3}$-20 mg $dm^{-3}$, in surface and ground waters. Trojanowicz[185] has designed a three-parameter system for simultaneous measurements of chloride, fluoride and nitrate in one stream.

An automatic system working in a modified Technicon Autoanalyzer scheme was proposed by Erdmann[186], who spiked the TISAB buffer with 0.2 mg $dm^{-3}$ background fluoride, and obtained for a 1.5 mg $dm^{-3}$ level of sample fluoride a standard deviation of 0.012 mg $dm^{-3}$. The electrode response was sufficiently rapid for 30 determinations per hour.

## 3.17. Chloride

The chloride concentration in natural water varies from fractions of mg $dm^{-3}$ to g $dm^{-3}$ in sea water and salines. Chloride may enter surface and ground waters from mineral deposits, or from industrial, municipal and agricultural pollution. The lowest levels are in boiler feed-water, where the determination of less than 100 µg $dm^{-3}$ is of interest because of the corrosive properties of water containing chloride.

The common occurrence of chloride requires special precautions for its determination at extremely low levels. Water used for solution preparation should be de-ionized to a conductivity of 0.1 $\mu\Omega^{-1}$ $cm^{-1}$. Reagents, vessels and skin contact may introduce contamination. Another source of contamination is to be found in reference electrodes or salt bridges normally containing potassium chloride solutions. Hence, in such cases, the mercury-mercury(II) sulphate reference electrode is recommended[187,188] or a double-junction reference electrode with a salt bridge with, say, 10 % potassium nitrate is preferred[189].

The commonly-used chloride-selective electrodes are the solid-state sensors, because the liquid-state electrode based on tetralkylammonium salt is prone to many interferences. The solid-state electrodes are based in most cases on single silver chloride crystals, on polycrystalline membranes, or on heterogeneous membranes. The first two are usually superior, and their characteristics resemble those of the silver-silver chloride electrode of the second kind. In several papers[190-192] it has been stated that the second kind of electrodes are equally good in performance, with the advantage of simplicity in preparation and low cost. Such electrodes have been used for the determination of chloride in water[188,193].

All the chloride electrodes are subject to similar interferences from anions forming slightly-soluble silver salts or of silver complexing ligands, but their level in waters is usually below that which can disturb the chloride determination. Despite the theoretical and operational similarity, it is assumed that the membrane electrode is somewhat more resistant to strong redox systems, while the second-kind electrode has a better detection limit, and is preferred in determinations where the chloride concentration is below that which is present owing to silver chloride solubility ($10^{-5}$ M = 0.35 mg $dm^{-3}$). Bardin[187] showed that in this range the potential varies linearly with concentration, and differential determinations are still possible. At $\sim$ 0.1 mg $dm^{-3}$, Torrance[188] found linearity of the calibration graph plotted in the coordinates of emf. versus $\log(c+1)$, where $c$ represents chloride concentration in mg $dm^{-3}$.

At low chloride level the potential of the electrode is influenced by the solubility of silver chloride, which in turn depends on temperature. One degree of difference in temperature between calibration and sample measurement at 25°C causes 4 % error, therefore both long and short term variations of temperature should be carefully avoided.

Another factor which ought to be considered is the effect of light on the electrode[194], but when screened from strong daylight, the crystal membrane electrode works well.

The ion-selective electrodes are better than the first or second kind electrodes in potentiometric titrations, but their most attractive use is for direct, single or continuous measurements.

Selmer-Olsen and Øion[195] determined 1.2 - 18 mg $dm^{-3}$ chloride in water using the Orion 94-17 crystal membrane electrode. The sample solution contained 0.5 M ammonium nitrate and 0.03 M nitric acid. Above 4.8 mg $dm^{-3}$ the errors were below 5 %, but for concentrations below 3 mg $dm^{-3}$, spectrophotometric procedures were performed. Similar accuracy was reported by Van Oort *et al.*[196] who determined 15 - 130 mg $dm^{-3}$ chloride under field conditions - the samples were merely acidified with a few drops of acetic acid.

More complex conditions were proposed by Weiss[197], who added to a 100 $cm^3$ sample 0.2 $cm^3$ sulphuric acid (1+1), 10 $cm^3$ 1 M potassium nitrate, and evaporated to half the initial volume. The samples were adjusted to pH 4, and the potential of the chloride electrode was measured and correlated to a calibration curve. In the range from 2 - 35 mg $dm^{-3}$ of chloride in surface and tap water the errors were from -3 to -13 %, compared with the data of titrimetric determinations.

Van Loon[198] used a heterogeneous electrode for the determination of 1 - 312 mg $dm^{-3}$ chloride in lake and stream water, without any sample pretreatment. In tap and river water analysis Hulanicki, Augustowska and Trojanowicz[199] also avoided pretreatment, and calibration was recommended with respect to water samples of which the chloride content had been determined titrimetrically. The electrode used was of the second kind. Several procedures have been compared by Mainka *et al.*[200] who used samples having and not having constant ionic strength. The standard addition method was considered the best. In order to remove hydrogen sulphide and some interfering organic substances, hydrogen peroxide in a nitric acid medium was added, and the potential was read before and after the addition of $10^{-5}$ M of sodium chloride. The limit of the determination was assumed to be 5 mg $dm^{-3}$.

The determination of chloride in sea water and brines[201] presents different problems, because of the high concentrations of analyte. Difficulties have been encountered because of changing liquid junction potentials and activity coefficients, especially when samples differ significantly in composition. In order to avoid this, samples were diluted with 10 % potassium nitrate solution, but bromide interferes. In this high concentration range the temperature effects were also significant. In sea water (11 - 26 g $dm^{-3}$) the standard deviation was 0.6 g $dm^{-3}$, and in brines (85 - 130 g $dm^{-3}$) it was 0.8 g $dm^{-3}$. When Takamatsu *et al.*[202] determined chloride in river and sea water, a linear calibration was obtained from 10 to $10^5$ mg $dm^{-3}$. Sulphide interference can be eliminated by the addition of lead acetate but most metal ions at 100 ppm do not disturb the determination, although copper and mercury should not exceed 10 and 0.1 mg $dm^{-3}$, respectively.

Several authors use titrimetric procedures which are slower, but, of course, offer better precision and accuracy. With the chloride ion-selective electrode at low analyte levels 18 % acetic acid was added to improve the potential break[200]. A similar electrode has been used in an automatic titrator for water analysis[203]. A continuous titration of chloride in natural waters may be carried out on the principle of electrolytic generation of silver ions by a programmed current source[204]. In this system (Fig. 5), at constant sample flow, the rate of titrant generation increases linearly with time. After the sample is overtitrated, the current decreases and a back-titration curve is obtained. The chloride content is read from these titration curves as a distance between the two curves at the equivalence point potential. The calibration curve is linear, and gives the relationship between the distance, measured in seconds, and the concentration of chloride. A single calibration curve is valid for a concentration range over less than one order of magnitude, but proper programming of the titrant generation current permits other concentration ranges to suit any sample.

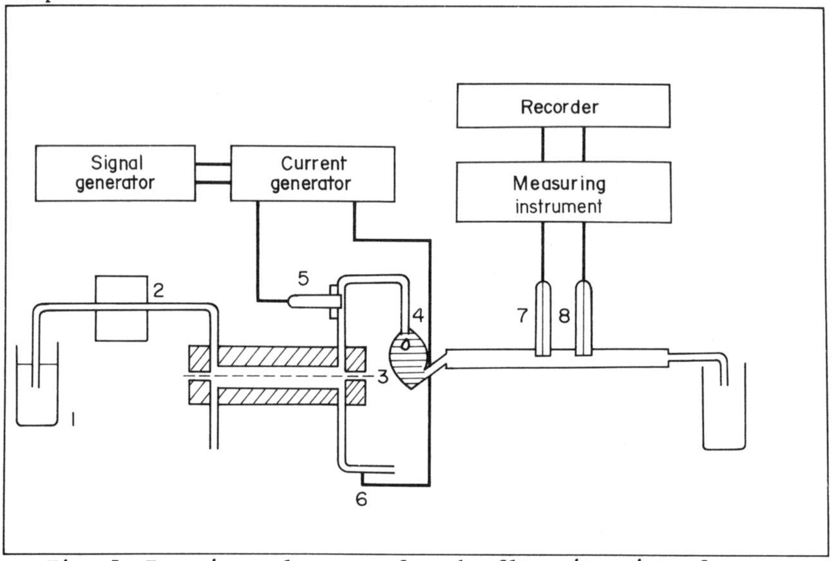


Fig. 5. Experimental set-up for the flow titration of chloride with electrolytically-generated silver ions[204]. Key: 1, solution reservoir; 2, peristaltic pump; 3, dialysis membrane; 4, drip vessel; 5, generating electrode; 6, counter electrode; 7, indicator electrode, 8, reference electrode[204].

A differential technique using two chloride electrodes was used by Florence[205] for determining 6 - 350 μg $dm^{-3}$ chloride in high-purity water. Despite the low concentrations, he claimed that membrane electrodes are more advantageous than those of the second kind. At the level of 100 μg $dm^{-3}$, the relative standard deviation was 3 %. The measurements were carried out in 1 M nitric acid, and phosphate and several metal ions (at 10 mg $dm^{-3}$) do not interfere, but traces of hydrogen sulphide exert a harmful effect. This procedure, although lengthy, was considered superior to spectrophotometric measurements.

Chloride titrations are carried out argentometrically, therefore for sensing the end point any silver-sensitive electrode is adequate. Mascini and Liberti[206] used a heterogeneous silver sulphide electrode for the successful titration of sulphide and chloride in waters, the amounts present being deduced from two distinct potential breaks. Jagner and Aren[207] tested several electrodes (silver metal, silver-sulphide membrane, silver bromide plus sulphide membrane) in titrations in 50 % ethanol. The most rapid response was observed for the bromide membrane electrode, which gave a precision of ± 0.2 %.

The silver-sulphide membrane electrode was used by Frost[208] for measuring chloride. This was done by adding on an excess of silver ions to samples containing

3 - 12 mg $dm^{-3}$ chloride after preliminary removal of sulphur compounds with charcoal and measuring the excess silver. Also, Reynolds[26] determined up to 1 mg $dm^{-3}$ chloride in waters by adding excess silver nitrate, adjusting pH to 5, and measuring the potential of the silver sulphide electrode. Precision was estimated to be 0.035 mg $dm^{-3}$.

Sekerka, Lechner and Wales[189] have developed a chloride ion-selective electrode based on insoluble mercury salts. They indicate that such an electrode, which has a membrane composed of mercury(I) chloride and mercury(II) sulphide gives more rapid response than the silver-based electrode. Its linear characteristics extend from 0.05 to 3500 mg $dm^{-3}$, and it is even possible to determine 0.01 mg $dm^{-3}$ chloride. The electrode was used by them in the analysis of tap water and some natural waters, the samples being mixed with a tenth of their volume of saturated potassium nitrate, and acidified to pH 2. Determinations in high-purity water and in heavy water have also been successfully performed. Sulphide is removed by the addition of bismuth nitrate[209]. At the other end of the scale, Sekerka and Lechner[210] have used the electrode for determining 1220 mg $dm^{-3}$ chloride in municipal waste-water, and 550 mg $dm^{-3}$ in industrial waste-water. A 100 $cm^3$ sample was tested with 1 $cm^3$ 50 % perchloric acid, 1 $cm^3$ saturated potassium persulphate, and 0.01 $cm^3$ 0.1 M copper nitrate, and the calibration curve technique equalled known addition of standard and analytic addition in precision[210].

An electrode working on the same principle has been obtained by rubbing the salt mixture into a graphite Selectrode[211]. Linear response was observed for concentrations above 0.35 mg $dm^{-3}$, but measurements on boiler were possible down to 0.01 mg $dm^{-3}$. Interference of iron(III) can be eliminated with fluoride, but several anions ($SO_4^{2-}$, $Br^-$, $I^-$, $S^{2-}$, $SO_3^{2-}$) interfere. The precision obtained at the 0.1 mg $dm^{-3}$ level seems to be better than that with the silver-based electrode. At lower concentrations, it is advantageous to reduce the temperature in order to improve sensitivity, although, of course, the response times become longer.

The importance of chloride determination has stimulated the development of several systems of continuous analysis. Tomlinson and Torrance[212] analysed high-purity power-station waters in a continuous system, using silver-silver chloride electrodes, for which the response was faster than for membrane electrodes. An improved detectability was observed at lowered temperature (10°C), and attention was paid to purity of reagents and to the water used for solution preparation.

The same type of electrode was used by Trojanowicz[185], who analysed tap water in a three-parameter system for fluoride, chloride and nitrate. Chloride was determined in the range below 100 mg $dm^{-3}$ as the first parameter in the one-stream system simply after addition of just an acetate buffer (Fig. 6). A double-function system was designed by Bertier[184] for determining fluoride and chloride (3.5 - 3500 mg $dm^{-3}$) in underground and surface waters.

Vanderputte, Dryon and Massart[213] analysed natural water in an automated continuous system. The chloride content was 2 - 350 mg $dm^{-3}$. The medium used consisted of sodium nitrate and an acetate buffer of pH 5.4. By using a relatively high flow-rate, it was possible to analyse 50 samples per hour without excessive noise in the measuring system, when it was properly earthed. Memory effects are present only below 10 mg $dm^{-3}$.

## 3.18. Chlorine

Chlorine content resulting from water chlorination or industrial pollution is determined in potable and waste waters as residual chlorine. This corresponds to free chlorine, hypochlorous acid, chloramines, etc. This can be performed with

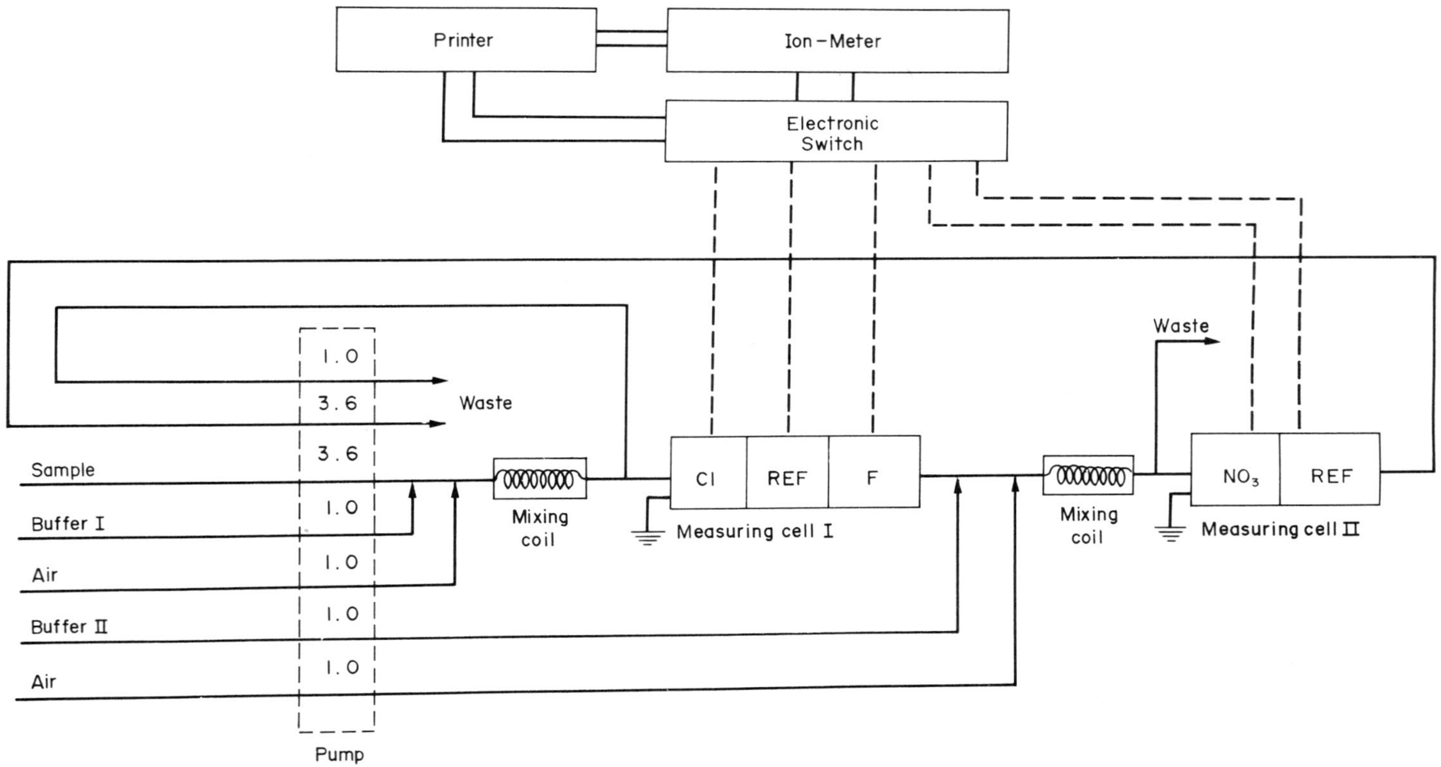


Fig. 6. Three parameter system for chloride, fluoride and nitrate measurements[185].
Buffer I = 0.1 M acetate buffer, 0.01 M DCTA, 0.1 % hydrogen peroxide.
Buffer II = 1 M orthophosphoric acid with 4 g $dm^{-3}$ silver carbonate.
Flow rates ($cm^2$ $min^{-1}$) are given on the diagram.

the iodide electrode[214] following on the change of iodide concentration, because of the oxidative properties of chlorine. A special chlorine electrode[215] is also based on the oxidation of iodide to iodine.

For iodide electrode determinations, the chlorine content should be in the range 0.1 - 1.0 mg $dm^{-3}$, and the concentration of iodide solution should be 5.6 x $10^{-5}$ M, which corresponds to 2 mg $dm^{-3}$ chlorine. Besides this, the solution contains 0.1 M acetate buffer and $10^{-3}$ M citrate for complexation of iron. The instability of chlorine in water requires rapid measurement after sampling.

Measurements with the Orion 97-70 chlorine electrode are simpler , cover the range from from 0.05 to 100 mg $dm^{-3}$, and are performed at pH 4. Rigdon, Moody and Frazer[216], by using the standard addition technique, determined 3 - 100 μg $dm^{-3}$ chlorine in waters with a computer-based automated system. Accuracy was found to be 2 μg $dm^{-3}$, but oxidants interfere in this procedure.

## 3.19. Bromide and iodide

Bromide in natural waters is usually low and does not exceed 1 mg $dm^{-3}$, except for some mineral waters and sea water. A higher content of bromide in surface water indicates industrial pollution.

Solid-state ion-selective electrodes are used for bromide determination. These have membranes of pressed silver-bromide and sulphide, or single crystals of silver bromide. Also used are silver-silver bromide electrodes of the second kind, and those based on mercury salts. Among the most harmful interferants are iodide, cyanide and sulphide. Chloride, however, may be present in molar excess of up to 400[217], but this probably depends also on absolute concentrations.

Harris and Williams[218] used the bromide membrane ion-selective electrode for direct measurement of bromide ( 5 - 500 μg $dm^{-3}$) in atmospheric waters, but below 100 μg $dm^{-3}$, preconcentration was found indispensible. In the similar range, bromide was determined by Weiss[197] in surface and tap water, using a similar sample preparation to that used for chloride determination. The results agreed within ±30 % with the spectrophotometric procedure based on bromination of phenol red. In the concentration range below 1 mg $dm^{-3}$, the method of standard addition is recommended.

Iodine occurs in trace amounts in natural waters in various forms as iodide, iodine or iodate, but its physiological activity is more important that that of bromine. Because of the availability of the iodide electrode, all forms of iodine are determined potentiometrically as iodide. The theoretical determination limit reaches down to a few μg $dm^{-3}$, but below 1 mg $dm^{-3}$ interferences occur due to redox processes and to adsorption of iodide at the electrode surface. These phenomena are more noticeable when the electrode membrane contains a larger proportion of silver iodide[219]. In careful measurements it is possible to determine down to 10 μg $dm^{-3}$, and the use of a reducing medium containing ascorbic acid is helpful.

Iodide has been determined in natural waters in the range 0.05 - 45 mg $dm^{-3}$ [220]. Rather large quantities of chloride, nitrate and perchlorate may be present, but the bromide quantity should not exceed 1000-fold excess of iodide.

When measuring the total content of iodine in natural water, Weiss[77] reduced iodine and iodate with sodium sulphite and hydrochloric acid. The final measurement was done in a slightly acid medium only, and on the basis of a calibration curve. The standard deviation was 0.006 - 0.02 mg $dm^{-3}$, while the amount determined ranged from 0.08 to 1.1 mg $dm^{-3}$. Similar pretreatment of the sample was applied in a titrimetric procedure, with the iodide ion-selective electrode. Otherwise, iodine and iodate were reduced with sodium arsenite in a nitric acid medium of pH 2 - 3[221]. Good results were obtained for about 200 mg $dm^{-3}$ iodine.

## 4. ON-LINE MONITORS

As mentioned in Section 3, several studies have been carried out on continuous or flow systems for laboratory use. Besides these, several manufacturers produce single- or multi-parameter apparatus, designed for industrial or routine monitoring of various types of waters. Such analysers are based on some common general principles, and their detailed operational layout is not very dependent on the species monitored in water. The most important part of the analysers comprise the supply system, usually containing a peristaltic pump with a system, of tubings, the measuring cell and the electronics for signal amplification and processing, meter display, chart recorder or printer. Provision is also normally included for automatic standardization and calibration.

The supply system is based on a multichannel peristaltic pump. The number of channels and the number of simultaneously-operated lines depends on the chemical complexity of the system. In most cases a two-channel system seems to be sufficient, because one leads the sample or standard, the second leads the reagent solution necessary for obtaining a proper medium for electrode measurements. An example of a more sophisticated system is that for cyanide, which has a four-channel unit (Fig. 7). The sample (channel 1) is initially mixed with a demasking solution containing EDTA (channel 2), and passes a reaction coil where it is heated, and after debubbling (channel 3) is mixed with an alkaline solution containing potassium dicyanoargentate(I) (channel 4), and enters the electrode compartment. In such systems the pump rotation is constant, and the rate of solution flow is controlled by the tubing diameter. In other systems, where the reagents cannot be mixed together before sample addition, or where special operations, such as prolonged heating, dialysis, etc., are needed, more complicated dosing systems are required. This also happens when the sample stream is segmented with air bubbles, as in the Technicon Autoanalyzer.

Operation of a monitor in a flow system with samples of natural waters containing inorganic, organic or biological particulate matter has the danger of fouling either the supply circuit or the electrodes in the measuring cell. Both may cause false indication. Occasionally, an ultrasonic or mechanical device can be used for cleaning electrodes. A great improvement would be an effective rapid by-pass filter.

The electrode cell should provide for trouble-free measurement, normally in thermostatted conditions. This requires proper earthing and small cell resistance, necessitating the location of both cell electrodes close together. The design of the cell significantly influences the stability of the potential, which on the other hand may be disturbed by an improperly-working pump and bad screening.

The electronics, besides its normal function in potentiometry, should have excellent long-term stability, and its additional aim is the control of standardization of the system, by switching on the stream of one or more standards at fixed intervals, e.g. every 6 or 12 h.

Because of the need to function over long periods, it is important that the consumption of reagent solutions by the monitor is as small as possible, thus lowering the cost of reagents and of labour by eliminating the need for frequent refilling of the solution reservoirs.

The basic scheme of the procedure for various ion-selective determinations is more or less similar. Therefore, the parent model of these monitors may be often used for different species by simply changing the sensor, and of the kind and rate of

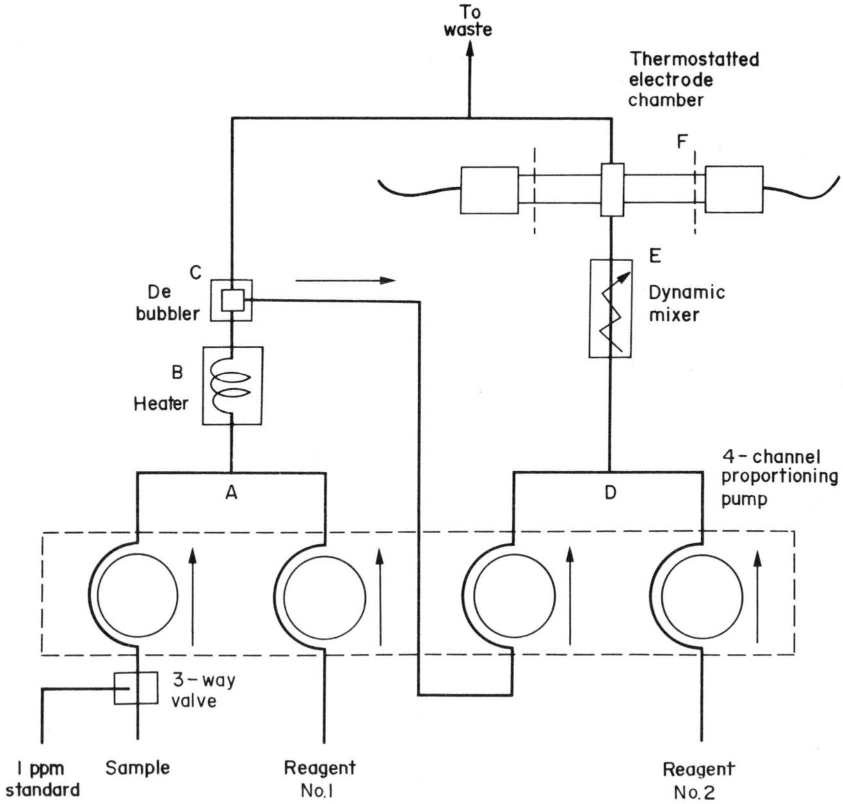


Fig. 7. Flow diagram of Orion Total Cyanide Monitor, Series 1000. Reagent No. 1 contains slightly acidic EDTA, and reagent No. 2 contains alkaline potassium dicyanatoargentate(I). (By courtesy of Orion Research, Inc.)

solution addition. In this way, e.g., the Series 1000 on-line analysers produced by Orion can be adapted for various determinations (Fig. 7). Thus, this monitor Series may be used in water analysis for the determination of carbonate, nitrate, fluoride, ammonia, sodium, cyanide, chloride, chlorine and water hardness. An extension of these applications is possible when heavily polluted waters are examined.

Other examples of monitors are the E.I.L. Model 8000 (Fig. 8), and the Condimat system made by Polymetron (Fig. 9). These monitors are designed for use with electrodes produced by the same manufacturer. They can be used for most of the applications in water analysis enumerated above. However, this does not complete the list of manufacturers supplying monitoring equipment, which also includes Beckman, Foxboro, Philips, Radiometer and Tacussel. These monitors differ only in technical details. It is difficult to consider any one of them superior to the others, but most of them, when properly adapted to a given programme, work efficiently.

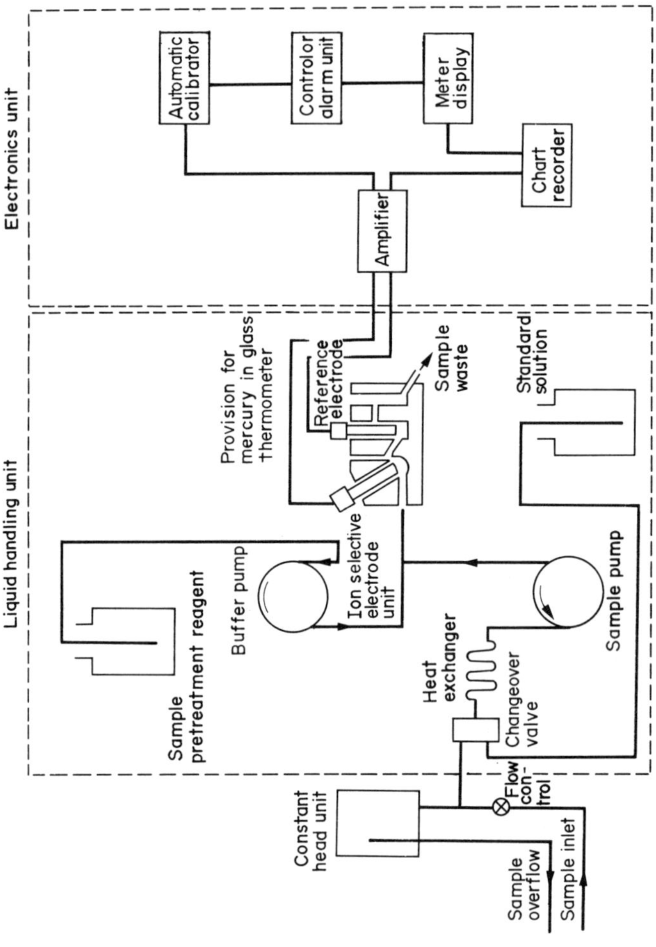


Fig. 8. Flow diagram of E.I.L. Automatic Industrial Monitor Model 8000.(By courtesy of Electronic Instruments Ltd.)

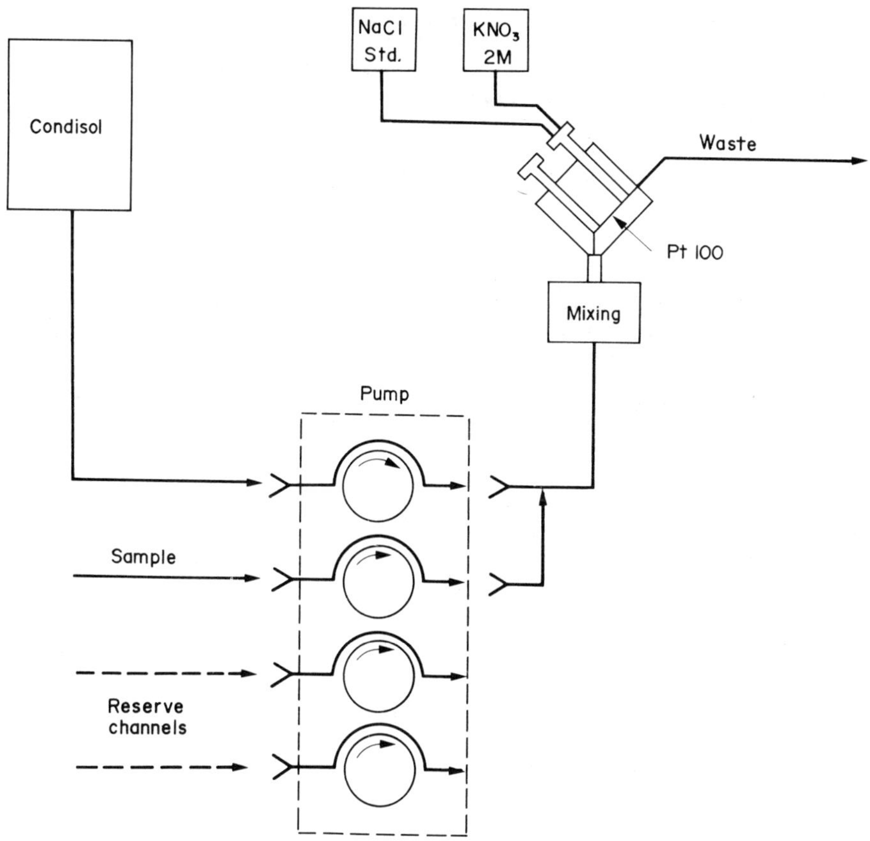


Fig. 9. Flow diagram of Polymetron Free Cyanide Monitor, CONDIMAT. (By courtesy of Polymatron.)

Among more recent developments, relatively simple but efficient devices are based on the principle of flow injection with non-segmented flow, invented by Růžička, Hansen and coworkers[49,52,118]. They are constructed of separate units, and may be adapted to a particular purpose by connecting the units into a complete thermostatted flow system. Different types of sensors may be used, among them ion-selective electrodes. Such systems are commercially available from BIFOK, Stockholm, Sweden. The progress in instrumentation and increasing needs and requirements for continuous analyses will progressively lead to more sophisicated and perfect sampling and measuring equipment.

## REFERENCES


1. J. N. Butler, in *Ion-Selective Electrodes* (Edited by R. A. Durst) p.143, NBS Special Publ. No.314 (1969).
2. F. Brudevold, E. Moreno and Y. Bakhos, *Arch. oral . Biol.* 17, 1155 (1972); *op. cit. C.A.*, 77, 105445s (1972).
3. T. N. Florence and G. E. Batley, *Talanta*, 24, 151 (1977).
4. L. Kamp-Nielsen, *Deep-Sea Res.* 19, 899 (1972).
5. M. J. Stiff, *Water Res.* 5, 585 (1971).
6. J. Gardiner, *Water Res.* 8, 23 (1974).
7. R. N. Sylva, *Water Res.* 10, 789 (1976).
8. P. G. Manning and S. Ramamoorthy, *J. inorg. nucl. Chem.* 35, 1577 (1973).
9. J. Buffle, F. L. Greter and W. Haerdi, *Analyt. Chem.* 49, 216 (1977).
10. A. Hulanicki and M. Trojanowicz, *Analytica Chim. Acta*, 68, 155 (1974).
11. M. Whitfield, *Proc. Analyt. Div. Chem. Soc.* 12, 56 (1975).
12. R. G. Bates, *Pure appl. Chem.* 36, 407 (1973).
13. A. K. Covington and R. A. Robinson, *Analytica Chim. Acta*, 78, 219 (1975).
14. J. B. Andelman, *J. Water Pollut. Control Federation*, 40, 1844 (1968).
15. R. M. Carlson and D. R. Keeney, *Instrumental Methods Analysis Soil Plant Tissue Proc. Symp.* p.39 (1970).
16. M. Mascini, *Inquinamento*, 13, 21 (1971).
17. R. J. Simpson, *Effluent Water Treat. J.* 11, 96 (1971).
18. S. J. Werber, *Int. Lab.* 28, (1971).
19. K. Slevogt, E. Seelos and W. Schraidt, *Vom Wasser*, 42, 1 (1974).
20. E. Pungor and K. Toth, *Vom Wasser*, 42, 43 (1974).
21. M. Trojanowicz and A. Hulanicki, *Chem. Analysis* (Warsaw), 22, 615 (1977).
22. E. Eicken, *Vom Wasser*, 49, 139 (1977).
23. G. J. Moody and J. D. R. Thomas, Applications of ion-selective electrodes, in *Ion-selective Electrodes in Analytical Chemistry*, (Edited by Henry Freiser) Vol.1, p.339, Plenum Press, New York, and London (1978)
24. D. Midgley and K. Torrance, *Potentiometric Water Analysis*, J. Wiley, Chichester (1978).
25. E. L. Eckfeldt and W. E. Proctor, Jr., *Analyt. Chem.* 43, 332 (1971).
26. R. C. Reynolds, *Water Res.* 7, 1333 (1971).
27. F. V. Ranzen and Z. Ya. Solovieva, *Radiokhimiya*, 15, 101 (1973).
28. I. Sekerka and J. F. Lechner, *Analyt. Lett.* 7, 463 (1974).
29. P. J. van der Winkel, J. Mertens, G. de Baenst and D. L. Massart, *Analyt. Lett.* 5, 567 (1972).
30. D. Hawthorn and N. J. Ray, *Analyst.* 83, 158 (1968).
31. H. M. Webber and A. L. Wilson, *Analyst*, 94, 209 (1969).
32. American Society for Testing and Materials, Test *ASTM: D* 2791-*T* (1969).
33. E. L. Eckfeldt and W. E. Proctor, Jr., *Analyt. Chem.* 47, 2309 (1975).
34. E. L. Eckfeldt, *Analyt. Chem.* 47, 2307 (1975).
35. R. W. Catrall, S. Tribuzio and H. Freiser, *Analyt. Chem.* 46, 2223 (1974).
36. T. Anfalt and D. Jagner, *Analytica Chim. Acta.* 66, 152 (1973).
37. K. L. Cheng and K. Cheng, *Amer. Chem. Soc. Div. Water Air Waste Chem.* General Papers, 11, 31 (1971); *op. cit. C.A.*, 77, 156206*x* (1972).
38. K. L. Cheng and K. Cheng, *Mikrochim Acta.* 385 (1974).
39. M. E. Thompson and J. W. Ross, Jr. *Science*, 154, 1643 (1966).
40. T. A. Hadjiioannou and D. S. Papastopoulos, *Talanta*, 17, 399 (1970).
41. G. J. Moody and J. D. R. Thomas, *Selective Ion Sensitive Electrodes*, Merrow, Watford (1971).
42. G. J. Moody, R. B. Oke and J. D. R. Thomas, *Analyst*, 95, 910 (1970).
43. G. H. Griffiths, G. J. Moody and J. D. R. Thomas, *Analyst.* 97, 420 (1972).
44. J. W. Ross, Jr. *Science*, 156, 1378 (1967).
45. J. Růžička, E. H. Hansen and J. C. T. Tjell, *Analytica Chim. Acta.* 67, 155 (1973).
46. A. Hulanicki and M. Trojanowicz, *Analytica Chim. Acta*, 87, 411 (1976).

47. A. Hulanicki and M. Trojanowicz, *Proc. Ion-Selective Electrodes Symposium* held at Matrafüred, Hungary, 18 - 21 October 1976 (Edited by E. Pungor and I. Buzás).
48. A. Craggs, G. J. Moody and J. D. R. Thomas, *Analyst.* 104, 412 (1979).
49. D. Ammann, M. Güggi, E. Pretsch, and W. Simon, *Analyt. Lett.* 8, 709 (1975).
50. A. Hulanicki and M. Trojanowicz, *Analytica Chim. Acta,* 68, 155 (1955).
51. A. Hulanicki and M. Maj-Zurawski and M. Trojanowicz, *Zh. analyt. Khim.* 32, 767 (1977).
52. E. H. Hansen, J. Růžička and A. K. Ghose, *Analytica Chim. Acta,* 100, 151 (1978).
53. R. C. Thurnau, U. S. Environment Protection Agency, 600/2-78,196 (1978).
54. M. Whitfield, J. V. Leyendekkers and J. D. Kerr, *Analytica Chim. Acta,* 45, 399 (1969).
55. A. Hulanicki and M. Trojanowicz, *Chem. Analysis* (Warsaw), 18, 235 (1973).
56. D. Jagner and J. P. Østergaard Jensen, *Analytica Chim. Acta.* 80, 9 (1975).
57. M. Mascini, *Analytica Chim. Acta,* 56, 316 (1971).
58. Orion Research Inc. *Application Bulletin A* 13.
59. T. F. Christiansen, J. E. Busch and S. C. Krogh, *Analyt. Chem.* 48, 1051 (1975).
60. C. Martin and A. M. Poudou, *Trav. Soc. Pharm. Montpellier,* 31, 371 (1971).
61. R. T. Oliver and R. F. Mannion, *16th Nat. Symp. Analysis Instrum. Div. Instr. Soc. Amer., Pittisburgh* (1970).
62. I. Sekerka and J. F. Lechner, *Analyst. Lett.* 7, 399 (1974).
63. I. Sekerka and J. F. Lechner, *Talanta* 22, 459 (1975).
64. G. J. Heijne and W. E. van der Linden, *Analytica Chim. Acta,* 93, 99 (1977).
65. D. J. Crombie, G. J. Moody and J. D. R. Thomas, *Talanta,* 21, 1094 (1974).
66. P. Lanza, *Analytica Chim. Acta,* 105, 53 (1979).
67. G. Johansson and K. Edström, *Talanta,* 19, 1623 (1972).
68. G. J. Moody, N. S. Nassory, J. D. R. Thomas, D. Betteridge, P. Szepesvary and B. J. Wright, *Analyst,* 184, 348 (1979).
69. M. J. Smith and S. E. Manahan, *Analyt. Chem.* 45, 836 (1973).
70. R. Jasinski and I. Trachtenberg, *Analyt. Chem.* 46, 364 (1974).
71. A. Hulanicki, M. Trojanowicz and T. Krawcyznski vel Krawczyk, *Water Res.* 11, 627 (1977).
72. R. M. Carlson and J. L. Paul, *Analyt. Chem.* 40, 1292 (1968).
73. E. Midgley, *Analyst,* 100, 386 (1975).
74. U. Fiedler, E. H. Hansen and J. Růžička, *Analytica Chim. Acta.* 74, 423 (1975).
75. C. W. Moore, C. E. Robertson and H. D. Nygren, *U. S. Geol. Survey Prof. Pap.* 450-*B* (1962).
76. F. G. Römer. L. Puyker, G. B. H. Brinkman and B. Griepink, *Z. analyt. Chem.* 289, 35 (1978).
77. D. Weiss, *Chem. Listy.* 66, 858 (1972).
78. Orion Research Inc. *Newsletter,* 6, 1 (1974).
79. J. H. Riseman, *Amer. Lab.* 4, 63 (1972).
80. C. Collombel, J. P. Durand and J. Bureau, *J. Eur. Toxicol.* 3, 291 (1970).
81. I. Sekerka and J. F. Lechner, *Water Res.* 10, 479 (1976).
82. R. Durst, *Analyt. Lett.* 10, 961 (1977).
83. G. Nota, *Analyt. Chem.* 47, 763 (1975).
84. Orion Research Inc., *Analytical Methods Guide*. 7th Edition, May (1975).
85. N. Shibata, *Analutica Chim. Acta,* 83, 371 (1976).
86. W. H. Evans and B. T. Partridge, *Analyst.* 99, 367 (1974).
87. M. J. Beckett and A. L. Wilson, *Water Res.* 8, 333 (1974).
88. I. R. McKenzie and N. N. W. Young, *Analyst.* 100, 620 (1975).
89. P. L. Bailey and M. Riley, *Analyst.* 100, 145 (1975).
90. W. L. Banwart, M. A. Tavatabai and J. M. Bremner, *Comm, Soil. Sci. Plant Analysis* 3, 449 (1972).
91. D. Midgley and K. Torrance, *Analyst,* 97, 626 (1972).
92. T. R. Gilbert and A. M. Clay, *Analst. Chem.* 45, 1757 (1973).

93. P. J. Le Blanc and J. F. Sliwinski, *Int. Lab.* 25, September/October (1973).
94. R. F. Thomas and R. L. Booth, *Environ. sci. Technol.* 7, 523 (1973).
95. L. Vandervenne and J. Oudewater, *Centre Belge d'Etude Doc. Eaux, Report No. 352* (1973).
96. B. W. Hawker, D. Midgley and K. Torrance, *Lab. Pract.* 22, 724 (1973).
97. J. Růžička, E. H. Hansen, P. Bisgaard and E. Reyman, *Analytica. Chim. Acta.* 72, 217 (1974).
98. J. Barica, *J. Fish Res. Bd. Canada,* 28, 759 (1971); *op. cit. C.A.* 75, 101149*j* (1971).
99. J. Barica, *J. Fish Res. Bd. Canada,* 30, 1389 (1973); *op. cit. C.A.* 79, 133577*t* (1973).
100. G. I. Goodfellow and H. M. Webber, *Analyst.* 97, 95 (1972).
101. J. Mertens, P. van der Winkel and D. L. Massart, *Analyt. Lett.* 6, 81 (1973).
102. D. Midgley and K. Torrance, *Analyst.* 98, 217 (1973).
103. J. Mertens, P. van der Winkel and D. L. Massart, *Bull. Soc. Chim. Belg.* 83, 19 (1974).
104. R. Leschber and H. Schumann, *DECHEMA Monogr.* 80, 737 (1976); *op. cit. C.A.* 86, 176996*u* (1977).
105. A. Hulanicki, R. Lewandowski and M. Maj, *Analytica Chim. Acta,* 69, 409 (1974).
106. A. Hulanicki, M. Maj-Zurawska and R. Lewandowski, *Analytica Chim. Acta,* 98, 151 (1978).
107. T. G. Sommerfeldt, R. A. Miln and G. G. Kozub, *Comm. Soil Sci. Plant Analysis,* 2, 415 (1971); *op. cit. C.A.* 76, 89864*z* (1972).
108. E. G. Shaw and P. Wiley, *Calif. Agric.* 5, 11 (1969).
109. W. Abbot, *J. Mississippi Acad. Sci.* 14, 11 (1968); *op. cit. C.A.* 72, 47249*g* (1970).
110. D. R. Keeney, B. H. Byrnes and J. J. Genson, *Analyst.* 95, 383 (1970.
111. D. Langmuir and R. L. Jacobson, *Enviro. sci. Technol.* 4, 834 (1970).
112. N. G. Bunton and N. T. Crosby, *Water Treat. Exam.* 18, 338 (1969).
113. D. J. Milham, A. S. Awad, R. E. Paul and J. H. Bull, *Analyst,* 95, 751 (1970).
114. D. Weiss, *Chem. Listy,* 69, 202 (1975).
115. A. W. M. Sweetsur and A. G. Wilson, *Analyst,* 100, 485 (1975).
116. J. Mertens and D. L. Massart, *Bull. Soc. Chim. Belg.* 82, 179 (1973).
117. K. Nowak and J. Luczak, *Roczniki PZH,* 26, 401 (1975).
118. E. H. Hansen, A. K. Ghose and J. Růžička, *Analyst.* 102, 705 (1977).
119. H. J. Nielson and E. H. Hansen, *Analytica Chim. Acta.* 85, 1 (1976).
120. Orion Research Inc. *Newsletter,* 2 44 (1970).
121. Orion Research Inc. *Methods Manual,* 93 Series Electrodes, (1975).
122. L. Weil and K. E. Quentin, *Vom Wasser,* 40, 125 (1973).
123. J. Mertens, R. van der Winkel and D. L. Massart, *Analyst. Chem.* 47, 522 (1975).
124. E. H. Hansen, J. Růžička and N. R. Larsen, *Analytica Chim. Acta,* 79, 1 (1975).
125. R. J. Stevens, *Water Res.* 10, 171 (1976).
126. G. K. Buckee, *J. Inst. Brew.* 80, 291 (1974).
127. G. A. Rechnitz, Z. F. Lin and S. B. Zamochnick, *Analyt. Lett.* 1, 29 (1967).
128. G. A. Rechnitz, G. M. Frickle and M. S. Mohan, *Analyt. Chem.* 44, 1098 (1972).
129. G. A. Rechnitz and M. S. Mohan, *Analyt Chem.* 45, 1323 (1973).
130. O. T. Takaishvili, J. P. Moconelidze, J. M. Karacheyeva and P. I. Davitaya, *Zh. analyt. Khim.* 30, 1629 (1975).
131. W. Miśniakiewicz and K. Raszka, *Proc. Ion-Selective Electrodes,* Conference held at Budapest, Hungary 5-9 September 1977, (Edited by E. Pungor and I. Buzás) 467 (1978).
132. J. W. Ross, Jr. and M. S. Frant, *Analyt. Chem.* 41, 967 (1969).
133. Orion Research Inc. *Application Bulletin.* 11.
134. J. O. Goertzen and J. D. Oster, *Soil Sci. Soc. Amer. Proc.* 36, 691 (1972).
135. M. Mascini, *Analyst.* 98, 325 (1973).
136. A. Hulanicki, R. Lewandowski and A. Lewenstam, *Analyst.* 101, 939 (1976).
137. E. P. Scheide and R. A. Durst, *Analyt. Lett,* 10, 55 (1977).
138. M. Trojanowicz, *Analytica Chim. Acta,* in press.

139. G. M. Shpeizer, E. A. Rubinchik and E. A. Atanova, *Tr. Irkutsk. Univ. Ser. Khim.* 3, 119 (1970); *op. cit. C.A.* 76, 49735*b* (1972).
140. R. Jasinski and I. Trachtenberg, *Analyt. Chem.* 45, 1277 (1973).
141. G. Ouzounian and Michard, *Analytica Chim. Acta,* 96, 405 (1978).
142. A. M. Y. Jaber, G. J. Moody and J. D. R. Thomas, *Analyst.* 101, 179 (1976).
143. Z. Balogh, M. Varhelyi and M. Kellermayer, *Rheumatol. Balneol. Allergol.* 17, 244 (1976); *op. cit. C.A.* 86, 176975*m* (1977).
144. Y. A. Demirjiam and K. H. Mancy, *Pittsburgh Conf. Analyt. Chem. Appl. Spectroscopy,* Cleveland (1969).
145. E. Mor, V. Scotto, B. Maricenaro and G. Alabiso, *Analytica Chin. Acta,* 75, 159 (1975).
146. Orion Research Inc. *Application Bulletin* 12.
147. E. W. Baumann, *Analyt. Chem.* 46, 1345 (1974).
148. G. M. Shpeizer, *Gidrokhim. Mater.* 53, 93 (1970).
149. J. Gulens and B. Labbate, *AECL* - 5542 (1976).
150. J. Gulens, K. Jessome and C. K. Macneil, *Analytica Chim. Acta.* 96, 23 (1978).
151. Electronic Instruments Ltd. *Instruction Manual,* $SO_2$ Probe (1976).
152. Orion Research Inc. *Instruction Manual* $SO_2$ (1975).
153. K. Nagy, *Kjemi,* 35, 15 (1975); *op. cit. C.A.* 86, 176935*g* (1977)
154. N. T. C. Crosby, A. L. Dennis and J. G. Stevens, *Analyst.* 93, 643 (1968).
155. S. J. Patterson, N. G. Bunton and N. T. Crosby, *Water Treat. Exam.* 18, 182 (1969).
156. J. A. Růžička and L. Mrklas, *Vod. Hospod.* B23, 107 (1973).
157. E. Bellack, *J. Amer. Water Works Assoc.* 64, 62 (1972).
158. *Standard Methods for the Examination of Water and Waste Water,* 13th Edition, p.168 Amer. Public Health Assoc. Washington, D.C. (1971).
159. T. Anfält and D. Jagner, *Analytica Chim. Acta.* 53, 13 (1971).
160. P. G. Brewer, D. W. Spencer and P. E. Wilkniss, *Deep-Sea Res.* 17, 1 (1970).
161. J. Bagg, *Analyt. Chem.* 48, 1811 (1976).
162. E. W. Baumann, *Analytica Chim. Acta,* 54, 189 (1971).
163. M. Trojanowicz, *Talanta,* 26, xxx (1979).
164. M. S. Frant and J. W. Ross, Jr. *Analyt. Chem.* 40, 1169 (1968).
165. T. B. Warner, *Analyt. Chem.* 41, 527 (1969).
166. I. Sekerka and J. F. Lechner, *Talanta,* 20, 1167 (1973).
167. T. B. Warner and D. J. Bressan, *Analytica Chim. Acta,* 63, 165 (1973).
168. C. J. Rix, A. M. Bond and J. D. Smith, *Analyt. Chem.* 48, 1236 (1976).
169. T. B. Warner, *Deep-Sea Res.* 18, 1255 (1971).
170. P. Pakalns and Y. J. Farrar, *Water Res.* 10, 1087 (1976).
171. J. E. Harwood, *Water Res.* 3, 273 (1969).
172. T. B. Warner, *Naval Res. Lab. Report* 7216 (1970).
173. Orion Research Inc. *Application Bulletin* 5A.
174. P. Garnier, R. Frank and J. Krembel, *Inform. Dent.* 16, 1 (1972).
175. O. Collombel, J. Bureau and J. Cotte, *Ann. Pharm. Franc.* 29, 541 (1971).
176. J. A. Růžička and P. Zajicek, *Vod. Hospod.* B19, 200 (1969).
177. P. Kauranen, *Analyt. Lett.* 10, 451 (1977).
178. T. B. Warner, *Water Res.* 5, 459 (1971).
179. T. S. Light, R. F. Mannion and K. S. Fletcher III, *Talanta,* 16, 1441 (1969).
180. P. J. Ke and L. W. Regler, *Analytica Chin, Acta,* 53, 23 (1971).
181. R. H. Babcock and K. A. Johnson, *J. Amer. Water Works Assoc.* 60, 953 (1968).
182. A. A. Diggens and D. E. Collis, *Water Treat, Exam.* 18, 192 (1969).
183. J. Mertens, P. van der Winkel and D. L. Massart, *J. Pharm. Belg.* 29, 181 (1974).
184. P. Bertier, *Analysis,* 2, 722 (1974).
185. M. Trojanowicz and A. Hulanicki, *Proc. Meeting Polish Chem. Soc.* Warsaw, p.1-368 (1977).
186. D. E. Erdmann, *Environ. sci. Technol.* 9, 252 (1975).
187. V. V. Bardin, *Zavod. Lab.* 28, 967 (1962).
188. K. Torrance, *Analyst.* 99, 203 (1974).
189. I.Sererka, J. F. Lechner and R. Wales, *Water Res.* 9, 663 (1975).
190. C. Harzdorf, *Z. analyt. Chem.* 270, 23 (1974).
191. M. R. Masson, *Talanta,* 22, 933 (1975).

192. M. Whitfield, *Ion-Selective Electrodes for the Analysis of Natural Waters*, Australian Marine Sci. Associ. Sydney (1971).
193. L. N. Moskvin, R. G. Leoneva, R. G. Fokina and E. P. Breshkina, *Zh. analyt. Khim*, 28, 2384 (1973).
194. G. J. Moody, R. B. Oke and J. D. R. Thomas, *Analyst*. 94, 803 (1969).
195. A. R. Selmer-Olsen and A. Ørien, *Analyst*. 98, 412 (1973).
196. W. J. van Oort, L. Gipon, C. Bognar, J. G. van Heyst, P. C. M. Frintrop and B. Griepink, *Z. analyt. Chem*. 270, 200 (1974).
197. D. Weiss, *Chem. Listy*, 65, 305 (1971).
198. J. C. van Loon, *Analytica Chim. Acta*, 54, 23 (1971).
199. A. Hulanicki, Z. Augustowska and M. Trojanowicz, *Chem. Anal*. (Warsaw), 22, 955 (1977).
200. E. Mainka, W. Coerdt, W. Koenig and A. von Beeckmann, *Gas-Wasser-fach. Wasser-Absawwer*, 114, 570 (1973),
201. N. Ogato, *Bunseki Kagaku*, 21, 780 (1972).
202. K. Takamatsu, K. Takahaski, K. Saito, and S. Kanno, *Eisei Kagaku*, 22, 286 (1976).
203. M. J. Fishman and O. J. Fiest, Jr. *U.S. Geol. Survey Prof. Pap*. 700-*C*, *C*-226, *C*-228 (1970); *op. cit. C.A.* 73, 123408*d* (1970).
204. E. Pungor, K. Toth and G. Nagy, *Hung. sci. Instr*. 35, 1 (1975).
205. T. M. Florence, *J. electroanal, Chem*. 31, 77 (1971).
206. M. Mascici and A. Liberti, *Analytica Chim. Acta*. 51, 231 (1970).
207. D. Jagner and K. Aren, *Analytica Chim. Acta*, 52, 491 (1970).
208. J. G. Frost, *Analytica Chim, Acta*, 48, 321 (1969).
209. I. Sererka, J. F. Lechner and L. Harrison, *J. A.O.A.C.*, 66, 625 (1977).
210. I. Sererka and J. F. Lechner, *J.A.O.A.C.*, 61, 1493 (1978).
211. G. B. Marshall and D. Midgley, *Analyst*, 103, 438 (1978).
212. K. Tomlinson and K. Torrance, *Analyst.*, 102, 1 (1977).
213. M. Vanderputte, L. Dryon and D. L. Massart, *Analytica Chim. Acta*, 91, 113 (1977).
214. Orion Research Inc. *Newsletter*, 2, 26 (1970).
215. Orion Research Inc. *Instruction Manual*, $Cl_2$ Electrode 97-70.
216. L. P. Rigdon, G. J. Moody and J. W. Frazer, *Analyt. Chem*. 50, 465 (1978).
217. Orion Research Inc. *Instruction Manual for Halide Electrodes*, (1971).
218. R. C. Harris and H. H. Williams, *J. Appl. Meteorol*, 8, 299 (1969).
219. A. Hulanicki, A. Lewenstam and M. Maj-Zurawska, *Analytica Chim. Acta*, 107, 121 (1973).
220. E. Pungor, *Analyt. Chem*. 39, 28*A* (1967).
221. Ye. A. Zeinalova, T. N. Morshina and M. M. Senyavin, *Zh. analit. Khim*. 30, 966 (1975).

# THE AUTHORS

Professor Adam Hulanicki

Dr Marek Trojanowicz

Adam Hulanicki was born in Warsaw in 1929, and holds the M.Sc., Ph.D. and Doctor Habilitus degrees of the University of Warsaw. In 1953, he commenced work as an assistant at the University of Warsaw, and in 1976 he was appointed Professor of Analytical Chemistry in its Institute of Fundamental Problems in Chemistry.

His early research work was on emission spectroscopy and trace analysis. Subsequently, he became interested in organic reagents in analytical chemistry, initially in dithiocarbamates, but later in organic reagents used as redox and compleximetric indicators. His present research field covers theory and application of ion-selective electrodes, coulometric titrations, and atomic absorption spectroscopy. His research and teaching activities are reflected in about 90 published scientific papers and two books - namely, *Emission Spectra Analysis*, published in 1956, and *Acid-Base Reactions in Analytical Chemistry*, published in 1973 and 1978 in Polish, and in 1975 in Russian.

Professor Hulanicki is Vice-Chairman of the Committee for Analytical Chemistry of the Polish Academy of Sciences, and a member of Polish Chemical Society. He serves on the Analytical Chemistry Division of I.U.P.A.C. and is a member of the Advisory Editorial Boards of *Talanta*, *Analytica Chimica Acta*, *Mikrochimica Acta*, and *Ion-Selective Electrode Reviews*.

Marek Trojanowicz was born in Warsaw in 1944, and holds the M.Sc. and Ph.D degrees of the University of Warsaw. In 1966-1967 he held a research fellowship from the Ministry of Science and Higher Education, and studied under the direction of Professor W. Kemula. In 1966 he joined the Chemistry Faculty of the University of Warsaw as an assistant.

He received his Ph.D. degree in 1974 for a thesis on "pM Electrodes as Sensors in Compleximetric Titrations".

His research interests include analytical techniques, with emphasis on ion-selective electrodes and their applications to environmental analysis, investigations of chemical equilibria, flow measurements, and data processing. He has published over 20 scientific papers.

In 1976 he spent one year as a research fellow in the Laboratory of Professor Nobuyuki Tanaka at Tohoku University, Japan, as research fellow. He is a member of the Polish Chemical Society.

*Ion-Selective Electrode Rev.*, Vol.1, 251-273.
Pergamon Press Ltd. Printed in Great Britain.

# RECENT TITLES—I

G. J. Moody and J. D. R. Thomas
Chemistry Department, U.W.I.S.T., Cardiff CF1 3NU, Wales

CONTENTS

## 1. ELECTRODE DEVELOPMENT AND NEW ELECTRODES, INCLUDING ASPECTS OF SELECTIVITY, RESPONSE TIMES, ETC.

### 1.1 Solid-state devices


Effect of pH on the response of a cyanide ion-selective electrode. M.Gratzl, F.Rakias, G.Horvái, K.Tőth and E.Pungor, *Analytica Chim.Acta*, 102, 85 (1978).

Dynamic response of the fluoride ion-selective electrode. R.C.Hawkings, L.P.V.Corriveau, S.A.Kushneriuk and P.Y.Wong, *Analytica Chim.Acta*, 102, 61 (1978).

Formation and properties of mixed lead sulphide-silver sulphide membranes for lead(II)-selective electrodes. G.J.M.Heijne, W.E. van der Linden and G. den Boef, *Analytica Chim.Acta*. 100, 193 (1978).

Investigation and development of selective membrane electrodes. XII Hydrogen chromate selective electrodes. I.Yu.Urusov, V.V.Sergievskii, A.F.Zhukov and A.V.Gordievskii, *Zh.Anal.Khim.*, 34, 156 (1979).

Ion-selective electrode for determining sulphate. M.Yu.Gorina and L.E.Ryvkina, *Zh.Anal.Khim.* 33, 2269 (1978).

Solid-state chloride-, bromide-, or iodide-selective electrodes. G. Papeschi, S. Bordi and M. Carla, *J.Electrochem.Soc.*, 125, 1807 (1978).

Behaviour of copper(II) ion-selective electrode in various ion-buffers. T. Hashizume, *Kitokyushu Kogyo Koto semmon Gakko Kenkyu Hokuku*, 11, 183 (1978).

Silver-selective solid membrane electrode. A.K. Jain, S.K. Srivastava, R.P. Singh and S. Agarawal, *Anal.Chem.*, 51, 1093 (1979).

Fluoride-sensitive electrode based on bismuth fluoride. V. Ebock and C. Neiser, *Z. Chem.*, 18, 343 (1978).

Behaviour of a polycrystalline fluoride-selective membrane electrode in aqueous-organic media. L.I. Manakova, N.V. Bausova, V.E. Moiseev, V.G. Bamburov and A.P. Sivoplyas, *Zh.Anal.Khim.* 33, 1517 (1979).

Response to calcium ions of the calcium fluoride crystal-membrane electrode. Malathi Ghosh, M.R. Dhaneshwar, R.G. Daneshwar and B. Gosh, *Analyst*, 103, 768 (1978).

Behaviour of iodide-selective electrodes at low concentrations of iodide. A. Hulanicki, A.Lewenstam and M. Maj-Zwawska, *Analytica Chim.Acta*, 107, 121 (1979).

Electrodes for the determination of (copper) ion activity. J. Vesely, J. Jindra and J. Gregr, *Czech.Patent* 176,359 (Cl G01N27/30) 15 January 1979.

Experimental study (selectivity towards $OH^-$ and potentiometric titration curve) of a specific membrane electrode: the fluoride ion electrode. R. Combes, P. Letellier and N. Baffier, *Actual.Chim.* (6), 27 (1979).

Research and development of selective membrane electrodes. Communication 13. Ion-selective electrodes without an internal reference solution. A.F. Zhukov, A.V. Vishnyakov, Yu.I.Urosov, A.V. Kopytin, L.I. Glasman and A.V. Gordievskii, *Zh.Anal.Khim..*, 34, 602 (1979).

## 1.2 PVC and other polymer matrix membrane devices, including liquid ion-exchanger and neutral carrier systems

Ion-selective liquid-membrane anionic-surfactant-sensitive electrode and its analytical applications. N. Ciocan and D.F. Anghel, *Fresenius Z. Anal. Chem.*, 290, 237 (1978).

Design of calcium-ion-selective electrodes. J.D.R. Thomas, *Lab. Practice*, 27, 857 (1978).

Siloxane compounds as active components in ion-selective electrode membranes. J.G. Schindler, G. Stork, H.J. Strueh and W. Schael, *Fresenius Z. Anal. Chem.*, 290, 45 (1978).

Film membrane electrode selective for ammonium ions. O.K. Stefanova and V. Rusina, *Electrokhimiya*, 14, 682 (1978).

Voltammetric ion-selective electrode for the determination of nitrate. J.A. Cox and G.R. Litwinski, *Anal. Chem.*, 51, 554 (1979).

Potassium-selective PVC membrane electrodes based on bis- and poly(crown ether)s. K. Kimwa, T. Maeda, H. Tamwa and T. Shono, *J.Electroanal. Chem. Interfacial Electrohem.*, 95, 91 (1979).

Uranyl ion-selective electrode based on uranyl bis-(2-ethylhexyl)phosphate. V.A. Mikhailov, V.V. Osipov and N.V. Serebrennikova, *Zh. Anal. Khim.*, 33, 1154 (1978).

Uranyl ion-selective electrode based on a new synthetic neutral carrier (N,N'-diheptyl-N,N',6,6-tetramethyl-4,8-dioxaundecanediamide). J. Senkyr, D. Amman, P.C. Meier, W.E. Morf, E. Pretsch, W. Simon, *Anal. Chem.*, 51, 786 (1979).

Selectivity rating of calcium ion-selective electrodes. G.J. Moody, N.S. Nassory and J.D.R. Thomas, *Proc. Anal. Div. Chem. Soc.* 16, 32 (1979).

Preparation of PVC-type calcium ion-selective electrode (bis(o-isooctylphenyl) phosphate as sensor) and determination of calcium in water. S-W. Sun, X. Wang and O-H. Ge, *Hua Hsueh Tung Pao,* (2), 128 (1979).

A new coated-wire cobalt(II)-selective electrode based on the benzalkonium tetrathiocyanatocobaltate(II) ion pair. K. Burger and G. Petho, *Analytica Chim. Acta,* 107, 113 (1979).

Improved liquid-membrane ion-selective electrodes. K.Koshicshi and T.Mizusakii, *Ger.Offen.*, 2, 845, 751 (Cl. GOIN 27/30) 26 Apr. 1979.

One-piece electrodes for determining the activity of ions. W. Simon, *Ger. Offen.*, 2, 739, 976 (Cl. GOIN 27/30 ) 18 Jan. 1979.

Study of a sodium ion-selective (neutral carrier) electrode. E.A. Materova, Z.S. Alagova, G.I. Shumilova and L.P. Vatlina, *Vestn. Leningrad. Univ., Fiz. Khim.* (4), 112 (1978).

The preparation of mono- and di-[4-(n-octyl)phenyl]- and -[4-(1,1,3,3-tetramethylbutyl)phenyl]phosphoric acids for use in PVC calcium ion-selective electrodes. A. Craggs, P.G. Delduca, L. Keil, B.J.Key, G.J. Moody and J.D.R. Thomas, *J. Inorg. Nucl. Chem.*, 40, 1483 (1978).

Preparation and properties of di-n-octyl(3-nitrophenyl)phosphonate and various di-n-alkylphenylphosphonates for use as ion-selective solvent mediators. A. Craggs, P.G. Delduca, L. Keil, G.J. Moody and J.D.R. Thomas, *J. Inorg. Nucl. Chem.*, 40, 1943 (1978).

Evaluation of calcium ion-selective electrodes based on di[(n-alkyl)phenyl]phosphate sensors and their calibration with ion buffers. A. Craggs, G.J. Moody and J.D.R. Thomas, *Analyst*, 104, 412 (1979).

Testing and characteristics of a clobutinol-sensitive electrode. K. Fukamachi and N. Ishibashi, *Yakugaku Zasshi,* 99, 126 (1979).

Combination ion-selective electrode based on solvent polymeric membranes. R.E. Dohner and W. Simon,*Anal. Letters*, 12, 205 (1979).

Response-time studies on neutral-carrier ion selective membrane electrodes. E.Linder, K. Tóth, E. Pungor, E. Morf and W. Simon, *Anal. Chem.*, 50, 1627 (1978).

Effect of a phenol derivative on the selectivity of organic sulphonate-selective electrodes. T. Fujinaga, S. Okazaki and H. Hara, *Chem. Letters.*, 11, 1201 (1978).

The influence of plasticizers on the characteristics of the potassium-valinomycin disc electrode. J.G. Schindler, G. Stork and H.-J. Strüh, *Fresenius Z. Anal. Chem.*, 292, 391 (1978).

Optimisation (with plasticisers) of a potassium-valinomycin disc electrode. J.G. Schindler, G. Stork and H.-J. Strüh, *Fresenius Z. Anal. Chem.*, 292, 396 (1978).

Nitrate ion-selective electrodes based on complexes of 2,2'-bipyridine and related compounds as ion-exchangers. T.L. Hwang and H.S. Cheng, *Analytica Chim. Acta.*, 106, 341 (1979).

Liquid membrane electrodes for perchlorate, thiocyanate, tetrafluoroborate and nitrite based on triphenyl-methane dyes. E. Hopirtean and E. Stefaniga, *Rev. Roum.Chim..*, 23, 137 (1978).

Ion-selective electrode responsive to chlorate ion. H. Kohara, *Kitakyushu Kogyo Koto Semmon Gakko Kenkyu Hokoku.*, 11, 159 (1978).

Effect of solvent in the nitrate-selective electrode. A. Hulanicki, M. Maj-Zurawska and R. Lewandowski, *Analytica Chim. Acta.*, 98, 151 (1978).

Coated-platinum sulphate-selective electrode. W. Baumann. *Analytica Chim. Acta*, 99, 247 (1978).

## 1.3 Enzyme and bacteria-based electrodes

Bacterial membrane electrode for L-cysteine. M.A. Jensen and G.A. Rechnitz, *Analytica Chim. Acta*, 101, 125 (1979).

(Amperometric) glucose and L-amino-acid electrodes based on enzyme membranes. G.J. Lubrano and G.G. Guilbault, *Analytica Chim. Acta*, 97, 229 (1978).

An enzyme electrode for acetylcholine. P.Durand, A. David and D. Thomas. *Biochem. Biophys. Acta*, 527, 277 (1978).

## 1.4 Field effect transistor (FET) systems

Ion-selective field transistors - a new type of electrode for chemical analysis and biomedical studies. Yu.G. Vlasov, *Zh. Prikl. Khim. (Leningrad)*, 52, 3 (1979).

The selectivity and temperature characteristics of ion-sensitive field effect transistors. O. Leistiko, *Phys. scr.*, 18, 445 (1978).

A new type of ion-selective device. J.S. Johannessen, T.A. Fjeldy and K. Nagy. *Phys. Ser.*, 18, 464 (1978).

Solid-state ion-selective electrodes with integrated electronics. T.A. Fjeldly, K.Nagy and J.S.Johannessen, *J. Electrochem. Soc.*, 126, 793 (1979).

### 1.5 Miscellaneous

Membrane with ion-selective properties (for pH). J. Schindler and W. Schael, *Ger. Offen.*, 2,750,807. (Cl. B01DB/04) 17 May 1979.

Reactivation of ion-selective electrodes. K. Nikei, S. Kawana, H. Miyagi and Y. Takada, *Japanese Patent*, 79, 26, 791 (Cl. G01N27/30) 28 Feb 1979.

Electrode (protected from penetration of the reference solution into the area of electric shielding) for measuring ions in solution. W. Ingold, *Ger. Offen.*, 2,805,088 (Cl. G01N 27/30) 3 May 1979.

Studies on ion-selective electrodes. Part III. On the preparation and testing of some ion-selective electrodes. R. Virtanen. *Kem.-Kemi.*, 6, 89 (1979).

Some aspects of asymmetry potentials in potentiometry. O.F. Schaefer, *Fresenius Z. Anal. Chem.*, 293, 30 (1978).

## 2. MECHANISTIC AND THEORETICAL ASPECTS

Use of a method of concentration cells for detecting the limiting stage of ionic transport through membranes based on neutral complexing agents. O.K. Stefanova, *Vestn. Leningrad Univ. Fiz. Khim.*, 4 94 (1978).

Interference films on the sensor membranes of solid-state copper(II) ion-selective electrodes. G.J. Moody, N.S. Nassory, J.D.R. Thomas, D. Betteridge, P. Szepesvary and B.J. Wright, *Analyst*, 104 348 (1979).

Limit of detection in analysis with ion-selective electrodes. D. Midgley, *Analyst*, 104, 248 (1979).

Ion transport in free and supported nitrobenzene Aliquat nitrate liquid membrane ion-selective electrodes. I. Bulk electrical properties including ion association and dielectric constant. D.E. Mathis and R.P. Buck, *J. Membrane Science*, 4, 379 (1979).

Ion transport in free and supported nitrobenzene Aliquat nitrate liquid membrane ion-selective electrodes. II. Interfacial kinetics and time-dependent phenomena. D.F. Mathis, F.S. Stover and R.P. Buck, *J.Membrane Science*, 4, 395 (1979).

Observations on the surface of cyanide ion-selective electrode by the use of energy dispersive X-ray fluorescence spectrometer. T. Aomi, *Denki Kagaku*, 45, 567 (1978).

Statistical approach for the selectivity of ion-selective membrane electrodes. C. Liteanu, E. Hopirtean, I.C. Popescu, I. Rica and E. Stefaniga, *Anal.Chem..*, 50, 1202 (1978).

Effects of surface heterogenicity on the sensitivity of sulphide-ion-selective electrodes. J. Gulens and B. Ikeda, *Anal. Chem.*, 50, 782 (1978).

Mechanism of the functioning of ion-selective membranes based on neutral complexing agents. Ion transport through the inner layers of a membrane containing valinomycin. V.E. Yurinskaya, O.K. Stefanova, and E.A. Materova, *Elektrokhimiya*, 15, 723 (1979).

Function (with respect to oxygen partial pressure) of the internal metallic lead of ion-selective disc electrodes. I. J.G. Schindler, G. Stork, H.J. Strüh, W. Schmid and K.D. Karaschinski. *Fresenius Z. Anal. Chem.*, 295, 248 (1979).

Mechanism of the functioning of ion-selective membranes based on neutral complexing agents. Distribution of ions of different charge signs between an aqueous solution and a membrane containing valimomycin. V.E. Yurinskaya, O.K. Stefanova, E.A. Materova and V.V. Glazunov, *Elektrokhimiya,* 15, 419 (1979).

A study of the mechanism of response of liquid ion-exchanger calcium selective electrodes. Part I. Zero current potential of commercial and modified electrodes. N.D. Van Mou and C. Gavach, *J. Electroanal. Chem. Interfacial Electrochem.*, 97, 151 (1979).

A study of the mechanism of response of liquid ion-exchanger calcium selective electrodes. Part III. Membrane behaviour under non-zero current conditions. W.D. Van Mou, and C. Gavach, *J. Electroanal. Chem. Interfacial Electrochem.*, 97, 171 (1979).

## 3. SOLUTION CHEMISTRY AND ION-SELECTIVE ELECTRODES

Comparison between measured and calculated concentrations of calcium ions in buffers. O. Scharff, *Analytica Chim. Acta.*, 109, 291 (1979).

Determination of free calcium concentrations with an ion-selective electrode in the presence of chelating agents in comparison with calculated values. Y.S. Kim and G.M. Padilla, *Anal. Biochem.*, 89, 521 (1978).

## 4. STANDARD ADDITION METHODS

The use of a microprocessor-based millivoltmeter in applications of ion-selective electrodes. G.J. Moody and J.D.R. Thomas, *Lab. Practice.*, 28, 125 (1979).

Novel computer evaluation of multiple standard addition with ion-selective electrodes. G. Horvái, L. Domokos and E. Pungor, *Fresenius Z. Anal. Chem.*, 292, 132, (1978).

Direct potentiometry with ion-selective electrodes. Method of double addition. R. Vadura and H. Pokorna, *Chem. Prum.*, 29, 194 (1979).

Studies on the standard addition method in ion-selective electrodes - an equation and a plot involving ΔpX. T-F. Chao, *K'o Hsenek T'ung Pao*, 24, 258 (1979).

Studies on the standard addition method in ion-selective electrodes - an equation a plot and a nomograph involving ΔE. T-F.Chao, *K'o Hsenek T'ung Pao,* 24, 212 (1979).

## 5. POTENTIOMETRIC TITRATIONS AND CONTINUOUS MONITORING

Novel titration technique for analysis of streamed sample - the triangle-programmed titration technique. III. Titrations with electrically generated bromine. G. Nagy, Z. Feher, K. Tóth and E. Pungor, *Analytica Chim. Acta,* 100, 181 (1978).

Electrochemical sensors for process-stream monitoring. Peter L.Bailey, *Anal.Chem.*, 50, 698A (1978).

Computer-controlled interference correction for ion-selective electrode measurements in a flowing system. P.D. Gaarenstroom, J.C. English, S.P. Perone and J.W. Bixler, *Anal.Chem.*, 50, 811 (1978).

Some practical considerations for the incorporation of ion-selective electrodes into continuous-flow analytical systems. D.C. Cowell, *Med. Lab. Sci.*, 35, 265 (1978).

Determination of the equivalence point in potentiometric titrations with Gran's first method used to test the electrode response. E. Still, *Analytica Chim. Acta,* 107, 377 (1979).

Apparatus and methods for continuous electrochemical analysis. U. Gruenke, P. Hartmann, U. Tietz, E. Keil, *Ger. (East) Patent,* 132,210 (Cl. GOIN 27/30), 6 Sept. 1978.

On-line use of ion-selective electrodes. A.A. Diggens and S. Lichtenstein, *Proc. Int. Water Conf. Eng. Soc., West Pa.,* 39, 83 (1978).

The acidimetric (potentiometric titration) estimation of fluoride. A comparison of the glass with the lanthanum fluoride electrode. M. Cartwright, and A.A. Woolf. *J. Fluorine Chem.,* 13, 501 (1979).

Single-point potentiometric titrations with ion-selective electrodes. G. Horvái, K. Tóth and E. Pungor, *Analytica Chim. Acta,* 107, 101 (1979).

Ion-selective electrodes in titrations using tetraphenylborate. K. Vytras, *International Lab.,* 35 (March 1979).

## 6. GENERAL ASPECTS OF METHODOLOGY WITH ION-SELECTIVE ELECTRODES

Ammonium-selective disc-electrode with the carrier-antibiotic nonactin. A method for the continuous enzymatic electrochemical determination of urea. I.J.G. Schindler, R.G. Schindler and O. Aziz, *J. Clin. Chem. Clin. Biochem.,* 16, 441 (1978).

Flow-through electrochemical cell with open liquid junction. W.J. Blaedel and Z. Yim, *Anal. Chem.,* 50, 1722 (1978).

Construction and behaviour of a micro flow-through copper(II)-selective electrode. W.E. van der Linden and R. Oostervink, *Analytica Chim.Acta,* 101, 419 (1978).

Ultramicroanalytical simultaneous determination of silver and cadmium with ion-sensitive electrodes. P. Hartmann, U. Gruenke and H. Berge, *Z.Chem.,* 19, 71 (1979).

Electrochemical assay. J.J. Kalinowski, *Ind. Res. Det.*, 21, 105 (1979).

Digital signal processing in the ion-selective electrode system. W.R. Tracey, *Diss. Abtr. Int. B*, 39, 3966 (1979).

Sources of error in ion-selective electrode potentiometry. R.A. Durst in *Ion-Selective Electrodes in Analytical Chemistry, Volume 1*, Ed. H. Freiser, Plenum, New York, 311 (1978).

Studies on ion-selective electrodes. Part II. Calibration of ion-selective electrodes with exponential dilution flask. R. Vistanen, *Kem.-Kemi*, 5, 614 (1978).

Computer automation of potentiometric analysis with ion-selective electrodes. J. Stanina, F. Bakker, J.J. Moels, J.E. Ordelman and A.G.M. Bruyn-Hes, *Analytica Chim. Acta*, 112, 45 (1979).

Microcomputer-controlled potentiometric analysis system. C.R. Martin and H. Freiser, *Anal. Chem.*, 51, 803 (1979).

Selection of suitable reference electrodes of pH and ion-selective measurement. W.C. Clark, *Beckman Rep.*, 22 (1978).

## 7. APPLICATIONS WITH ION-SELECTIVE ELECTRODES

### 7.1 Coordination complexes and reaction kinetics

Stability of alkaline earth monofluoride complexes in methanol. J.W. Bixler and A.M. Bond, *J. Inorg. Chem.*, 17, 3684 (1978).

Kinetic study of the iron(II)-induced perbromate-iodide reaction with an iodide ion-selective electrode and kinetic determination of iron, perbromate, ethylene diamine-tetraacetic acid, diethylenetriaminepentaacetic acid, and ethylene glycol-bis[2-aminoethyl ether]-N,N',N',N'-tetraacetic acid. L.A. Lazaron and T.P. Hadjiioannou, *Anal. Chem.*, 51, 790 (1979).

Metal [iron(III); copper(II)]-EDTA complex formation studies in ethanol-aqueous mixtures with cupric ion-selective electrode. N. Radic, *Anal. Letters*, 12, 115 (1979).

Experiments (involving calcium complexes) with the PVC matrix membrane calcium ion-selective electrode. B.J. Birch, A. Craggs, G.J. Moody and J.D.R. Thomas, *J. Chem. Educ.*, 58, 740 (1978).

### 7.2 Mineralised tissue and dental materials

Rapid determination of fluoride in bovine jaws. P. Bourbon, J. Alary, M. Segui and J.F. Esclassan, *Analysis*, 6, 364 (1978).

Fluoride content of buccal surface enamel from caries-free mediaeval subjects. A. Richards, M.J. Larsen and O. Fejerskov, *Archs oral Biol.*, 24, 83 (1979).

The kinetics of inorganic phosphate in human dental plaque and saliva. A. Tatevossian and C.T. Gould, *Archs oral Biol.*, 24, 461 (1979).

Distribution and kinetics of fluoride ions in the free aqueous and residual phases of human dental plaque. A. Tatevossian, *Archs oral Biol.*, 23, 893 (1978).

Fluoride concentration in tooth germs of permanent teeth of the domestic pig. R.L. Speirs, *Archs oral Biol.*, 23, 1019 (1978).

Binding of fluoride in oral streptococci. S. Kashket and F.J. Bunick, *Archs oral Biol.*, 23, 993 (1978).

Incorporation of a dietary fluoride supplement into bone and developing teeth of the domestic pig. R.L. Speirs, *Archs oral Biol.*, 23, 1013 (1978).

Effect of a cariogenic challenge on fluoride uptake by enamel of rats receiving fluoridated drinking water. J.R. Mellberg and R.H. Larson, *Caries Res.*, 12, 137 (1978).

Calcium fluoride and fluoridated hydroxyapatite formation in relation to the acid dissolution of enamel mineral. S. Duke and G.C. Forward, *Caries Res.*, 12, 12 (1978).

Supplemental fluoride recommendations for infants based on dietary fluoride intake. S.M. Adair and S.H.Y. Wei, *Caries Res.*, 12, 76 (1978).

Relationship between fluoride in the drinking water and the plasma fluoride concentration in man. J. Ekstrand, *Caries Res.*, 12, 123 (1978).

Effects of fluoride on the initial colonization of teeth *in vivo*. M. Kilian, M.J. Larsen, O. Fejerskov and A. Thylstrup, *Caries Res.*, 13, 319 (1979).

Effect of fluoride on the growth of hydroxyapatite and human dental enamel. Z. Amjad and G.H. Nancollas, *Caries Res.*, 13, 250 (1979).

Enhanced fluoride penetration and retention in enamel. R.T.Koritzer and J.S.Levy, *Caries Res.*, 13, 259 (1979).

Surface enamel fluoride and strontium in relation to caries prevelance in man. P.C. Spector and M.E.J. Curzon, *Caries Res.*, 13, 227 (1979).

A preliminary study of the microelement composition of the outer layer of dental enamel. T.W.Cuttress, *Caries Res.*, 13, 73 (1979).

A laboratory evaluation of fluoride dentrifices in Greece. T.M. Athanassouli and D.S. Papastathopoulis, *Comm. Dent. Oral Epidemiol.*, 6, 185 (1978).

The release of fluoride from alginate impression material. F. Hattab and G. Frostell, *Comm. Dent. Oral Epidemiol.*, 6, 273 (1978).

Clinical appearance of dental fluorosis in permanent teeth in relation to histologic changes. A. Thylstrup and O. Fejerskov, *Comm. Dent. Oral Epidemiol.*, 6, 315 (1978).

The calcium selective electrode - a possible tool in dental research. K. Bjorvatin and T. Mørch, *Acta Odont. Scand.*, 37, 259 (1979).

Dental caries and fluorosis in Arussi province, Ethiopia. B. Olsson, *Comm. Dent. Oral Epidemiol.*, 6, 338 (1978).

Dental findings in high fluoride areas of Ethiopia. B. Olsson, *Comm. Dent. Oral Epidemiol.*, 7, 51 (1979).

Kenhardt revisited - a study in a high fluoride area. D.H. Retief, E.L. Bradley, F.H. Barbakow, M. Friedman, J.I. Bischoff and E.H.M. van der Merwe, *J. Dent. Ass. S.Africa,* 34, 427 (1979).

Effects of a single dose of fluoride on calcium metabolism. M.J. Larsen, F. Melsen, L. Mosekilde and M.S. Christensen, *Calcif. Tiss. Res.,* 26, 199 (1978).

Effects of fluoride on *in vitro* calcification of tendon matrix. C.L. Wadkins and R.A. Luben, *Calcif. Tiss. Res.,* 26, 51 (1978).

Mineral substance of bone tissue and of experimental cutaneous calcinosis in rats: chemical analysis and ESR study. H.J. Tochon-Danguy, M. Geoffroy, G. Boivin and C.A. Baud, *Calcif. Tiss. Res.,* 26, 259 (1978).

Localised fluoride release from fluorine-carrying polyphosphonates. M. Anbar, E.P. Farley, D.D. Denson and K.R. Maloney, *J. Dent. Res.,* 58, 1134 (1979).

Titanium and fluoride concentrations in titanium tetrafluoride and APF treated enamel. B. Clarkson and J. Wefel, *J. Dent. Res.,* 58, 600 (1979).

The effect of metal/fluoride complexes on fluoride uptake and caries-like lesion formation in enamel. B.H. Clarkson, *J. Dent. Res.,* 57, 784 (1978).

The influence of fluoride on apatite formation from unstable supersaturated solutions at pH 7.4. E.D. Eanes and J.L. Meyer, *J. Dent. Res.,* 57, 617 (1978).

Comparison of macro- and micromineral concentrations in the serum and drinking water of healthy children in South Finland. A. Helle, *Proc. Finn. Dent. Soc.,* 74, 46 (1978).

Four chemically distinct stages in developing enamel from bovine incisor teeth. C. Robinson, P. Fuchs, D. Deutsch and J.A. Wetherell, *Caries Res.,* 12, 1 (1978).

Release of fluoride from fluoride-containing chewing gum. C. Bruun and H. Givskov, *Comm. Dent. Oral Epidemiol.,* 6, 27 (1978).

Fluoride absorption from subcutaneous sites: influence of volume and concentration. G.M. Whitford, K.E. Reynolds, D.H. Pashley and R.S. Callen, *J. Dent. Res.,* 57, 452 (1978).

Effect of temperature of $SnF_2$ solution on tin and fluoride uptake by bovine enamel. M.S.Putt, J.F.Beltz and J.C.Muhler, *J. Dent. Res.,* 57, 772 (1978).

Fluoride uptake and caries rate in hamster molars after topical fluoride application followed by acidulation. A. Nathaniel, I. Gedalia, J. Anaise and L. Daphni, *J. Dent. Res.,* 57, 534 (1978).

Effect of long term administration of fluoride on plasma fluoride and calcium in relation to forming enamel and dentin in rats. O. Fejerskov, M.J. Larsen, K. Josephsen and A. Thylstrup, *Scand. J. Dent. Res.,* 87, 98 (1979).

Determination of fluoride and chlorhexidine from chlorhexidine/fluoride-containing dentrifices. O.K. Dolles, P. Bonesvoll, O.N. Gamst and P. Gjermo, *Scand. J. Dent. Res.,* 87, 115 (1979).

Accumulation and clearance of fluoride in human saliva after different topical fluoride treatments. U. Heintze and L.G. Petersson, *Swed. Dent. J.,* 3, 141 (1979).

Surface adsorption of fluoride and ionic exchange reactions on hydroxyapatite. G. Rölla and W.H. Bowen, *Acta Odont. Scand.*, 36, 219 (1978).

Studies of prophylaxis pastes containing sodium-potassium aluminium silicate and fluoride. M.S. Putt, C.J. Kleber and J.C. Muller, *J.Dent.Res.*, 58, 1659 (1979).

Influence of milk products on fluoride bioavailability in man. J. Ekstrand and M. Ehrnebo, *Eur. J. Clin. Pharmacol.*, 16, 211 (1979).

## 7.3 Biomedical fluids

### 7.3.1 Blood fluids

Ion-selective discelectrodes with neutral cation carriers for automated electro-analysis of serum. J.G. Schindler, H.K. Dürr, W. Riemann, H.-E. Braun and V. Kellner, *Biomedizinische Technik*, 23, 45 (1978).

An ion-selective electrode for the determination of calcium activity. D.M. Band, C.H. Fry and T. Treasure, *J. Physiol.*, 276, 1P (1978).

Determination of ionised calcium with the Orion SS-20 and AMT Electrion Auto compared. F.H. Loken, S.B. Arnaud and S.J. Rehfeld, *Clin. Chem.*, 24, 2066 (1978).

Higher values for ionised calcium with a new type of electrode for Orion SS-20. S. Ohman and L. Larsson, *Clin. Chem.*, 24, 2070 (1978).

Simplified method for measurement of calcium-ion activity in serum using Serum-Nonovettes and a calcium-ion-sensitive disc-electrode. H. Maier, A. Roeckel, G. Schmid and A. Heidland, *Res. Exp. Med.*, 173, 297 (1978).

Determination of blood-ionised calcium in a large segment of the normal-adult population. L.J. Drop, C. Fuchs and P.M. Stulz, *Clin. Chem. Acta*, 89, 503 (1978).

Serum ionized calcium and corrected total calcium in borderline hyperparathyroidism L. Larsson and S. Ohman, *Clin. Chem.*, 24, 1962 (1978).

Measurement of free calcium ion in capillary blood and serum. N. Fogh-Andersen, T.F. Christiansen, L. Komarmy and O. Siggaard-Andersen, *Clin.Chem.*, 24, 1545 (1978).

An analysis of measured and calculated calcium quantities in serum. K.O. Pedersen, *Scand. J. Clin. and Lab. Invest.*, 38, 659 (1978).

Relationship between duration of Q-T interval and plasma ionized calcium concentration. D. Scheidigger and L.J. Drop, *Anesthesiol.*, 51, 143 (1979).

Effect of thiazoles on serum calcium. M. Mahamadi, L. Bivins and K.L. Becker, *Clin. Pharmacol. and Therapeutics*, 26, 390 (1979).

Some effects of pH on plasma calcium in the human and cat. D.M. Band and J.W. Broadway, *J. Physiol.*, 289, 30P (1979).

Serum calcium and potassium determination by the split membrane ion-selective electrode technique. R.W. Cattrall and K-T. Jong, *Talanta*, 25, 541 (1978).

Analysis for potassium in human erythrocytes by use of a standard-addition method and an ion-selective electrode. E.A. Mangubat, T.R. Hinds and F.F. Vincenzi, *Clin. Chem.*, 24, 635 (1978).

The application of potassium selective electrodes in the intensive care unit. T. Treasure, *Intens. Care Med.*, 4, 83 (1978).

Potassium-selective haemoanalysis using the valinomycin-disc electrode. J.G. Schindler, *Biomedizinische Technik*, 24, 203 (1979).

On-line continuous potentiometric measurement of potassium concentration in whole blood during open heart surgery. H.F. Osswald, R. Asper, W. Damai and W. Simon, *Clin. Chem.*, 25, 39 (1979).

Normal range for serum sodium by ion-selective electrode analysis exceeds that by flame photometry. W. Annan, N.A. Kirwan and W.S. Robertson, *Clin. Chem.*, 25, 643 (1979).

Normal values for sodium with ion-selective electrodes. G.B. Levy, *Clin. Chem.*, 25, 1516 (1979).

Concentrations of ionic, total and bound fluoride in plasma. L. Singer and R.H. Ophaug, *Clin. Chem.*, 25, 523 (1979).

Chloride-selective liquid-membrane electrodes based on lipophilic methyltri-n-alkylammonium compounds and their applicability to blood-serum measurements. K. Hartman, S. Luterotti, H.F. Osswald, M. Oehme, P.C. Meier, D. Ammann and W. Simon, *Mikrochim Acta*, 235 (1978).

### 7.3.2 Sweat and saliva

Modified ion-selective-electrode method for measurement of chloride in sweat. P.R. Inley, J.A. Dye, D.A.Lichti, J.M.Byers and Jane R. Williams, *Amer. J. Clin. Pathol.*, 69, 615 (1978).

Comparison of the chloride electrode and gravimetric chloride titration sweat tests. W.J.Warwick, L.G. Hansen, I. Viela and J. Matheson, *Amer. J. Clin. Path.*, 72, 142 (1979).

Measurement of chloride in sweat with the chloride selective electrode. W.J. Warwick and L.G. Hansen, *Clin. Chem.*, 24, 2050 (1978).

Measurement of chloride in sweat by use of a selective electrode and strip-chart recorder. W.J. Warwick and L.G. Hansen, *Clin. Chem.*, 24, 381 (1978).

Quality of sweat test performance in the diagnosis of cystic fibrosis. H. Shwachman and A. Mohmoodian, *Clin. Chem.*, 25, 158 (1979).

The flow rate dependent excretion of ionized calcium in human parotid saliva. H. Maier, M.J. Coroneo, G. Antonczyk and A. Heidland, *Archs oral Biol.*, 24, 225 (1979).

Sweat testing to cystic fibrosis: Diagnostic screening with a combination ion-selective electrode. P.T.Bray, G.C.F.Clark, G.J.Moody, G.Thomas and J.D.R.Thomas, *Archs.Disease in Childhood*, 53, 483 (1978).

7.3.3 Urine

Urinary fluoride concentration in 6- to 9-year old children drinking water with a very low fluoride level. I. Shannon and D.M. Saunders, *Caries Res.*, 13, 18 (1979).

7.3.4 Intracellular fluids

Carbon dioxide decreases the intracellular potassium activity in frog muscle. F. Huguenin and T. Zeuthen, *J. Physiol.*, 287, 11P (1979).

Non-passive chloride distribution in mammalian heart muscle: microelectrode measurement of the intracellular chloride activity. R.D.Vaughan-Jones, *J.Physiol.*, 295, 83 (1979).

Ion-sensitive intracellular microelectrodes. How to make and use them. R.C. Thomas, *Academic Press, London, 1978*. Pp.110.

7.3.5 Miscellaneous

Evaluation of the Space Stat 30 sodium/potassium Ion Analyzer. S. Patel and P.O'Gorman, *Clin. Chem.*, 24, 1856 (1978).

Evaluation of the (Orion) Space-Stat 20 Ionized Calcium Analyser. L. Larsson and S. Ohman, *Clin. Chem.*, 24, 731 (1978).

Measurement of sodium in albumin solutions with ion-selective electrodes. M.S. Mohan, R.G. Bates, J.M. Hiller and M.J.D. Brand, *Clin. Chem.*, 24, 580 (1978).

Dorsal root potentials and changes in extracellular potassium in the spinal cord of the frog. R.A. Nicoll, *J. Physiol.*, 290, 113 (1979).

Ionized calcium during plateletpheresis. I.O. Szymanski, *Transfusion*,,18, 701 (1978).

Calcium binding of Troponic C. I. A potentiometric study. S. Iida, *Jap. J. Biochem.*, 86, 733 (1979).

Relationship of free and total calcium in hypercalcemic conditions. J.H. Ladenson, J.W. Lewis, J.M. McDonald, E. Slatopolsky and J.C. Boyd. *J. Clin Endocrinol. and Metabolism*, 48, 393 (1979).

The regulation of extramitochondrial free calcium ion concentration by rat liver mitochondria. D.G. Nicholls, *Biochem. J.*, 176, 463 (1978).

Calcium ions and the regulation of $NAD^+$-linked isocitrate dehydrogenase from the mitochondria of rat heart and other tissues. R.M.Denton, D.A. Richards and J.G. Chin, *Biochem. J.*, 176, 899 (1978).

Flexible valinomycin electrodes for on-line determination of intravascular and myocardial $K^+$. J.L. Hill, L.S. Gettes, M.R. Lynch and N.C. Hebert, *Amer. J. Physiol.*, 235, H455 (1978).

Regulation of chloride in quiescent sheep-heart Purkinje fibres studied using intracellular chloride and pH-sensitive microelectrodes. R.D. Vaughan-Jones, *J. Physiol.*, 295, 111 (1979).

Halothane hepatotoxicity and fluoride production in mice and rats. B.H. Gorsky and H.F. Cascorbi, *Anesthesiol.*, 50, 123 (1979).

Is enflurane defluorination inducible in man? J.R.Dooley, R.I. Mazze, S.A. Rice and J.D. Borel, *Anesthesiol.*, 50, 213 (1979).

Enflurane and methoxyflurane metabolism at anesthetic and at subanesthetic concentrations. A.E. White, W.C. Stevens, E.I. Eger, R.I. Mazze and B.A. Hitt, *Anesthesiol. Analg.*, 58, 221 (1979).

Hepatotoxicity and halothane metabolism in an animal model with application for human toxicity. M.J.Cousins, J.H. Sharp, G.K. Gourlay, J.F. Adams, W.D. Haynes and R. Whitehead, *Anaesth. Intens. Care*, 7, 9 (1979).

Methoxyflurane-induced nephroxicity: influence of food intake on some biochemical indications of renal function in Fischer 344 rats. L.E. Arthaud and T.A. Loomis, *Toxicol. and App. Pharmacol.*, 45, 845 (1978).

Acute fluoride toxicity: the influence of acid-base status. K.E. Reynolds, G.M. Whitford and D.H. Pashley, *Toxicol. and App. Pharmacol.*, 45, 415 (1978).

## 7.4 Enzyme systems and bacteria

An "antibody electrode", preliminary report on a new approach in enzyme immunosasay. J-L. Boitieux, G. Desmet and D. Thomas, *Clin. Chem.*, 25, 318 (1979).

Determination of L-amino-acids and alcohols with oxidase enzymes and a tubular iodide-selective electrode. M. Mascini and G. Palleschi, *Analytica Chim. Acta*, 100, 215 (1978).

Enzyme electrode for measurement of penicillin fermentation broth; step toward application of enzyme electrodes in fermentation control. H. Nilsson, K. Mosbach, S.O. Enfors and N. Molin, *Biotechnol. Bioeng.*, 20, 527 (1978).

Inhibiting effects of copper on bacteria related to the free ion concentration. L.P.T.M. Zevenhuizen, J. Dolfing, E.J. Eshuis and I.J. Scholten-Koerselman, *Microbial. Ecol.*, 5, 139 (1979).

## 7.5 Medicinal and pharmaceutical preparations

Determination of fluoride in solutions by means of an ion-selective electrode. J.E. Wassenaar and W.T. Binnerts, *Voeding*, 39, 18 (1978).

Determination of chloramphenicol in pharmaceutical preparations by the cadmium ion-selective electrode, spectrophotometry and atomic absorption spectrometry. S.S.M. Hassan and M.H. Eldesouki, *Talanta*, 26, 531 (1979).

The determination of penicillins by desulphurisation with lead and EDTA titration. S.S.M. Hassan, M.T.M. Zaki and M.H. Eldesouki, *Talanta*, 26, 91 (1979).

A $NH_4^+$-selective-enzymatic flow-through system. A method for the continuous enzymatic, electrochemical determination of urea. II. J.G. Schindler, R.G. Schindler and O. Aziz, *J. Clin. Chem. Clin. Biochem.*, 16, 447 (1978).

Potentiometric titration of trace concentrations of penicillamine using ion-selective electrodes. S.M. Donahe, G.E. Janauer and T.D. Zucconi, *Anal. Lett. Part B*, 11, 721 (1978).

Membrane electrodes with tetraphenylborate for certain basic drugs and their practical use. K. Selinger, and R. Staroscik, *Pharmazie*, 33, 208 (1978).

Novel penicillin enzyme electrode. C.J. Olliff, R.T. Williams and J.M. Wright, *J. Pharm. Pharmacol.*, 30, 45P (1978).

Determination of some thiobarbiturates with ion-selective membrane electrodes. V.V. Cosofret and A.A. Bunaciu, *Anal. Letters*, 12, 17 (1979).

Calcium determination in some pharmaceutical products using an ion-selective electrode. Z. Pranjic-Anusia, *Acta Pharm. Jugosl.*, 29, 29 (1979).

Electrochemical sensors for the determination of some pharmaceutical products. C. Luca, C. Baloescu, G. Semenescu, T. Tolea and E. Semeuescu, *Rev. Chim. (Bucharest)*, 30, 72 (1979).

## 7.6 Beverages and foods

Measurement of fluoride, with special reference to milk, using a fluoride ion-selective electrode, and an investigation of interferences caused by certain ionic species. G.D. Smith, G. Beswick and D.A. Rosie, *Fluoride*, 12, 142 (1978).

Determination of iodide in nutritional beverage products using an ion-selective electrode. Preston Miles, *J. Assoc. Off. Anal. Chem.*, 61, 1366 (1978).

Potentiometric determination of iodide in milk by the use of an ion-selective electrode. E. Sucman, M. Sucmanova and O. Synek, *Z. Lebensm.-Unters. Forsch.*, 167, 5 (1978).

The ingestion of fluoride in tea. S.C. Duckworth and R. Duckworth, *Brit. dent. J.*, 145, 368 (1978).

## 7.7 Vegetation, vegetables, fruits and juices

No titles

## 7.8 Air and stack gases

Fluorine-selective electrode and its use for analysing waste gases and waste water. P.V. Rakcheev, T.G. Repenkova, G.I. Kravchenko, A.G. Gordievskii and A.A. Vlazneva, *Tr.- Mosk. Khim.-Tekhnol. Inst. im D.I.Mendeleeva*, 93, 56 (1977).

Method and apparatus for the continuous measurement of gas traces with ion-sensitive electrodes. U. Fritze and H. Hershinger, *Ger. Offen.*, 2,723,310 (Cl. GOIN 27/46) 07 Dec. 1978.

## 7.9 Water

Analysis of the nitrate content of surface water and effluent water. Application of a plastic-membrane nitrate-selective electrode in water-quality monitors. H.C. Brinkhoff, *Environ. Sci. Technol.*, 12, 1392 (1978).

Determination of fluoride in solutions by means of an ion-selective electrode. J.E. Wassenaar and W.T. Binnerts, *Voeding*, 39, 18 (1978).

Ion-selective electrode for determination of chloride ion in biological materials food products, soils and waste water. I. Sekerka and J.F. Lechner, *J. Assoc. Off. Anal. Chem.*, 61, 1493 (1978).

In-plant methods for fluoride determination in water using ion-selective electrodes. D. Weil and K.E. Quentin, *Z. Wasser Abwasser Forsch.*, 11, 133 (1978).

Determination of chloride in high-purity waters in the range 0 to 20 μg $l^{-1}$ of chloride using ion-selective membrane electrodes incorporating mercury (1) chloride. G.B. Marshall and D. Midgley, *Analyst*, 104, 55 (1979).

On-line use of ion-selective electrodes (for water analysis). A.A. Diggens and S. Lichtenstein, *Proc. Int. Water Conf. Eng. Voc. West Pa.*, 39, 83 (1978).

Potentiometric determination of cyanides in waste waters using ion-selective electrodes. T.B. Andreeva, V.A. Krasnozhea and Yu.M. Trofimov, *Fosfornaya Prom-st.*, (5), 40 (1978).

Analytical measurement of ammonium ion in water. Compagnie Europeene de Troitement des Eaux Laboratoire A. Chemie des Eaux et de l'Environnement, Ecole Nationale de Chimie de Rennes, *Fr. Demande*, 2,382,009 (Cl. G01N 27/26) 22 Sept. 1978.

Practical methods for fluoride determination by ion-selective electrodes in water. D. Weil, *Hydrochem. Hydrogeol. Mitt.*, 247 (1978).

Determination of fluorides in river water. R. Gallego, J.L. Bernal and R. Pardo, *Afinidad*, 35, 333 (1978).

Use (for NaCl activity in sea water) of a cell with a glass electrode with sodium function and a chloride-sensitive membrane electrode at 1-600 kg $cm^{-2}$ and 0.25°. P.A. Kryukov and S.Ya. Tarasenko, *Izv. Sib. Otol. Akad. Nauk SSSR, Ser. Khim. Nauk*, 2, 37 (1979).

Ion-selective electrodes in water quality analysis. R.C. Thurnou, *Report*, (1978) EPA/600/2-78/106; Order No. PB-285724, 44 pp.

Sulphonated alizarin fluorine blue (AFBS). Part IV. A critical comparison of the use of AFBS against alizarin fluorine blue (AFB) and the fluoride electrode for the determinaticn of low fluoride concentrations:interferences with the AFBS method and their removal. S.F. Deane, M.A. Leonard, V. McKee and G. Svehla, *Analyst*, 103, 1134 (1978).

Determination of residual chlorine in water with computer automation and a residual chlorine electrode. L.P. Rigdon, G.J. Moody and J.W. Frazer, *Anal. Chem.*, 50, 465 (1978).

Determination of total chlorine residual in treated waste waters by electrode. R.L. Jenkins and R.B. Baird, *Anal. Letters*, 12, 125 (1979).

## 7.10 Rocks and Soils

Use of a silver sulphide electrode in geochemical investigations. N.N. Kolpakova and G.D. Mironova, *Geokhimiya*, 12, 1873 (1978).

Ion-selective electrode for determination of chloride ion in biological materials, food products, soils and waste water. I. Sekerka and J.F. Lechner, *J. Assoc. Off. Anal. Chem.*, 61, 1493 (1978).

Analysis of fluoride in acetic acid leach solution of fluorspar with ion-selective electrode. R.L. Zimmerman, Jun., and H.G. Bertrand, *Anal. Lett. Part A*, 11, 569 (1978).

Use of selective electrodes in soil and analysis. J. Glinski, W.S. Stepniewski and Z. Stepniewska, *Pr. Kom. Nauk. (Pol.Tow.Glebozn)*, 2, 23 (1977).

Direct potentiometric determination of fluorine in mineral raw materials of different composition by a fluoride-selective electrode. G.I. Bebeshko, V.P. Roze, V.A. Khalizova, *Zh. Anal. Khim.*, 34, 507 (1979).

Improving a method for fluorine determination in phosphogypsum. T.N. Vladimitskaya, A.P. Gorskaya and I.S. Tishchenko, *Metody Anal. Kontrolye Kach. Prod. Khim. Prons-sti*, (9), 26 (1978).

Determination of chloride ion concentration in aqueous extracts from soils using an ion-selective electrode. S.G. Samokhvalov, A.A. Shaimukhametova, A.L. Erinov and I.V. Alekhina, *Khim. Sel'sk. Khoz*, 17, 58 (1979).

## 7.11 Industrial

Comments on "Determination of fluorine in petroleum and in petroleum-process catalysts with a fluoride-selective electrode". J.N. Wilson and C.Z. Marczewski, *Anal. Chem.*, 50, 1584 (1978).

Electrochemical sensors for process-stream monitoring. P.L. Bailey, *Anal. Chem.*, 50, 698A (1978).

Application of ion-selective electrodes to environmental problems in the steel industry. D.S. Macintyre, B.G. Cooksey and J.M. Ottaway, *Proc. Anal. Div. Chem. Soc.*, 16, 18 (1979).

Determination of chlorine in concentrated sulphuric acid solution with a residual chlorine electrode. K.D. Brown and G.A. Parker, *Anal. Chem.*, 51, 1332 (1979).

Use of ion-selective electrodes for monitoring chemical industry waste water. A.A. Sakharov and E.K. Mikhaevich, *Metody Anal. Kontrolya Kach. Prod. Khim. Prom-sti.*, (9), 30 (1978).

Analysis of trivalent chromium plating solution by volumetric analysis and ion-selective electrode method. S. Konishi and G. Shuichi, *Kimzoku Hyomen Gijutsu*, 30, 38 (1979).

Application of ion-selective electrodes in metal finishing industries, G. Subramanian, N. Chandra and G.P. Rao, *J.Electrochem. Soc. India*, 27, 37 (1978).

Continuous monitoring of a heavy water plant effluent with a sulphide-selective electrode. J. Gulens, K. Jessome and C.K. Macneil, *Analytica Chim.Acta,* 96, 23 (1978).

## 7.12 Environmental and public health

A rapid method of the determination of fluoride in sewage sludges. R.E. Rea, *Water Pollut. Control (Maidstone, England),* 78, 139 (1979).

Determination of total chlorine residual in treated waste waters by electrode. R.L. Jenkins and R.B. Baird, *Anal. Letters,* 12,125 (1979).

Isocyanurate specific electrode and method of analysis (of swimming pool water) using quaternary ammonium isocyanuric acid salts. M.L. Pinsky, *U.S. Patent,* 4,134,798 (Cl. 204-1T; G01N 27/56) 16 Jan. 1979.

Determination of iodine by solvent extraction-ion-selective electrode method. N. Fukuzaki, T. Suzuki, R. Sugai and T. Oshina, *Bunseki Kagaku,* 28, 60 (1979).

Culture studies on the effects from fluoride pollution on the growth of marine phytoplankters. L. Oliveira, N.J. Antia and T. Bisalputra, *J. Fish Res. Board Can.,* 35, 1500 (1978).

## 7.13 Applications in non-aqueous media

Potentiometric determination of fluoride ion by means of a fluoride selective electrode using organic solvents. V.P. Novak, V.I. Bogovina, L.A. Vasilava, *Nov. Metody* Instrument. Analiza Materialov,M., 125 (1979).

Ion-selective electrodes in non-aqueous solvents. E. Pungor and K.Toth, in (Editor: J.J. Lagowski) "Chem. Non-Aqueous Solvents", Academic Press, New York, 1978, p.145.

Method for the determination of methanol in binary methanol-water mixtures by use of ion-selective electrodes. G.J. Kakabadse, A.H. Maleila, M.N. Khayat G. Tassopoulos and A. Vehdati, *Analyst,* 103, 1046 (1978).

Stability of alkaline earth monofluoride complexes in methanol. J.W. Bixler and A.M. Bond, *J. Inorg. Chem.,* 17, 3684 (1978).

## 7.14 Miscellaneous

Ion-selective electrodes in organic functional group analysis: Microdetermination of hydrazines with the copper electrode. S.S.M. Hassan and M.T.M. Zaki, *Mikrochim. Acta,* (I), 137 (1979).

Ion-selective electrodes in organic functional group analysis: Microdetermination of quinones with halide electrodes and Gran's plot. S.S.M. Hassan and M.B. Elsayes, *Mikrochim. Acta,* (II), 333 (1979).

Determination of sodium and thiols in alkali solutions. V.I. Nazarova, *Koks Khim.,* 51 (1978).

Some analytical applications of a new silver-ion-selective membrane electrode. V.V. Cosofret, *Rev. Roum. Chem.*, 23, 1489 (1978).

Ion-selective electrode for determinations of chloride ion in biological materials, food products, soils and waste water. I. Skerka and J.F. Lechner, *J. Assoc. Off. Anal. Chem.*, 61, 1493 (1978).

Method for the determination of methanol in binary methanol-water mixtures by use of ion-selective electrodes. G.J. Kakabadse, A.H. Maleila, M.N. Khayat, G. Tassopoulos and A. Vahdati. *Analyst*, 103, 1046 (1978).

Indirect potentiometric determination of hydrazine, isoniazid, sulphide and thiosulphate with chloramine-T ion-selective electrode. M.A. Koupparis and T.P. Hadjiioannou, *Talanta*, 25, 477 (1978).

Dual ion-selective electrode detector for simultaneous detection of bromine- and chlorine-containing compounds in gas chromatography. T. Kojima, M. Ichse and Y. Seo, *Analytica Chim. Acta*, 101, 273 (1978).

Determination of gold in cyanide solutions with a liquid-membrane ion-selective electrode. Yu.V. Shavnya, A.S. Bychkov, O.M. Petrukhin, V.A. Zarinskii, L.V. Bakhtinova and Yu.A. Zolotov, *Zh. Anal. Khim.*, 33, 1531 (1978).

Determination of cyanide present in solution. R. Nakamura and K. Takahara, *Jpn. Kokai Tokkyo Koho* 79 52, 592 (Cl GOIN 31/22), 25 Apr 1979, 3 pp.

Research and development of selective membrane electrodes. Communication 14. Determination of mercury(II) and thallium(III). A.V. Kopytin, A.F. Zhukov, Yu.I.Urosov, L.A.Kopytina and A.V. Gordievskii, *Zh. Anal. Khim.*, 34, 465 (1979).

Potentiometric titration of ionic surfactants. D.E. Anghel, N. Ciocan and G. Popescu, *Tr.-Mezhdunar. Kongr. Poverkln.-Akt. Veshchestvam 7th*, 1, 524 (1976) (Publ. 1979)

Potentiometric microdetermination of lead(II) with an ion-selective lead electrode and its application to the analysis of organic lead compounds. A. Campiglio, *Mikrochim.Acta*, I, 267 (1979).

Combination chloride-selective microelectrode for corrosion study. K. Kitamura and E. Sato, *Boshoku Gijutsu*, 27, 517 (1978).

Microdetermination of organic halogens using a sodium ion-selective glass electrode as an endpoint sensor for the argentometric titration. K. Hozumi, K. Kawakami and N. Akimoto, *Buneski Kagaku*, 28, 5 (1979).

Some analytical applications (for halogen and sulphur containing organic substances) of a new silver ion-selective membrane electrode. *Rev. Rom. Chim.*, 23 1489 (1978).

## 8. GLASS pH ELECTRODES: DEVELOPMENT, APPRAISAL, ETC.

Accurate measurements of concentration of hydrogen ions with a glass electrode: calibrations using Prideaux and other universal buffer solutions and a computer-controlled automatic titration. A. Avdeef and J.J. Bucher, *Anal. Chem.*, 50, 2137 (1978).

Accuracy of the hydrogen-ion-selective glass electrode. E.P.Serjeant and A.G. Warner, *Anal. Chem.*, 50, 1724 (1978).

Glass micro-electrode for pH measurement. N.A. Afedotov, *Zavod. Lab.*, 44, 1309 (1978).

Error and error propagation in determination of pH values. S. Ebel, E. Glaser and H. Mohr, *Fresenius Z. Anal. Chem.*, 293, 33 (1978).

Assessment of glass electrodes for determining pH in boiler-feed water. D. Midgley and K. Torrance, *Analyst*, 104, 63 (1979).

Intramyocardial pH as an index of myocardial metabolism during cardiac surgery. F.J.M. Walters, G.J. Wilson, D.J. Steward, R.J. Domenech and D.C. MacGregor, *J. Thoracic and Cardiovascular Surg.*, 78, 319 (1979).

Enzymatic determination of urea in serum based on pH measurements with the flow -injection method. J. Růžička, E.H. Hansen, A.K. Ghose and H.A. Mottola, *Anal. Chem.*, 51, 199 (1978).

## 9. GAS-SENSING ELECTRODES: DEVELOPMENT AND NEW SYSTEMS

### 9.1 Gas sensors dependent on conventional ion-selective electrodes

Improved air-gap electrode and its application to measurement of carbonate in interstitial water. F.G. Romer, L. Puyker, G.B.H. Brinkman and B. Griepink, *Fresenius Z.Anal. Chem.*, 289, 35 (1978).

Comparison of three methods for determination of sulphur dioxide in food and drink. N. Jennings, N.G. Bunton, N.T. Crosby and T.G. Alliston, *J. Assoc. Public Anal.*, 16, 59 (1978).

Limits of detection of gas-sensing probes. Application to the ammonia sensor. F. van der Pol, *Analytica Chim. Acta*, 97, 245 (1978).

### 9.2 Other types of electrochemical gas sensors

Development of a miniature rapid response carbon dioxide sensor (based on differential complementary metal oxide semiconductor). G.E. Bingham, C.H. Gillespie and J.H. McQuaid (Lawrence Livermore Laboratory, California), *Report*, (1978) UCRL-52440. Pp 25.

Sintered zirconium for oxygen sensors. N. Miwa and K. Tanaka, *Japanese Patent*, 78, 149, 204 (Cl C04336/48) 26 Dec. 1978)

Oxygen sensor for exhaust gas system. T. Hori, K. Uchida, Y. Ohtsuka, T. Furutani and H. Wakisaka, *Japanese Patent*, 79, 12, 887 (Cl G01N 27/58) 30 Jan. 1979.

Gas sensor(s) (especially for iso-$C_4H_{10}$) with improved sensitivity and durability. Y. Yokomizo, T. Takahashi, M. Inai and Z. Tanno, *Japanese Patents*, 79, 32, 395 to 79, 32, 399 (Cl G01N 27/12) 9 March 1979.

Gas sensor with electrochemical cells. A. Alberry, M.D. Crook and G.S. Vizard, *Ger. Offen.*, 2, 832, 328 (Cl G01N 27/56) 8 Feb. 1979.

Electrochemical cell for measuring oxygen in gases. F.J. Rohr and H. Holick, *Ger. Offen.*, 2738755. (Cl. G01N 27/50) 8 March 1979.

Solid electrolyte gas sensor having a protective bonding layer. J.D. Bode and S.K. Rhee, *U.S. Patent*, 4,107,018 (Cl 204-195s; G01N 27/46) 15 August 1978.

Improved solid electrolyte oxygen gas sensors. D.C. Davis, D.J. Romine, P.R. Woodruff, J.D. Bode, T.Y. Tien and C.T. Young, *U.S. Patent*, 4,136,000 (Cl 204-195s, Cl 23 G1/02) 23 January 1979.

Oxygen sensor, T. Akatsuka, *U.S. Patent*, 4, 116, 797, (Cl. 204-195s; G01N 27/46) 26 September 1978.

Electrochemical measuring probe for determining the oxygen content in exhaust gases. H. Maurer, F. Rieger, E. Linder and R. Schuessler, *Ger. Offen.*, 2, 732, 743 (Cl. G01N 27/50) 8 February 1979.

Oscillations in the emf of solid-state electrochemical oxygen sensors. R.E. Hetrick and E.M. Logothetis, *Appl. Phys. Lett.*, 34, 117 (1979).

Electrochemical probe for determining the oxygen content in automobile exhaust gases. H. Maurer, F. Rieger and E. Linder, *Ger. Offen.*, 2, 729, 475. (Cl. G01N 27/50) 11 January 1979.

Electrochemical oxygen-measuring apparatus with internal reference system and solid (Pb-Pbo) electrolyte. M. Henault and G. Vitter, *Ger. Offen.*, 2, 812, 921 (Cl. G01N 27/52) 30 November 1978).

Gas sensors with improved sensitivity (for CO and $H_2$). N. Ichinose, H. Ohkuma and Y. Yokomizo, *Japanese Patent*, 78 141 093 (Cl G01N 27/12) 8 December 1978.

## 10. APPLICATIONS OF GAS SENSORS

Enzymic-potentiometric determination of urea in human serum. Ellen M. Berman, S.J. Yao, S.K. Wolfson, Jun. and J.M. Tokarsky, *Bioelectrochem. Bioenerg.*, 5, 63 (1978).

Determination of ammonium-nitrogen in silage samples by an ammonia (-sensitive) electrode. E. Byrne and S. McCormack, *Commun. Soil Sci. Plant Anal.*, 9, 667 (1978).

Nitrogen determination in coal and coke with an ammonia electrode. K.J. Doolan and C.B. Belcher, *Talanta*, 25, 398 (1978).

Continuous determination of 'hydrogen extractable' nitrogen from silicon-iron by an ammonia-gas-sensing electrode. R.G. Hirst and C.M. Maucione, *Anal. Chem.*, 50, 1046 (1978).

Microelectrode determination of pH and $pCO_2$ in cat proximal tubule after benzolamide: evidence for hydrogen ion secretion. T.D. DuBose, L.R. Pucacco, D.W. Seldin, N.W. Carter and J.P. Kokko, *Kidney Internat.*, 15, 624 (1979).

Use of an ammonia electrode for the determination of nitrogen in oximes, melamine-formaldehyde resin and ethylenimine-added-wood meal. Y. Hiroboryashi, *Ringyo Shikenjo Kenkyu Hokoku*, 304, 1 (1979).

## 11. REFERENCE ELECTRODES

Field-effect transistor as solid-state reference electrode. P.A. Comte and J. Janata, *Analytica Chim. Acta,* 101, 247 (1978).

Electrochemistry in anisole. I. Reference electrode and electroactivity range. G. Dabosi, M. Martineau and G. Durand, *Analusis,* 6, 289 (1978).

## 12. BOOKS AND GENERAL REVIEWS

Ion-selective electrodes in analytical chemistry. Volume 1. Ed. Henry Freiser Plenum Press, New York, 1978. Pp 439. Contains articles on
"Theory and principles of membrane electrodes" (R.P. Buck),
"Precipitate-based ion-selective electrodes" (E. Pungor and K. Tóth),
"Ion-selective electrodes based on neutral carriers"(W.E.Morf and W. Simon),
"Poly(vinyl chloride) matrix membrane ion-selective electrodes (G.J.Moody and J.D.R. Thomas),
"Sources of error in ion-selective electrode potentiometry" (R. A. Durst), and
"Applications of ion-selective electrodes" (G.J. Moody and J.D.R. Thomas).

Developing the commercial potential of macrocyclic molecules. R.A. Schwind, T.J. Gilligan and E.L. Cussler, in (Editors: R.M. Izatt and J.J. Christensen) "Synth. Multident. Macrocyclic Compd", Academic Press, New York, 1978, p.298. (Includes ion-selective electrode applications of macrocyclic molecules).

Ion-selective electrodes. D.S. Veselinovic, *Glas. Hem. Drus. Beograd.,* 43, 817 (1978).

Ion-sensitive Intracellular Microelectrodes. How to Make and Use Them. R.C. Thomas. Academic Press, London, 1978. Pp 110.

Ion-specific electrodes. M. Yiran Vayreda, *Circ. Farm.,* 36, 325 (1978).

Ion-selective electrodes responsive to anions. M. Kataoka and T. Kamichara, *Dojin Nyusu,* 8, 1 (1978).

Ion-selective electrodes and trace analysis. N. Ishibashi, A. Jyo and K. Kine, *Kagaku Zokam (Kyoto),* 78, 35 (1978).

Analysis with ion-selective electrodes. J. Veselý. D. Weiss and K. Štulik. Ellis Horwood, Chichester 1978, Pp.245.

Ion-selective electrodes. Conference held at Budapest, Hungary, 5-9 September 1977. Ed. E. Pungor, Akadémiai Kiadó, Budapest, 1978, PP.613.

From conventional membrane electrodes to ion-sensitive field-effect transistors. P. Bergveld and N.F. de Rooij, *Med. and Biol. Eng. and Computing,* 17, 647 (1979).

# FORTHCOMING CONFERENCES

July 20-26 1980

SAC 80. An International Conference at The University of Lancaster, United Kingdom organised by The Analytical Division of The Chemical Society. The Conference will have sessions on various aspects of analytical chemistry, including electroanalysis with ion-selective electrodes. Presentations will be by lectures and posters and there is to be an exhibition of scientific equipment and "Do it Yourself" workshop sessions. There will also be an accompanying persons programme.

Registration details are now available from

The Secretary, Analytical Division,
The Chemical Society,
Burlington House,
LONDON W1V OBN,
United Kingdom.

October 13-15 1980

Third Scientfic Session on Ion-Selective Electrodes at Mátrafüred, Hungary organised by the BTU Research Group for Industrial Chemical Analysis of the Hungarian Academy of Sciences. The Session will be devoted to new developments in the theory and investigation of ion-selective electrode mechanism and new applications of ion-selective electrodes.

Preliminary registration details and further information from

The Organising Committee of the Third Scientific Session on Ion-Selective Electrodes,
Institute for General and Analytical Chemistry,
Technical University,
H-1521 BUDAPEST XI.,
Gellért tér 4,
Hungary.

April 13-16 1981

International Symposium on Electroanalysis in Clinical, Environmental and Pharmaceutical Chemistry at UWIST, Cardiff, United Kingdom organised by The Electroanalytical Group of the Analytical Division of The Chemical Society. The programme will be devoted to all aspects of electroanalysis as it can apply to clinical, environmental and pharmaceutical chemistry.

The programme of the Symposium will emphasise methodology and applications of electroanalytical methods, especially with regard to the use of ion-selective electrodes, gas sensors and polarography. Aspects of development, operation and mechanisms as well as new areas, such as piezoelectric crystals will be included where these are likely to have relevance to progressing electroanalysis in biomedical and environmental fields.

The First Circular seeking an expression of interest and a call for papers is now available from

Short Courses Section (Electroanalysis Symposium),
UWIST,
CARDIFF CF1 3NU,
Wales, United Kingdom.

# SUBJECT INDEX

# AUTHOR INDEX

# ION-SELECTIVE ELECTRODE REVIEWS

VOLUME 2

## Other Related Pergamon Titles of Interest

ALBAIGES: Analytical Techniques in Environmental Chemistry
BOUMANS: Line Coincidence Tables for Inductively Coupled Plasma Atomic Emission Spectrometry
COSOFRET: Membrane Electrodes in Drug Substances Analysis
JEFFERY & HUTCHISON: Chemical Methods of Rock Analysis, 3rd edition
MEITES: An Introduction to Chemical Equilibrium & Kinetics
PATAKI & ZAPP: Basic Analytical Chemistry
YINON & ZITRIN: The Analysis of Explosives

## Related Journals of Interest

*Electrochimica Acta*
*Progress in Analytical Atomic Spectroscopy*
*Talanta*

Full details of all Pergamon publications/free specimen copy of any Pergamon journal available on request from your nearest Pergamon office.

# ION-SELECTIVE ELECTRODE REVIEWS

## Volume 2

*Editor-in-Chief*
J. D. R. THOMAS
*UWIST, Cardiff, Wales*

PERGAMON PRESS
OXFORD · NEW YORK · TORONTO · SYDNEY · PARIS · FRANKFURT

| | |
|---|---|
| U.K. | Pergamon Press Ltd., Headington Hill Hall, Oxford OX3 0BW, England |
| U.S.A. | Pergamon Press Inc., Maxwell House, Fairview Park, Elmsford, New York 10523, U.S.A. |
| CANADA | Pergamon Press Canada Ltd., Suite 104, 150 Consumers Rd., Willowdale, Ontario M2J 1P9, Canada |
| AUSTRALIA | Pergamon Press (Aust.) Pty. Ltd., P.O. Box 544, Potts Point, N.S.W. 2011, Australia |
| FRANCE | Pergamon Press SARL, 24 rue des Ecoles, 75240 Paris, Cedex 05, France |
| FEDERAL REPUBLIC OF GERMANY | Pergamon Press GmbH, 6242 Kronberg-Taunus, Hammerweg 6, Federal Republic of Germany |



First edition 1981

**British Library Cataloguing in Publication Data**

Ion-selective electrode reviews.
Vol. 2
1. Electrodes, Ion selective
I. Thomas, J. D. R.
541.3'724 QD571

ISBN 0 08 028434 5

First published in *Ion-Selective Electrode Reviews*, Volume 2, Nos 1 and 2, 1980, and supplied to subscribers as part of their subscription. Also available to non-subscribers.

*Printed in Great Britain by A. Wheaton & Co. Ltd., Exeter*

# CONTENTS

*Ion-Selective Electrode Rev.* Vol.2, p.1. 0191-5371/80/0901-0001$05.00/0


# EDITORIAL

Articles devoted to field effect transistors, microelectrodes, calcium ion-selective electrodes, industrial applications, roles in relation to dental research and materials, and applications in water analysis graced the pages of the first volume of ION-SELECTIVE ELECTRODE REVIEWS. They set the scene for the wide scope and appeal of ion-selective electrodes- not only to dedicated electrochemists, but to all chemists as well as scientists in the life sciences and various branches of technology.

The articles of the present issue emphasise the policy of covering important areas of development and application in divers fields. In this respect, it is timely to have the article by Professor M. Mascini on known addition, Gran's plots and related methods; these are often essential in the effective deployment of ion-selective electrodes.

Professor Mascini wrote his article whilst on an assignment at the University of Somalia. This widens the geographical area of origin of articles which, to date, have come from Europe and America, but this was not the objective of going to Somalia and Professor Mascini is pleased to be back in his native Rome!

Selective ion-sensors are the essence of ion-selective electrodes, but many of the sensors can be useful in other sensing devices. It is for this reason that Professor G. Guilbault was invited to contribute his article on piezoelectric devices. Furthermore the article is a prelude to further articles on gas sensors, and especially covering those areas where ion-sensors used in ion-selective electrodes have a significant role.

"All are but parts of a stupendous whole," (Alexander Pope).

J.D.R. Thomas

July 1980 J.D.R. THOMAS.

*Ion-Selective Electrode Rev.* Vol.2, pp.3-16. 0191-5371/80/0901-0003$05.00/0


# USES OF THE PIEZOELECTRIC CRYSTAL DETECTOR IN ANALYTICAL CHEMISTRY

George G. Guilbault

Department of Chemistry, University of New Orleans, New Orleans, Louisiana 71022, U.S.A.

CONTENTS



KEYWORDS: Piezoelectric crystal detectors; air pollution monitors; gas chromatography detector; water vapour detector; gas detectors.

# USES OF THE PIEZOELECTRIC CRYSTAL DETECTOR IN ANALYTICAL CHEMISTRY

George G. Guilbault

## 1. INTRODUCTION

Piezoelectricity was first discovered in 1880 by Pierre and Jacques Cure [1] who discovered that a pressure exerted on a small piece of quartz caused an electrical potential between deformed surfaces, and that the application of a voltage to a quartz crystal effected physical distortions. A crystal activated by a concise electrical charge vibrates mechanically for a short time until a physical equilibrium is attained.

Alpha quartz is most widely for piezoelectric crystal detectors. Its properties of water insolubility and resistance to temperatures up to 579°C with no loss of piezoelectric properties make it particularly desirable. The frequency of the quartz crystal is dependent upon the physical dimensions of the quartz plate and upon the thickness of the electrodes placed on it. For use as a piezoelectric crystal detector, only AT or BT cut quartz plates are useful [2]. These crystals are two high-frequency mode plates which vibrate in a shear mode about an axis parallel to the major surface; these plates have low or zero temperature coefficients at the temperature of use, and have surfaces that are antimodal in displacement. The coefficients are a critical function of the angle of cut. The AT cut is superior in temperature coefficient and in mass sensitivity. The 9 MHz crystals used most frequently are 10 to 16 mm discs, squares and/or rectangles which are approximately 0.19 mm thick. The metal electrodes are 0.3 µm to 1 µm thick, 3 to 8 mm in diameter and are made of gold, nickel, silver or aluminum. The resonant frequency of a crystal is normally obtained using a frequency meter attached to the output of an oscillator circuit.

In 1959, Sauerbrey [3,4] developed a relationship between the mass of metal films depostied on quartz crystals and the change in frequency. This relationship was derived for quartz (AT-cut) crystals vibrating in the thickness shear mode.

$$\Delta F = -2.3\ 10^{6}\ F^{2}\ \frac{\Delta M}{A}$$

where $\Delta F$ = the change in frequency due to the coating [Hz]
$F$ = the frequency of the quartz plate [MHz]
$\Delta M$ = mass of deposited coating (g)

A = area coated ($cm^2$)

This equation predicts that a commercially available 9 MHz crystal would have a mass sensitivity of about 400 Hz $\mu g^{-1}$ or a 15 MHz crystal a sensitivity of 2600 Hz $\mu g^{-1}$. It is therefore apparent that the vibrating quartz crystal can be an extremely sensitive mass indicator. The detection limit is estimated to be about $10^{-12}$ g [5].

Sauerbrey successfully tested equation (1) by evaporating thin metal films on quartz crystals, weighing them on a microbalance and monitoring the change of frequency electronically. Metals and many other solids change the frequency, but do not greatly affect the crystal's ability to vibrate. However, when liquids are deposited on the crystal surface the ability to vibrate is often impaired, because the vibrating crystal surface dissipates energy in the liquid. If a gas is allowed to adsorb onto the liquid coating, the amplitude of vibration changes proportionally to concentration.

Piezoelectric devices have been used for many years for controlling frequency in the communications field, as selective filters in electronic networks and recently for the detection and determination of micro concentrations of substances. In this article the applications of piezoelectric crystal detectors in analytical chemistry will be reviewed.

## 2. APPARATUS

The experimental set-up with the piezoelectric quartz crystal detector used by the authors is shown schematically in Fig. 1. The cell design is the most sensitive one for use in a flow system.

Fig. 1. Experimental apparatus. 1 = recorder, 2 = digital-to-analog converter, 3 = frequency counter, 4 = oscillator, 5 = power supply.

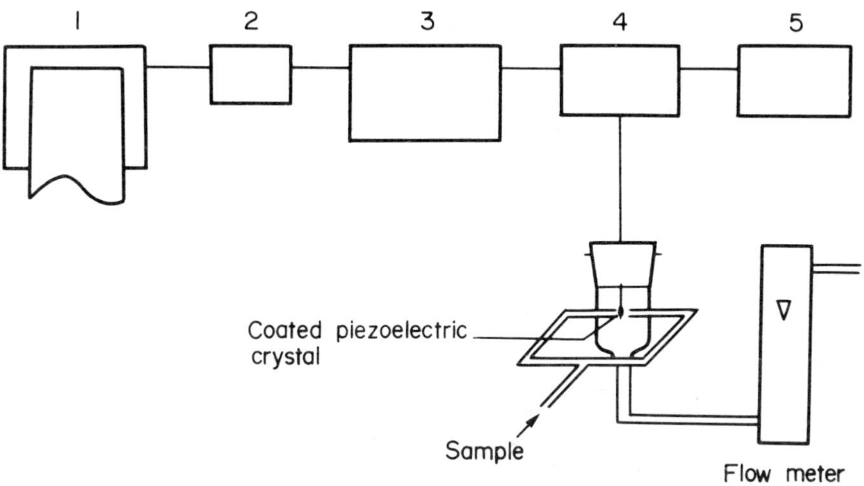

The piezoelectric crystals used were 9-MHz quartz crystals with silver plated metal electrodes on both sides. The crystals were AT-cut and mounted in HC 25/U holders (JAN Crystals Mfg. Co.). The instrumentation consisted of a low frequency OX transistor oscillator (International Crystal Mfg. Co., Oklahoma City) powered by a regulated power supply (Heath Kit model 1P-28). The applied voltage was kept constant at 9V-dc. The frequency output from the oscillator was measured by a frequency counter (Health-Schlumberger model SM-4100), which was modified by a digital-to-analog converter, so that the frequency could be recorded. The frequency (peak maximum) could be read on either the frequency counter or the recorder.

Air as the carrier gas was supplied by a vibrating diaphragm air pump. An empty gas chromatographic column of stainless steel was placed between the pump and cell in order to minimize a change in flow rate. The flow rate of air was maintained at $30 \pm 2$ $cm^3$ $min^{-1}$, according to previous works [6,7,8,9].

For field measurements of organophosphorus compounds the apparatus shown in Fig. 1 is inadequate. A self-contained small, portable detector was designed for this using the basic components shown in Fig. 1, but with a 6 V battery, a mini pump and digital readout. [8]

## 3. DETECTORS DEVELOPED FOR ANALYSIS

### 3.1 Detector for gas chromatography

King [10,11] described a coated piezoelectric crystal for use as a sensitive and selective detector for gas chromatography. Since the frequency of a vibrating crystal can easily be measured to $\pm$ 1.0 Hz, any frequency change greater than this can be easily detected. King, by coating the crystals with different substrates used in gas chromatographic columns, proposed an interaction of the coating phase with the same components of a gas stream while on the crystal surface. The frequency of the crystal depended on the mass of the vapor adsorbed by the coating. Applying equation (1) King estimated that detection limits of $10^{-9}$ g could be realized. Furthermore, this detection limit was independent of carrier gas provided the carrier gas did not partition in the substrate. King called his device the "piezoelectric sorption detector" since the interaction "is probably a combination of adsorption and absorption."

The high sensitivity, selectivity and ruggedness of the sorption detector makes it ideal for use in analytical chemistry, and several excellent articles on applications of the piezoelectric crystal detector as a GC detector waer published by Karasek et al. [12,13,14].

### 3.2 A piezoelectric detector for water vapor

King described a coated crystal analyzer which is sensitive to as little as 0.1 ppm water [10,11,15]. Since the adsorption isotherms of many materials are known, the performance of a sorption detector is readily predicted. The relative adsorption of water on crystals coated with molecular sieves, hygroscopic polymer, silica gel, and polar liquids were investigated by King. The liquid coated crystals exhibited a rapid response and their response is quite linear. The solid adsorbents were outstanding for assay of water at very low concentrations.

Since the adsorption characteristics of uncoated crystals are of only limited usefulness, the response of coated gold, nickel, and aluminum crystals to water vapor were investigated [15]. When deliquescent salts were used as coatings for crystals, interesting water detectors resulted. The response of the detectors is rapid at normal conditions, but at low temperatures and low humidities the response can be very slow due to a lack of driving force for the reaction, slower diffusion and existence of a solid state reaction. A crystal coated with 0.05 μm of lithium chloride for example, had a time constant of 18 min at $-32^{o}C$ and less than 1 s at room temperature.

Water detectors can be constructed from a wide variety of materials and each has its own range of usefulness [16]. A selective water detector based on a hygroscopic polymer coated crystal has been commercially available since 1964 [17]. The unique qualities of the instrument are ppm detection in 30 s, high selectivity and long life time.

Several studies have demonstrated that the piezoelectric crystal detector can be used to measure the moisture content of the Martian atmosphere. The atmospheric environment on the Martian surface differs significantly from the earth's atmosphere. For example, at least 80% carbon dioxide can be found in the Martian atmosphere. Recent Mariner probes have shown that the total pressure of the Martian atmosphere is approximately 600 to 800 $Nm^{-2}$. In addition, evidence suggests that atmospheric water vapous pressure may range from 0.007 to 120 $Nm^{-2}$ [18]. A solution to these problems were found by using a King piezoelectric sorption hygrometer with several different polymer coatings [19,20]. The quartz crystal detector is extremely sensitive to mass changes and the water vapour is detected by measuring the change in mass of an hygroscopic film on the crystal. Since the wind-blown Martian dust could accumulate on the crystal, and thus interfere with vapour measurements, a need developed for an electronic method of removing dust particles from the crystal surface. Methods were developed for quantitatively determining water under different experimental conditions (temperature, pressure, etc.).

### 3.3 Piezoelectric crystal detectors for detection and determination of air pollutants

3.3.1 Detector for sulphur dioxide. The analysis of sulphur dioxide in air has long been a problem. Oil refineries, pulp mills, and effluents from a number of other industrial stacks constitute a prime source of sulphur dioxide discharge into the atmosphere. Burning of high sulphur fuels in automobiles is another cause of pollution at ground level. Therefore, an ever-increasing demand has arisen for new, simple and inexpensive methods for measurement and control of sulphur dioxide pollution.

There have been several research reports describing the use of coated piezoelectric quartz crystals as highly sensitive detectors for sulphur dioxide and many coating materials have been investigated.

Trace amounts of sulphur dioxide can be detected using a new detector cell design with triethanolamine and quadrol as coating materials [21]. One important feature of this design is that the column effluent is split into two equal streams which directly and simultaneously fall on the opposite faces of a coated crystal. This arrangement is expected to improve the sensitivity since the amount of the sample gas reacting with the coating at any moment is appreciably increased.

Effects of changes of temperature have been investigated in several experiments. Guilbault and Lopez-Roman [22] pointed out that the frequency of the crystals increased with increases in temperature, especially from 100°C to 200°C. From 25°C to 40°C the effect of the temperature is very small, an increase of about 40 Hz. These experiments show that temperature must remain constant during the reading, but a change of 10°C is not critical.

Triethanolamine was used as coating material by Cheney and Homolya [23,24], who measured the temperature dependence of an uncoated 9 MHz crystal and found 71 Hz of change when the temperature increased from 10°C to 35°C. Adsorption and desorption of the sulphur dioxide on the coating material at different temperatures were also investigated.

The method of coating the crystal with various substrates has been shown to be very important. Several different techniques were evaluated for application of the substrate to the piezoelectric crystal; dropping, dipping and spraying methods. Earp [25] had shown in a previous study that the sensitivity of the sorption detector is inversely proportional to the area coated and to the distance from the center of the crystal. As the area of coating on the electrode surface decreased, the sensitivity increases to a maximum and then decreased again due to less surface area being available for sorption. It is essential, therefore, that the coating be placed on the crystal in the exact same manner each time. In the drop method, the substrate is dissolved in a volatile solvent and then a drop of this solution is placed at the center of the crystal electrode using a micro-syringe or glass bar; the solvent is then allowed to evaporate. In the dip method, the crystal is dipped into a solution of the substrate in a volatile solvent and then allowed to dry. In the spray method the crystal is coated by spraying a solution of the different compounds (contained in a suitable solvent) onto a crystal surface. The critical factor involved in coating the crystal is not so much the amount of coating as the ability to reproduce the coating operation, as proposed by Hartigan [26].

Cheney et al. [27] used cotton swabs to coat a crystal with the substrate ethylenedinitrotetraethanol for detection of sulphur dioxide. The technique of coating the crystal was checked for repeatability of both coating amount and sensitivity. A greater sensitivity to sulphur dioxide was obtained by using a 9 MHz crystal coated in the centre (340 Hz) rather than a totally coated crystal (260 Hz). The authors also found that the frequency change due to a varying coating is predictable and consistent for a centre-coated crystal, but not for a fully coated crystal.

Considerable interferences were caused by nitrogen dioxide and moisture in the assay of sulphur dioxide using all coatings. An accurate quantitative determination cannot be obtained for sulphur dioxide in the presence of nitrogen dioxide using coated piezoelectric crystals. Guilbault et al. [28] showed that a hydrophobic membrane filter (pore size 0.45 µm) was successful in greatly reducing the response of a quadrol coated crystal to atmospheric moisture. Each additional filter layer further reduced the effect of moisture response, and with 4 layers the response for moisture was completely suppressed.

The use of piezoelectric crystals for the detection of sulphur dioxide in stack-gas and automobile exhaust analysis has been described [29].

3.3.2 Ammonia detectors. The detection of ammonia gas in the parts per $10^9$ range was effected using coated piezoelectric crystal detectors. When Ucon 75-H-90,000 and Ucon-LB-300X were used as coatings [30], good sensitivity to ammonia was obtained, and nitrogen dioxide reacts only after a 5 minute exposure of the coating. The nitrogen dioxide reacted with the Ucon coatings, and new compounds were formed on the crystal when a great sensitivity for both ammonia and nitrogen dioxide resulted. The formation of new compounds was evident from the appearance of new bands and the shift of some bands observed in the infrared spectra of these substances. Some problems were caused by atmospheric moisture, and by high concentrations of organic compounds, which dissolve the coatings.

Several materials such as Capiscum annuum pods, ascorbic acid, and ascorbic acid with silver nitrate were applied as coatings for the selective detection of ammonia in the atmosphere [31]. These specific coatings were not dependent on activation, and could be used at ambient conditions. The ascorbic acid and an extract of Capiscum annuum react reversibly with ammonia and this type of coated crystal can be used to detect ammonia over a wide concentration range. In order to determine the type of reaction which occurs between ammonia and the coating substrates, infrared spectra were obtained under various conditions. These spectra suggested that the reaction of ammonia with ascorbic acid is a simple acid-base reaction, and this was assumed to be the mechanism of the Capiscum-ammonia reaction also. The nature of the increased sensitivity caused by silver nitrate is not known. However, it undoubtedly results from some type of interaction between ascorbic acid and silver. This interaction is probably an oxidation of ascorbic acid by silver ion, which is in turn reduced.

Hlavay and Guilbault [32] have described two new coatings for the detection of ammonia in ambient air: L-glutamic acid-hydrochloride and pyridoxine-hydrochloride. These two coatings give greater sensitivity and better selectivity than with previously described substrate coatings. The response time for both coatings is less than 7 min, and complete reversibility of response is observed in 5 min. No significant interferences were found from other gases. The effect of moisture on the coating material was eliminated by using a gas chromatographic precolumn packed with silica gel. With these coatings, ammonia was detected at less than parts per $10^9$ concentrations.

The reaction between L-glutamic acid hydrochloride and ammonia was investigated by infrared methods. According to the infrared spectra a reaction between the carboxyl group and ammonia could have occurred, yielding the ammonium salt of the carboxylic acid. With pyridoxine hydrochloride and ammonia the infrared spectra provides evidence that the reaction between the hydroxyl group and ammonia has occurred, yielding the ammonium salt of pyridoxine.

3.3.3 Detector for hydrogen sulphide. Hydrogen sulphide is a dangerous gas which has presented a safety problem in a number of industries in America and elsewhere. This is especially true since hydrogen sulphide at a dangerous level is not noticed by workers and levels can increase suddenly. A method for selective detection of the gas in the atmosphere has been developed [33]. This method is based on the adsorption of hydrogen sulphide on to the surface of a quartz crystal coated using an acetone extract of various soots resulting from the burning of several organochlorine compounds. The extract of a soot prepared from chlorobenzoic acid provided the best substrate material, and the method is most useful in the concentration range 1 to 60 ppm. Since the Occupational Safety and Hygiene Authority limit is 20 ppm as a ceiling value, this detector provides measurement within the range of interest.

Other coatings materials, such as lead acetate, metallic silver, metallic copper and anthraquinone-disulphonic acid for detection of hydrogen sulphide with coated piezoelectric crystal have been proposed by King [10].

3.3.4 Detector for hydrogen chloride gas. Hydrogen chloride gas is a noxious byproduct of many industrial processes. The detection of this gas has been accomplished using a crystal coated with different tertiary amines. A successful technique for placing these amines on the surface of a piezoelectric quartz crystal was found to be vapor deposition of the amines on the crystal surface in a flask using vacuum and heating. The amount of amines deposited could be monitored during this process, and reproducible coatings were produced.

The adsorption of hydrogen chloride on a crystal coated with triphenylamine was investigated as a function of time [34]. The response in the field was demonstrated during the Titan-Centaur launch of the first U.S. Mars landing vehicle. Coupled with preliminary laboratory data, a concentration level of 2 parts per $10^9$ was found.

Other amines were investigated for detection of hydrogen chloride gas in the parts per $10^9$ range. Among these the most sensitive were trimethylamine hydrochloride and triphenylamine [6]. These two coatings give great sensitivity and enable the detection of hydrogen chloride in the ppm to parts per $10^9$ concentration range. A new coating technique for deposit of the substrates on the surface of the crystals was also investigated. The response time observed is only a few seconds, and a complete reversibility of response was observed in less than 1 min. No serious interferences for any gases at 1000 ppm concentration, except ammonia and moisture was observed using trimethylamine hydrochloride as the coating material. A gas chromatographic column packed with silica gel was used to eliminate the effect of moisture.

3.3.5 Detector for organophosphorus compounds and pesticides. In recent years great concern has arisen concerning the effect of using pesticides in the environment. Some pesticides are powerful cholinesterase inhibitors and as such are toxic to wildlife and humans, as well as to pets. The organophosphorus insecticides are all structurally related and undergo similar reactions. Since almost all of the organophosphorus pesticides contain either phosphoryl or thiophosphoryl groups, and the thiophosphoryl pesticides readily undergo oxidation reactions to produce phosphoryl containing compounds, di-isopropylmethyl phosphonate (DIMP) was chosen as a model compound in studies by Guilbault and Scheide [35].

Applying different inorganic salts ($FeCl_3$, $CuCl_2$, $NiCl_2$, $CdCl_2$) as coatings for the crystal, the detection of DIMP in the ppm range was effected. It was established that the $FeCl_3$-DIMP complex, which was used as the substrate in the determination of low concentrations of DIMP, could not be used in detection of other organophosphorus compounds of similar structure [35,36]. Therefore, a detector was built using the $FeCl_3$-paraoxon complex as the substrate on the piezoelectric crystal for the specific determination of paraoxon.

A special instrument was developed from a modified design of Karasek's apparatus [12] in the Shackelford-Guilbault experiments [37]. The instrument consisted of two modified Clapp oscillators, the signals from which were mixed, and the resultant frequency difference fed through a pump diode circuit to either an ammeter or a recorder. A cobalt-isonitrosobenzoylacetone(IBA)-diethyl-p-nitrophenylphosphonate (Paraoxon) complex was evaluated for use as a substrate on the quartz crystal. Three different compounds: parathion, DDVP (dimethyl-dichlorovinyl phosphonate) and DIMP, were studied; parathion showed the strongest interaction with the coating used. The recovery time for the crystal when parathion was the pesticide detected, was longer than that for DDVP in the ppb range, but recovery to the original frequency was complete. Various interferences were also investigated, and water vapour was compensated for by use of a mixer circuit in the instrument.

Several experiments have been done in order to determine the most optimum conditions for detection of pesticides with a coated crystal [35]. Among these, the cut-off points and the optimization of coatings were investigated in detail. The cut-off is important in determining the optimum amount of substrate to be applied to the crystal. The frequency of oscillation of a crystal decreases with increasing mass according to the Sauerbrey equation, until the crystal is overloaded and ceases to resonate. It was established that a slightly larger amount of substrate can be added before cut-off occurs, if the crystal was coated on two sides instead of one.

Guilbault and Tomita [7] found a piezoelectric quartz coated with 3-PAD, Triton X-100 and sodium hydroxide, to be an excellent detector for assay of organophosphorus compounds, possessing properties of high sensitivity, excellent selectivity, fast response and long life time .

Several coatings have been screened for response to organophosphorus compounds: 2-and 3-PAD (pyridinium aldoxime dodecyl), 2-PAM (pyridinium aldoxime methyl), histidine hydrochloride and succinyl choline salts were the most reactive and selective. Of these, 3-PAD was the best coating for compounds with G agent structure, and histidine hydrochloride for compounds of the malathion type.

3.3.6 Detector for aromatic hydrocarbons. Since aromatic hydrocarbons are very common in gasoline and are naturally also found in auto exhausts, a fast and selective method for detection of the aromatic hydrocarbons in air is needed. A nujol mixture of trans-chlorocarbonyl-bis-(triphenylphosphine)irridium (I) [trans-IrCl (CO) $(PPh_3)_2$] was used as a coating material by Karmarkar and Guilbault [38] in a piezoelectric device. A benzene solution of the nujol + Ir-complex (1% m/m) was prepared and the crystal was coated on both sides with a tiny brush. It was found that this crystal coating was reactive to aromatic hydrocarbons but was not as sensitive to aliphatic and ordinary olefines. Hence, aromatic compounds such as xylenes, benzaldehyde, 1,3,5-trimethylbenzene, anisole, n-butyl benzene could be detected at low concentration (1-120 ppm), but compounds such as hexane, heptane, octane, cyclohexane, etc., could be detected only at high concentrations. No pronounced effect of atmospheric moisture on the detection limit was observed. The method seems to be fast, reproducible and inexpensive.

A coated piezoelectric quartz crystal for detection and assay of toluene in the working place has been developed by Guilbault et al. [8]. Carbowax 550 was used as a coating substrate, and toluene vapor can be detected in the linear range 30 to 300 ppm with a reproducibility better than 4%. The response time was 30 s, and a complete reversibility was obtained in less than 40 s. No interferences were observed at a 5% volume/volume level. The lifetime of the detector is more than two months. Also a portable monitoring device for toluene, which is 20 cm x 14.7 cm x 9 in dimension and less than 7kg in mass, has been developed.

3.3.7 Detector for mercury in air. In recent years considerable attention has been devoted to the development of methods for detecting low concentrations of mercury in air because the presence of this substance in the environment is very dangerous for human beings. A quartz peizoelectric crystal detector with gold evaporated onto the crystal surface has been shown to be a selective and sensitive substrate [39,40]. By placing of the crystal into a variable oscillator circuit, measurements of the change in frequency of the crystal due to the increase in mass permits a sensing of the amount of mercury vapour present in air down to lower than parts per $10^9$ concentrations. Calibration curves were obtained from ppm to below parts per $10^9$ concentrations of mercury. Reversibility was achieved by placing the detector in an oven held at a temperature of 150°C and flow switching a stream of clean as well as dry air over the detector. The effect of flow rates of air, temperature and interferences on the detection of mercury was investigated. The detector designed by Scheide and Taylor has potential for use both as an air pollution sensor and in industrial hygiene applications.

3.3.8 Detector for mononitrotoluene. Two new coating materials have been evaluated as to selectivity and sensitivity for the detection of mononitrotoluene (MNT), which can serve as a reliable indication of the presence of its less volatile parent (trinitrotoluene, TNT). These are Ucon XMP-1018 and Carbowax 1000. These two coatings were found to be very sensitive to MNT vapor. Carbowax 1000 was selective to MNT, while Ucon XMP-1018 suffers interference from organic vapours, such as toluene. With Carbowax 1000, no interference was observed from any of the inorganic gases and organic vapors which were tested, except some perfumes. Some kinds of perfumes, which were saturated at room temperature or higher, gave a response with the coatings. However, the interference due to perfumes could be considered negligible when the concentrations of these vapors were one-tenth saturation. Since concentration of all perfumes in air can be usually considered less than one-tenth of sample saturated at room temperature, these compounds would not interfere in the assay of MNT. [41]

The coating was also shown to be active toward atmospheric humidity. This problem could be corrected by using air as the carrier gas.

Other parameters that affect the efficiency of the detector were also studied and evaluated. The flow rate of the carrier gas was observed to affect the sensitivity of detection. As the flow rate was increased, the change in frequency decreased, because the adsorption of the MNT vapour become incomplete on the surface of the coatings. On the other hand, the sensitivity also became lower with decrease in flow rate, because MNT vapour diffuses in the tube between the injection port and the crystal, resulting in a spread of the signal peak. The optimum flow rate of carrier gas observed was 30 $cm^3\ min^{-1}$. The temperature of the measuring cell had an effect on the sensitivity, which decreased with increasing cell temperature. However, the cell temperature also affected the response time of the detector. The response become faster as the cell temperature was increased. At a 70°C cell temperature, the response time observed was only a few seconds and a complete reversibility of response was obtained in less than 20 s, while at room temperature the response time was about 20 s or more and a complete reversibility took more than 2 min. For rapid analysis, 50°C is the optimum cell temperature, with respect to both the sensitivity and response time. Under these optimum conditions, the change in frequency ranged from 200 Hz for 10 ppm MNT to 20 Hz for 10 ppb when 2 $cm^3$ samples were injected. The coatings were very stable and the coated crystal gave almost the same sensitivity even after a month. The detector has potential use as a simple device for the assay of explosives. [39]

3.3.9 Carbon monoxide detector. A piezoelectric crystal detector for carbon monoxide has been developed using mercury (II) oxide, which reacts with carbon monoxide to produce $Hg^o$. The $Hg^o$ produced is then detected by using a gold coated piezoelectric crystal.

The optimum temperature for the reaction:

$$HgO + CO \longrightarrow Hg^o + CO_2$$

has been found to be 210°C. Mercury (II) oxide decomposes at 500°C; however, at this temperature (210°C), some mercury (II) oxide will be decomposed and gives a slightly changing background , affecting the baseline. By using a reference cell, this background is completely eliminated. In the reference cell, the carbon monoxide sample will first pass through silver oxide and is quantitatively oxidized before it reaches the mercury (II) oxide. Therefore, only the Hg background is measured in the reference cell.

The gold coated piezoelectric crystal adsorbs $Hg^o$ at room temperature. Reversibility is achieved by desorption at 150°C. A straight line calibration plot, that is ΔF vs concentration, has been obtained in the ppm range with a 10 $cm^3$ sample. By varying the sample size, a 10 fold or 100 fold increase in sensitivity in the linear range can be achieved [42].

## 4. SUMMARY AND FUTURE APPLICATIONS

Piezoelectric crystal detectors have been used in many areas of analytical chemistry, although mainly as air pollution sensors. Instruments which are portable for field use are already in the development stage. The instrumental design and fabrication of piezoelectric detecotrs has recently been improved to such an extent that there are no major problems to be anticipated in using these devices. It should be possible to develop many new coating materials that are selective or specific for certain pollutants, and then a group of many of these sensors could be placed at strategic positions throughout a city or area with information relayed back to a computer or central station for evaluation. Individual personal piezoelectric crystal monitors can be developed for workplace monitoring. It is likely that the piezoelectric crystal detector will play an important role in the near future for air pollutant control monitoring.

The reader is invited to other reviews written by the author in the field of piezoelectric detectors for additional information [9,43,44].

## 5. REFERENCES


1. F. W. Maarsen, M. C. Smit and J. Matze, Recueil, 76, 713 (1957).

2. R. A. Heising, Quartz Crystal for Electrical Circuits, Van Nostrand, New York, (1946), p. 24.

3. G. Z. Sauerbrey, Z. Physik, 155, 206 (1959).

4. Idem., ibid., 178, 457 (1964).

5. A. W. Warner, and C. D. Stockbridge, The Measurements of Mass using Quartz Crystal Resonators, Symposium on Vacuum Microbalance Techniques, Los Angeles, Calif. (1962).

6. J. Hlavay and G. G. Guilbault, Anal. Chem., 50, 965 (1978).

7. G. G. Guilbault and Y. Tomita, Anal. Chem., July 1980.

8. G. G. Guilbault, M. Ho and B. Rietz, Anal. Chem., Aug., 1980.

9. G. G. Guilbault, Final Report, ARO Grant DAHC-74-G-0119, Oct. 31, 1977.

10. W. H. King, Jr., Anal. Chem., 36, 1735 (1964).

11. W. H. King, Jr., U. S. Patent 3,164,004 (Jan. 5, 1965).

12. F. W. Karasek and K. R. Gibbins, J. Chromatogr. Sci., 9, 535 (1971).

13. F. W. Karasek, P. Gruy, H. H. Hill and J. M. Tiernay, Chromatog., 124, 179 (1976).

14. F. W. Karasek and J. M. Tiernay, J. Chromatog., 89, 31 (1974).

15. W. H. King, Jr., Res. Dev., 20(4), 28 (1969).

16. W. H. King, Jr., Ger. Offen. 1, 901, 845 (August 13, 1970).

17. H. M. Crawford, Analysis and Instrumentation, Plenum Press, New York, (1964), pp. 105-108.

18. W. H. King, Jr., and G. M. Varga, Jr., Final Report for Martin-Marietta Corp., Contract No. RCO-446067.

19. W. H. King. Jr., F. E. Steidler and G. M. Varga, Jr., Phase I-II Report for Martin-Marietta Corp., Contract No. RC1-201037.

20. Idem., ibid., Contract No. RC1-201107.

21. K. H. Karmarkar and G. G. Guilbault, Analytica Chim. Acta 71, 419 (1974).

22. G. G. Guilbault and A. Lopez-Roman, Environ. Letters, 2, 35 (1971).

23. J. L. Cheney and J. B. Homolya, Anal. Letters, 8, 175 (1976).

24. J. L. Cheney and J. B. Homolya, Sci. of the Total Environ., 5, 69 (1976).

25. R. B. W. Earp, Ph.D. Dissertation, University of Alabama (1966).

26. M. J. Hartigan, Ph.D. Dissertation, University of Rhode Island (1970).

27. J. L. Cheney, T. Norwood, and J. B. Homolya, Anal. Letters, 9, 361 (1976).

28. K. H. Karmarkar, L. M. Webber and G. G. Guilbault, Environ. Letters, 8, 345 (1975).

29. K. H. Karmarkar, L. M. Webber and G. G. Guilbault, Analytica Chim. Acta, 81 265 (1976).

30. K. H. Karmarkar and G. G. Guilbault, Analytica Chim. Acta, 75, 111 (1975).

31. L. M. Webber and G. G. Guilbault, Anal. Chem., 48, 2244 (1976).

32. J. Hlavay and G. G. Guilbault, Anal. Chem., 50, 1044 (1978).

33. L. M. Webber, K. H. Karmarkar and G. G. Guilbault, Anal. Chem., 97, 29 (1978).

34. NASA Tech. Brief, MFS-23357.

35. E. P. Scheide, and Guilbault, Anal. Chem., 44, 1764 (1972).

36. G. G. Guilbault, Analytica Chim. Acta, 39, 260 (1967).

37. W. M. Shackelford and G. G. Guilbault, Analytica Chim. Acta, 73, 383 (1974).

38. K. H. Karmarkar and G. G. Guilbault, Environ. Letters, 10, 237 (1975).

39. E. P. Scheide and J. K. Taylor, Environ. Sci. Tech., 8, 1097 (1974).

40. E. P. Scheide and J. K. Taylor, Am. Chem. Soc. Div. of Environ. Chem., 14, 329 (1974), presented, Los Angeles, California, April, 1974.

41. Y. Tomita, M. Ho and G. G. Guilbault, Anal. Chem., 51, 1475 (1979).

42. G. G. Guilbault and M. Ho, Anal. Chem., in the press.

43. L. M. Webber, J. Hlavay and G. G. Guilbault, Mikrochimica Acta, 1978, 351.

44. J. Hlavay and G. G. Guilbault, Proceedings, 4th Joint Conference on Sensing Environmental Pollutants, (1978) 205.

# THE AUTHOR

Dr. Guilbault received his BS degree from Loyola University in 1958 and his Ph.D. from Princeton University in 1961. After serving 5 years at Edgewood Arsenal, he joined the staff of the University of Louisiana in New Orleans (now the University of New Orleans) in 1966. During 1973-75, he was a visiting Professor at the Technical University of Denmark. His fields of interest are analytical, clinical, physical and biochemistry, especially in the use of enzymes as analytical reagents, enzyme electrodes, immobilized enzymes and fluorometric methods of analysis. He received the 1977 American Chemical Society's award in analytical chemistry.

*Ion-Selective Electrode Rev.* Vol.2, pp.17-71. 0191-5371/80/0901-0017$05.00/0


# USES OF KNOWN ADDITION, GRAN'S PLOTS AND THE RELATED METHODS WITH ION-SELECTIVE ELECTRODES

Marco Mascini

Istituto Chimica Analitica, Università di Roma, Rome, Italy

CONTENTS

KEYWORDS: Analate addition; analate subtraction; computer methods with ion-selective electrodes; Gran's plots; known subtraction; standard addition; titrations with ion-selective electrodes; known addition.

## 1. INTRODUCTION

The large impact produced by ion-selective electrodes around 10 years ago was based on their promising feature of determining by a direct potentiometric measurement the amount of a certain species in solution without the need for any tedious pretreatment (1-3).

Although by the Nernst law a potentiometric measurement is related to the "activity" of an ionic species in solution, many people are more accustomed to concentration terms, such as M, g $dm^{-3}$ ppm, etc. Hence, several methods of measuring concentration with ion-selective electrodes were proposed.

Two more classical techniques were first used, namely, the direct method and potentiometric titration. However, in the last 10 years two other techniques have been developed and discussed, namely, the known (or standard) addition (or subtraction) and the Gran titration; these are the subject of this review.

The four methods differ in precision, accuracy and in the time required per analysis. Generally the time used is compensated by better precision, but in some applications one technique may be more appropriate than others. This is because in any measurement technique three major effects must be considered, namely, ionic strength, complexation and the interfering species present. The various measurement techniques are affected differently by these effects and depending on the particular application, the analyst can choose the best approach.

Ionic strength affects only the activity coefficient of the ion, but because of it the activity and concentration calibration curves are separated. This divergence can be overcome by diluting standards and samples with an electrolyte which does not interfere and gives a high and constant ionic strength to all solutions; but such a remedy is not always possible.

Ion binding or complexation is a solution effect. Generally ion-selective electrodes measure one ionic species in solution such as, $F^-$ but not HF, $Cu^{2+}$ but not $Cu(EDTA)^{2-}$, and $S^{2-}$ but not $HS^-$ or $H_2S$. Such specificity is the basis for the-end point detection in some potentiometric titrations. However, when the ion in the sample is partly associated it is often necessary to extract it for the measurement or assume the free fraction to be constant during the measurement for both standards and samples.

Finally, some chemical species may interfere and react with the electrode material (by precipitation or ion exchange reaction) and alter the potentiometric response. This effect is not easy to overcome. If an ion interferes it should be removed, for example, $OH^-$ is removed by adjusting the pH, or in some special cases the interference effect can be computed and thus cancelled.

In the following sections features of the direct method, known addition or subtraction, and Gran plots are discussed.

## 2. THE DIRECT METHOD OF ION-SELECTIVE ELECTRODE MEASUREMENTS

The direct method is simply derived from the procedure pH measurement (4). The sample to be analysed is pretreated when necessary and the electrodes are immersed in it. The potential measurement is recorded and related to a calibration curve obtained with standard solutions. The technique is generally recommended for routine measurements because of the short time involved (two or three minutes) and with flow through systems up to 100 samples per hour can be monitored.

All ion-selective electrodes which have been described, have generally been used by this technique, for the method is undoubtedly attractive and the results are sufficiently precise for most purposes.

The direct method is particularly suitable for those samples in which the ion is in a free uncomplexed state or can be decomplexed by appropriate pretreatment. It can also be used when the ion can be obtained free in a constant fraction provided that the fraction is constant for standards and samples.

An example of the decomplexation technique is the direct determination of fluoride in water samples with "TISAB" solution. In fact HF is a weak acid with a $pK_a$ of 3.14 and at low pH values some of the $F^-$ is bound as HF which is not sensed by the fluoride ion-selective electrode. As for all weak acids the higher the pH the greater is the fraction of total fluoride that is dissociated and in order to have 99% or more dissociated, the pH must be higher than the $pK_a$ by 2 units or more. In this case the pH should be greater than 5.14.

Other examples of species that can bind fluoride are aluminium and iron (III) ions. In the case of aluminium the thermodynamic formation constant, K, for the complex, $AlF^{2+}$, formed from $Al^{3+}$ and $F^-$ is approximately $10^6$. In order to use this value to determine the fraction of fluoride complexed by a given concentration of $Al^{3+}$, it is necessary to remember that with increasing pH hydroxy complexes of $Al^{3+}$, such as $Al(OH)^{2+}$ are formed. At a relatively low pH of 5, however, a large fraction of aluminium is not complexed with $OH^-$. Thus, for example, taking a value of $10^{-4}$M as the concentration of the $Al^{3+}$

form and neglecting the formation of HF or any activity coefficient, it is calculated that about 99% of the fluoride is complexed as $AlF^{2+}$.

Similar considerations apply with other metal ions like iron, beryllium, zirconium, thorium, uranium and calcium which complex fluoride. Such complexations led to the development of the reagent known as TISAB (total ion strength adjustment buffer) for use when measuring fluoride in water samples (5-9). The buffer is prepared in various ways and one recipe is:

> Acetic acid (glacial) 57 $cm^3$, sodium chloride 58 g and CDTA 1 g. Dissolve and mix these constituents in about 500 $cm^3$ distilled water and add strong NaOH solution (about 5M) with continuous stirring until the pH reaches $5.5 \pm 0.3$. Dilute to 1 $dm^3$.

By diluting standard and sample 1:1 with TISAB the fluoride is uncomplexed because of the formation of stable metal CDTA complexes and the pH ensures that all fluoride is free.

A similar technique has been developed for calcium determination in water samples. Calcium in fact forms complexes with a wide variety of ligands (10). Using a calcium ion-selective electrode it has been shown that approximately 16% of the calcium in sea water is not ionized, and is probably complexed with sulphate, carbonate and bicarbonate (11).

In other systems the binding of calcium by sulphate also can be significant and can lead to erroneous results when electrode measurements do not take this into account. For example, using a value of $2x10^2$ for the thermodynamic constant, $K_f$, for the formation of soluble $CaSO_4$ from $Ca^{2+}$ and $SO_4{}^{2-}$ (10) and using Debye-Hückel value for activity coefficients, in a $5x10^{-3}M$ solution of calcium sulphate, approximately 22% of the calcium is bound as $CaSO_4$. This indicates that calcium sulphate is not a good choice as calibrating solution. A better choice would be calcium chloride because it remains ionized. In any case, calcium ion-selective electrodes have been used for measuring $K_f$ for soluble $CaSO_4$, a value of 24.5 being obtained (12).

In typical treated municipal water with I = 0.004, $[Ca^{2+}]$ = $6.5x10^{-4}M$ and $[SO_4{}^{2-}]$ = $2.7x10^{-4}M$ the problem is less serious. The calibration indicates that only 1.5% of the calcium is bound as $CaSO_4$. However, if one considers such a water with high concentration of these species (13), namely, $6x10^{-3}M$ $[SO_4{}^{2-}]$ , $3.5x10^{-3}M$ $[Ca^{2+}]$ and I = 0.04, then the calculation indicates that approximately 20% of the calcium and 10% of the sulphate are bound as $CaSO_4$.

Similar calculations can be made for the binding of calcium by carbonate to form soluble $CaCO_3$, using value of $K = 1.6x10^3$. With typical values in municipal waters of $[HCO_3^-]=7.5x10^{-4}M$, $[Ca^{2+}]$ = $6.5x10^{-4}M$, pH = 7.3 and I = $5x10^{-3}$, only 0.1% of the calcium is bound as $CaCO_3$. In this case, as the pH increases so will the extent of calcium binding due to conversion of bicarbonate to carbonate. The calculation with unusually high concentration of these species, namely, $[CO_3{}^{2-}]$ = $5x10^{-4}M$, $[Ca^{2+}]$ = $3.5x10^{-3}M$ and I = 0.04 indicates that approximately 8% of the calcium is bound as $CaCO_3$.

Hulanicki and Trojanowicz (14) have suggested a useful direct method for the determination of calcium in water supplies; they developed a "constant complexation buffer" for addition at a volume ratio of 1:1 to both samples and standards. This ensures that the correct pH is achieved (pH = 9.1), magnesium ions are masked, the ionic strength is fixed and a constant proportion of the calcium ion is complexed. The buffer consists of:

0.02 M ammonium hydroxide, 0.02 M ammonium chloride, 0.04 M potassium nitrate, 0.04 M acetylacetone and 0.02 M iminodiacetic acid.

With this buffer, determinations of calcium in the range 20-800 mg $dm^{-3}$ were performed (14).

The above examples show how careful has to be the choice of the pretreatment solution and how difficult it is to generally apply the direct method.

The final consideration on the direct method concerns the precision. Typical measurements in open beakers and at room temperature are within ± 1 mV and this corresponds to a precision of ± 4% on the concentration value of monovalent ions and ± 8% of bivalent ions.

Thus, the direct method of using ion-selective electrodes has advantages and disadvantages. Another advantage is that precision is constant over all the calibration curve and over several orders of magnitude, so that at high and low concentration the precision is the same, at least over the linear portion of the calibration curve. A disadvantage is that the precision is relatively low but it can be improved as long as great care is spent in temperature control, stirring rate, regulation sample pretreatment, choice of reference electode, junction potential control, but as sson as the technique is elaborated to give greater accuracy and precision the speed is lost and the technique may often lose its advantage over alternative methods.

## 3. KNOWN ADDITION, KNOWN SUBTRACTION AND RELATED METHODS

### 3.1 Theory

3.1.1. The known addition method. This technique consists of measuring the potential in a sample before and after the addition of a known concentration of the substance that is being determined.

In any sample solution containing an unknown total concentration of determinand, $C_A$, the electrode will respond to the activity of the free ionic form of the determinand and its potential with respect to a reference electrode and recorded with a potentiometer can be expressed by:

$$E_1 = E^o + S \log (f\, C_A\, x) \qquad (1)$$

where f is the activity coefficient, x the fraction of the total concentration of the determinand present as free ions, S is the slope of the electrode calibration curve, and $E^o$ takes into account the overall standard potential of the ion selective electrode, the reference electrode and the junction potential.

If a volume, V, of a solution containing a known concentration, $C_S$, of determinand is added to the original volume, $V_O$, of the sample in which the electrodes are immersed the new potential is

$$E_2 = E^{o'} + S' \log \left[ f' \, x' \frac{(V_o C_A + V C_s)}{(V_o + V)} \right] \qquad (2)$$

where the primed terms $E^{o'}$, $S'$, $f'$ and $x'$ denote values of the previously defined terms in the second solution.

If the electrode is responding in the linear portion of the calibration curve, $S' = S$ and if the junction potential can be considered unchanged, i.e., $E^o = E^{o'}$, then the difference between the two potential readings can be arranged to

$$E_2 - E_1 = S \log \left[ \frac{f' \, x' \, (V_o C_A + V C_s)}{f \, x \, C_A \, (V_o + V)} \right] \qquad (3)$$

If the conditions are arranged that the ionic strength does not change then $f' = f$, and if the free fraction of the ion remains unchanged it follows that $x' = x$. Equation (3) can then be simplified to

$$E_2 - E_1 = S \log \left[ \frac{V_o C_A + V C_s}{C_A (V_o + V)} \right] \qquad (4)$$

and rearranged to give the expression below, that is, the total concentration in the original sample is expressed as

$$C_A = \frac{C_s}{10^{\Delta E/S} \left[ 1 + \frac{V_o}{V} \right] - \frac{V_o}{V}} \qquad (5)$$

If, finally, the volume of the reagent added is less than or equal to approximately 1-2% of that of the sample, the volume change can be ignored and equation (5) becomes

$$C_A = \frac{C_s \; V}{V_o} \times \frac{1}{(10^{\Delta E/S} - 1)}$$

or

$$C_A = \frac{C_s \, V}{V_o} \cdot K \qquad (6)$$

where K values are as tabulated in Table I for a range of millivolt differences ($\Delta = \Delta E$ for monovalent ions and $\Delta = \Delta 2 \times \Delta E$ for bivalent ions) for theoretical slopes of 59 mV decade$^{-1}$ for monovalent ions and 29.5 mV decade$^{-1}$ for bivalent ions. Hence, the total concentration $C_A$ may be obtained by mulipIying the value obtained from the Table by the nominal change in the determinand concentration.

3.1.2. <u>The known subtraction method.</u> This method is different in that the standard solution added to the sample is not the determinand solution but a solution of a species which reacts with the determinand quantitatively to produce a decrease in the determinand concentration with corresponding change in the electrode potential.

TABLE I. Values of K for the known addition method.

$\Delta$ = mV difference, $\Delta = \Delta E$ for monovalent ions, $\Delta = 2 \times \Delta E$ for bivalent ions.

| $\Delta$ | K | $\Delta$ | K |
|---|---|---|---|
| 1 | 25.6 | 16 | 1.15 |
| 2 | 12.3 | 17 | 1.06 |
| 3 | 8.06 | 18 | 0.98 |
| 4 | 5.92 | 19 | 0.91 |
| 5 | 4.63 | 20 | 0.84 |
| 6 | 3.79 | 21 | 0.79 |
| 7 | 3.18 | 22 | 0.74 |
| 8 | 2.72 | 23 | 0.69 |
| 9 | 2.38 | 24 | 0.64 |
| 10 | 2.10 | 25 | 0.60 |
| 11 | 1.87 | 26 | 0.57 |
| 12 | 1.68 | 27 | 0.54 |
| 13 | 1.51 | 28 | 0.50 |
| 14 | 1.38 | 29 | 0.48 |
| 15 | 1.26 | 30 | 0.45 |

In this case the theoretical treatment is the same and instead of equation (6) one obtains

$$C_A = \frac{C_s\ V}{Z\ V_o} \cdot (K + 1) \qquad (7)$$

where Z represents the combination ratio between the species added and the ion in solution. For example, if the ion to be determined is $S^{2-}$ and $Ag^+$, Z = 2, if the ion to be determined is $Cu^{2+}$ and EDTA is added, then Z = 1.

3.1.3. Analate addition and analate subtraction methods. Another variation of the technique can be obtained when the sample is added to a solution containing a known concentration of the ion to be measured. Often the technique is called "analate addition" or "analate subtraction". The latter term is used when a decrease of the ion to be measured results.

The equation obtained for analate addition is

$$C_A = \frac{C_s\ V}{V_o} (10^{\Delta E/S} - 1) \qquad (8)$$

and for the analate subtraction

$$C_A = \frac{C_s\ V}{V_o} (1 - 10^{-\Delta E/S}) \qquad (9)$$

with the same meaning for symbols as defined above.

The technique of analate addition, otherwise known as sample addition is a variation of the standard addition method. For a known volume of a standard solution of the ion to be determined a known volume of the sample is added. The technique is used when the concentration of the ion in the sample is relatively high and a dilution is necessary in order to bring the concentration into the linear range. The technique is also used in other cases, that is, when the sample is very acid or very alkaline and it should be buffered before the measurement, or when the sample is a nonaqueous solution, or when the sample has a very high ionic strength.

An example of the use of analate addition is the determination of nitric acid in stainless steel pickling baths (15). The pickling bath contains nitric and hydrofluoric acids and $Fe^{3+}$ ion and by the analate addition method extra information about the pickling process may be obtained.

The analate subtraction technique is useful for applying the ion-selective electrode measurements to chemical species for which a selective membrane electrode is not available but where the sought species can quantitatively react with ions for which ion-selective electrode measurements can be made. The ion-selective electrode and the reference electrode are kept in a standard solution to which a small volume of the sample is added and the reaction between the sought species and the ion in the standard solution takes place. The disappearance of the standard ion is monitored and related to the concentration of the sought species in the sample.

The analate addition method has been applied to the determination of oxidants in air with an iodide-selective electrode (16,17). Air was bubbled for a limited time through an iodide solution buffered at pH 7 in which iodide was $10^{-5}$M; the ozone and other oxidants in the air oxidize the iodide to iodine and the change of concentration of iodide brought about by the oxidant was determined through electrode potential measurement.

### 3.2. Practical examples

3.2.1. Known addition. To determine the concentration of iodide ions in a sample by the known addition method, i.e., by adding iodide ions to the sample, take 50 $cm^3$ of the sample ($V_o$ =50) and using an iodide ion-selective electrode with a reference electrode measure the e.m.f. Such a result was -180 mV ($E_1$ = -180). Add 1 $cm^3$ (V = 1) of a standard solution, $10^{-2}$ M, of iodide ions ($C_s = 10^{-2}$) and again measure the e.m.f., namely -194 mV ($E_2$ = -194). From Table I compute for $\Delta$= (-180 + 194) = 14 mV for n = 1 a value of K = 1.38. Now, by equation (6) the unknown concentration of the solution can be calculated:

$$C_A = \frac{10^{-2} \times 1}{50} \times 1.38 = 2.7 \times 10^{-4} M \qquad (10)$$

3.2.2. Known subtraction. To determine the concentration of $Cu^{2+}$ ions in a sample by the known subtraction method, i.e., by adding EDTA to a buffered solution take 50 $cm^3$ of sample ($V_o$ = 50) and using a copper membrane ion-selective electrode with a reference electrode measure the e.m.f., namely, +130 mV ($E_1$ = +130 mV). Add 1 $cm^3$ (V = 1) of a standard solution of $10^{-2}$ M EDTA ($C_s = 10^{-2}$) and again measure the e.m.f., namely, + 120 mV ($E_2$ = +120). From Table I compute for $\Delta$ = 2 x [+130 -(+120)] = 20 (because copper is bivalent) K = 0.84.

The total concentration of copper can be computed using equation (7):

$$C_A = \frac{10^{-2} \times 1}{50} \times 1.84 = 3.7 \times 10^{-4} \text{ M} \qquad (11)$$

The examples outlined in this section and in section 3.2.1. have been published (18) and demonstrate the simplicity of the operation. The forte of the method lies in obtaining the total concentration of a species without reference to an electrode calibration curve and by means of a very simple operation sequence.

### 3.3 Assumptions of the known addition/subtraction methods

The reliability of the known addition method depends on the validity of the assumption described in the derivation of equation 6, and it is worth-while discussing the various features of fixing the conditions of measurement. The various assumptions are

(i) the change of volume during the addition is negligible;

(ii) the degree of complexation remains unchanged;

(iii) the activity coefficient of the ion does not change, i.e., the ionic strength remains constant;

(iv) the junction potential at the reference electrode does not change, i.e., the ionic strength remains constant;

(v) the ion-selective electrode has a constant response slope and (for using the data of Table I) the slope has a theoretical Nernstian value;

(vi) no interferents are present in the sample; and

(vii) in the known subtraction method the final concentration is within the limit of linear Nernstian response of the electrode.

For (i), the assumption of negligible dilution is reasonable as long as the concentration of the addition component is sufficiently high. Also, the increasingly common use of micropipetes which repoducibly dispense volumes under 100 $mm^3$ make this very feasible.

Assumption (ii) is completely valid when the ion is completely uncomplexed. However, when it is partly complexed it is important that the concentration of the complexing agent is at least 20-100 fold in excess to the concentration of the ion to be determined to ensure a constant degree of complexation. Often it is advisable to add a complexing agent to the sample solution in order to ensure this requirement. Often the complexation degree depends on pH, and it may be necessary to buffer the sample in order to satisfy the conditions. Also, some buffers, such as, acetate, phosphate, tris and ammonia behave like complexing agents so the addition of these substances serves a dual purpose, namely, buffering the pH and making constant the complexation fraction of the ion. This assumption depends on the composition of the sample which should be at least approximately known by using the addition method.

Assumption (iii) and (iv) can become valid merely by adding an inert electrolyte to make the ionic strength constant. The standard solution can be prepared without the inert electrolyte because the variation of volume is negligible and hence there will be a negligible effect on the ionic strength during the standard addition. The need for the precaution of adding an inert electrolyte in order to maintain the ionic strength is emphasised by illustrative measurements on calcium containing solutions whereby $10^{-1}$M calcium chloride is the known addition standard added to $10^{-3}$M calcium chloride (19). The associated variation in ionic strength with the repercussions on activity coefficients demonstrate that under such conditions, the activity coefficient needs to be considered in order to arrive at the correct deduction (19).

Assumption (v) is especially important when using the values of Table I. In fact, very few electrodes have a truly Nernstian value of the calibration slope, but all electrodes have a very large range in which the slope is constant. Non-Nernstian slopes can sometimes be controlled by using samples with known concentration and by adding a standard solution, in which case Table I is still valid if the Δ is computed in the following way

$$\Delta = \frac{(E_2 - E_1) \times 59.1}{S} \qquad (12)$$

where $E_2 - E_1$ is the millivolt difference between the two electrode potential measurements, 59.1 the theoretical Nernstian value and S the slope of the experimental system.

Assumptions (vi) and (vii) are often practically related to assumption (v) because when an interfering species is present or when the concentration of the determinand species is below a certain value the slope of the electode does not remain constant at the linear or near-Nernstian value. The manipulation of the new slope, although operative over a limited range, can be handled as discussed for assumption (v).

### 3.4 Precision of the known addition method

Optimization of precision of the known addition measurement can be discussed in terms of equation (6),

$$C_A = \frac{C_s V}{V_o} \times \frac{1}{(10^{\Delta E/S} - 1)} \qquad (6)$$

The standard deviation of $C_s$, V and $V_o$ can be considered negligible, while the standard deviation of $C_A$ will depend on ΔE ($E_2$ and $E_1$) measurments and on the value of S, the electrode slope.

By differentiating the equation (6) and rearranging the following equation representing the percent of standard deviation of $C_A$ propagated from the standard deviation of experimental potential measurements is obtained

$$\% \sigma_{C_A} = \frac{2 \sigma_E}{0.43 x S} \left(1 + \frac{C_A V_o}{C_s V}\right) \times 100 \qquad (13)$$

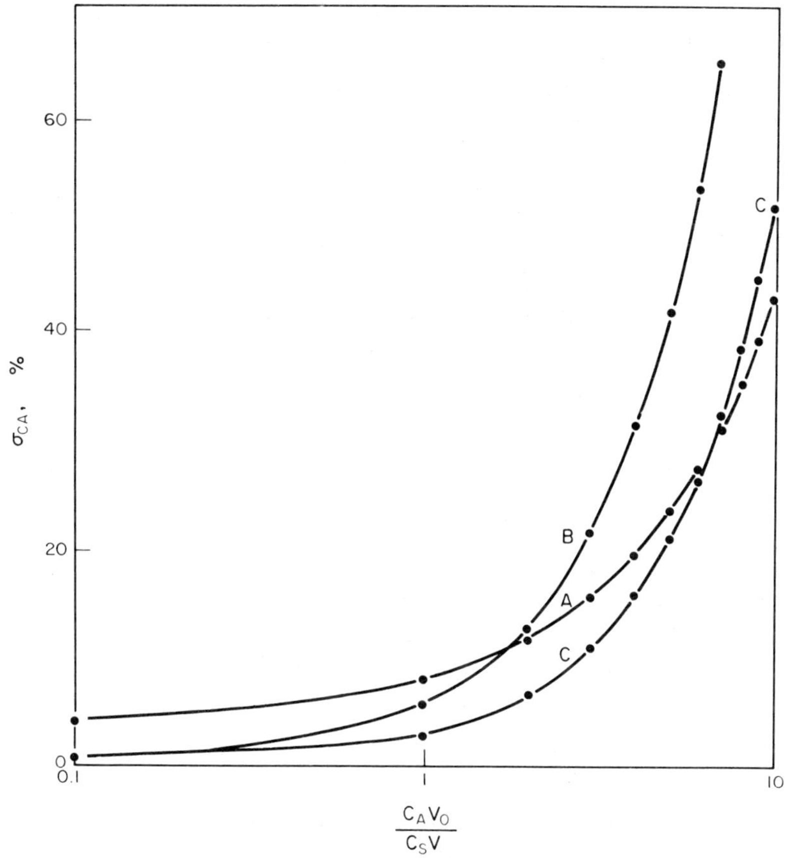


Fig.1. Standard deviation of concentration, $\sigma$ , as propagated from the standard deviation of potential measurement, $\sigma_E$, and of the slope, $\sigma_S$, value. Curve A refers to $\sigma_E$ = 0.5 mV and for a theoretical slope of 59.1 mV, $\sigma_S$ = 0. Curves B and C are for $\sigma_S$ = 0.5 mV and 1.0 mV respectively, $\sigma_E$ = 0.

$\sigma_E$ represents the standard deviation of the potential measurement, the factor 2 taking into account that two potential measurements are made. S is the slope, $0.43 = \frac{1}{\ln 10}$ and $C_s$ V represents the amount of ion added and $C_A$ $V_o$ the amount of the ion in the sample solution.

Equation 13 is depicted in Fig.1 as curve A for a slope of 59.1 mV and a $\sigma_E$ = 0.5 mV as function of the ratio $\frac{C_A\ V_o}{C_s\ V}$ on semi-log paper.

From Fig.1 it can be seen that when the amount added (V $C_s$) is 1/10 of the amount of the sample ($C_A\ V_o/C_s\ V$ = 10), the relative standard deviation $\sigma_{C_A}$ propagated is 44%. It must be kept in mind that the ratio $C_AV_o/C_sV$, must be around 1 or greater than 1. When the ratio is 1 the relative standard deviation is 8% and when the ratio is 0.1, the relative standard deviation becomes 4%. This value corresponds to a potential variation of 60 mV.

The standard deviation propagated by an uncertainty on the S measurement can also be computed. By differentiating equation (6) and rearranging the following equation is obtained

$$\%\sigma_{C_A} = \frac{\sigma}{0.43xS}\left[1 + \frac{C_A\ V_o}{C_s\ V}\right] \ln\left[1 + \frac{C_A\ V_o}{C_s\ V}\right] x\ 100 \qquad (14)$$

$\sigma$ is the standard deviation of the slope S and with the other letters of equation 14 having the same meaning as previously.

Equation (14) is depicted in Fig.1 as Curve B and computed for $\sigma$ = 1 mV and S = 59.1 mV, and also as Curve C for $\sigma$ = 0.5 mV, in each case as a function of the ratio $\frac{C_A\ V_o}{C_s\ V}$.

It is evident that the slope uncertainty is less important if the ratio $C_A\ V_o/C_s$ V, is around or below 1, but it becomes extremely important for higher ratios in 13. If the ratio is $\leqslant 1$ the uncertainty of the slope can be neglected in comparison to uncertainty of the potential values. When the ratio is 1, the variation of potential for monovalent ions is 18 mV and for bivalent ions it is 9 mV.

From these considerations, the practical rule was derived (20) that the variation of potential in the known addition method should be between 10 and 30 mV for monovalent ions and between 5 and 15 mV for bivalent ions. These potential differences correspond respectively to ratios of 2 and 0.5 and from Fig.1 the relative standard deviations can be computed.

Furthermore, because of considerations concerning complexation and activity (assumptions (ii) and (iii) above) it is preferable to maximize $C_s$ and minimize V for a particular value of $C_s$ V.

## 3.5 Applications

The precision of the measurement of the known addition method and the time spent to perform analysis is similar to that of direct method outlined in section 2. The method is more appropriate when a rapid and approximate analysis of occasional samples is needed and for which the calibration curve is not ready or the electrode potential is not stable or the electrode is not properly conditioned. In fact with this procedure, only one standard solution is required and calibration drift is unimportant.

The method is also appropriate when the ion to be determined is partially complexed, in fact often the ion cannot be decomplexed for direct measurement, or because there is not a decomplexing agent available or because the addition of another species interferes with the electrode material.

Some comparisons between the two techniques have been made in the literature. Different conclusions have been reached on the relative merits of the direct and known addition methods.

In the analysis of tap water (21) for nitrate (this ion is not complexed) the accuracy and precision of a known addition procedure are worse than those of the direct method with a calibration curve. In an analysis of river water for ammonia (22) with an ammonia gas electrode the results were comparable for the two procedures. The determination of sulphur in petroleum by a sulphur dioxide probe is reported to be successful by the known addition method (23). Here, the sulphur dioxide produced by combustion of the petroleum is collected in a solution containing a large excess of tetrachloro-mercurate (II) ions at pH 6.9 which complex nearly all the gas as di-sulphato-mercurate (II) ions. After the collection period the pH of the solution is altered to 1.2, when 20% of the sulphur dioxide is still complexed. The total sulphur dioxide is determined by a known addition procedure.

The subtraction method is used when it is not possible to prepare a standard solution of the determinand for addition to the sample, either because no pure reagents are available or because the determinand ion is unstable. This is the case for sulphide determination in waste water (24). Many samples have been processed by adding silver nitrate to polluted water samples in order to determine sulphide concentration. For example, to 50 $cm^3$ of sample add 1 $cm^3$ aliquots standard silver nitrate solution (0.1M) and record the difference in potential values. Table I may be used for evaluating the concentration. The subtraction procedure has also been used for polluted sea-water samples, with reliable results (25-26).

For black liquor of the pulp and paper trades Orion Research Incorporated (27) have developed a procedure with a cadmium ion-selective electrode. Cadmium was chosen instead of silver because the latter could be reduced to the metallic form by polysulphides and aromatic polyhydroxy groups in the black liquor. Because of this, the silver electrode could give erroneously high sulphide values, since the formation of metallic silver as well as the precipitation of silver sulphide would affect the extent of reduction in the level of free silver ion in the silver nitrate subtraction reagent. For the cadmium electrode method the procedure of standard subtraction is inverted to an analate subtraction. Thus, a measured amount of the black liquor is added to a reagent containing a known concentration of cadmium ions, and a cadmium ion-selective electrode is used to follow the decrease in the cadnium ion concentration, since cadmium reacts with the sulphide in the sample. The original sulphide concentration can be calculated from the amount of cadmium removed from the reagent by means of the following equation which is a variation of equation (6):

$$C_A = \frac{C_S V}{V_o}\left[1 - \frac{1}{10^{\Delta E/S}}\right] \qquad (15)$$

In this case the reagent contains three additional chemicals; acetic acid, sodium hydroxide and Cellosolve. The acetic acid and sodium hydroxide serve to fix the pH and total ionic strength of the reagent so that changes do not occur when the sample is added. Cellosolve prevents the black liquor from coating the electrodes during the measurement. The Orion procedure states that it is not necessary to add any antioxidant to the reagent since, in the presence of excess cadmium ion, the reaction between sulphide and cadmium ion takes place rapidly. In the short time needed for the determination there is no significant oxidation of the sample sulphide.

Another interesting example of the successful application of this rapid technique is the determination of silver in fixing solutions for photographic development in the range $10^{-4}$-$10^{-1}$M and using the silver ion-selective electrode (28). The fixing solution typically contains a large excess of thiosulphite in a sulphite-bisulphate buffer which serves to prevent the precipitation of silver chloride. The silver that accumulates in the bath is mostly complexed as $[Ag(S_2O_3)]^-$ and the level of free silver ions present is low even at high concentrations of silver. The determination is performed by adding a known quantity of silver ions to a measured volume of the fixing bath. Most of the added silver is immediately complexed by the excess thiosulphate ion present in the bath. The change in electrode potential is used to calculate the original total silver concentration of the sample using equation (6).

The determination of fluoride in vegetation due to emissions from aluminium smelting factories is based on extraction of fluoride from dried plant material and subsequent analysis by a lanthanum fluoride membrane electrode (29). The extraction should involve alkali fusion, but to improve speed and simplicity direct extraction with perchloric acid (0.1 M) has been successfully used. In this medium, $F^-$ is strongly associated with protons but it has been shown that accurate results can be obtained by the standard addition procedure, that is, by measuring a stable potential value before and after adding a known amount of fluoride.

Finally, Table II collects some other examples of known additions and known subtractions taken from the literature. They are still indicative; the precision and the time required for analysis in fact plays a substantial role in choosing a method. Normally, the time required is around two minutes and the relative precision is not better than 4%.

Any analytical problem relevant to ion-selective electrodes can be solved by a known addition, known subtraction or related procedure provided that the relative precision is sufficient for the particular problem. Table II gives an idea of the various analytical problems which have been solved with this technique in recent years. Many other analytical problems which have been solved with ion-selective electrodes through other techniques may also be solved by the addition/subtraction and similar procedures.

### 3.6 Meters for standard addition and subtraction measurements

The standard addition and subtraction methods have become very popular and in some instruments specially designed for ion-selective electrode measurements, a new scale has appeared by which computation of the total concentration of the ion by equation (6) or (7) becomes immediate. Such a scale for an Amel Model 328 Ion Meter is shown in Fig.2.

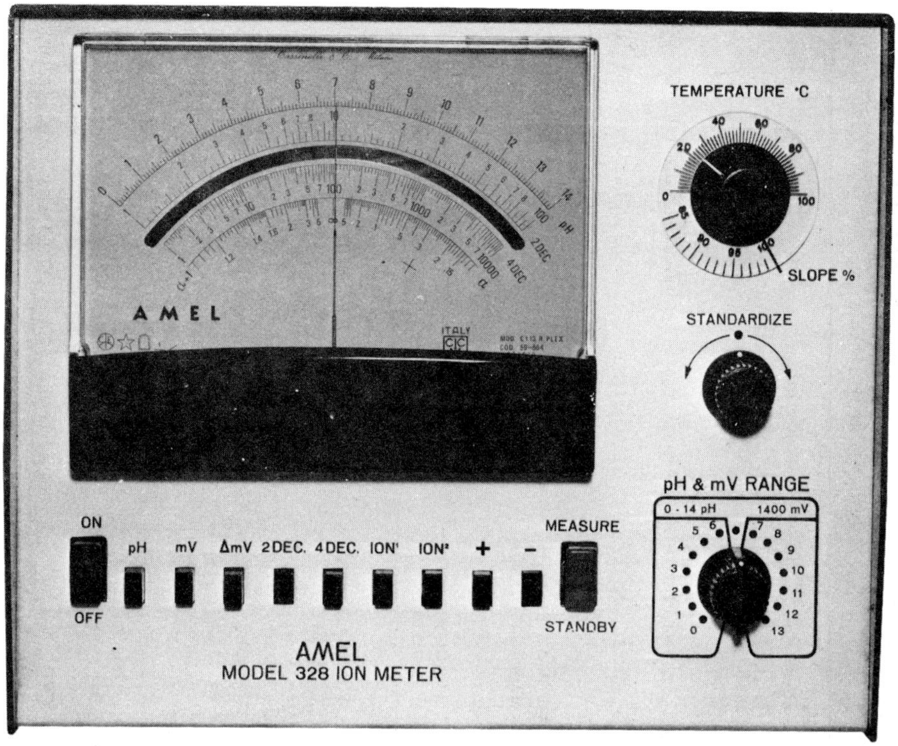


Fig.2. The front panel of an Amel Model 328 Ion Meter. The standard addition/subtraction scale is the bottom one.

To the right of $\infty$ on the mid point of the lowest scale is the standard addition scale, and to the left is the standard subtraction scale. The electrode is placed in a sample of volume $V_O$, and the meter pointer adjusted to read $\infty$. After a standard addition of reagent (addition or subtraction) is added to the sample (the value now being V) the value $C_AV_O/C_SV$ is read directly from the scale. Thus, a standard solution of concentration $C_S$, equal 1.00 times some power of 10, and if the volume ratio, $V/V_O$, is a power of 10, i.e., 1:100 or 1.5:150, the value of $C_A$ divided by (the power of the standard solution, times the power of the volume ratio) can be read directly from the scale. For example, for a $10^{-3}$M sample solution of volume 100 $cm^3$ with the addition of 1 $cm^3$ standard of $10^{-1}$M concentration the reading on the scale would be $10^{-3}/(10^{-1} \times 10^{-2}) = 1$.

This addition or subtraction scale makes the assumption that $V_O$ is about 50 times greater than V or greater, that is, the volume change upon the addition is negligible. In cases where this condition cannot be fulfilled, for instance when the sample volume is very small then the volume of the sample and the standard should be nearly equal.

TABLE II. Analytical problems solved by the known addition, known subtraction and related methods

| Species | Sample | Pretreatment | Electrode or Sensor | Technique | Ref. |
|---|---|---|---|---|---|
| Ammonia | Musts and wines | Adjust to pH 11 with NaOH | Ammonia | Addition of $NH_4^+$ | 30 |
| | Sea water | Spike sample with $NH_4Cl$, add concentrated NaOH | Ammonia | Addition of $NH_4^+$ | 31,32 |
| | Beer | Adjust to pH 11 with NaOH | Ammonia | Addition of $NH_4OH$ | 33 |
| | Plasma | Dilute 1 + 3 with pH 12 buffer | Ammonia | Addition of $NH_4OH$ | 34 |
| Bromide | Wine | Dilute sample with solution containing $H_3PO_4$, $KNO_3$, $CuSO_4$ | Bromide | Addition of $Br^-$ | 35 |
| Bromine | Aqueous solution | Add sample to known concentration of iodide, bromine reduced to bromide | Iodide | Analate subtraction, measure loss of iodide | 36 |
| Hypobromite | Aqueous solution | Add sample to known concentration of iodide | Iodide | Analate subtraction, measure loss of iodide | 36 |
| Cadmium | $CN^-$ plating bath | Add excess hypochlorite, if necessary adjust pH to 10 with NaOH | Cadmium | Addition of $Cd^{2+}$ | 37 |
| Calcium | Beer | Filter sample, adjust to pH 5.5-6.0 with NaOH using methyl red indicator | Calcium | Addition of $Ca^{2+}$ | 38 |
| | Saliva | - | Calcium | Analate addition | 39 |
| | Serum | - | Calcium | Addition of $Ca^{2+}$ | 40 |

| | | | | | |
|---|---|---|---|---|---|
| | Sugar solutions | Prepare standards in sucrose background | Calcium | Addition of calcium to sucrose solution | 41,42 |
| | Urine | - | Calcium | Analate addition | 39 |
| | Wine | Dilute sample 1+ 9 | Calcium | Addition of $Ca^{2+}$ | 42 |
| | Aqueous solution | - | Calcium | Addition of $Ca^{2+}$ | 19 |
| Chloride | Boiler cleaning solution | Add fixed concentration of silver to each sample | Silver | Subtraction of $Cl^-$ | 44 |
| | Copper(II) sulphate (plating bath | - | Chloride | Addition of $Cl^-$ | 37 |
| | Food seasonings | Add 1g sample to 100 $cm^3$ standard | Chloride | Analate addition | 39 |
| | Ice-cream | Dilute ice-cream, mix with sodium nitrate/nitric acid buffer containing silver nitrate to precipitate most of the chloride | Chloride | Addition of $Cl^-$ | 39 |
| | Tomato juice | Add sample to chloride standard solution | Chloride | Analate addition | 39 |
| | Platinum catalyst | Dissolve in conc. $H_2SO_4$, fume, neutralize and dilute | Chloride | Analate addition | 39 |
| | Serum and biological fluids | Add to acidified chloride standard | Chloride | Analate addition | 39 |
| | Saliva | Add to acidified chloride standard | Chloride | Analate addition | 39 |
| | Urine | Add to acidified chloride standard | Chloride | Analate addition | 39 |

continued overleaf

TABLE II (continued)

| Species | Sample | Pretreatment | Electrode or sensor | Technique | Ref. |
|---|---|---|---|---|---|
| Chlorine | Aqueous solution | Add sample to reagent containing known concentration of iodide | Iodide | Analate subtraction, measure loss of iodide | 36 |
| Hypochlorite | Aqueous solution | Add sample to known concentration of iodide | Iodide | Analate subtraction | 36 |
| Chromium (+6) | | Acidity sample with $H_2SO_4$ add tc known concentrat-ion of iodide | Iodide | Analate subtraction, measure loss of iodide | 39 |
| Citrate | Aqueous samples | Add sample to copper nitrate solution at pH 6 | Copper | Analate subtraction | 39 |
| | Cyanide plating baths | Add excess hypochlorite, adjust pH to 4-7 with HCl | Copper | Addition of $Cu^{2+}$ | 37 |
| Copper (+2) | Natural waters | Dilute filtered sample 1+1 with acetic acid/KOH/NaF/ formaldehyde buffer | Copper | Addition of $Cu^{2+}$ | |
| | Pyrophosphate bath | Add excess of pyro-phosphate | Copper | Addition of $Cu^{2+}$ | 39 |
| | Silicon | Dissolve in HF | Copper | Addition of $Cu^{2+}$ | 46 |
| | Microsample | To 25 $mm^3$ of sample add standard increments | Copper | Addition of $Cu^{2+}$ and subtraction with EDTA | 47 |
| Cyanate | Water | Heat sample with 50% $H_2SO_4$ for thirty minutes to hydrolyse cyanate to ammonia, cool, add NaOH | Ammonia | Addition of ammonia | 39 |

| | | | | | |
|---|---|---|---|---|---|
| Cyanide | Silver plating bath | Add 100 $cm^3$ 0.25 M NaOH | Silver | Addition of $Ag^+$ | 37,48 |
| Fluoride | Detergents | Dilute with citrate buffer | Fluoride | Addition of $F^-$ | 49 |
| | Feeds | Sample dried and ground then extracted after one hour in HCl, citrate buffer added | Fluoride | Addition of $F^-$ | 50,52 |
| | Glass | Fuse in $ZnO/Na_2CO_3$, dissolve neutralize | Fluoride | Addition of $F^-$ | 53 |
| | Sea water | To 100 $cm^3$ of sea water add standard fluoride | Fluoride | Addition of $F^-$ | 54 |
| | Uranyl nitrate | Separate fluoride from uranium using Dowex 50W-X4 cation exchange column, buffer with ammonium acetate | Fluoride | Addition of $F^-$ | 55 |
| | Wine | Add TISAB to sample | Fluoride | Addition of $F^-$ | 56,57 |
| | Vegetation | Dry the foliage, grind, mix and weigh 1.0 g, add 50 $cm^3$ $HClO_4$ (0.1M) | Fluoride | Addition of $F^-$ | 29 |
| Formaldehyde | | Add 5 $cm^3$ sample to 50 $cm^3$ reagent containing $Ag^+$ | Silver | Analate subtraction | 39 |
| Iodide | Selenium | Melt sample with NaOH, add water and boil, neutralize to pH 6-7 with 50% $H_2SO_4$ filter | Iodide | Addition of $I^-$ | 58 |
| | Milk | To 50 $cm^3$ of milk, KCl is added | Iodide | Addition of $I^-$ | 59,61 |

continued overleaf

TABLE II (continued)

| Species | Sample | Pretreatment | Electrode or sensor | Technique | Ref. |
|---|---|---|---|---|---|
| Iodide | Rocks and soils | Fuse with $NaCO_3/K_2CO_3$ at 700°, dissolve in $NH_4OH$, reduce | Iodide | Addition of $I^-$ | 62 |
| Lead | Paint | Dissolve paint chips in nitric acid, neutralize with excess sodium acetate to pH 5-6, add MeOH 1+1 | Lead | Addition of $Pb^{2+}$ | 39 |
| Nitrate | Pickling baths | 0.5 $cm^3$ of pickling bath containing $HNo_3$ and HF to 50 $cm^3$ of standard solution | Nitrate | Analate addition of $NO_3^-$ | 15 |
| | Aqueous solution | Standard solution | Nitrate | Addition of $No_3^-$ | 19 |
| Nitrate | Foods | Extract ground sample with $H_2O$, add buffer | Nitrogen oxide | Addition of $NO_2^-$ | 63 |
| Nitrogen | Sewage | Digest dried sample in $H_2SO_4$, dilute, add NaOH/NaI solution | Ammonia | Addition of $NH_4^+$ | 39,64, 65 |
| Nitrogen oxides | Air | Absorb in 0.1 M NaOH, add buffer | Nitrogen oxide | Addition of $NO_2^-$ | 63 |
| Oxidants | Air | Scrub with $5x10^{-5}M$ $I^-$ at phosphate buffer pH 7 | Iodide | Analate subtraction | 16,17 |
| Potassium | Felspar | Dissolve in HF and m-benzene disulfonic acid, add water, CDTA adjust to pH 8.4 | Potassium | Addition of $K^+$ | 66 |

| | | | | | |
|---|---|---|---|---|---|
| | Fertilizers | Extract fertilizer in acetate buffer | Potassium | Addition of $K^+$ | 39 |
| | Saliva | - | Potassium | Addition of $K^+$ | 39 |
| | Urine | - | Potassium | Addition of $K^+$ | 39 |
| | Wines | Dilute 1+9 with water | Potassium | Addition of $K^+$, 1% alcohol | 67 |
| Proteins | Serum | Split disulphide bonds with NaOH/urea, add $AgNO_3$ | Sulphide | Analate subtraction | 68 |
| Silver | Cyanide plating baths | Dilute 1 + 99 or 1 + 999 | Silver | Addition of potassium silver cyanide complex | 37, 69 |
| | Photographic fixing solution | To 50 $cm^3$ of sample add 1 $cm^3$ of $AgNO_3$ (0.01 M) | Silver | Addition of $Ag^+$ | 28,70 |
| | Gold cyanide plating bath | Dilute 1:10 or 1:100 with distilled water | Silver | Addition of $K[Ag(CN)_2]$ | 37 |
| Sodium | Felspar | Dissolve in HF and 3-benzene-disulphonic acid, dry, add water and CDTA, adjust pH to 8.4 | Sodium | Addition of $Na^+$ | 66 |
| | Pulping liquors | Dilute sample, mix aliquot with ammonium carbonate or add undiluted sample to $NaCO_3/NH_4LO_3$ | Sodium | Analate addition | 71 |
| | Sea water | Calibrate electrodes in artificial sea water, no preparation necessary for known addition | Sodium | Addition of $Na^+$ | 39 |
| Sulphate | Watts nickel baths | Add sample to reagent of water and 1.4-dioxane and known concentration of lead | Lead | Analate subtraction, measure loss of lead | 37 |

continued overleaf

TABLE II (continued)

| Species | Sample | Pretreatment | Electrode or sensor | Technique | Ref. |
|---|---|---|---|---|---|
| Sulphide | Pulping liquors | Dilute cadmium standard reagent with Cellosolve, add known amount of sample to reagent | Cadmium | Analate subtraction measure loss of cadmium | 72, 73 |
| | Waste water | To 50 $cm^3$ of waste waters add 1 $cm^3$ of $AgNO_3$ (0.1M) (15-300ppm of sulphide) | Silver | Subtraction of silver | 24 |
| Sulphur | Fuel oils | Combust sample on cotton wick, absorb combustion products in tetrachloromercurate(II) solution, add sulphamic acid, and add $SO_2$ buffer | Sulphur dioxide | Addition of $SO_2$ | 74,23 |
| Sulphur dioxide | Air and stack gases | Absorb sample in tetrachloromercurate(II) solution, add sulphamic acid and dilute, add $SO_2$ buffer 10+1 | Sulphur dioxide | Addition of $SO_2$ | 74 |
| | Air and stack gases | Absorb samples in iodine solution and measure iodide increment | Iodide | Analate addition | 17 |
| | Wine | Add HCl/glycerol acid reagent to diluted sample | Sulphur dioxide | Addition of $SO_2$ | 74 |
| Thorium | Uranium alloys | Dissolve, add to buffered fluoride solution, mask other metals | Fluoride | Analate subtraction | 75 |

| | | | | | |
|---|---|---|---|---|---|
| Uranium | Uranium alloys | Dissolve, add to buffered fluoride solution, mask other metals | Fluoride | Analate subtraction | 75 |
| Zirconium | Uranium alloys | Dissolve, add to buffered fluoride solution, mask other metals | Fluoride | Analate subtraction | 75 |

In such circumstances, it is still possible to use the direct reading standard addition/subtraction scale provided that initially instead of setting the meter at midscale ($\infty$), the meter should be set to read a value R (right hand) on the standard addition scale calculated from the formula:

$$R = \frac{V_o}{V} \qquad (16)$$

Having set the meter to read R with the electrodes in the sample solution, the addition of the standard is made and the total concentration is read on the standard addition or subtraction scale as explained above. Note that an initial setting of R is made on the standard addition scale regardless of whether an addition or subtraction is made. The simplest value of R is 1 where the volume of the sample and the standard are equal.

Lastly, the Orion Ionalyzer Model 901 Meter incroporates a high impedence amplifier and analog-to-digital converter whose output is fed to a microprocessor pre-programmed with the appropriate equations for calculating pH and concentration. Thus, the Model 901 can perform all calculations of standard addition or subtraction and analate addition or subtraction, with easy manual correction for electrode slope, volume variation, monovalent or bivalent ion. The micro-computer built up inside takes into account all parameters (procedure, slope value, positive or negative ion, monovalent or bivalent) and gives a display of the total concentration. Apart from the cost the instrument seems very flexible and up to date for ion-selective electrode measurements. The instrument has been evaluated (76) for various analyses using fluoride, bromide and chloride solid state ion-selective electrodes and for a PVC matrix membrane barium ion-selective electrode. It was concluded that the results for known standards and analytical samples were of acceptable accuracy and precision (76).

## 4. GRAN'S PLOTS

Generally, the equivalence point of a potentiometric titration is calculated by plotting potential values against the volume of titrant and locating the point of maximum slope of the titration curve. In this generalized way all information resident in the points relatively far away from the inflexion point is wasted.

Gran (77, 78) and later others (79, 80) devised a way for treating potentiometric titration data which instead of using the points around the maximum slope of the titration curve, utilizes all the points, and especially those far from the equivalence point. The method consists of linearizing the titration curve, i.e., in computing from the titration data a function which when plotted against the volume of titrant gives a straight line and which can be extrapolated to the equivalence point. The plots, called Gran's plots, were originally devised for titration with a pH indicator electrode in acid-base titrations and for use with certain other potentiometric sensors, such as the silver-silver chloride electrode. After the great expansion of analytical potentiometry with the advent of ion-selective electrodes, gas sensor probes and enzyme electrodes, the Gran's plot principle has been applied to all potentiometric titrations. Included in the principle is the so-called multiple standard addition procedure or "addition titration".

The basic theory is quite simple and is now illustrated. The response 'E' of an ion-selective electrode to a monovalent cation with respect to a suitable reference electrode can be written as follows, without taking into account liquid junction potential, activity coefficient, complexation with other ions and interfering effects:

$$E = E' + S \log C_A \qquad (17)$$

$C_A$ is the concentration of the free cations and E' incorporates the standard and reference electrode potentials. By rearranging equation (17) becomes

$$C_A = K \cdot \text{antilog } E/S$$

Thus, antilog E/S is a measure of $C_A$. By plotting the antilog E/S, proportional to $C_A$ against the volume of the titrant, V, added, a linear plot called a Gran's plot is obtained.

The plot is linear because the decrease of $C_A$ is linearly dependent on the addition of volume of the titrant, V. The greater the volume of titrant B added, the smaller the free concentration of the A species. The plot can be thus extropolated to $C_A \doteq 0$, i.e., to antilog E/S = 0 to intercept on the titrant volume axis and locating a point, $V_e$, which is the equivalence point and where the concentration of the A cation is zero. However, it is widely appreciated that at this point an equilibrium is reached where the concentration of the cation A is never true zero, and this is the reason that the extrapolation should be made since the points in this region do not follow a straight line.

Because of the relationship between [A] and [B], $[A]^x [B]^y = K_e$ where $K_e$ is an equilibrium constant and *x* and *y* are the stoichiometric ratios it follows that if $C_A$ = K antilog E/S

then $$C_B = K' \text{ antilog } (- E/S') \qquad (19)$$

Thus, after the equivalence point, by plotting antilog (- E/S') *vs* volume added another straight line is obtained and showing the increase of the titrant concentration after the equivalence point; this second straight line can be also extrapolated to find a $C_B = 0$, where the antilog (- E'/S) is zero, to the intercept on the volume of titrant axis locating the same $V_e$ where the level of excess titrant is zero.

The Gran's plot is, therefore, composed of two straight lines, one before the equivalence point showing the analate concentration decreasing and with negative slope, the other after the equivalence point showing the titrant concentration increasing and with positive slope.

With most ion-selective electrodes, only one straight line can generally be obtained, when the ion for which the electrode is considered selective is in excess. This happens because the limit of detection of the electrode is generally higher than the concentration of the ion at the equivalence point or on the other side of the titration when the reacting ion is in excess. This restriction does not happen with the glass electrode for pH measurement nor with some ion-selective electrodes, like silver or copper, or cadmium in several complexometric or precipitation titrations.

A difficulty arises because of volume changes during the titration. In fact, a correction is needed because of the increase of volume accompanying the addition of titrant. The correction factor is

$$C_A = \frac{C_{A_o}(V_o + V)}{V_o} \qquad (20)$$

where $C_A$ is the corrected concentration, $C_{A_o}$ the initial concentration, $V_o$ the initial volume and V the volume added. From the practical point of view, it is therefore necessary to plot the function

$$(V_o + V) \cdot \text{antilog } E/S \qquad (21)$$

which incorporates the volume correction, and which is linearly proportional to the concentration of the ionic species in solution, $C_A$. For the case following the equivalence point the function

$$(V_o + V) \text{ antilog } (- E/S') \qquad (22)$$

is plotted on the ordinate.

The above expressions hold for cations; for anions the plus and minus signs should be reversed for before and after the equivalence point.

The advantages of Gran's plots are as follows:

(i) Points located near the equivalence point can be discarded, while in conventional titrations such points have to be taken in greater numbers in order to avoid overshooting of the end point. Such points are difficult to evaluate in potentiometric titrations because in that region the electrode potentials are usually very unstable and drift readily.

(ii) It is easier to solve those problems when the end point break is poor, i.e., in cases where the solubility of the precipitate is high or when the complex is too dissociated or when the starting concentration of the ion is near the limit of detection of the electrode or when other ions in solutions are interfering at the electrode surface.

Since a straight line is defined by only two points, it will be sufficient to take just a few points far away from the equivalence points, i.e., when the addition is really proportional to the increase or decrease of ion concentration. Such points will be taken when the concentration of the ion is sufficiently high to determine the electrode potential in the range where there is Nernstian or near-Nernstian behaviour, i.e., far from the limit of detection, far from the effect of interference by other ions and when the solubility or the dissociation of the complex does not influence the increase or decrease of the ion concentration.

## 5. TITRATIONS OF ACIDS AND BASES

Before the commercial availability of the wider range of ion-selective electrodes, the glass electrode was widely used as pH indicator and Gran's plots had many applications in titrations of acids and bases. These are briefly reviewed.

## 5.1 Strong acid -strong base

Tn a strong acid/strong base titration, two straight lines are obtained, the first by plotting on the ordinate on the acid side of the equivalence point the function:

$$F = (V_o + V)\ 10^{\ E/S},\ \text{that is,}\ (V_o + V)\ 10^{-pH} \qquad (23)$$

such a function being related to the hydrogen ion concentration and as depicted in Fig.3 shows how the relation decreases linearly upon the addition of hydroxide ions. On the alkaline side of the equivalence point is plotted the function:

$$F = (V_o + V)\ 10^{\ E/S},\ \text{that is,}\ (V_o + V)\ 10^{\ pH} \qquad (24)$$

Here, Fig.3 depicts an increase in the relation upon the addition of hydroxide ions. These functions have been widely discussed in the literature as well as their deviations from linearity (80-88).

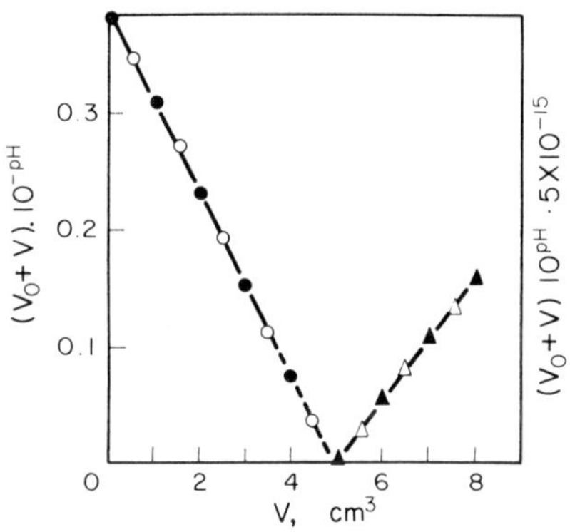


Fig.3. Gran's plots for the titration of a strong acid with a strong base (80).

Deviations due to one or more of the following factors may arise:

(i) If curvature happens at values of V remote from $V_e$ the activity coefficient and/or the junction potential cannot be neglected at the extreme values of hydrogen or hydroxide ion concentrations. If however the plots are linear near the equivalence point, the value of $V_e$ must be obtained from this region alone (Fig.4)

(ii) If the acid solution contains a metal ion which starts to hydrolyse at a hydrogen ion concentration in the range $10^{-4} < [H^+] < C_A$, the function on the acidic side of the equivalence point may become curved as the acidity is decreased. If the plots are linear at the highest acidity used, the equivalence point of the strong acid may again be determined by extrapolation (Fig.5).

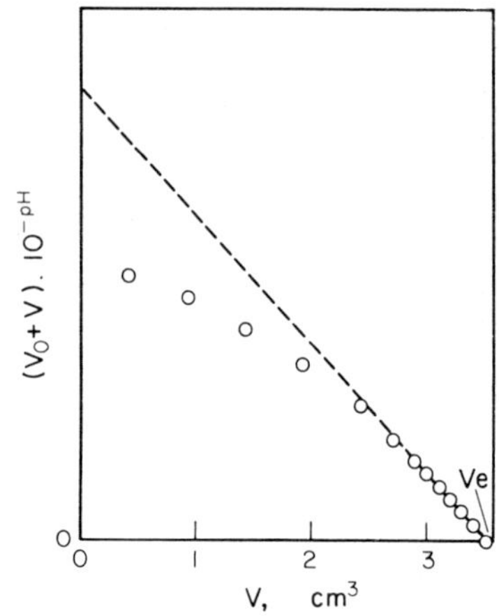


Fig.4. Gran's plot (80) for a system in which $E_j$ varies with $[H^+]$ : 35.0 $cm^3$ of nitric acid (0.210 M) was titrated with lithium hydroxide (2M) in 50% v/v aqueous dioxane containing 0.5M $(Li^+, H^+)$ $NO_3^-$.

(iii) If the strong base is contaminated by carbonate, the functions on the alkaline side of the equivalence point are curved around $V_e$. The value may be found by using only measurements in the acid region (Fig.6). If, however, the function on the alkaline side is linear over an appreciable range of V, they may be extrapolated to $V_e'$ and

$$V_e' - V_e = \frac{V_o C_A}{C_B} \qquad (25)$$

where $C_B = [OH^-]_o + 2[CO_3^{2-}]$ and $V_e = V_oC_A/[OH^-]_o$. $V_o$ is the sample volume, $C_A$ the concentration of hydrogen ion in the sample, $C_B$ the analytical concentration of the basic solution, and $[OH^-]_o$ the concentration of hydroxide ions in the basic titrant solution. An estimate of the carbonate ion concentration may therefore be obtained.

## 5.2. Weak acid - strong base

In the titration of a weak acid with a strong base the function after the equivalence point remains the same, but that before it is

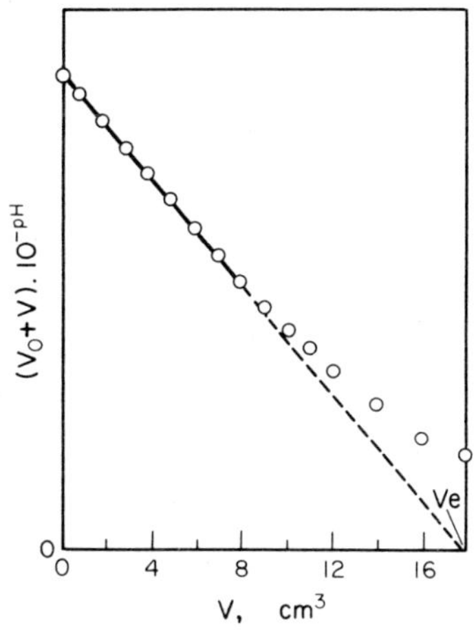


Fig.5. Gran's plot (80) for the standardization of a strong acid containing hydrolysable metal ions. 40,0 cm$^3$ of perchloric acid (0.01M) containing vanadyl perchlorate ($5.0 \times 10^{-2}$M) was titrated with sodium bicarbonate (0.02118 M) in a sodium perchlorate (3M) medium.

slightly altered. In fact, Gran devised the relation:

$$F = V \cdot 10^{E/S} \text{ or } V \cdot 10^{-pH} \qquad (26)$$

Since for a weak acid $[H^+]$ is low even at the beginning of a titration, $E_j$ will probably be negligible. If, moreover, the initial solution is not too concentrated, the ionic strength will not vary and the activity coefficient as well as the conditional constants of the acid and the water product will remain approximately constant (Fig.7). Function (26) is based on the assumption that the reaction:

$$OH^- + HA = A^- + H_2O$$

is quantitative, i.e., that $A^- = V \cdot C_B$ and $HA = C_A - VC_B$, by neglecting an initial dissociation of the acid and the equilibrium near the equivalence point.

Deviations from linearity arise because of the following:

(i) At the beginning of the titration the acid can be dissociated if only moderately weak and $[A^-] > V \cdot C_B$. The value of the $V_e$ must then be found using only that part of the function which is linear.

(ii) If the acid is very weak as V approaches $V_e$, an equilibrium condition occurs and $[HA] > C_A - VC_B$. The equivalence point may then be found by extrapolation of the upper linear region.

(iii) As for the case of the titration of a strong acid with a strong base, curvature in the alkaline region will be observed if the base contains carbonate.

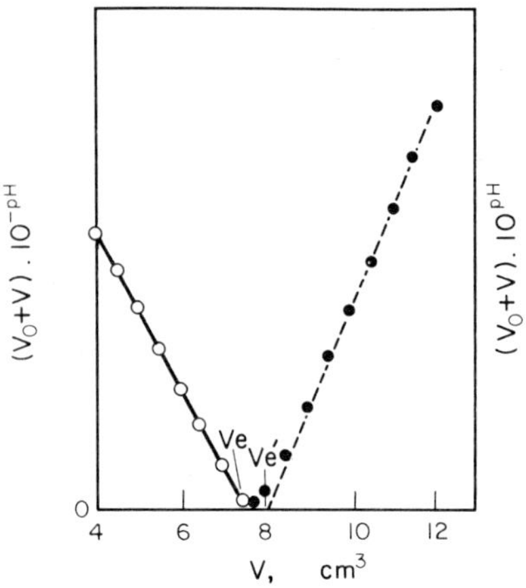


Fig.6. Portions of Gran's plots (80) for the titrations of a strong acid with a solution containing both carbonate and hydroxide.

## 5.3 Very weak acid - strong base

Gran's plots give excellent results for acids with stability constants $<10^8$: when the constant is $\geq 10^8$ the plots obtained are no longer straight but slightly curved. Moreover, the asymptotes do not intersect at the V axis and the point where they intersect does not coincide with the equivalence point. These deficiencies originate in the assumptions of the Gran's plot. In order to overcome these an extension of Gran's method has been presented by Ingman and Still (81) for very weak acids.

Ingman and Still (81) did not make any assumptions but the value of the stability constant must be determined at the prevailing ionic strength. The function, plotted on the ordinate axis against the volume is:

$$F = V \cdot 10^{K-pH} + \frac{V_o + V}{C_{NaOH}}\left([H^+] - [OH^-]\right)\left(1 + 10^{K-pH}\right) \quad (27)$$

where K is the stability constant, $[H^+] = 10^{-pH}/10^{-0.08}$ and $[OH^-] = 10^{-pOH}/10^{-0.12}$ at ionic strength 0.1M. V is the volume of the titrant $V_o$ the volume of the sample. Ingman and Still (81) show how to calculate the function manually, but thereafter solution was by computer.

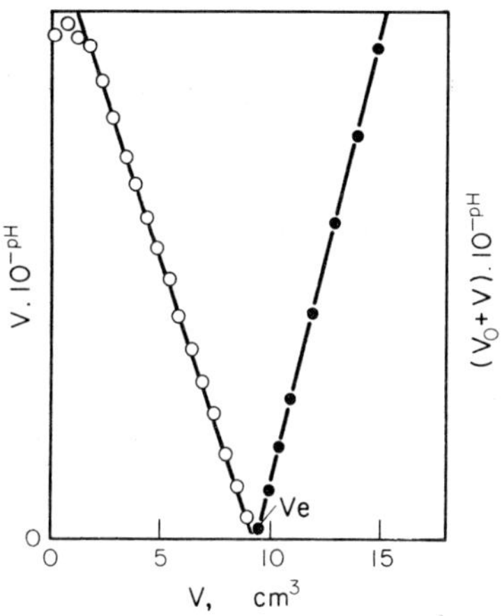


Fig.7. Gran's plots for the standardization of a weak acid with a strong base (80). 20.0 $cm^3$ of acetic acid ($\sim$0.05M) titrated with 0.106M sodium hydroxide (0.106M).

### 5.4. Other cases

Various other acid-base titrations have been discussed in the literature, including that of a weak base with a strong acid,titration of a diprotic acid, a poly-protic acid, a polymeric acid, mixtures of strong acids and weak acids and mixtures of weak acids. These are not discussed in this review because of their specialised nature and limited practical utility.

## 6. PRECIPITATION TITRATIONS

In a precipitation reaction the concentration of an ion, A, originally $C_A$ disappears as the titrant, B, is added to the sample solution, and the decrease in the concentration of ion, A, is linearly proportional to the volume of the titrant. In the reaction:

$$xA + yB = A_x B_y$$

as long as A is in excess

$$[A] = \frac{V_o}{(V_o + V)} \cdot C_A - \frac{x}{y} \cdot \frac{V}{(V_o + V)} \cdot C_B$$

$$= \frac{x}{y} \cdot \frac{(V_e - V)}{(V_o + V)} \cdot C_B \qquad (28)$$

Also, if the potential given by

$$E = E^o + S \log [A] , \quad [A] = K.\text{antilog } E/S \qquad (29)$$

By plotting the function

$$F = (V_o + V) \text{ antilog } E/S \qquad (30)$$

against V a straight line which intersects the V axis (F = 0) at V = $V_e$ will be obtained.

After the equivalence point, when B is present in excess and rememberinf that $[A]^x \ [B]^y = K_s$, where $K_s$ is the solubility product, the function:

$$F' = (V_o + V) \text{ antilog } -\left(\frac{x}{y} \ \frac{E}{S}\right) \qquad (31)$$

is obtained and is linear with V, the volume added.

The above can happen if the concentration of the ion for which the electrode is selective is higher than the detection limit of the electrode, or if the equilibrium is easily established and the relationship between A and B ions is stoichiometric.

By way of illustration, in the case of the titration of calcium with fluoride by use of the fluoride ion-selective electrode as the indicator electrode only one straight line is obtained corresponding to fluoride being in excess (89) (Fig.8). In Fig.8 the superiority of the Gran's plot with respect to the conventional plot is well demonstrated.

The function, on the ordinate,

$$F = (V_o + V)\ 10^{-E/S} = (V_o + V)\ 10^{-pF} \qquad (32)$$

is linearly proportional to the concentration of fluoride ions and after the equivalence point the function increases linearly with addition of the fluoride standard solution as titrant.

In some cases, a single electrode may sense both ions, the analate and the titrant, A and B; for instance in the case of formation of AgCl the A species can be $Cl^-$ and the B species can be $Ag^+$; and in the titration of chloride ions with silver ions, the silver chloride electrode follows the decrease in chloride concentration before the equivalence point and the increase of silver concentration after it. By plotting before the equivalence point the function

$$F = (V_o + V)\ 10^{-E/S} = (V_o + V)\ 10^{-pCl} \qquad (33)$$

which is proportional to the concentration of chloride and after the equivalence point the function

$$F = (V_o + V)\ 10^{E/S} = (V_o + V)\ 10^{-pAg} \qquad (34)$$

which is proportional to the concentration of silver ion, two straight lines are obtained as in Fig.9.

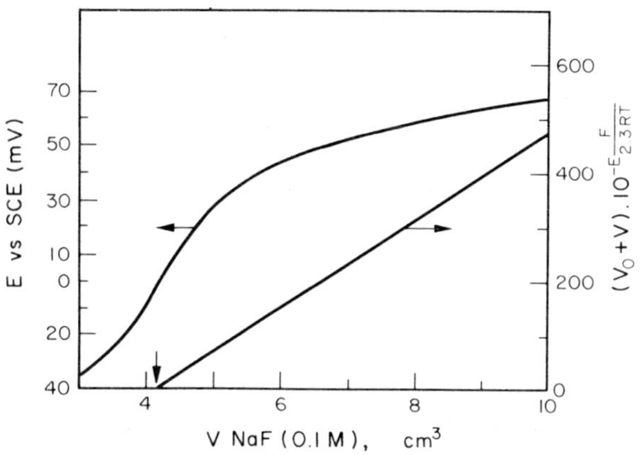


Fig.8. Potentiometric titration of 20 cm$^3$ of $Ca^{2+}$(0.01M) with sodium fluoride (0.100M). Upper curve is the usual potential/volume plot and the lower one is Gran's plot (89).

In practice, as can be seen from the experimental plots, the curves depart from linearity near the equivalence point and the extent depends on the reaction involved. The phenomenon arises because the concentration of A is never reduced quite to zero by reaction with B and also the concentration of B is not zero at the equivalence point. Thus, in the strong acid/strong base reaction the method does not take into account the autoprotolysis of the water and neither in the silver ion/chloride precipitation reaction does it take into account the limited but definite solubility of silver chloride or in the case of the fluoride reaction the solubility of calcium fluoride. The further away is the titration from the equivalence point, the less the effect of autoprotolysis of water on the hydrogen or hydroxide ion concentration, or of the solubility of silver chloride on the silver ion or chloride ion concentration, or of the solubility of calcium fluoride on fluoride ion concentration. In practice, as already emphasised, the further away the points are from the equivalence point the better the linearity.

Other characteristics of precipitation reactions, like adsorption or coprecipitation, are generally less important in Gran's plots

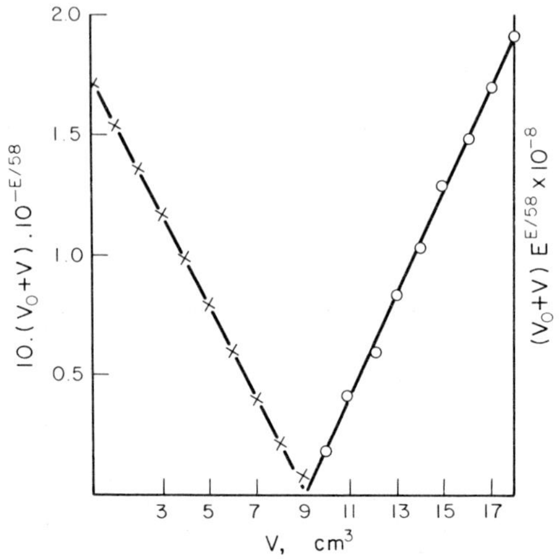


Fig.9. Gran's plot of chloride titration with silver ions 50 $cm^3$ $Cl^-$ ( $\approx 2 \times 10^{-3}$M) with silver nitrate ($10^{-2}$M) using a chloride ion-selective electrode.

because they become relevant when the primary ion concentration is low i.e., around and after the equivalence point. Hence, as above, by taking into account only the first part of the plot, when the primary ion is in excess the plot gives a more accurate result.

The formation of soluble complexes between either the primary ion or the titrant and other ions present in the sample will cause a curvature in the plots. It is sometimes possible to de-mask the ion or the titrant by adding a substance which will selectively precipitate or complex the masking agent without reacting with either the determinand or the titrant. The pH of a solution can often affect the result by changing the proportion of free anion present as in phosphate titrations, or of the proportion of free metal by the formation of metal hydroxy complexes, or by changing the composition of the substance precipitated as in the formation of basic salts. Thus, the addition of an excess of strong acid, or strong base, or of a suitable buffer solution may be required.

The presence of ion-selective electrode interfering ions will limit the use of the electrode in conventional titrations. However, in most cases Gran's plots can give accurate results by taking into account only the measurements where the primary ion is in excess, i.e., far from the equivalence point where the primary ion concentration is related to the potential values. Liquid ion-exchange electrodes are particularly prone to interference errors, as discussed by Schultz (90).

In conclusion, errors resulting in an under-estimate of the equivalent volume, increase with decrease in the concentration of the primary ion and with increase in the concentration of interfering ion. The errors also increase as the solubility product increases and as the degree of dilution during the titration increases (91). Similar problems are encountered in complex formation ititrations.

## 7. COMPLEX FORMATION TITRATIONS

In a complex formation titration, the following reaction takes place:

$$xA + yB = A_x B_y \qquad (35)$$

The conditional complexation constant, $K_c$, is given by:

$$K_c = \frac{[A_xB_y]}{[A]^x [B]^y} \qquad (36)$$

As long as A is present in excess, i.e., before the equivalence point, the concentration of A decreases with the addition of the titrant, B, according to:

$$[A] = \frac{V_o [A]_o}{(V_o + V)} - \frac{x}{y} \cdot \frac{[B]_o V}{(V_o + V)} \qquad (37)$$

where $[B]_o$ is the concentration of titrant B, V the volume of B added and $V_o$ the volume of the sample of concentration, $[A]_o$

Because $[A] \propto$ antilog E/S, $(V_o + V)$ antilog E/S represents the concentration of A which decreases linearly with V as long as $\frac{x}{y}$ remains constant throughout the titration. The function will be zero when $[A]$, given by equation (37) is zero, i.e., when

$$V_o [A]_o = \frac{x}{y} [B]_o V_e \qquad (38)$$

By extrapolating $(V_o + V)$ antilog E/S to the abscissa axis, it will be possible to locate $V_e$.

After the equivalence point,

$$[A_x B_y] = \frac{1}{y} \frac{[B]_o V_e}{(V_o + V)} \qquad (39)$$

and

$$[B] = \frac{[B]_o (V - V_e)}{(V_o + V)} \qquad (40)$$

Because $[A] \propto$ antilog E/S, the following relation is obtained:

$$(V_o + V)^{(1 - 1/y)} \cdot 10^{\frac{x}{y}(-E/S)} = K_c (V - V_e) \qquad (41)$$

In the usual case where x = y = 1 equation (41) becomes

$$10^{-E/S} = K_c(V - V_e) \quad (42)$$

Deviations from linearity are expected in this case

(i) if a variation of the ratio of complex formation happens during the titration,

(ii) when the reaction is not complete, i.e., around the equivalence point, and

(iii) where the electrode is not selective, i.e., at very low concentration of the primary ion when it is usual to plot where the ion, for which the electrode is selective, is in excess, and as discussed above, before the equivalence point.

Whitfield and Leyendekkers (92-94) described the use of the Orion calcium electrode and the Orion calcium-magnesium electrode for the complexometric titration of EGTA and DCTA; a Gran's plot procedure was used to locate the equivalence points for $Ca^{2+}$ and $Mg^{2+}$ ions in several water and sea water samples with $Na^+$ ions as interfering ions.

The pH of solution should be kept constant during such complexometric titrations, especially since the strength of the complexing agent is often pH dependent. If the solution is too acid the hydrogen ion competes with the cation for the ligand and the complex may not be formed. Thus, with EDTA, which reacts with hydrogen ions in solution it is imperative to use a buffer, but it must be remembered that often a buffer is a weak complexing agent for metal ions when the following equilibrium must be borne in mind:

$$AC + B \rightleftharpoons AB + C$$

AC is the weak complex and AB is the strong complex formed in the solution. It has been shown (95) how in this case there is no need for correction for the volume variation, i.e., plotting antilog (E/S) vs V is adequate for obtaining a straight line.

## 8. MULTIPLE STANDARD ADDITION

The multiple standard addition procedure, or so-called addition titration, is the technique similar to the standard addition procedure described in Section 3, but as the name implies several additions are made and the potentiometric data are linearized through Gran's function.

The concentration of the ion, for which the electrode is selective, increases in the sample as additions are made and the function:

$$F = (V_o + V) \text{ antilog } E/S \quad (43)$$

is proportional to the increasing concentration (89). By plotting the function (43) against the volume, V, added, a straight line is obtained which does not start from zero concentration as that from the equivalence point in the titrations of Section 4. Instead, it starts from the sample unknown concentration. The sample concentration can be obtained from the intercept of the line on the abscissa for a zero value of function (43). $V_e$, taken from the intercept corresponds to an apparent volume of the standard solution in the sample before the additions. The unknown concentration, $C_A$, can be calcultated by the following formula:

$$C_A = - \frac{C_s V_e}{V_o} \quad (44)$$

where $V_o$ is the volume of the sample, $C_s$ the concentration of the standard solution and $V_e$ the volume obtained by extrapolating function (43) to the abcissa. The numerical value of $V_e$ is negative since it lies to the left of the zero value corresponding to no addition of standard. $V_e$ can be regarded as the volume of standard which needs to be removed from the sample in order for the concentration of species A to be zero.

Several advantages are characteristic of addition titration or multiple standard addition technique. Firstly, an equilibrium is not involved in the additions, hence there are no slow processes of salt solubilities which can bend the linear plot. Again, because the result is dependent upon several readings the effect of random errors in individual readings on the result is reduced and the accuracy and precision are, therefore, greatly increased. Finally, the technique may be used in some cases even if interfering ions are present in the sample for by adding a solution containing the primary ion, A, its concentration increases eventually to a value in which the interfering effect on the electrode response is negligible and the result is then calculated by extrapolation from these interference-free readings.

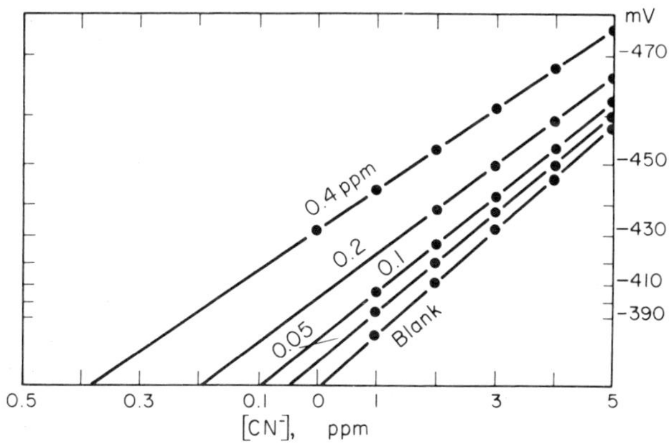


Fig.10. Known additions of 0.01 mg $dm^{-3}$ cyanide to known samples containing cyanide in the range zero to 0.4 mg $dm^{-3}$. The results are plotted on Gran's plot paper without volume correction (96).

A variation of the addition titration has been developed by Frant, Ross, Riseman (96) for the determination of low levels of cyanide. Instead of using a cyanide ion-selective electrode and simply adding cyanide ions in solution, they proposed adding a small amount of an indicator solution of $[Ag(CN)_2]^-$. In this case the concentration of free silver ion is related to the cyanide in the sample through the relationship:

$$[Ag^+] = \frac{[[Ag(CN)_2]]}{\beta_2 [CN^-]^2} \qquad (45)$$

where $\beta_2$ is the stability constant for the formation $[Ag(CN)_2]^-$ ions. A silver sulphide membrane electrode becomes an indicator of cyanide in the solution with a slope of 118 mV decade $^{-1}$ (Fig.10). This sensor is more accurate for low concentrations. By adding cyanide solution and plotting the Gran Function the results are more accurate than with cyanide electrode (97).

## 9. ASSUMPTIONS FOR GRAN'S PLOT PROCEDURES

The reliability of Gran's plot procedures depends on a series of assumptions which the analyst must keep in mind:

(i) the activity coefficient of the ion is assumed to be constant throughout the titration, i.e., the ionic strength is presumed to remain constant.

(ii) the junction potential at the reference electrode solution interface remains constant.

(iii) the ion-selective electrode has a constant response slope at least when measurements are far from the equivalence point.

(iv) the medium is free from interfering ions, or at least their contribution to electrode response is negligible for measurements far from the equivalence point.

(v) the reaction taking place during the titration is unique and stoichiometric

(vi) the reaction cannot be considered to be quantitatively complete.

The above assumptions include some phenomena related to the concentration of the ion in the sample and phenomena related to the concentration of the ion in the titrant.

Assumption (i) is related to the fact that electrode potential is related to a measure of the activity and not concentration; a high and constant ionic strength can ensure constancy of the activity coefficient. The simplest way to achieve this is to add an inert electrolyte to the sample and add a titrant with the same high concentration of inert electrolyte. Orion Research Incorporated and general literature references refer to this electrolyte as an ISA, i.e., an ion strength adjustor. This procedure also makes the junction potential constant, thus satisfying the second assumption. However, because the titration is generally carried out within a stirred solution, the junction potential can become very sensitive to stirring if an adequate ionic strength is not realised. As long as high ionic strength is maintained the Gran's plot titration procedure permits the use of a stirred solution and hence results in a rapid titration.

Assumptions (iii) and (iv) which in other procedures must be rigorous, are less imperative in the Gran's plot method. In fact, for a sample in which the ion of interest is even below the limit of detection of the electrode, addition of the standard solution will lead to a sufficiently high concentration of the ion where a Nernstian or near-Nernstian relationship, or a constant slope between the anti-log E/S and the volume of the titrant will soon be obtained. Extrapolation should then be from the high values of added standard or titrant. In relation to this, chloride concentrations near the solubility of silver chloride (which is the electrode material) cannot be determined by the direct method, but can be successfully determined by a Gran's plot by adding ions to the solution (98). Also, in the determination of nitrate in water samples the interference of chloride ions on the nitrate electrode can be overcome by a plot procedure based on adding nitrate ions. Addition of the nitrate makes the electrode potential less sensitive to the chloride interference, and by extrapolation from the points of sufficiently high added nitrate it is possible to obtain the initial nitrate concentration which cannot be sensed directly by the electrode (98).

Assumptions (v) and (vi) concern relationships in titrations and do not apply in the case of addition titrations where no reaction takes place. If, for instance an acid to be titrated is too weak a correction must be applied in order to obtain a linearity as discussed in Section 5; similar situations can arise for precipitation and complexometric formation reactions have very high solubility products and low stability constants, respectively.

In relation to assumptions (v) and (vi), Fig.8 presented the Gran's plot for the determination of calcium with fluoride but without correction for high solubility of the salt. If the concentration of the sample was lower, for instance $10^{-3}$M, and the Gran's plot did not give a straight line then a more elaborate function to linearize the titration data and which takes into account the exact value of the stability constant could be applied in a manner similar to that discussed in Section 5.3 for titrations of a very weak acid with strong base. However, such elaborate functions can be handled better with a computer (99-101).

Finally, in Table III several cases taken from the literature are presented and in which Gran's plot procedures have been used.

## 10. PRECISION OF GRAN'S PLOT PROCEDURES

The standard deviation of the volume, $V_e$, obtained by extrapolation in Gran's plot procedures depends on several factors. In the simplest case of the addition titrations, or multiple standard addition procedure, Buffle, Monnier and Parthasarathy (115-117) have carried out an interesting analysis. Fig.11 show how the percentage error, $\frac{\Delta V_e}{V_e}$, varies with a function, log R, where R is given by the following:

$$R = \frac{C_s + C_{sol}}{C_A + C_{sol}} \cdot \frac{V_{max}}{V_o} \qquad (46)$$

TABLE III. Analytical problems solved by the Gran's plot procedures

| Species | Sample | Preparation procedure | Electrode or sensor used | Technique | Ref. |
|---|---|---|---|---|---|
| Aluminium | Alloys | Dissolve metal or alloy in HCl, dilute, adjust to pH 3.5 - 4.0 with acetate buffer | Fluoride | Addition of sodium fluoride | 102-105 |
| Arsenate | Aqueous solution | Add buffered $La(NO_3)_3$ to sample, dilute with distilled water | Fluoride | Back titration of excess lanthanum with sodium fluoride | 106 |
| Azide | Wines and foods | Add NaOH, boil, add $H_2SO_4$, and distil, analyse distillate | Silver | Addition of silver | 107 |
| Bromide | In presence of chloride | - | Bromide | Addition of $AgNO_3$ | 108 |
| Calcium | In presence of $Na^+$ | Titrate with complexones | Calcium | Addition of DCTA | 93 |
| Calcium | In aqueous solution (illustrative experiment) | - | Calcium | Addition of $Ca^{2+}$ | 19 |
| Chloride | Low levels | Add $KNO_3$ for ionic strength adjustment | Silver | Addition of $AgNO_3$ | 39 |
| Copper | In presence of $Zn^{2+}$ | Titrate with EDTA | Copper(II) | Addition of EDTA | 109 |
| | Natural waters | Add a complexing anti-oxidant buffer, add standard $Cu^{2+}$ | Copper | Addition of $Cu^{2+}$ | 110 |

| | | | | | |
|---|---|---|---|---|---|
| Cyanide | Waste waters | Decomplex with acid and EDTA, adjust to pH 11, add $K[Ag(CN)_2]$ indicator | Silver | Addition of cyanide | 96,97 |
| Fluoride | Polluted air | Air bubbled into NaOH (0.1M) | Fluoride | Addition of NaF | 89 |
| | In presence of complexing metals | Several buffers depending on the metal | Fluoride | Addition of $F^-$ | 111 |
| | Sea water | - | Fluoride | Addition of $F^-$ | 71 |
| | Vegetation | Extract with $HClO_4$ for 20 min at 20°C | Fluoride | Addition of $F^-$ | 29 |
| Hydrogen ion | Acidic sample | Dilute sample with $NH_4NO_3$(5M) | pH electrode | Addition of acid | 112 |
| Nitrate | Pickling baths | Add increments of 0.5 $cm^3$ sample to standard solution | Nitrate | Analate addition | 15 |
| | Aqueous solution (illustrative experiment) | - | Nitrate | Addition of $NO_3^-$ | 19 |
| Phosphate | | Add known excess of $La(NO_3)_3$ to sample to form a precipitate | Fluoride | Addition of NaF | 39 |
| Pyro-phosphate | Instant mashed potato | Add sample to hot neutral $NaClO_4$ cool, add excess $La(NO_3)_3$ | Fluoride | Addition of NaF | 39 |
| Selenite | Aqueous solution | Add $La(NO_3)_3$ solution to the sample, and titrate with NaF | Fluoride | Addition of $F^-$ | 113 |
| Sulphide | Sediments | Mix with antioxidant buffer, disperse ultra-sonically | Cadmium | Addition of $Cd^{2+}$ | 114 |

$C_{sol}$ is the solubility of the electrode material, i.e., a kind of detection limit, $C_S$ is the concentration of standard solution, $C_A$ is the unknown concentration of the ion in the sample of volume, $V_o$, and $V_{max}$ is the maximum volume used for the addition titration. From the practical point of view, R, is the ratio between the amount added and the amount present before the titration.

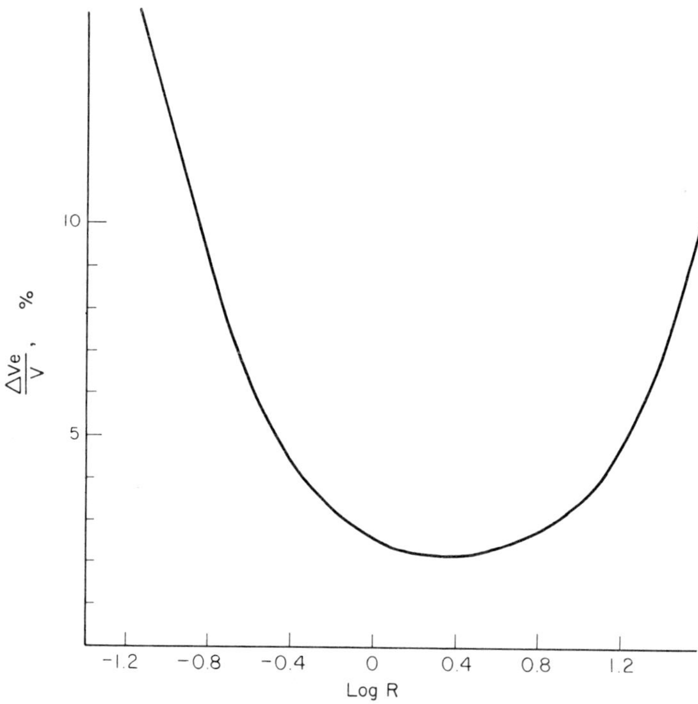


Fig.11. A plot of $\Delta V_e/V_e$ vs log R (see text). $V_o/V_o$ = 0.1%, $\sigma V/V_o$ = 0.05%, $\sigma E$ = 0.1 mV, $V_{max}$ = 10 $cm^3$ $V_o \geq 10$ $cm^3$.

The conclusion of Buffle, Monnier and Parthasarathy was that in order to keep the error of $V_e$ minimal, the value of R should lie between 1 and 3, irrespective of the errors in V, $V_o$, E and in the plot itself. This means that the maximum potential variation should be in the range between 18 and 36 mV for monovalent ions and 9 and 18 mV for bivalent ions. Fig.12 shows other plots. Thus, Fig.12(a) shows the relative error in $V_e$ as a function of the number of points taken with the emergence of the second rule that there should be at least ten points. Fig.12(b) depicts the relative error as a function

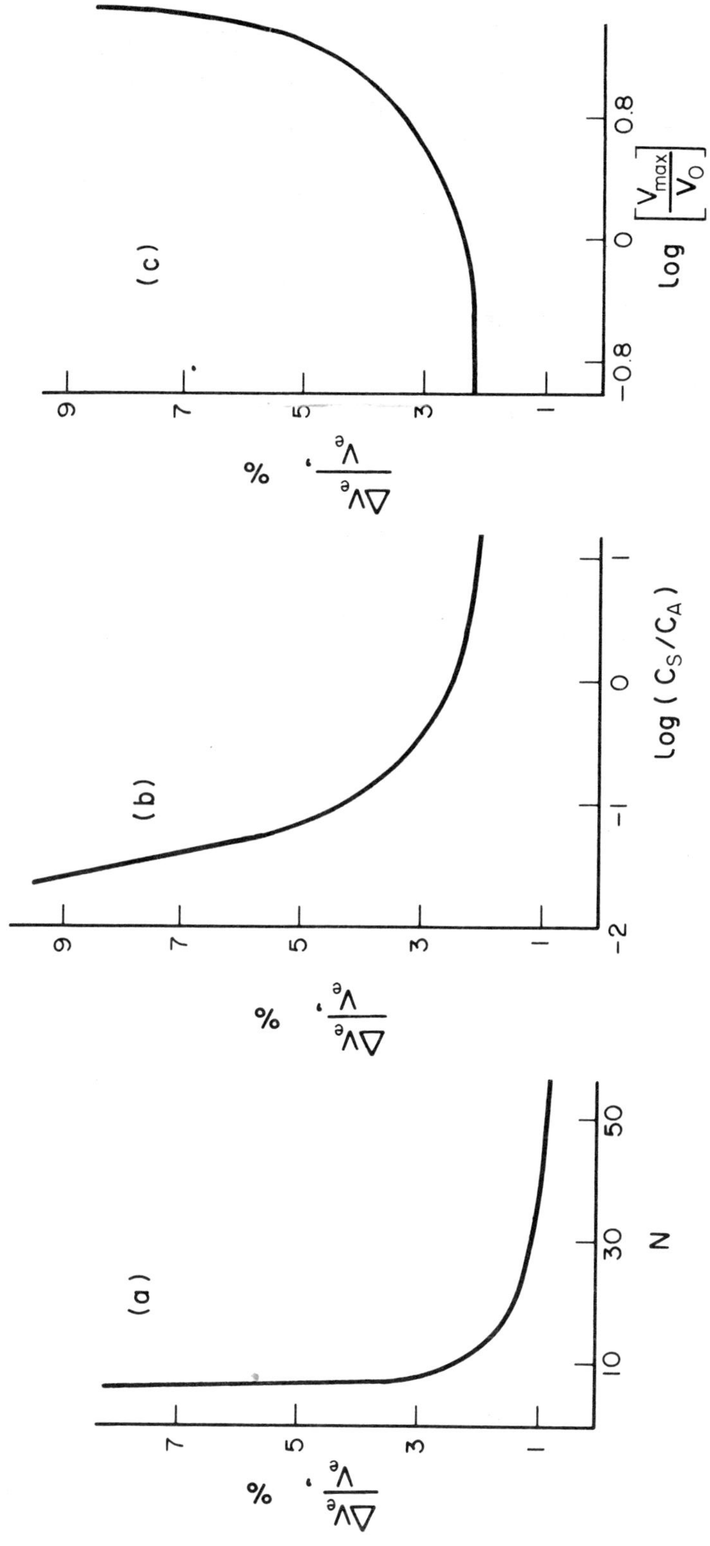


Fig. 12. Plots of $(\Delta V_e/V_e)_{min}$. (a) Versus N; (b) versus log $(C_S/C_A)$; (c) versus log $(V_{max}/V_O)$. (See text).

of the ratio of the concentration of the standard solution added, $C_S$, and the unknown concentration, $C_A$, of the sample. Thus, $C_S/C_A$ should be >1, and this is usual. Fig.12(c) depicts the basis for another rule. Here, the relative error is plotted against the ratio between the maximum volume added, $V_{max}$, and the volume of the sample, $V_o$. The $V_{max}/V_o$ ratio should be around or <1.

Finally, the effect of the error caused by the electrode slope on the $V_e$ value has been determined theoretically. Significant errors can occur in the measurement of concentration if the slope varies more than 2%, that is, 1 mV from 59.1 mV for monovalent ions, or 0.5 mV from 29.6 mV for bivalent ions.

The above rules taken for the simplest of the titrations, namely, the addition titration are readily extended to other cases. In fact, in the other types of titrations, other problems, such as high solubility for precipitation titrations, or complex formation which is not too stable, or a variation in the ratio of complexation can further increase the calculated standard deviation (118, 119).

## 11. DEVICES FOR DRAWING GRAN'S PLOTS

In order to plot the Gran titration curve it is necessary to make calculations in order to relate the electrode potentials to concentrations. The familiar pocket scientific calculators now commonly used make the calculations very rapid but other ways have been suggested to eliminate the otherwise tedious computions:

(i) A calibration curve can be used to convert millivolt readings to concentrations. It is not important to draw the real electrode calibration curve for it is adequate to draw a calibration curve with the true electrode slope value on semilog paper. The fact that the curve is displaced on the mV axis will influence the slope of the Gran plot, but the intercept on the horizontal axis will be the same. In this case, however, the volume correction has to be computed.

(ii) The log scale on an ion meter can be used to read concentration values directly. In this case, the slope can generally be regulated by a potentiometer which is more or less accessible to the panel of the ion meter. Often this can be regulated as a percentage of the ideal Nernstian slope. However, it is also necessary in this case to compute the volume correction.

(iii) Gran's Plot Paper or antilog - linear paper, first introduced by Orion Research Incorporated (120) and now distributed by several ion-selective electrode manufacturers is a very convenient way to plot titration and addition data. In the discussion subsequent to Gran's presentation in 1952 of his method to linearise titration curves it was pointed out (78) that the calculations could be simplified considerably if antilogarithmic linear paper was used. Alas, at that time such paper was not generally available, and it was only in 1970 that it was introduced by Orion Research Incorporated as a result of the wide interest and commercial development of ion-selective electrodes.

The special graph paper is available in two forms, with and

without in-built correction for volume change. The more common is the type with an allowance for a 10% volume change; thus, if the sample volume is 100 $cm^3$ the total volume of added increments must finally be 10 $cm^3$. The ordinate axis is antilogarithmic, so that values of E plotted on it become converted to antilog (E/S) which avoid the need for calculations. The slope, S, is built into the paper at 58 mV for monovalent ions or 29 mV for bivalent ions. This is the major drawback of the method if the electrode slope is less than 58 mV, say 55 or 56 mV as frequently happens, the intercept will be in error, as shown in Section 10.

Fig.10 and 13 give examples of titrations performed with this special antilog-linear paper available from ion-selective electrode suppliers. Note that the horizontal lines divert from the abscissa in order to take into account the volume correction.

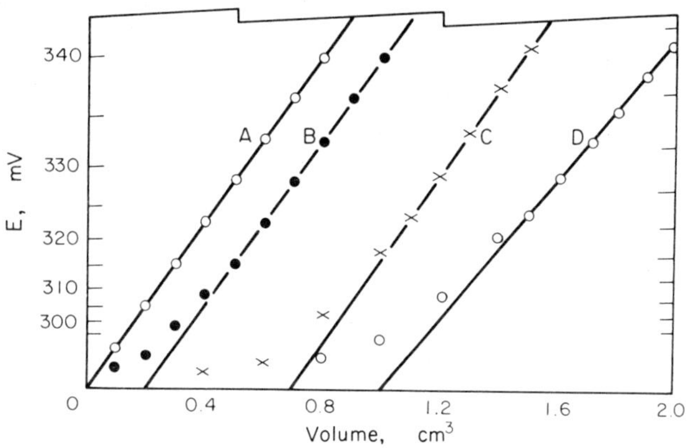


Fig.13. Potentiometric titration of azides, $N_3^-$, with silver nitrate (0.01 M) in a 20 $cm^3$ sample (107). A, blank titration $N_3^-$ = 0; B, $N_3^-$ - $10^{-4}$ M; C, $N_3^-$ - 3.5 x $10^{-4}$ M; and D, $N_3^-$ = 5 x $10^{-4}$ M.

(iv) A Gran ruler has been described (121). This device is very similar in principle to Gran's plot paper. The ruler, illustrated in Fig.14, permits the transformation of standard millimetre graph paper into Gran's paper. The advantage over the Gran's plot paper is that the normal graph paper which is always available can be used by sliding the ruler along the abscissa according to the volume addition printed at the foot of the ruler the $\Delta E$ variation of potential, converted into antilog units can be directly plotted on the ordinate. In order to allow for the volume correction provided by the ruler, the base of the ruler is positioned according to the volume variation and so introduces a correction factor. The slope error is, however, still present because the theoretical Nernstian slope is built into the ruler as in Gran's plot paper. This is unfortunate since this error is often found in practice for rarely do ion-selective electrodes show a truly theoretical response.

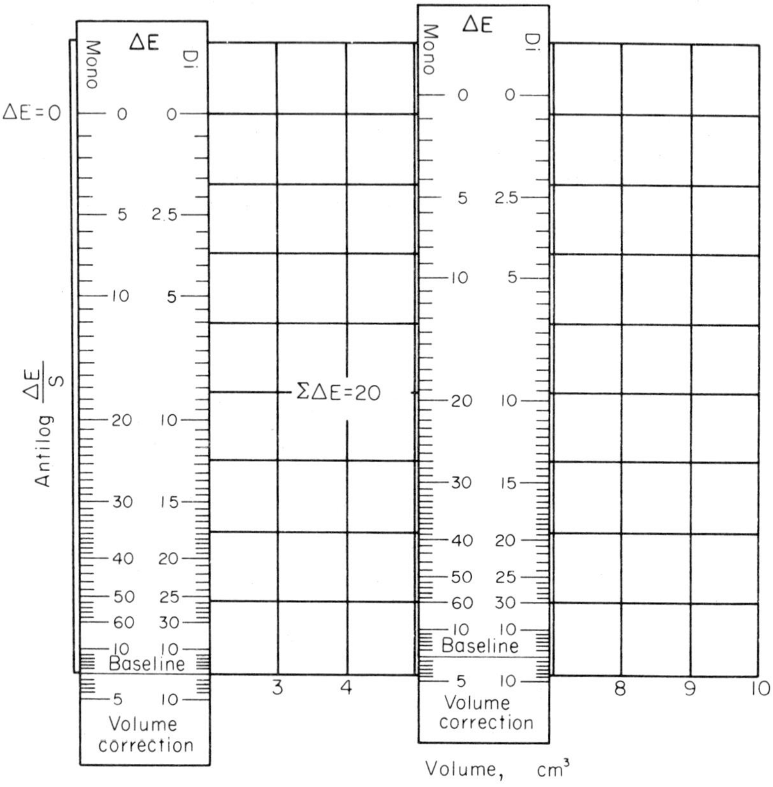


Fig.14. The Gran Ruler (121).

(v) An electronic antilog converter has been proposed (109). This can be connected to the output of a pH meter and its function is to convert the potential value measured by the meter to the concentration of an ion, say $[Cu^{2+}]$, in the solution according to the relationship:

$$\text{Antilog}\left[(E - E')/S\right] = \left[Cu^{2+}\right] \quad (47)$$

E' and S are obtained by calibration and must be adjusted on the device.

The antilog converter shown in Fig.15 performs the following mathematical operation: $U_{out}$ = antilog $[(E - E')/S]$ where E is the analog input signal of the converter. During the titration the input signal changes by S mV for each decade charge in the concentration of the metal ion in solution. In the first stage a value E' is subtracted from the input signal, E; the value of E' can be adjusted from 0 to 400 mV by means of potentiometer R4. (E - E') is multiplied by a constant 1/S. 1/S can be adjusted by potentiometer R6 for bivalent ions from 0.05 to 0.025 corresponding to S values from 20 to 40 mV, but a change in R7 makes it possible to use another range for S, say for monovalent ions.

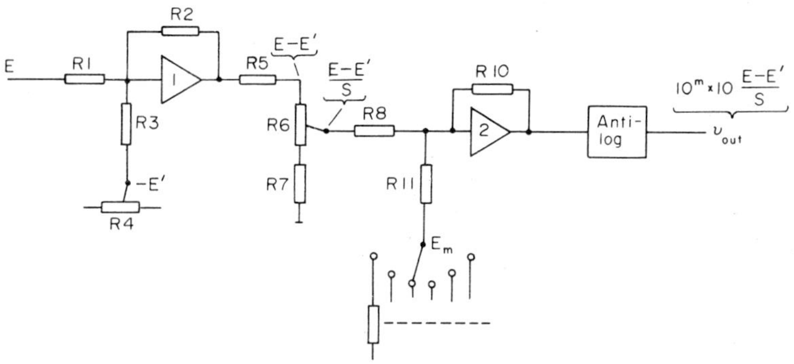


Fig.15. Diagram of an "antilog converter" (109) for use in Gran-type addition titrations.

To provide optimal use of the converter, the output signal can be amplified stepwise by a factor $10^m$, where m is an adjustable integer from 0 to 5. This is achieved by adding [(E - E')/S] a potential value equal to m:

$$U_{out} = \text{antilog} \left[(E - E')/S + m\right] \quad (48)$$

In this case there is no volume correction but the volume change was kept within 2% so that its influence was negligible (109). However, it has been suggested (109) that the converter output be connected to a volume corrector of the kind described by Johansson (122).

## 12. USE OF COMPUTERS

Computer evaluation of experimental data of Gran's plots and the use of computers to control titrations through automatic burettes and continuous evalution of experimental data have been realised in several ways in the past few years.

Brand and Rachnitz (123) proposed the first computer approach to the standard addition method using ion-selective electrodes. To handle the results it is necessary to find $E_o$ and slope, S, of the electrode, in order to arrive at the concentration of the unknown. However, prior knowledge of $E_o$ and S is not necessary. Basically, the procedure follows from the familiar known addition relationship of equation (49) valid for any number, ( j - 1), of additions of $V_i$ $cm^3$ of solution of known concentration, $C_S$, to $V_o$ $cm^3$ of solution of unknown concentration, $C_A$, ($V_1$ = 0). The best values of $E_o$, S, and $C_A$ are found to fit the experimental data.

$$E_i = E_o + S \log \frac{C_A V_o + \sum_{i=1}^{j} C_S V_i}{V_o + \sum_{i=1}^{j} V_i} \qquad (49)$$

The problem has been solved by use of a non linear least squares curve fitting procedure by using Taylor's formula. By differentiating equation (49) the following is obtained:

$$\frac{\partial E_i}{\partial E_o} = 1 \qquad (50)$$

$$\frac{\partial E_i}{\partial S} = \log \frac{C_A V_o + \sum_{i=1}^{j} C_s V_i}{V_o + \sum_{i=1}^{j} \cdot V_1} = x_i \qquad (51)$$

$$\frac{\partial E_i}{\partial C_A} = \frac{S\ V_i \log e}{C_A V_o + \sum_{i=1}^{j} C_s V_i} = y_i \qquad (52)$$

and by Taylor's formula:

$$E_i = E_o + S x_i + \Delta E_o \frac{\partial E_i}{\partial E_o} + \Delta S \frac{\partial E_i}{\partial S} + \Delta C_A \frac{\partial E_i}{\partial C_A} \qquad (53)$$

a function error may then be established:

$$Z_i = E_i - E_o - S\ x_i \qquad (54)$$

and values of $\Delta E_o$, $\Delta S$ and $\Delta C_A$ calculated by a least squares fit to the equation:

$$Z_i = \Delta E_o + \Delta S\ x_i + \Delta C_A y_i \qquad (55)$$

The initial estimates for $E_o$, S and $C_A$ are corrected by addition of $\Delta E_o$, $\Delta S$ and $\Delta C_A$, respectively, and the cal-culations repeated until the values of $E_o$, S and C converge generally in 3 or 4 cycles. The program, called ADDFIT, is written in Fortran IV and it is available from Prof.G.A.Rechnitz (Department of Chemistry, University of Newark, Delaware).

The method was used by another group (128) when the treatment was extended to locate the equivalence point in precipitation and complexation titrations and in which either the analyte or titrant was electroactive. Furthermore, by using data points in a region where the activity ratio of electroactive and interfering ions are favourable, misleading data are excluded. The program, called TITRATE, has also been written in Fortran IV and is available from Dr.Isbell (Department of Chemistry, University of Hawaii, Honolulu, Hawaii 96822).

The multiple standard addition technique has been simplified for use with a desk computer (125) and similar results to those of Brand and Rechnitz (123) were obtained.

A computer evaluation of titrations by Gran's method has been described (126) as a basis for an undergraduate analytical chemistry experiment. Subroutine programs are included for least squares analysis of data and for construction of the Gran's plot. Gran's plots are illustrated for the evaluation of data over the entire titration and for the last 50% of the titration.

A laboratory minicomputer system (127) has been applied to the monitoring of several ion-selective electrodes in a flowing stream. A standard addition analytical approach was used with a least squares fit to the data used to obtain electrode response slope and the unknown analyte concentration for each electrode. The computer was programmed so that three different modes of data monitoring were possible, namely,

(i) operator-controlled selection of data regions to be collected for subsequent analytical computation,

(ii) computer-controlled selection of data collection regions based on a pre-selected time-delay after each addition of standard, and

(iii) selection of data collection regions base on realtime computer identification of successive voltage plateaus in the potentiometer output during a series of standard additions.

Frazer and co-workers (128) proposed the use of an "on line computer" to process and display the data using Gran's method. This, they called an interactive experiment between the computer and the scientist. The computer will perform all routine operations which can be pre-defined, and the scientist will observe the unusual events to interpret the phenomena. Several examples are given. They introduced an "error function" employed as a means for testing the linearity of the Gran plot. A minimum in the error function plot therefore indicates the region of maximum linearity of the Gran function. A later paper (129) presented an expanded discussion of the error function technique for locating the equivalence points of potentiometric titrations.

Ariano and Gutknecht (130) have described a computer-controlled potentiometric analysis system. During a typical analysis, a series of standard additions are made to an unknown solution in a cell containing an ion-selective electrode and a reference electrode. These additions are optimized in that the volume of each addition is automatically adjusted so as to yield an even distribution of the resulting cell voltages. The cell voltages which are acquired by using a computer-optimizing sampling technique, and the standard addition data are fitted to the Nernst equation using the non-linear least squares procedure.

Finally, a paper published by Fleet and Ho (131) outlines a procedure for monitoring cyanide with an ion-selective electrode based on a Gran's plot standard addition technique in a continuous flow system. The use of a computer program facilitates the calculation of the results.

## 13. CONCLUSION

Known addition, Gran's plot and related procedures can greatly widen the scope of application of ion-selective electrodes. Furthermore, the present trends towards computerisation and the use of microprocessors take the drudgery out of the associated calculations and can cope with overcoming certain of the limitations of the addition methods.

## 14. REFERENCES


1. R.A.Durst, (Editor). "Ion Selective Electrodes", N.B.S.Spec.Publ. No.314, Washington D.C. (1969).
2. G.J.Moody and J.D.R.Thomas, "Selective Ion Sensitive Electrodes", Merrow, Watford (1971).
3. K.Camman, "Das Arbeiten mit Ionen Selektione Elektroden", Springer-Verlag, Berlin (1973).
4. R.Bates, "Determination of pH: Theory and Practice", J.Wiley, New York and London (1973).
5. J.B.Andelman, J.Water Pollut.Control. Part I, 1844 (1968).
6. M.S.Frant and J.W.Ross jr., Anal.Chem., 40, 1169 (1968).
7. ASTM D 1179-70.
8. I.Sekerka and J.F.Lechner, Talanta, 20, 1167 (1973).
9. J.E.Harwood, Water Res., 3, 273 (1969).
10. L.G.Sillén and A.Martell, "Stability Constants of Metal-Ion Complexes", Spec.Publ.No.17, Chemical Society, London (1964).
11. M.E.Thompson and J.W.Ross, Science, 154, 1643 (1966).
12. A.Craggs, G.J.Moody and J.D.R.Thomas, Analyst, 104, 961 (1979).
13. C.N.Durfor and E.Becker, J.Amer.Water Works Assoc., 56, 237 (1964).
14. A.Hulanicki and M.Trojanowicz, Analytica Chim.Acta, 68, 155 (1974).
15. J.O.Burman and G.Johansson, Analytica Chim.Acta, 80, 215 (1975).
16. A.F.Trachtenberg and I.H.Suffet, "65th Annual Meeting of the Air Pollution Control Association" June 18-22 1972.
17. M.Mascini and M.T.Muratori, Ann.Chim., 287, (1975).
18. AMEL, Milan, Italy, "Uso degli Electrodi Sens-Ion", Commercial leaflet, 1.3.1972.
19. A.Craggs, G.J.Moody and J.D.R.Thomas, J.Chem.Educ., 51, 541 (1974).
20. P.L.Bailey, "Analysis with Ion-Selective Electrodes", Heyden and Son Ltd., London, 1976, p.189.
21. A.Hulanicki, R.Lewandowski and M.Maj, Analytica Chim.Acta 69, 409 (1974).
22. R.F.Thomas and R.L.Booth, Environ.Sci.Technol., 7, 523 (1973).
23. J.A.Krueger, Anal.Chem., 46, 1338 (1974).
24. Amel, "Application Bulletin. Determination of Sulphide in Waste Waters", Milan, Italy.
25. Amel, Milan, Italy, private communication, 1979.

26. H.Z.Balog, M.Varhelyi and M.Kellermayer, Rheumatol. Balneol. Allergol., 17, 244 (1976).

27. Orion Research Incorporated, Application Bulletin, No.12. "Determination of Total Sulphide content in water". Cambridge, Ma., U.S.A. (1969).

28. Orion Research Incorporated, Application Bulletin, No.14. "Silver in Photographic Fixing Solutions". Cambridge, Ma., U.S.A. (1970).

29. A.E.Villa, Analyst, 104, 545 (1979).

30. D.J.McWilliam and C.S.Ough, Amer.J.Enol.Viticult., 25, (2), 67 (1974).

31. R.F.Thomas and R.L.Booth, Environ.Sci.Technol., 7, 523 (1973).

32. T.R.Gilbert and A.M.Clay, Anal.Chem., 45, 1757 (1973).

33. F.Dawert and T.Nitsche, Brauwissenschaft, 29, 299 (1976).

34. A.F.Attili, D.Autizi and L.Capocaccia, Biochem.Med., 14, 109 (1975).

35. J.E.Graf, T.E.Vaughan and W.H.Kipp, J.AOAC, 59, 53 (1976).

36. Orion Research Incorporated, Newsletter, 2, 26 (1970). Cambridge, Ma., U.S.A.

37. M.S.Frant, Plating,686 (1971).

38. Orion Research Incorporated, Application Bulletin No.8, "Determination of Total Calcium Levels in Beer by the Known Addition Method". Cambridge, Ma., U.S.A (1970).

39. Orion Research Incorporated, Analytical Methods Guide. Ninth edition, Cambridge, Ma., U.S.A. (1978).

40. G.J.Moody, R.B.Oke and J.D.R.Thomas, Proc.Soc.Anal.Chem., 7, 180 (1970).

41. M.A.Clarke, Proc.1970 Technical Session on Cane Sugar Refining Research, Oct. 12-13 (1970) p.19.

42. T.Henscheid, K.Scheonrock and P.Berger, J.of the A.S.S.B.T., 16, 482 (1971).

43. Orion Research Icorporated, Newsletter, 2, 33 (1970).

44. J.G.Frost, Analitica Chim.,Acta, 48, 321 (1969).

45. M.J.Smith and S.E.Manahan, Anal.Chem., 45, 836 (1973).

46. M.Taddia and P.Lanza, Ann.Chim., 65, 719 (1975).

47. J.M.van der Meer, G.den Boef and W.E.van der Linden, Analytica Chim.Acta, 85, 317 (1976).

48. L.N.La Patnick, Analytica Chim.Acta, 72, 430 (1974).

49. A.L.Schick, J.AOAC, 56, 798 (1973).

50. J.Tusl, J.AOAC, 53, 267 (1970).

51. J.R.Melton, W.L.Hoover and J.L.Ayers, J.AOAC, 57, 508 (1974).

52. L.Torma and B.E.Ginther, J.AOAC, 51, 1181 (1968).

53. V.H.Gebhardt, H.Horn and L.Stephan, Glastech.Berichte, 48, 63 (1975).

54. C.J.Rix, A.M.Bond and J.D.Smith, Anal.Chem., 48, 1236 (1976).

55. H.H.Moeken, H.Eschrich and G.Willeborts, Analytica Chim.Acta., 45, 233 (1969).

56. C.Martin and S.Brun, Trav.Soc.Pharm., Montpellier, 29, 161 (1969).

57. G.De Baenst, J.Mertens, P.van den Winkel and D.L.Massart, J.Pharm. Belg., 28, 188 (1973).

58. U.Westerlune-Hemerson, Anal.Chem., 43,1120 (1971).

59. E.A.Crecelius, Anal.Chem., 47, 2034 (1975).

60. G.S.Craven and M.C.Griffith, Aust.J.Dairy Tech., 75 (1977).

61. Orion Research Incorporated, Application Bulletin, A19. Cambridge, Ma., U.S.A.

62. W.H.Ficklin, J.Res.U.S.Geol.Survey, 3, 753 (1975).

63. Orion Research Incorporated, Nitrogen Oxide Electrode Instruction Manual, Cambridge, Ma., U.S.A.

64. J.Mertens, P.van den Winkel and D.L.Massart, Internat.J.Environ. Anal.Chem., 4, 25 (1975).

65. E.F.Eastin, Commun.in Soil Sci. and Plant Anal., 7, 477 (1976).

66. J.H.Puffer and R.S.Cohen, Chem.Geol., 15, 217 (1975).

67. Orion Research Incorporated, Potassium Electrode Instruction Manual, Cambridge, Ma., U.S.A.

68. P.W.Alexander and G.A.Rechnitz, Anal.Chem., 46, 860 (1974).

69. L.N.La Patnick, Analytica Chim.Acta., 72, 430 (1974).

70. Amel, Application Bulletin, "Determination of $Ag^+$ in photographic solutions". Milan, Italy.

71. B.L.Lenz and J.R.Mold, T.A.P.P.I., 54, 2051 (1971).

72. G.Bronow, T.Ilus and G.E.Miksche, Acta.Chem.Scand. 26, 1117 (1972).

73. Orion Research Incorporated, Application Bulletin, A 15. Cambridge, Ma., U.S.A.

74. Orion Research Incorporated, Sulphur Dioxide Electrode Manual, Cambridge, Ma., U.S.A.

75. Fu Chung Chang, Hui-TUh Tsai and Shw-Chii Wu, Analytica Chim.Acta., 71, 477 (1974).

76. G.J.Moody and J.D.R.Thomas, Lab.Practice, 28, 125 (1979).

77. G.Gran, Acta Chem.Scand., 4, 559, (1950).

78. G.Gran, Analyst, 77, 661 (1952).

79. D.Dyrssen, Acta Chem.Scand., 19, 1265 (1965).

80. F.J.C.Rossotti and H.Rossotti, J.Chem.Educ., 42, 375 (1965).

81. F.Ingman and E.Still, Talanta, 13, 1431 (1966).

82. A.Johansson, Analyst, 95, 535 (1970).

83. D.Dyrssen, D.Jagner and F.Wengelin, "Computer Calculations of Ionic Equilibria and Titration Procedures". Wiley, London. (1968).

84. C.McCallum and D.Midgley, Analytica Chim.Acta, 65, 155 (1973).

85. C.McCallum and D.Midgley, Talanta, 21, 723 (1974).

86. C.McCallum and D.Midgley, Analytica Chim.Acta, 78, 171 (1975).

87. C.McCallum and D.Midgley, Anal.Chem., 1232 (1976).

88. T.Anfalt and D.Jagner, Analytica Chim.Acta, 57, 165 and 177 (1971).

89. A.Liberti and M.Mascini, Anal.Chem., 41, 676 (1969).

90. F.A.Schultz, Anal.Chem., 43, 1523 (1971).

91. P.Carr, Anal.Chem., 43, 425 (1971).

92. M.Whitfield and J.V.Leyendekkers, Analytica Chim.Acta, 45, 383 (1969).

93. M.Whitfield, J.V.Leyendekkers and J.D.Kerr, Analytica Chim.Acta, 45, 399 (1969).

94. M.Whitfield and J.V.Leyendekkers, Analytica Chim.Acta, 46, 63 (1969).

95. Orion Research Incorporated, Newsletter, 3, 17 (1971).

96. M.S.Frant, J.W.Ross and J.H.Riseman, Anal.Chem., 44, 2227 (1972).

97. H.Clysters, F.Adams and F.Verbeek, Analytica Chim.Acta, 83, 27 (1976).

98. M.Mascini, unpublished work.

99. I.Hansson and D.Jagner, Analytica Chim.Acta, 65, 363 (1973).

100. M.Igri, W.Kakolowicz, L.G.Sillen and B.Warnqvist, Talanta, 14, 126 (1967).

101. T.Anfalt and D.Jagner, Analytica Chim.Acta, 47, 57 (1969).

102. B.Jaselskis and M.K.Bandemer, Anal.Chem., 41, 855 (1969).

103. E.W.Baumann, Anal.Chem., 42, 110 (1970).

104. Orion Research Incorporated, Newsletter, 3, 5 (1971).

105. N.Radic, Analyst, 101, 657 (1976).

106. W.Selig, Mikrochim.Acta, 3, 349 (1973).

107. C.Botré, M.Mascini, B.Bencivenga and G.Pallotti, Il Farmaco, 28, 218 (1973).

108. Orion Research Incorporated, Newsletter, 3, 3 (1971).

109. J.M.van der Meer and J.C.Smit, Analytica Chim.Acta, 83, 367 (1976).

110. M.J.Smith and S.E.Manahan, Anal Chem., 45, 836 (1973).

111. W.Selig, Mikrochimica Acta, 3, 87 (1973).

112. Orion Research Corporation, Newsletter, 2, 35 (1970).

113. J.L.Bernal, E.Barrado and R.Pardo, Analytica Chim.Acta, 111, 71 (1969).

114. E.J.Green and D.Schnitker, Marine Chemistry, 2, 111 (1974).

115. J.Buffle, N.Parthasarathy and D.Monnier, Analytica Chim.Acta, 59, 427 (1972).

116. J.Buffle, Analytica Chim.Acta, 59, 439 (1972).

117. N.Parthasarathy, J.Buffer and D.Monnier, Analytica Chim.Acta, 59, 447 (1972).

118. T.Anfalt and D.Jagner, Analytica Chim.Acta, 57, 165 (1971).

119. I.Hansson and D.Jagner, Analytica Chim.Acta, 65, 363 (1973).

120. Orion Research Incorporated, Newsletter, 2, 49 (1970).

121. C.C.Westcott, Analytica Chim.Acta, 86, 269 (1976).

122. A.Johansson, Talanta, 21, 1269 (1974).

123. M.J.D.Brand and G.A.Rechnitz, Anal.Chem., 42, 1172 (1970).

124. A.F.Isabell, R.L.Pecsok, R.H.Davies and J.H.Purnell, Anal.Chem., 45, 2363 (1973).

125. G.Horvái, L.Domokos and E.Pungor, Z.Anal.Chem., 292, 132 (1978).

126. T.J.MacDonald, B.J.Barker and J.A.Caruso, J.Chem.Educ., 49, 200 (1972).

127. J.J.Zipper, B.Fleet and S.P.Perone, Anal.Chem., 46, 2111 (1974).

128. J.W.Frazer, A.M.Kray, W.Selig and R.Lim, Anal.Chem., 47, 869 (1975).

129. J.W.Frazer, W.Selig and L.P.Rigdon, Anal.Chem., 49, 1251 (1977).

130. J.M.Ariano and W.F.Gutknecht, Anal Chem., 48, 281 (1976).

131. B.Fleet and A.Y.W.Ho, Talanta, 20, 793 (1973).

Acknowledgement: Grateful thanks are extended to the following authors for permission to reproduce material from the named papers:

F.J.C. Rossotti, H. Rossotti, J. Chem. Educ., 42, 376 (1965). (Figures 1-5). Ref. 80.

A. Liberti, M. Mascini, Anal. Chem., 41, 676 (1969). (Figure 3). Ref.89

M.S. Frant, J.W. Ross, J.H. Riseman, Anal. Chem., 44, 2227 (1972). (Figure 1). Ref. 96.

C. Botrè, M. Mascini, B. Bencivenga, G. Pallotti, Il Farmaco, 28, 218 (1972). (Figure 2). Ref. 107.

C.C. Westcott, Analytica Chim. Acta, 86, 269 (1976). (Figure 1). Ref.12

J.M. Van der Meer, J.C. Smit, Analytica Chim. Acta, 83, 367 (1976). (Figure 1). Ref. 109.

# THE AUTHOR

Professor Marco Mascini

Marco Mascini graduated at the University of Rome in 1963 and is now Professor of Quantitative Analysis in the Faculty of Science at the University of Rome. He has worked in several research establishments in Europe and in the United States of America, that is, in Professor Pungor's laboratory at the Technical University, Budapest, and Professor Guilbault's laboratory at the University of New Orleans, and in Professor Rechnitz's laboratory at Newark, University of Delaware.

Professor Mascini has been engaged on ion-selective electrode research since 1968, having been engaged on their preparation, analytical evaluation and application in several areas ranging from air pollution monitoring to industrial applications and clinical analysis. Presently, his research interests are on the development of biosensors obtained by coupling ion-selective probes with enzymatic, bacterial and antibody systems.

*Ion-Selective Electrode Rev.* Vol.2, pp.73-115. 0191-5371/80/0901-0073$05.00/0


# THE ROLE OF ION-SELECTIVE ELECTRODES IN CYSTIC FIBROSIS STUDIES

G. J. Moody and J. D. R. Thomas
Chemistry Department, Redwood Building, UWIST, Cardiff CF1 3NU, Wales

CONTENTS

KEYWORDS: Chloride ion-selective electrodes; cystic fibrosis; sodium ion-selective electrodes; sweat analysis.

## 1. INTRODUCTION

Cystic fibrosis (fibrocystic disease of the pancreas, muscovi-diosis) is a generalised inherited disease of children, adolescents and young adults involving a malfunction of exocrine glands. In patients with all the clinical manifestations of the disorder a characteristic triad of symptoms, namely, chronic pulmonary disease, pancreatic deficiency and elevated levels of sodium and chloride in sweat is evident [1]. The clinical picture can be greatly complicated by other symptoms such as intestinal obstruction and hepatic cirrhosis. The abnormality is present at birth, persists throughout life and is unrelated to the severity of the condition.

Various estimates ranging from 1:100 to 1:90 000 live births have been made of the incidence of cystic fibrosis [2-24] but the genuine incidence in Caucasians is thought to range from 1:1000 to 1:2000 live births. It is inherited and 1 in 25 of the population carries the abnormal gene and can transmit the disease [25]. When two carriers marry, there is a one in four chance of a CF child at each pregnancy.

TABLE I. Some estimated incidences of cystic fibrosis

| Country | Location | Incidence Relative to Live Births | Reference |
|---|---|---|---|
| Denmark | Entire country for the period 1945-1949 | Initially 1:20 206, rising to 1:4530 | 2 |
| | Entire country for the period 1963-1967 | 1:4530 | 2 |
| | Copenhagen for the period 1958-1962 | 1:2893 | 2 |
| Hawaiian Islands | Caucasian group | 1:3800 | 3 |
| | Mongolian group | 1:90 000 | 3 |
| India/ Pakistan | "The disease had not been suspected before death as we were under the impression that cystic fibrosis does not occur in this part of the world." | | 4 |
| Sweden | Sweden, 1950-1957 | 1:7700 to 10 000 | 5 |
| | Sweden | 1:3000 | 6 |
| | Sweden, 1974 | 1:3500 | 7 |
| Turkey | Istanbul Maternity Clinic | 1:3000 | 8 |
| United Kingdom | West Midlands (Asian immigrants) | 1:10 000 | 9 |
| | Wales, 1970-1975 | 1:1654 | 10 |
| | South Wales, 1971-1973 | 1:1850 | 11 |
| | South Wales/North Staffs | 1:1850 | 12 |
| United States | Hartford County, Conn. | 1:489 to 1:1863 | 13 |
| | Hartford County, Conn. | 1:500 to 1:2630 | 14 |
| | Indiana | 1:1370 | 15 |
| | Massachussets | 1:100 to 1:10 000 | 16 |
| | Massachussets | 1:500 to 1:1000 | 17 |
| | Massachussets, Vermont and New Hampshire | 1:2300 | 18 |
| | Minnesota | 1:1000 to 1:1400 | 19 |
| | Minnesota | 1:970 | 20 |
| | Ohio | 1:3500 to 1:4000 | 21 |
| Italy | Milan, 1974 | 1:2650 | 22 |
| Cuba | Havana, 1975 | 1:5065 | 23 |
| New Zealand | Hospitals in Auckland 1974-1975 | 1:2070 | 24 |

The lower incidences in Table I are generally attributed to poor diagnoses in earlier assessments. Countrywide assessments for Denmark are lower because convenience and opportunity give a better chance of diagnosis in a large city like Copenhagan.

Cystic fibrosis (CF) is the most common autosomal recessive disease in Caucasians. Compared with other congenital or hereditary childhood diseases, CF is about one half as common as mongolism; about as common as club foot, dislocation of the hip or cleft lip, and about ten times as common as phenylketonuria [26]. However, the incidence of CF in Negro infants is about one seventh [18] to one ninth [26] that in Caucasian infants (Table I).

Until recently, the mortality rate for CF children was high and one third died by one year of age and one half by four years [27]. Fortunately, this disturbing trend was substantially altered with intensive therapy. Thus, while treatment with the earlier types of antibiotics, diet or pancreatic replacement did little to lower the mortality rate, the picture greatly improved with intensive pulmonary or prophylactic pulmonary cleansing therapy when the mortality rate dropped to about one seventh of normal [28]. Indeed, from 1962 at the famous Minnesota CF Centre the rate fell from about 20% per year to about 2% per year [26]. This clearly indicates the importance of early diagnosis of neonates [29] in order to commence prompt medical treatment and possibly counselling [30].

A recent survey, however, did not support the hypothesis that effective treatment following earlier non-symptomatic diagnosis prolonged survival [31]. Nonetheless, the desirability of early diagnosis is emphasised by continuing advances such as the correction of the deficiency of proteolytic enzymes in the gastro-intestinal tract (the so-called pancreatic replacement) and antibiotic therapy for the recurrent chest infections associated with CF cases. This is emphasised by the fact that in 1964 only 12% of CF children survived beyond adolescence; by 1978 the proportion reaching adulthood had risen to 75% [32].

Despite the many criteria for diagnosis, doubt has been expressed about CF being a screenable disease [33]. This is largely because the criteria for screenable diseases with a specific biochemical lesion and primary treatment cannot be met. However, this criticism is counterbalanced by the obvious benefits already outlined regarding treatment and counselling. Warwick [34] has given a resumé of the background thinking to newborn screening for CF, especially since prophylactic approaches to therapy can provide an obligation to diagnose every patient.

## 2. DIAGNOSIS OF CYSTIC FIBROSIS

Few diseases are so readily diagnosed as CF and a wide variety of tests have been devised. These include high electrolyte content of finger and toe nails [35,36], specific stool enzyme assay [24,37], high albumin content of meconium [6,7,10,12,23,38-48], high sodium content of parotid saliva [49,50], sweat electrolyte tests based on high chloride levels determined with chloride ion-selective electrodes [2,8,26,51-65] or chloridometer [66] and similarly high sodium levels determined with sodium-sensitive glass electrodes [55,58,67], high electrical conductivity [68,69], and neutron activation analysis for sodium in nail (and possibly hair) clippings [8,70-80].

The absence of pancreatic enzyme activity in duodenal secretions, chronic pulmonary involvement and family history of the disease are also useful diagnostic tests. However, certain other tests, such as the agar plate finger/palm print, the filter paper print, the film patch test and "Quantabs" are unreliable and have been largely discarded since they tend to measure an amount of salt rather than concentration.

Convenience, simplicity, reliability and discrimination are very important diagnostic criteria. Measurements of the electrolytes in finger and toe nails can be discounted on these scores, for while the salt in nails originates in the sweat glands, the amount of salt will not only be proportional to the salt concentration of the sweat but also to the rate of sweating, frequency of bathing, and other aspects, e.g. diet, of the subject's activities for some time before nail sampling.

Similar criticisms have been made regarding neutron activation analysis of nails and since sample washing is the only non-mechanical step followed, it is more prone to error. Washing may, therefore, account for the considerable variation reported by various investigators [8,71,74,78,80] as well as the significant differences apparent for controlled variations in the washing procedures in just a single study [8]. For example, in ten neonates clippings were obtained from three fingers of the same hand, one group was washed *in vivo* with ethanol ([Na] = $145.17 \pm 30.61$ mmol $kg^{-1}$); the second group washed with ethanol after cutting ([Na] = $147.92 \pm 21.95$ mmol $kg^{-1}$) and the third group washed with water and rinsed with acetone ([Na] = $93.87 \pm 29.40$ mmol $kg^{-1}$). The diagnostic utility of neutron activation analysis of nail clippings is, therefore, dependent on a standard washing procedure, but nail clipping analyses are of no value in the newborn period [77]. The charge of >£3 per sample for 100 sample batches alone (private communication, Analytical R and D Unit, Harwell, February, 1975) mitigates against the method for routine screening.

For practical purposes, screening by means of the Boehringer-Mannheim test for albumin in meconium is simple but unfortunately restricted to the newborn. The sodium content of saliva from the parotid gland is normally determined with a glass sodium ion-selective electrode, but the desirability for early diagnosis precludes this procedure until the fourth month of life. The same remark applies to the assay of the elevated (total) calcium levels in parotid saliva using a calcium ion-selective electrode [81].

It is not surprising, therefore, that a great deal of attention for universal CF diagnosis has focussed on sweat electrolyte tests and which constitutes the theme of this review. All these tests are centred on the fact that sodium and chloride levels are almost always elevated in CF sweat samples [82,83]. These are probably the most important for screening purposes using, for example, sodium ion-selective electrodes [55,58,67], and chloride ion-selective electrodes [2,6,8,25,26,51-65,84-90]. The alternative scheme, based on laboratory assays and derived from the Gibson and Cooke procedure [66] also depends on sweat production but much larger amounts of sweat (ideally ~100 mg) are necessary.

## 3. THE BASIS OF SWEAT ELECTROLYTE TESTS

The abnormally high concentration of electrolytes in the secretions of eccrine sweat glands of CF patients was first recognised [82,83] in 1953. The abnormality is present at birth and persists throughout life but there is no relationship between electrolyte levels and severity of the disease or of the organs involved [91]. It is found in at least 98% of individuals homozygous for the CF gene and in only a very few other well-demarcated cases [92], such as hypoadrenocorticalism, ectodermal dysplasia and rare genetic defects; each being capable of precise differential diagnosis.

The "Sweat Test" has, therefore, become a cornerstone for diagnosing cystic fibrosis, although false negative CF cases, i.e., with normal electrolyte concentrations in the sweat, must always be kept in mind [93-96].

Any test comprises three quite distinct steps:
(i) preparation of the skin of the chosen test site,
(ii) stimulation of sweating (the most intricate step), and
(iii) measurement of the chloride and/or sodium levels in that sweat.
Many improvements have been effected at each step since 1959, especially regarding the means of sweat stimulation and its actual analysis from an *in situ* standpoint.

The fundamental steps of the sweat test may be summarized as follows:

SKIN-SITE PREPARATION

↓

SWEAT STIMULATION AND ACCUMULATION

↓

SWEAT ANALYSIS

| LABORATORY | | *IN SITU* |
|---|---|---|
| Titration or Chloridometer | chloride | Chloride ion-selective electrode |
| and/or | | and/or |
| Flame photometer or Atomic absorption spectroscopy | sodium | Sodium ion-selective electrode |

The sweat abnormality is limited to monovalent ions including sodium, potassium and chloride [91]. It can be demonstrated unequivocally by the Gibson-Cooke test using pilocarpine iontophoresis for sweat stimulation but at least 100 mg of sweat needs to be collected for subsequent laboratory analysis [66]. While the more aesthetic physiological heat-induced sweating should be considered, at least for research purposes [97], the technique based on the parasympathomimetic agent, pilocarpine, obviates the danger of heat stress and claustrophobia of plastic bag methods [98-100], and the discomfort following sub-cutaneous injection of cholinergic drugs [100].

Pilocarpine(I) occurs naturally in the leaflets of South American shrubs, *Pilocarpus jaborandi* and *Pilocarpus microphyllus* and with muscarine and arecholine is among the three major cholinomimetic alkaloids. Its major action is to stimulate autonomic effector cells

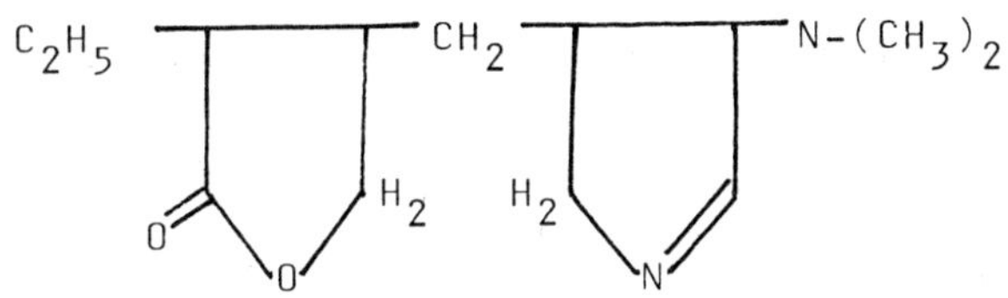


I

in a fashion analogous to cholinergic post-ganglionic nerve impulses.

Few drugs readily penetrate the intact skin but percutaneous ion transfer can be increased by iontophoresis. This is the basis of the Gibson-Cooke procedure wherein the said transfer is effected by passage of a galvanic current through an aqueous pilocarpine solution applied to the skin site underneath an electrode [66]. The drug is thus locally concentrated in the dermis and subcutaneous tissues and stimulates the sweat glands without producing general systemic effects.

Pilocarpine iontophoresis requires only five to ten minutes to complete and over a subsequent thirty minutes generally produces sufficient sweat for taking to the laboratory. Gibson and Cooke determined chloride in small sweat samples by polarography [66]; larger samples were titrated with mercury(II) nitrate using S-diphenylcarbazone indicator [101] or a chloridometer [102], while sodium was determined with an ion-selective electrode or flame photometer. More recently, chloride in sweat has been analysed directly on the skin with a combination ion-selective electrode [51,86]. Some duplicate results for chloride in the sweat of 740 normal and 196 known CF subjects, respectively, obtained by two different routine analytical techniques at an established CF centre are shown in Table II. Four sweat tests were conducted in one sitting on each subject - one electrode and one Gibson-Cooke on each arm - the correlation coefficient being 0.91.

TABLE II. The mean chloride values of normal and CF sweat samples analysed by two different methods [60]

| Method | Subject | No. | Chloride/mM | | |
|---|---|---|---|---|---|
| | | | Mean | sd | Range |
| Orion 417 Skin Meter [a] | Normal | 740 | 13.7 | 10.7 | 1-80 |
| Pilocarpine Iontophoresis Sweat Test [b] | Normal | 740 | 11.6 | 9.1 | 1-68 |
| Orion 417 Skin Meter [a] | CF | 196 | 90.6 | 20.3 | 40-200 |
| Pilocarpine Iontophoresis Sweat Test [b] | CF | 196 | 94.6 | 21.9 | 21-180 |

a Using pilocarpine iontophoresis technique for sweat stimulation [86]

b The term "pilocarpine iontophoresis sweat test" used at the 1975 GAP conference [89], and by some workers, embraces the pilocarpine iontophoresis sweat stimulation step and the collection and removal of sweat to the laboratory for subsequent chloride/sodium assays. This can be confusing [88] for the pilocarpine iontophoresis sweat stimulation step is also part of the *in situ* electrode sweat test [86]. In this review "pilocarpine iontophoresis" refers solely to the stimulation step, the purposes of this Table being that it includes the complete Gibson-Cooke procedure.

The 1975 Cystic Fibrosis Foundation GAP conference recommended that diagnosis of cystic fibrosis be based on a quantitative pilocarpine iontophoresis sweat test (See Table II). The chloride should be determined by titration or using a chloridometer and sodium by flame photometry [89]. Furthermore, the test should be duplicated or repeated at least once and the results interpreted by an experienced clinician. While this is a sound basis for making a diagnosis, screening methods ought ideally to be simpler, faster and cheaper to perform with the minimum possible number of false negatives and, at worse, only a few false positives. In principle, a test based on the Orion 470 Skin Meter largely meets such requirements.

## 4. SWEAT SECRETION AND COLLECTION

Since the sweat test has progressively assumed cumulative importance as a diagnostic method, the collection of sweat is an important factor in both laboratory and screening methods. It is, therefore, pertinent to consider some anatomical and physiological aspects of sweat [103].

### 4.1 The eccrine glands

The number of eccrine sweat glands in human skin is estimated to lie between $2 \times 10^6$ and $5 \times 10^6$. They occur in almost all skin areas but the density is greatest in the palms of the hand and soles of the feet and least on the thighs, neck and back. The glands develop in the fourth month of uterine life but there is no post-natal development. The nerve supply is based on sympathetic post-ganglionic fibres but is cholinergic. Although sweat glands respond to adrenergic drugs, there is no apparent adrenergic nerve supply and no sweating occurs in a denervated zone of skin.

## 4.2 Stimulation of sweating

Sweat composition varies with time and rate of sweating [104, 105]. Sweat collection for diagnostic sweat tests should, therefore, be standardised to ensure reliable comparative data and it is futile to analyse unstimulated sweat [106]. However, some analyses of unstimulated sweat with sodium ion-selective electrodes have been reported [55,58,67] but even the original investigators [55] now favour the analysis of stimulated sweat.

4.2.1 Local heat stimulation. Local heating of skin does not cause sweating except at 43 to 46°C when skin damage may arise [107]. Nevertheless, local heat stimulation may be used with an aluminium cylinder set at 48°C under carefully controlled conditions [20,26, 51]. The interscapular area of the back, despite the low eccrine gland concentration, is a recommended test site for newborns, the test site being prewashed with distilled water and thoroughly dried with gauze [26,51,54]. The area is then covered with a Parafilm sheet (10 x 10 cm) which serves as an inert barrier and also prevents evaporation and contamination during the five minutes when the aluminium heating block is held firmly in position on the area. In air-conditioned rooms it is recommended that the baby be wrapped in a blanket [54]. After the five-minute stimulation period the aluminium block and Parafilm are removed from the site where the skin temperature is 36.1 to 37.2°. Chloride is then measured *in situ* with a chloride ion-selective electrode calibrated at 37°C although the mean skin temperature is 36.7°C [51,54]. An alternative procedure is to use a water bath heater set at 50°C which gives a mean skin temperature of 37.0°C ranging from 36.8 to 38.7°C [54].

Local heat stimulation produces sufficient sweat for an *in situ* screening test, but is not recommended for individuals who are cold, dehydrated or who have skin complaints [54]. It has been used to screen 20 071 children up to eleven years old in twenty six Minnesota communities using the anterior forearm skin areas [26]. No CF cases were found, thus emphasising high mortality rates and the need for early screening. Later, three CF cases were diagnosed but none were covered in the initial programme, thus vindicating its effectiveness despite some convern over spurious results (since corrected) and caused by fluctuations in the line voltage [26].

4.2.2 The plastic bag. Apart from keeping the subject in a room at 32°C for one to two hours [79,82,83], the first simple, reliable practical sweat stimulation/collection method employed a plastic bag [98]. However, the long confinement period, frequently needed for sufficient sweat, the danger of hyperpyrexia and the constant vigil to prevent forceful exit of the infants from the bag posed procedural difficulties. This method is therefore not frequently used and was ignored by Kopito and Shwachman in a review [55] which stated a preference for pilocarpine iontophoresis.

4.2.3 Use of cholinergic drugs. Before discussing sweat collection following pilocarpine iontophoresis it is pertinent to consider stimulation by injection of cholinergic drugs. One such procedure involves the intradermal injection of β-methylacetylcholine hydrochloride (Mecholyl) [107-109] with collection periods of ten to thirty minutes [109]. Mauer and West [100] preceded the mecholyl injection by physostigmine, but in view of occasional side reactions - flushing and bradycardis - they adopted a single injection of furtrethonium [Furmethide]. However, this drug has been withdrawn and replaced by bethanechol (Urecholine) [57,100,110] although the sweat collection period may last up to one hour.

4.2.4 Pilocarpine iontophoresis. Pilocarpine iontophoresis was originally devised to stimulate sweating for laboratory test samples [66] and is the method favoured by participants at the 1975 GAP conference for stimulating sweat production [89]. The dangers of heat stress and the pain of injection need not be endured and a correct pilocarpine iontophoresis technique largely avoids such discomforts. Any sign of unease caused by poor electrode/skin contact is easily alleviated by pressing momentarily on the electrode. However, in view of the risk, however slight, of ventricular fibrillation and respiratory arrest, it is recommended that iontophoresis equipment should not give currents >10 mA on short circuiting [111].

The Orion Research Incorporated skin chloride unit is an important development and is extensively used as an *in situ* screening test for cystic fibrosis [8,55-57,59-61,64,65,111-114]. The present 417 model conveniently incorporates the iontophoresis and combination chloride ion-selective electrode in a single unit. The method currently employed for screening in Cardiff is detailed in Section 10. After completion of the five-to-ten-minute iontophoresis stage sufficient sweat has usually collected on the skin surface in just two or three minutes to complete the chloride analysis with the chloride ion-selective electrode.

4.3 Sweat collection

The physical collection of sweat for subsequent laboratory analysis is usually based on sorption by pre-weighed filter paper or gauze sorbent. Thus, the overnight sweat collection reported by Shwachman and Mahmoodian [115] involves placing a weighed 5 cm square gauze pad on the subject's washed back in the evening. The pad area is sealed with plastic, taped and then removed on the next morning, reweighed and analysed for sweat. Although 100 mg sweat is desirable for laboratory analysis [89] samples of just 40 mg were considered satisfactory for measuring sodium and chloride levels in 29 CF patients and 47 controls [115]. Comparable results were obtained using sweat collected after pilocarpine iontophoresis [115]. The quantities of sweat obtained after pilocarpine iontophoresis [66] and furtrethonium injection [100] respectively, are shown in Table III.

Pilocarpine iontophoresis is now the accepted stimulation procedure for collecting sweat for laboratory analysis of sodium and chloride [89]. The collection procedure is similar to the Shwachman-Mahmoodian method described above except that the collection period following the actual stimulation is just thirty minutes. Specimens

TABLE III. Sweat sample collection data following stimulation by pilocarpine iontophoresis or furtrethonium injection

| Stimulation Details | Collection Details | Number of Tests | Sweat Collected /mg | Remarks (Reference) |
|---|---|---|---|---|
| Filter paper (2.5cm) placed on washed area. $0.2cm^3$ of 0.2% pilocarpine nitrate added to positive electrode followed by few $cm^3$ of 0.7M sodium bicarbonate to negative electrode. Iontophoresis for 5 min at 2mA | Preweighed filter paper (2.5cm) covered with plastic (3 x 3 cm) for 30 min. Filter paper then quickly placed in a preweighed bottle | 55 | 18-135 | Only 4 samples < 40 mg [66] |
| Square gauze ($7.5cm^2$) on washed site. $4cm^3$ of 0.05% or 0.1% pilocarpine nitrate added to positive electrode. Iontophoresis for 15 min at 4mA | Preweighed gauze ($7.5cm^2$) covered with plastic for 15-90 min. Gauze then placed in preweighed bottle | 34 | 140-1620 (Mean=692) | Longer time provided little extra sweat [66] |
| Furtrethonium iodide ($0.2-0.4cm^3$) containing 0.66-1.3 mg of active agent injected intradermally into centre of washed skin site | Preweighed gauze ($5cm^2$) covered with vinyl sheet for 60 min. Gauze then placed in weighing dish | 90 | $0.1-1.4cm^3$ (Mean=$0.3cm^3$) | Sweat samples rarely $<0.1cm^3$ Bethanechol is less effective [100] |

<100 mg should be discarded mainly because small sweat volumes lead to considerable errors from both evaporation losses and analytical difficulties [89]. Duplicate samples may be collected following the pilocarpine stimulation step on each arm (or the legs of small infants). The second iontophoresis step is started on the second arm when the sweat collection on the first arm is under way. However, infants under six weeks old may still provide insufficient sweat for analysis [89].

## 5. SWEAT COMPOSITION

Sweat is essentially a clear, watery secretion. Since abnormal sweat electrolyte levels constitute a major diagnostic basis for cystic fibrosis the principal inorganic components as well as lactate and urea in normal sweat and plasma are shown in Table IV [116].

TABLE IV Selected components in "normal" sweat and plasma (mM except for pH)

| Species | Sweat | Plasma |
|---|---|---|
| $Na^+$ | 10-80 | 137-148 |
| $K^+$ | 7.0 | 3.9-5 |
| $Cl^-$ | 10-65 | 101-106 |
| $HCO_3^-$ | Trace | 0-10 |
| $Ca^{2+}$ | 0.5-1.5 | 2.5 |
| $Mg^{2+}$ | 0.0005 | 0.5 |
| $SO_4^{2-}$ | <0.5 | 0.5 |
| $HPO_4^{2-}$ | 0.005 | 0.5 |
| Urea | 5-10 | 2.5-5 |
| Lactate | 4-40 | 0.5-10 |
| pH | 4.2-7.5 | 7.5 |

Some comparative data on the chloride and sodium levels measured in the sweat of normal control subjects and diagnosed CF subjects are shown in Table V.

The variation in sweat composition during sweating and with sweating rate is an interesting phenomenon. Thus, Schwartz and Thaysen [109] found that the sodium and potassium in the initial sweat collected following mecholyl injection was invariably higher than in any subsequent collection period. This was the case for sodium even when the sweating rate was lower in the initial than in the subsequent period, and neither was it related to the fact that the sodium level rose to a maximum with increased sweating rate, that is, from 1.92 mM at a flow rate of 35 g $m^{-2}min^{-1}$ to 103.8 mM at a flow rate of 25.3 g $m^{-2}min^{-1}$. In no case did the sodium concentration approach that in plasma whereas the potassium sweat:plasma ratio remained generally constant at 2.2 with variable sweating rates.

TABLE V Chloride and sodium contents of sweat from normal and cystic fibrosis subjects

| Ion | Series | 'Normal' Control Subjects | | | C.F. Subjects | | | Stimulation Method | Analysis Method [Reference] |
|---|---|---|---|---|---|---|---|---|---|
| | | Number | [Ion]/mM Range | [Ion]/mM Mean | Number | [Ion]/mM Range | [Ion]/mM Mean | | |
| $Cl^-$ | 1 | 8 | 11-60 | 45.4 | 14 | 84-136 | 110.1 | Pilocarpine iontophoresis | Mercurimetric titration[74] |
| | 2 | 39 | 7-49 | 21.1 | 11 | 80-122 | 94.8 | | Polarography [66] |
| | 3 | 34 | 12-39 | 23.0 | 68 | 52-148 | 116.0 | | Mercurimetric titration[55] |
| | Mean of subjects in each group | | - | 24.3 | - | - | 112.8 | - | - |
| $Na^+$ | 4 | 36 | 17-40 | 26.8 | 26 | 54-200 | 131.1 | Spontaneous | Sodium ion-selective electrode on washed skin [67] |
| | 5 | 28 | 10-80 | 36.0 | 27 | 20-360 | 135.0 | | Sodium ion-selective electrode on washed skin [55] |
| | 6 | 17 | 6.2-85.2 | 39.5 | - | - | - | Mecholyl injection | Flame photometry [109] |
| | Mean of subjects in each group | | - | 32.5 | - | - | 133.0 | - | - |

The first sample phenomenon has been studied in some detail by Schwarz, Sutcliffe and Style [111]. Pilocarpine iontophoresis stimulation is suitable for such studies since additional readings may be taken *in situ* with an ion-selective electrode as long as sweating continues. Most normal subjects show relatively high sodium values in the first one to three samples but fortunately this is not the case in CF subjects where sodium concentrations are uniformly high in all samples. Schwartz and Thaysen [109] concluded that the sodium and potassium cations are delivered into a precursor solution, and that the sodium, but not the potassium, is reabsorbed by a subsequent process of limited capacity. The first sample phenomenon can be attributed to a lag in the action of the ion transport factor and, of course, the regulating factor is deficient in cases of cystic fibrosis.

Sex, age and health of the subject condition sweating rates and hence sweat electrolyte levels [112]. Thus, sweating rates are lower in normal children than in adults: an abrupt increase occurs in the ten to twelve years age range. Sex differences become apparent at puberty and females are only 25 to 50% as responsive as males to sweat stimulation [17].

## 6. LABORATORY AND SCREENING SWEAT ELECTROLYTE TESTS

The quantitative analysis of sweat is a valid diagnostic test for cystic fibrosis. Laboratory analysis for sodium and chloride with more than 60 mM in conjunction with clinical symptoms like recurrent respiratory disease, malabsorption or a family history of cystic fibrosis, is consistent with the diagnosis of cystic fibrosis [89]. However, there are borderline areas where interpretation is difficult (Table V) and while values of 45 to 60 mM of chloride and sodium generally warrant repetition of the tests, values exceeding 60 mM have been found in some normal adults. Only careful clinical and repeated analyses can resolve such cases.

### 6.1 Laboratory sweat electrolyte analysis

Following stimulation, the laboratory test requires that sufficient sweat be collected for analysis either *in situ* or in the laboratory.

6.1.1 Elution of sweat from the collection matrix. One method involves adding, say distilled or deionised water (2 $cm^3$), to the weighing bottle containing the filter paper or gauze with its absorbed sweat and squeezed to obtain a proper extract. A sample (1 $cm^3$) of the fluid is taken for chloride estimation and the remainder centrifuged to remove any troublesome fibres before sodium estimation by atomic absorption spectroscopy.

The above technique often gives poor recoveries and Schwarz, Sutcliffe and Style [111] preferred eluting the sweat-soaked supports by clamping between two perspex plates dipping into a reservoir of distilled or deionised water. Between 2 and 3 $cm^3$ of eluate is adequate for removing sweat and which is fibre-free.

Two of the three analytical methods approved by the GAP Conference [89] depended on elution of the sweat from the gauze matrix with 5 $cm^3$ of distilled water for samples >150 mg and 3 $cm^3$ for samples

> 100 mg. Aliquots (1 $cm^3$) of the eluate were used for the mercury (II) titration technique for chloride [101] and aliquots of suitable dilution for determining sodium (and, if necessary, potassium) by say flame photometry.

6.1.2 Analytical methods. Titrimetry [101], polarography [66], amperometric titration [102,118], flame photometry [89], etc. have been variously used to measure chloride and sodium in sweat samples after collection from the skin site.

The combination chloride ion-selective electrode can also measure chloride in laboratory samples but at present its principal role is for rapid *in situ* monitoring on the skin itself. With regard to precision, the electrode should be used carefully. For example, when taking the threshold value of 60 mM of univalent chloride or sodium it should be remebered that a 10% higher value alters the emf response by just 2.45 mV and by 2.72 mV for a ten percent lower level of such ions. Thus, readings of 54 and 66 mM of the above ions differ by only 5.17 mV at 25°C. Sodium, and potassium if necessary, are normally determined by flame photometry with lithium as a convenient internal standard. The virtue of potassium values is that the sodium:potassium ratio is normally <4 in children and <3 in neonates [84,119] but can be depressed by intake of mineralocorticoids [120]. The necessary dilution of sweat for sodium analysis is easily judged from the chloride concentration which is measured first.

It is important that sodium and chloride concentrations should not differ appreciably from the criterion [121] that $[Na^+] = 1.12 \times [Cl^-]$. Certainly, a result where chloride exceeds sodium is suspicious and when sodium exceeds chloride by 10 mM, or more, merit a repeat analysis on fresh sweat.

## 6.2 Sweat electrolyte screening tests

Early screening tests were unreliable; indeed the excessive number of false positives and false negatives obtained with the chloride-plate test led to its demise [122]. Despite its low cost and simplicity, experience is required to interpret the imprint of the reaction between skin chloride and silver nitrate/potassium chromate in the agar plate.

6.2.1 Electrical conductivity. Recent screening methods are quantitative rather than the qualitative pass/fail criterion of the agar plate and analogous tests. Thus, the measurement of electrical conductivity of sweat [68] depends on enhanced conductivity readings in CF cases due to the associated high sweat electrolyte levels. The test can be performed directly on the skin surface on sweat obtained either spontaneously or by stimulation. It can be completed quickly without the need for highly trained staff and there is no need for frequent calibration. However, the method has been criticised notably because no large series of patients or control subjects has been tested [106] and because of its inability to detect known CF cases [123]. Considerable quantities of sweat are also needed in order to cover the conductivity electrodes [124].

6.2.2 Ion-selective electrodes. In the development of diagnostic screening methods considerable interest has centred on the scope of chloride and sodium ion-selective electrodes, especially for direct measurement on the sweat-skin surface. In their most convenient form as 'combination electrodes' these potentiometric ion sensors have a built-in second reference electrode to complete the working cell.

Direct skin measurements on sweat stem from early work on skin pH with glass electrodes [125]. Apart from the silver billet electrode, the new sensors arose from the development of glass membrane electrodes selective to sodium and later other types of membranes selective to a substantial range of cations and anions [126,127] which have found important biomedical applications [128].

The general procedure is to position the precalibrated combination electrode on the well-washed skin area where sweating has already been induced. The chloride or sodium level corresponding to the millivolt meter reading is equated to an emf-concentration calibration graph but the Orion 417 skin chloride unit permits a direct read-out [86]. The data for a modified procedure obtained with this commercial unit now used in Cardiff are discussed in Section 10.

## 7. AN APPRAISAL OF ION-SELECTIVE ELECTRODE SWEAT ANALYSIS DATA

The results of chloride measurements found by direct skin-electrode contact after sweat stimulation by localised heating, pilocarpine iontophoresis and also dry skin are summarised in Tables VI to VIII. Similar work with sodium ion-selective electrodes is summarised in Table IX.

Naturally, most attention has centred on the convenient Orion chloride combination ion-selective electrode (Models 96-17 and 97-17). Some screening has also been done with the 94-17 version [8,54,112] but the need for a complementary reference electrode is really most inconvenient.

There is discrimination in the various studies between chloride and sodium levels in the sweat of normal subjects and CF patients (Tables VI to IX). Certain spurious values are, however, apparent, for example, the high chloride value (45.9 mM) found using the Beckman 39187 silver billet electrode compared with 22.3 mmol $dm^{-3}$ obtained using the Orion 94-17 electrode (Table VII). However, other workers [20,26,51] using the same Beckman model 39187 electrodes found 'normal' chloride levels of <30 mM (Table VI). A photoelectric response is always possible with this billet-type electrode and it has now been superseded by the Beckman silver-silver chloride combination model 78527. Incorporation of silver sulphide into the silver chloride matrix of the solid state chloride crystal membrane ion-selective electrode provides a mechanically strong sensor matrix and essentially eliminates photoelectric effects [132].

The use of rapid-reading sweat test devices has been criticised since the amount of sweat produced is not measured [106]. This factor, together with the variation in sweat composition, sweat evaporation and the first sample phenomenon [111], can account for non-reproducible results, e.g. chloride values of 34, 50, 86 and 50 mM in the sweat of the same patient using a combination ion-selective electrode [131]. Kopito and Shwachman [55] suggest that the chloride levels for two tests on each subject must agree to within 10 mM for <50 mM or to within 25 mM for readings >50 mM.

TABLE VI Sweat chloride measurements made with chloride ion-selective electrodes on the skin after local heating

| Number of Subjects: Normal | Number of Subjects: C.F. | [Chloride]/mM: Range | [Chloride]/mM: Mean(±sd) | Combination Electrode/Meter System Employed | Washing Scheme Before: Stimulation | Washing Scheme Before: Measurement | Remarks [Reference] |
|---|---|---|---|---|---|---|---|
| 28 | - | 5-30 | $13\pm6.5$ | | Site washed with distilled water, dried with gauze | | [51] |
| 49 | 49 | 55-130 | $89.2\pm10.6$ | | | | [51] |
| 20 071 | - | 93% < 30. Retesting of 57 cases > 30, gave persistently high chlorides. | | Beckman 39187[a] silver billet with expanded scale Beckman 76 pH meter | Site on forearm washed with distilled water, dried with gauze. | None. Electrode (after washing with distilled water) placed on site after blotting sensor | Values > 30mM are suspect. One or more values > 40mM are retested using pilocarpine iontophoresis. [20] |
| 348 | - | 95% < 30. All but one case of > 30 related to poor fluid intake, fever or poor initial test. | | | Intersapular site of neonate's back washed with distilled water, dried with gauze. | | Cost per test 52 cents over 10 000 tests. Good correlation with alternative pilocarpine Coltlove chloridometer technique (r = 0.935). [26] |
| 30 | - | 10-60 | $21\pm9.7$ | Orion 96-17 chloride standardised at room temp. with Orion 400 meter | Site wiped with gauze (moistened with distilled water) and dried. | Electrodes washed with distilled water, placed quickly on sweating site. | Compensatory corrections for difference between room and skin temperature held to be unnecessary. Gel packs used for sweat stimulation. [55] |
| 21 | 21 | 40-150 | $109\pm35$ | | | | |
| Assessment of method, breakdown of data on subjects not given but ogive based on 1156 sweat chloride values | | | | Beckman 78527V silver-silver chloride, Orion 96-17 chloride or Radiometer PKS67011 chloride electrodes standardised at 37°C used with Beckman 76pH, Orion 401 or Radiometer 27pH | Interscapular area washed with distilled water and dried with gauze | None. Electrode (after washing with distilled water) placed on site after blotting sensor | Two tests on each subject to agree within 10mM for < 50mM for > 50mM. Precision comparable with that of Gibson-Cooke method. Gel pack used for sweat stimulations. [54] |

a This is not a combination electrode.

TABLE VII Sweat chloride measurements taken with chloride ion-selective electrodes on the skin after pilocarpine iontophoresis

| Number of Subjects | | [Chloride]/mM | | Combination Electrode/ Meter System Employed | Washing Scheme Before | | Remarks (Reference) |
|---|---|---|---|---|---|---|---|
| Normal | C.F. | Range | Mean(±sd) | | Stimulation | Measurement | |
| 1 | - | - | 19.9±4.8[a] | Beckman 78527V silver-silver chloride or Orion 96-17 chloride at 33.5°C with Beckman 76004 "Expandomatic" pH meter | Not detailed | Dried electrode with salt-free gauze after rinsing with water applied to skin | Useful for comparing ionic activities in presence and absence of functional sweat glands. [58] |
| 43 | - | 8-42 | - | | | | Children(1-4 yrs old).Component activities from sources other than sweat obscured on dry skin.[58] |
| 55 | 55 | - | 75[b] | Orion 94-17[c] chloride with Orion 401 ion meter | Not detailed | Not detailed | [12] |
| 536 | - | - | | | | | [12] |
| 3000 | - | - | 45.9±19.9 (46.0;45.8;48.4;40.8)[d] | Beckman 39187[c] silver billet(3000) or Orion 94-17[c] (2061)used with Beckman"Expandomatic pH meter) | Reference made to Gibson and Cooke[66] | Not detailed | Values greater than mean plus 2sd regarded as suspect for CF.[8] |
| 2062 | - | - | 22.3±12.0 (22.6;22.1;21.3;17.2)[d] | | | | |
| 47 | - | 10-50 | 25±8.5 | Orion 96-17 chloride standardised at room temperature with Orion 400 meter | Not detailed Surface of forearm used. Back for infants and neonates.) | Gauze-pad moistened with ion-free water.Site blotted with gauze.Electrode rinsed with distilled water and quickly placed on sweat site. | Compensatory corrections for temperature differnces between skin and room felt to be unnecessary.Kopito and Shwachman prefer pilocarpine agency for sweat stimulation [55] |
| 75 | 75 | 45-200 | 101±32 | | | | |
| 4 | 4 (15 measurements) | 10-86 | - | Orion chloride but model not specified | Not detailed | Not detailed | Wide variations for individuals, eg,34,50,86 and 50mM for one subject attributed to variable sweating.[129] |

Table continued on next page.

| | | | | | | | |
|---|---|---|---|---|---|---|---|
| 55 | 55 | - | 40[b] | | Orion instructions followed.[86] | | [112] |
| 754 | - | - | | | | | |
| 24 940 | 11 | - | - | | | | False positive rate is 5%. CF subjects comprised 9 confirmed "true"positives and 2 known "false"negatives.[130] |
| 1 | 1 | - | 14.5 | | | | Low chloride( 10mM)also obtained in laboratory after pilocarpine iontophoresis, but sodium was high at 100mM. Discrepancy attributed to cloxacillin. [96] |
| 1 | 1 | 72-90 | 82 | | | | |
| 1205 | - | 5-100 | 34.4±12.7 | Orion 97-17 chloride with Orion 417 skin chloride meter unit. | Procedure discussed in Section 9 | | Age 0-20 days. Test is best after day 2. [64] |
| 409 | - | 7-57 (10-29 in 79%cases) | - | | Skin washed with gauze (moistened with distilled water). Repeated 3 times. | Test area rapidly wiped with moist gauze and blotted dry. | All 40-50mM cases routinely checked by urecholine injection and Cotlove chloridometer analysis. This reference method on 32CF patients gave 99±11mM compared with 80±30mM for Orion system.[57] |
| 58 | 58 | 40-200 (85% with 60) | ] | | | | |
| 400 | 1 | 10-90 (in babies <1wk old) | - | | | | Charge of $4.50 per test proposed in 1971. [131] |
| 4 | - | 11-115 | - | | | | |
| 5 61 tests) | - | 6-39 | - | Orion system modified: felt pads saturated with pilocarpine nitrate/$NaHCO_3$;extended iontophoresis period; sweat allowed to build up under plastic cap for 10 min. | Skin washed twice with distilled water on fresh gauze. Site finally wiped dry. | Electrode put on sweat site after removing plastic cap. | Orion system modified to combat poor precision and occasional "false" positive cases. Revision might give precision of reference methodologies. [84] |

a Eight repeat measurements on same skin site.
b Normal upper limit except in neonatal period.
c These are not combination electrodes.
d Consecutive mean values for first 4 days of life.

TABLE VIII Sweat chloride measurements taken with chloride ion-selective electrodes on dry skin surfaces

| Number of subjects: Normal | CF | [Chloride]/mM: Range | Mean($\pm$SD) | Combination Electrode/ Meter System Employed | Washing Scheme before Measurement | Remarks (Reference) |
|---|---|---|---|---|---|---|
| 1 | - | - | 34.4 $\pm$ 2.3 | Beckman 78527V Silver-silver chloride(silver billet) or Orion 96-17 chloride at 33.5°C with Beckman 76004 "Expandomatic" pH meter | Wet electrode tip, after washing with deionised water, applied to volar surface of mid forearm with gentle pressure | Useful in comparing ionic activities in presence and absence of functional sweat glands<br>8 repeat measurements made on same skin site [55] |
| 1 | - | - | 22.0 $\pm$ 2.2 | | | |
| 1 | - | - | 10.9 $\pm$ 2.0 | | | |
| 40 | - | 8-70 | - | | | Children (1-4 years old). Component activities from sources other than sweat obscured on dry skin [55] |
| Not specified | | | | Chloride. Type unspecified but probably solid state crystal variety | Not specified | Local Urecholine injection to stimulate sweating in cases being checked. Otherwise, stimulation method not specified [40] |

TABLE IX Sweat sodium measurements taken with sodium ion-selective electrodes on the skin

| Number of Subjects: Normal | Number of Subjects: C.F. | [Sodium]/mM: Range | [Sodium]/mM: Mean($\pm$SD) | Combination Electrode/ Meter System Employed | Washing Scheme before Measurement[a] | Sweat Stimulation Method | Remarks (Reference) |
|---|---|---|---|---|---|---|---|
| 36 | - | 17-40 | 26.8 $\pm$ 5.2 | Beckman 31142 sodium with expanded scale pH meter | Rinsed electrode placed directly on washed skin area | None | [67] |
| 26 | 26 | 54-200 | 131.1 $\pm$ 42.5 | | | | [67] |
| 28 | - | 10-80 | 36 $\pm$ 9.6 | | | | [55] |
| 27 | 27 | 20-360 | 135 $\pm$ 54 | | | | [55] |
| 1 | - | - | 50.7 $\pm$ 6.3 | Beckman 78420V sodium or Orion 96-11 sodium at 33.5°C with Beckman 76004 "Expandomatic" pH meter | Wet electrode tip, following washing with deionised water, applied directly to the volar surface at mid forearm with gentle pressure (A) | Dry skin measurements | Useful in comparing ionic activities in presence and absence of functional sweat glands. 8 repeat measurements made at same time [58] |
| 1 | - | - | 27.6 $\pm$ 3.4 | | | | |
| 1 | - | - | 10.7 $\pm$ 1.4 | | | | |
| 1 | - | - | 59.8 $\pm$ 16.7 | | Dried electrode with salt-free gauze after rinsing with water applied to moist surface (B) | Pilocarpine iontophoresis | |
| 40 | - | 14-120 | - | | As in A above | Dry skin measurements | Children (1-4 years old) Component activities from sources other than sweat obscured on dry skin [58] |
| 43 | - | 5-40 | - | | As in B above | Pilocarpine iontophoresis | |

[a] Scheme before sweat stimulation not specified.

The emf values for readings of 45 mM and 55 mM differ by only 5.17 mV at 25°C whereas at lower levels a 10 mM difference leads to a wider spread of emf readings; there is, of course, a smaller spread at high levels.

If the ion-selective electrode techniques are to have low incidences of false positive and false negative results, detailed attention is required of all the factors likely to cause variations in readings. Evaporation can be very rapid and from a collection matrix produce up to a two-fold rise in the concentration of sweat electrolytes in thirty seconds [111] and is a source of serious error [63].

The principal sweat stimulation techniques, that is, local heating by aluminium cylinder or gel pack, or pilocarpine iontophoresis, do not seem to lead to overall differences in results. The individual variations mentioned above can well relate to the quantity of sweat produced and skin washing procedures (Tables VI to IX). There is no indication, however, as to how far skin electrolytes can contribute to falsify results when sweat volumes are small [58].

## 8. DEFECTS AND HAZARDS OF ION-SELECTIVE ELECTRODE SWEAT TESTS

All sweat tests are fraught with difficulties and which arise at all the three steps outlined in Section 3.

### 8.1 General aspects

The European Working Group on Cystic Fibrosis reported a false-positive rate of five percent based on 25 000 tests with chloride ion-selective electrodes [113] while Brimblecombe and Chamberlain recorded a false-negative rate of at least fifteen percent [114]. This is disconcerting but does not detract from the fact that it is a simple, cheap, and, when used carefully, accurate method for the direct measurement of chloride in sweat on the skin. The results are, however, often irreproducible and apart from the quantity and nature of the sweat produced and skin washing procedure, technical problems arise with the care and maintenance of the electrodes. Thus, although the method is simple, conscientious attention must be given to ritual, while personnel, if not trained in laboratory techniques, must be closely supervised by those who fully appreciate the technology of ion-sensors.

### 8.2 Dietary and related factors

The net effect of false-negative sweat test readings is that the patient, unless otherwise diagnosed, misses early treatment. This might have happened with the four cases of low chloride ion-selective readings (mean 37 mM) noted by Fontaine and Ferriaux [131] even though the subsequent laboratory data were unequivocally high (127 mM). In 1972 Griffiths and Bull [96] reported the case of a child with proven CF with raised sodium yet normal sweat chloride being treated with cloxacillin. This has obvious implications regarding screening procedures and it was suggested that in at least some CF children the cloxacillin radical might be replacing some of the chloride in the sweat. Mothers on a course of such drugs could inadvertently administer the drug to infants either via breast milk or transplacental passage [133]. These non-electrode defects are disturbing in such cases [133,134] and it is considered prudent to establish the nature of any course of drugs by nursing mothers or children [133].

Intramuscular injection of aldosterone has been observed to lower sweat electrolyte levels of CF patients to the normal range [120]. Peripheral oedema has also been cited as a cause of false-negatives and possibly false-positives [92].

There is some evidence that false-positives can arise through excessive salt intake in the diet and a high laboratory test [83] result (51 and 54 mM) fell to the definitely normal range (33, 36, 29 and 35 mM) when the salt in the subject's food was reduced [20].

## 8.3 Hazards

Pilocarpine iontophoresis is the most commonly used sweat stimulation method for monitoring electrolytes with ion-selective electrodes [Tables VII and IX] and is also an integral part of the Orion 417 skin chloride test unit [86]. The safety aspects of the iontophoresis step have been questioned [111] and care is needed in the design of equipment in order to avoid burns [118].

An electric current passing through the body carries the risk of ventricular fibrillation, respiratory arrest, asphyxia, burns and blistering [111]. Blanc reported a fatality with an alternating current of 60 cycles and 46V [125]. However, the chance of a serious effect on the heart, respiratory centre or muscles by the direct current used in pilocarpine iontophoresis is regarded as unlikely [111]. Nevertheless, the intensity of an electric shock depends on the current passing through the vital body organs (Table X). The resistance of wet skin is typically $10^3$ ohm $cm^{-1}$ and a person standing on a wet floor and "holding" a direct current terminal at >50V can receive a current around 50 mA and experience a most unpleasant sensation [135].

TABLE X. Human physiological effects of electric currents [117]

| Current (mA) | Comments |
|---|---|
| 3 | Sensation threshold |
| 10 | Mild sensation |
| 100 | Severe shock, muscular paralysis and breathing problems |
| 200 | Death |

The situation is rather reassuring with pilocarpine iontophoresis since the recommended currents are normally 2 mA and only exceptionally 4 mA [66,118]. However, if the iontophoresis electrode is not in full contact with the skin site area, or the pads are not adequately saturated with the recommended solutions, or if the pads become dry, then skin burns may occur. Thus, if the patient shows any sign of discomfort, the iontophoresis blocks and pads should be checked and in cases of unusual discomfort the current should be reduced.

Blisters can be caused by the products of electrolysis formed near the skin surface even from pilocarpine nitrate at 1 mA current [111]. It is recommended that a stack of 10 to 12 filter papers saturated with pilocarpine nitrate be placed between the skin site and the iontophoresis block to prevent blister formation [111]. Blisters can be avoided by injections of chlorinergic drugs but are unfortunately often painful unless administration can be effected with a jet injector.

## 8.4 Ion-selective electrode factors

Ion-selective electrodes have found many medical applications, especially regarding fluoride and calcium measurements [128,136]. In 1965 Warwick and Hansen proposed the use of a chloride ion-selective electrode for the direct measurement of chloride in sweat while still on the skin site [51]. Apart from problems of skin washing and sweat stimulation the combination ion-selective electrode has to be used in a manner which is dramatically different from probably any other application of a macro ion-selective electrode to date. Thus, the sweat samples are extremely small (few $mm^3$) and the temperature profile between the sensor/skin interface, and also the reference electrode located some distance above the interface, is unknown. As with all ion-selective electrodes, interferences from other ions or, more significantly in this case, by substances excreted in the sweat, or by contamination of the skin site by soaps or cleansing emulsions containing, say, hexachlorophene, is always a possibility. The relative merits of pilocarpine nitrate and pilocarpine hydrochloride must be fully considered since the latter is recommended for the much used Orion 417 skin chloride system. The bromide salt must never be used.

The chloride ion-selective is free from the pH sensitivity of sodium glass electrode that might occur at sweat pH ranges and less prone to photoelectric effects than the silver billet electrode.

Leakage of the internal reference electrode solution from the cone and socket joint concentric to the chloride ion-selective sensing crystal in the 96-17 Orion combination electrode model is inevitable and indeed essential to its functioning. Thus, although the reference solution has been devised to provide a junction solution low in chloride (20 mM), a leakage rate of 0.5 to 1 $mm^3 min^{-1}$ arises through the cone-socket joint [62]. This, compared with a sweating rate of 0.17 to 0.5 $mm^3 min^{-1}$ in the iontophoresis area, can lead to significantly low chloride results and hence false negative decisions - a considerably more serious defect than false positive readings. Therefore, there is a strong case for allowing the sweat to actually accumulate before placing the electrode on the site as suggested by Szabo, Kenny and Lee [84].

# 9. A MODIFIED SWEAT CHLORIDE SCREENING TEST PROGRAMME

Sibenga has commented that laboratory errors undoubtedly cause more inaccuracies in diagnosis of cystic fibrosis than in other medical conditions [92]. The sweat test has become a corner-stone for CF diagnosis but odd cases are known with normal salt levels in the sweat [95,96]. The test based on a chloride ion-selective electrode comprises three quite distinct steps (Section 3). Some fundamental problems are associated with these steps:

(i) Chloride contamination from the potassium chloride in the reference compartment of the combination electrode.
(ii) Interference from Hibitaine or Cetavlon in antiseptic skin swabs.
(iii) Uncertainty of skin temperature in relation to an electrode pre-calibrated at, say 35°C.
(iv) Pressure variations at the skin-sensor interface and contact time.
(v) Care and maintenance of the electrode and iontophoresis unit.
(vi) Inadequate quantities of sweat - a problem pertaining to any method of sweat analysis.
(vii) Evaporation of sweat - an effect which overestimates chloride.

If the ion-selective electrode test is working properly, then the result should be the same as if the world-wide standard Gibson and Cooke method had been used. This section reviews recent work aimed at locating and measuring the various errors which cause deviation from that standard. With few exceptions [60-63,65,84] most determinations using a chloride ion-selective electrode have been undertaken using the method originally developed by Orion Research Incorporated [86]. In the same context virtually no attention has been directed towards an evaluation of skin site preparations, long-term care and maintenance of electrodes, and associated interference effects or the statistical treatment of analytical data [60-65].

## 9.1 Leakage of electrode filling solution

The leakage rates of filling solutions for four Orion 96-17 chloride ion-selective electrodes ranged from 0.12 to 0.83 $mm^3min^{-1}$ with a mean value of 0.56 $mm^3min^{-1}$ (sd = 0.26). The electrodes were dismantled and recharged between each leakage run.

Examples of calculated leakage effects at 15s and 30s electrode/skin contact times following a 1-minute sweat accumulation at an imaginary (but realistic) sweating rate are shown in Table XI. Leakage is thus usually insignificant when the electrode-skin contact time is <15 s but assumes importance if there is little sweat to measure of if sweat residues on the sensor surface from previous run(s) are not removed. The concentration of chloride in the

TABLE XI Calculated effects of leakage of internal filling solution on skin chloride readings [62]

| Skin contact time/min | [Chloride] in sweat/mM | | |
|---|---|---|---|
| | Expected[a] | Calculated for 2 Leakage Rates($mm^3min^{-1}$) | |
| | | 0.5 | 1.0 |
| 0.25 | 10.0 | 10.4 | 10.7 |
| | 20.0 | 20.0 | 20.0 |
| | 50.0 | 48.8 | 47.8 |
| | 100.0 | 96.9 | 94.1 |
| 0.50 | 10.0 | 10.6 | 11.2 |
| | 20.0 | 20.0 | 20.0 |
| | 50.0 | 48.1 | 46.5 |
| | 100.0 | 95.0 | 90.6 |

a Assumed chloride concentration on skin 1 minute before positioning electrode and sweating rate 2.5 $mm^3 min^{-1}$.

reference electrode barrel (20 mM) is about the middle of the normal range for sweat chloride. Thus, any error associated with its inevitable leakage will tend to increase the number of false-negative sweat test results [62].

## 9.2 Interference effects

No pH interference was observed over the range 4 to 9, the usual pH range for sweat being about 4.2 to 7.5 (Table IV). Neither urea, sodium sulphate, potassium nitrate, sodium citrate, hexachlorophene nor pilocarpine interfered with the chloride ion-selective electrode response. However, bromide and iodide would constitute serious interferences owing to rapid, metathetic interaction with silver chloride in the sensor membrane [127,137].

$$AgCl(s) + Br^- \rightleftharpoons AgBr(s) + Cl^-$$

This is especially important since certain skin antiseptics like Savlon and Cetavlon contain Cetrimide (cetyl trimethyl ammonium bromide) and which are occasionally impregnated in swabs. This fact must be considered in relation to possible false-positive results. The interference effect of Cetavlon disappeared after conversion of the Cetrimide component to the acetate form.

The general effect of Hibitaine (chlorhexidinegluconate) is to deposit a Hibitaine-sensitive coating on the chloride sensor membrane and response times lengthen. In any situation where Hibitaine is likely to have been used, then the chloride measured with chloride ion-selective electrodes must be seriously suspect since Hibitaine-induced intercept shifts of up to 100 mV can occur [62]. Even when chlorhexidine levels are not high enough to seriously affect an individual measurement, there is a cumulative effect in that the sensor surface becomes dull. This could well be an important cause of apparent "ageing" of sweat-testing chloride ion-selective electrodes.

Clearly, it is improper to use any of these antiseptic materials when washing the intended sweat stimulation site. Isopropanol, followed by two batches of distilled water proved to be a very satisfactory alternative for skin washing purposes [62,64]. Medi Swab$^R$ supplied by Pharmax, Bexley, Kent, UK, were used to apply these washes and also to wipe the skin areas [64].

## 9.3 Reproducibility and stability of electrodes

The precision of replicate measurements for three sodium chloride standards at 25°C are presented in Table XII in mV rather than concentrations - an error of 1 mV representing about four percent error in concentration. Thus, the standard deviations for sweat chloride levels would represent a reproducibility of about 1%.

The long term stability of these electrodes is excellent and those examined withstood repeated dismantling and reassembly with <1 mV differences in the emf readings of a given test solution. This is especially relevant to any screening programme since it is imperative that the electrode be dismantled and washed after any batch of subjects has been examined.

TABLE XII. Replicate measurements with an Orion 96-17 chloride ion-selective electrode at 25°C [62]

| [Sodium chloride] /mM | Potential/mV Mean(N=50) | Range | sd |
|---|---|---|---|
| 1 | +67.01 | 66.00-67.90 | 0.43 |
| 10 | +10.71 | 10.50-11.00 | 0.14 |
| 100 | -44.18 | 44.00-44.50 | 0.20 |

Transfer between test solutions of equal concentration consistently gave reproducible results, for example, 50 readings at the 10 mM chloride level fell into a range of just 0.50 mV [62].

### 9.4 Temperature effect

Skin temperatures are commonly 33 to 35°C but 28°C is not unusual in adults. Thus, the Orion combination electrode, for example, will be subjected to a range of temperature conditions yet will have been calibrated at, say 37°C. During the *in situ* measurement the electrode membrane sensor surface will be a degree or so below skin temperature and in turn the rest of the combination body including the reference electrode a few degrees below that of the sensor. Therefore, in the absence of a completely thermostatted electrode maintained at skin temperature, errors seem inevitable. However, there will be some self-compensatory feature since for chloride levels in sweat >20 mM, the effect of reference filling solution leakage opposes that of temperature [62]. This problem, of course, does not arise in the Gibson-Cooke technique [66].

No allowance has usually been made for sweat on the skin being at a temperature greater than the ambient surroundings. This is generally considered unimportant because the period of pressure-skin contact (10 to 20 s) is too brief to appreciably affect the temperature of the electrode crystal membrane [55].

The results presented in Table XII were conducted with bulk, thermostatted standards. More realistic results obtained with very small sample volumes and no conscious temperature control are shown in Table XIII using an electrode calibrated at 20°C. Measurements were then taken on two standard chloride solutions, namely, in bulk, as 2 $mm^3$ drops on Nescofilm on the bench, as 2 $mm^3$ drops on Nescofilm strapped to an adult forearm, and as 1 $mm^3$ drops on the well washed skin site of an adult forearm [63]. Accuracy and precision are not significantly worse on 1 $mm^3$ control samples placed on skin than in bulk solution (Table XII).

TABLE XIII. The performance of the Orion 94-17 combination chloride ion-selective electrode with various sodium chloride situations [63]

| Form of Sample | Sodium Chloride Sample | | Measured [Chloride][a] /mM | | Error (%) |
|---|---|---|---|---|---|
| | Concentration /mM | Volume /mm$^3$ | Mean | sol(N) | |
| Bulk solution in beaker at room temperature(17°C) | 20 | - | 21.5 | 0.36 (5) | + 7.5 |
| | 100 | - | 105.1 | 1.04 (5) | + 5.1 |
| Resting on Nesco-film on bench at 17°C | 20 | 2 | 22.0 | 0.22 (5) | +10.0 |
| | 100 | 2 | 106.2 | 2.86 (5) | + 6.2 |
| Resting on Nesco-film on top of skin | 20 | 2 | 17.6 | 0.26 (5) | -17.0 |
| | 100 | 2 | 91.9 | 0.84 (5) | - 8.1 |
| Resting on skin (33°C) | 20 | 1 | 18.0 | 0.45 (7) | -10.0 |
| | 100 | 1 | 106.1 | 6.83 (9) | + 6.1 |

a Electrode calibrated at 20°C with standard chlorides

### 9.5 Pressure effect on the electrode at the sweat site

Virtually no information has been available regarding the effect of pressure on the electrode at the skin site [61,63,138]. Clinical cases are known where pressure variations altered the skin chloride meter readings by 20 mM but independent evidence established that insufficient sweat had collected at the sweat site for reliable readings to have been taken.

The effect of varying pressures on the electrode readings has been investigated on sweating skin sites and standard 20 and 100 mM sodium chloride solutions (1 mm$^3$) placed on well washed adult skin sites [63]. When readings were steady using light pressure, the electrode was then forced downwards until the skin was depressed by about 1 cm. Typically, this operation caused about 1 mV changes in the recording taken as shown on a Servoscribe 1S Recorder. On releasing the pressure the original potential was rapidly and largely restored. Thus, a mean value of 18.0 mM (sd = 0.45) was obtained in seven separate runs with 20 mM sodium chloride (1 mm$^3$), the range being 17.5 to 18.9 mM [63].

More recently, Warwick and Hansen [61] conducted similar chart recorder runs but where the Orion 97-17 chloride electrode was deliberately top-loaded with weights ranging from 50 to 2000g. The supposed pressure sensitivity of such electrodes was again shown to be false, that is, the electrode is slightly and temporarily sensitive to a pressure change but will give the same result with any constant pressure - at least within their 40 to 1670 g cm$^{-2}$ range [61].

## 9.6 Care and Maintenance of the Orion 417 Sweat chloride measuring system

It is imperative that the chloride ion-selective electrode be immediately dismantled, thoroughly washed and dried after every measuring session and only reassembled just prior to the next measuring session. Unless this routine chore is conscientiously followed, serious corrosion of the sensor membrane is likely to arise.

Close attention must also be given to both pilocarpine iontophoresis electrode blocks. The batteries of the skin meter also need to be checked and charged as appropriate [64].

## 9.7 Sweat stimulation and accumulation

The original method for conducting iontophoresis was inadequate in several respects and has therefore been modified by several workers [64,65,84]. Thus, the original concentration of pilocarpine nitrate (3 mM) has been increased to 9.5 [65], 23.8 [84] and 100 mM [63], respectively, and the original potassium sulphate (57 mM) increased to 600 mM [63] or replaced by 0.12 mM sodium bicarbonate[65].

The iontophoresis anode and cathode blocks have also been modified so that the metal surfaces lie flush with the skin and a groove cut between them to prevent short circuiting [63]. Cellulose phosphase filter papers were used instead of the original, rather expensive, Orion felt pads which tended to dry out in storage. The anode

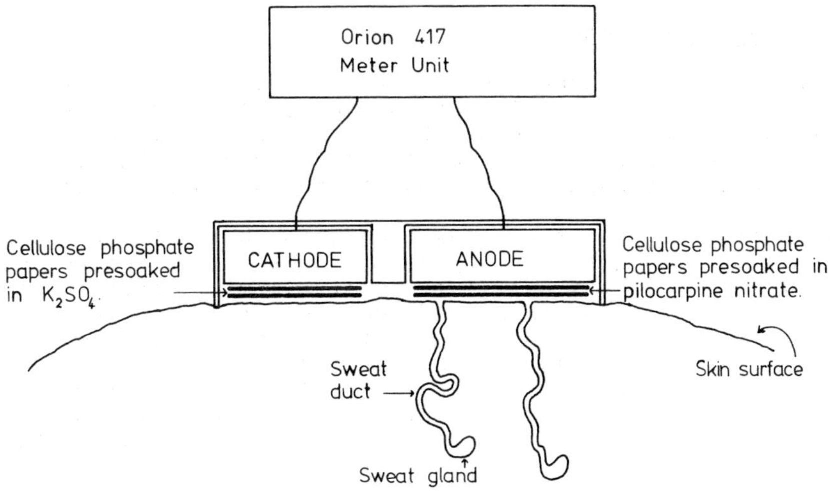


Fig.1 Schematic presentation of a pilocarpine iontophoresis run

pads (1.5 x 2 cm) were soaked with 100 mM pilocarpine nitrate (0.1 $cm^3$) and dried. The smaller cathode pads (1 x 2 cm) were similarly treated with 600 mM potassium sulphate (0.1 $cm^3$) [63].

Sweating was then induced at 1.5 mA supplied by the Orion 417 meter for 5 minutes and which in principle should transfer about 1 mg of pilocarpine to the subcutaneous areas for sweat production (Fig. 1). Pilocarpine hydrochloride supplied with the Orion 417 meter kit has been replaced by the pilocarpine nitrate in order to avoid possible chloride contamination [63,65].

This modified pad system was so satisfactory that a baby asleep at the start of a test usually remained so throughout [64]. Patients treated with a similarly modified iontophoresis system are reported to suffer no discomforture, nor do they develop burns [65]. Ordinary filter papers are not recommended because slight discomfort was reported by adult volunteers [64].

In the zero-time sweat collection period recommended by Orion Research Incorporated, the user is advised to wipe off the skin-iontophoresis site and just apply the electrode. Unfortunately, in many cases the sweat may not really have accumulated at that stage and incorrect readings may be obtained [65]. In an attempt to overcome this serious drawback, the Ionto-block is removed, the site washed twice with gauze soaked in deionised water to remove iontophoresis debris, dried with a tissue and covered with transparent, impervious Nescofilm. After three minutes this taped Nescofilm is removed and the electrode measurement taken immediately on the visible sweat site with just light downward pressure [64]. Finley *et al.* [65] employ a ten-minute period for sweat accumulation, rather than a zero-time Orion type collection, and they used a Kimble disposable plastic cap cover instead of Nescofilm.

### 9.8 Evaporation effects

Evaporation from uncovered sweat samples before positioning the ion-selective electrode will increase the real chloride concentration to an extent dependent on temperature and exposure time. The problem, of course, cannot be completely eliminated but is certainly reduced by allowing sweat to accumulate in the covered site [64,65].

Pairs of micro electrodes inserted through the taped Nescofilm on to sweating skin sites show promise for continuous studies throughout the sweat period [139]. Suitable micro chloride ion-selective electrodes and micro reference electrodes have been fabricated by dipping the ends of drawn out melting point tubes into Corning chloride liquid ion-exchanger (0.4 g) PVC (0.17 g) tetrahydrofuran (6 $cm^3$) and 2 percent agar gel - 0.1 mol $dm^{-3}$ potassium nitrate respectively. Each electrode was then backfilled with internal 0.1M potassium chloride solution and an internal silver-silver chloride wire sealed in with plasticine.

## 10. RECENT SCREENING PROGRAMMES

The results of three recent sweat programmes based on chloride ion-selective electrodes and/or the Gibson and Cooke reference technique are outlined in the following sections.

## 10.1 The Cardiff chloride test programme

The procedural modifications discussed in Section 9 have been adopted for screening neonates at the University of Wales Hospital, Cardiff. Routine screening was attempted for 1218 newborn infants but five were incomplete because the infants were removed for paediatric checks. A further eight cases failed to sweat so the ultimate programme was based on 1205 chloride ion-selective electrode tests [64].

The results of screening gave a mean sweat chloride of 34.4 mM (± 0.36 SE). There was little difference between the 594 males (mean 33.4 mM ± 0.51 SE) and the 611 females (mean 34.9 mM ± 0.52 SE), the t-distribution value of 2.06 being close to that of 1.96 for P = 0.05. A similar result also applies to breast feeding (482 cases: mean 34.4 mM ± 0.50 SE) compared with bottle feeding (492 cases: mean 32.0 mM ± 0.50 SE). The procedure used is described in the Appendix.

The results were classified according to age and sweat chloride (Tables XIV and XV), while the relation between a subjective estimate of sweating, sweat chloride and age is given in Table XVI.

TABLE XIV. Sweat chloride levels based on age distribution [64]

| Age Group /day | Number in Group | Mean [Chloride][a] /mM (±SE) |
|---|---|---|
| 0-0.5 | 62 | 38 (1.3) |
| 0.5-1.0 | 157 | 39 (1.1) |
| 1.0-1.5 | 160 | 39 (0.85) |
| 1.5-2.0 | 174 | 37 (0.88) |
| 2.0-2.5 | 179 | 34 (0.90) |
| 2.5-3.0 | 118 | 33 (1.1) |
| 3.0-3.5 | 176 | 32 (1.5) |
| 3.5-4.0 | 56 | 30 (1.4) |
| 4.0-4.5 | 46 | 29 (1.8) |
| 4.5-5.0 | 46 | 27 (1.8) |
| 5.0-20 | 131 | 27 (1.0) |

a Range = 3 to 100 mM

TABLE XV. Distribution of chloride level in sweat samples [64]

| Chloride group/mM | Number in Group | Mean age/day (±SE) |
|---|---|---|
| < 5 | 2 | 0.9 (0.06) |
| 5-10 | 24 | 3.7 (0.50) |
| 10-15 | 62 | 4.3 (0.42) |
| 15-20 | 115 | 3.5 (0.40) |
| 20-25 | 134 | 3.1 (0.17) |
| 25-30 | 136 | 2.7 (0.14) |
| 30-35 | 138 | 2.7 (0.15) |
| 35-40 | 136 | 2.7 (0.19 |
| 40-45 | 160 | 2.0 (0.13) |
| 45-50 | 241 | 1.8 (0.08) |
| 50-55 | 34 ) | 1.7 (0.19) |
| 55-60 | 13 ) b | 1.9 (0.26) |
| 60-100[a] | 10 ) | - |

a The individual values with ages (in days) in parentheses are:
65* (1.04)
70* (5.48); 70* (3.25); 70* (0.85); 70 (20); 70 (1.25);
90 (3.17); 90 (0.58)
100*(0.96); 100 (3.04)
*Trace sweating only
b False positives

TABLE XVI. Mean sweat chloride levels according to degree of sweating [64]

| Sweating class[a] | Number in class | [Chloride] mM | | Age/day | |
|---|---|---|---|---|---|
| | | Mean(±SE) | Range | Mean(±SE) | Range |
| 1 | 3 | 45 (14) | 10-65 | 1.6(0.33) | 1.0-2.3 |
| 2 | 850 | 36(0.41) | 3-100 | 2.3(0.07) | 0.1-20 |
| 3 | 320 | 30(0.68) | 7-70 | 3.3(0.13) | 0.1-20 |
| 4 | 32 | 30(2.3) | 8-70 | 3.3(0.31) | 0.8-7.6 |

a Sweating class:
1 = No sweating; 2 = trace; 3 = moderate; 4 = excessive.

Sweat was assessed subjectively on a four point scale ranging from 'no sweat' to 'excessive' and it is relevant that it was generally more difficult to obtain sufficient sweating in babies about one day old.

Attempts were made with some babies to actually collect sweat after the non-selective electrode measurement by washing the skin site twice with gauze pads soaked in deionised water and drying with a fresh tissue. This was intended for a comparative chloride assay with a chloridometer but the samples of about 15 mg were below the minimum 100 mg recommended [66]. The planned correlation study was, therefore, undertaken with sufficient sweat from 23 adult volunteers when a correlation coefficient of 0.96 was obtained between chloride measured with an Orion chloride ion-selective electrode and the Radiometer Model CMT 10 Chloridometer. This is to be compared with $r = 0.89$ for 47 normal subjects and $r = 0.86$ for 50 cystic fibrosis cases reported by the Tucson group using virtually the same two techniques [68].

Instances of high sweat chloride, the difficulty in obtaining enough sweat and the extent to which the Orion 417 Meter Unit can be used in conjunction with meconium analysis for reducing the incidence of false positive results are important factors in any screening programme. The fiftyseven cases (5 percent) in Table XV with sweat chloride levels >50 mM may be considered false positives. However, in no instance did the Boehringer-Mannheim meconium test strip, the electro-immunoassay of albumin or the determination of the albumin: $\alpha_1$-antitrypsin ratio give a positive indication of CF. Conversely, three of the four cases included in this test programme, and independently found to have meconium albumin >5 mg $g^{-1}$ dry mass, gave 23, 26 and 48 mM sweat chloride respectively; the fourth case (<1 day old) did not give sweat.

Sweat chloride fell with age (Table XV) and sweat induction was generally difficult in very young babies (Table XVI). Furthermore, the eight cases where no sweating was possible were all <1.3 days old (six <0.8 day) and this indicates that screening should be delayed until at least the second day after birth.

No cases of CF were detected among the 1205 babies screened. However, it was possible to examine 2 proven CF cases born in other local hospitals. In Case 1 (25 days old) the chloride ion-selective electrode test gave 90 mM chloride compared with 90 mM using the chloridometer. The meconium data, namely the Boehringer-Mannheim test; the albumin (469 mg $g^{-1}$ dry mass) and the albumin:$\alpha_1$-antitrypsin ratio (7.2) were all in the positive category while the Gibson and Cooke sodium was 106 mM.

Case 2 (21 days old) presented with meconium ileus was severely jaundiced and dehydrated. A confirmatory Gibson and Cooke test [66] was not possible owing to insufficient sweating, but the chloride ion-selective electrode test gave 70 mM and necropsy examination also indicated a positive CF case.

## 10.1 The Tucson chloride test programme

The Tucson test programme covered 172 normal subjects and 75 patients with cystic fibrosis and each individual was studied in duplicate using left and right forearm skin sites. The original method described for the application of an Orion chloride ion-selective electrode was also modified (see Section 9.7) using two iontophoresis cells on left and right arms simultaneously and compared with the reference quantitative method of Gibson and Cooke.[66] The modified pilocarpine iontophoresis procedure was studied for reproducibility in normal individuals (Table XVII), reproducibility between left and right arm sites in individuals (Table XVIII), and compared with the quantitative Gibson and Cooke technique both in normal subjects and in patients with cystic fibrosis (Table XIX).

TABLE XVII. Sweat chloride levels (mM) determined with an electrode for three normal individuals over five days[65]

| Normal Subject[a] | Day | | | | | Mean | s.d. |
|---|---|---|---|---|---|---|---|
| | 1 | 3 | 3 | 4 | 5 | | |
| 1 | 15 | 13 | 14 | 16 | 15 | 14.6 | 1.14 |
| 2 | 8 | 6 | 7 | 6 | 9 | 7.2 | 1.3 |
| 3 | 22 | 20 | 23 | 24 | 21 | 22.0 | 1.58 |

a Ages not given but probably <18 years

The reproducibility of measurements using the modified Orion procedure was considered satisfactory for normal individuals whose age, though not specified, is probably greater than one month (Table XVII). Much greater variations have been reported for the chloride and sodium sweat levels for normal neonates in the first four days of life but which settled down thereafter in the first month to 26 mM for chloride [119]. The slight statistical differences between the means for 125 normal and 25 cystic fibrosis subjects measured by the modified Orion procedure (Table XVIII) was not

TABLE XVIII Statistical data for the analysis of sweat from left and right arm sites using an ion-selective electrode [65]

| Parameter | Subject type | |
|---|---|---|
| | Normal | Cystic Fibrosis |
| Right arm (mean)/mM[Cl] | 13.752 | 109.000 |
| Left arm (mean)/mM[Cl] | 13.224 | 105.000 |
| Difference of means | 0.528 | 4.000 |
| sd of difference | 2.586 | 8.421 |
| SE of difference | 0.231 | 1.684 |
| Correlation | 0.931 | 0.899 |
| SE of estimate | 2.479 | 8.156 |
| Intercept | 0.971 | 9.246 |
| Slope | 0.910 | 0.907 |
| t value (P = 0.05) | 2.010 | 2.123 |
| Number of subjects | 125 | 25 |
| Age of subjects | 1month-18years | 3months-25years |

TABLE XIX Statistical data for the analysis of sweat from left and right arm sites using two different methods [65]

| Parameter | Subject type | |
|---|---|---|
| | Normal | Cystic Fibrosis |
| Mean right arm(Gibson and Cooke)/mM[Cl] | 14.149 | 98.720 |
| Mean left arm(Chloride electrode)/mM[Cl] | 15.234 | 103.600 |
| Difference of Means | -1.085 | -4.880 |
| SD of difference | 2.880 | 8.482 |
| SE of difference | 0.420 | 1.696 |
| Correlation | 0.894 | 0.864 |
| SE of estimate | 2.890 | 8.649 |
| Intercept | 2.006 | 8.654 |
| Slope | 0.934 | 0.962 |
| t value (P = 0.05) | -2.241 | -2.88 |
| Number of subjects | 47 | 50 |
| Age of subjects | 14months-17years | Not listed |

medically important. Again the slight differences evident between the electrode and the reference quantitative techniques were not considered to be clinically important (Table XIX).

Finley *et al.* [65] concluded that the proposed modification is simple, reliable, reproducible and accurate and which compares well with the reference method and has not been clinically misleading [65].

### 10.3 The Minnesota chloride test programme

Another test series has been conducted [60] with 740 healthy subjects and 146 cystic fibrosis patients using both the original Orion technique and a gravimetric chloride titration technique [140]. Four sweat tests were performed on each subject in one sitting - one ion-selective electrode and one gravimetric on each arm. The statistics for both groups of subjects (Table II) are for 1480 pairs of tests on healthy subjects and 292 pairs of tests on cystic fibrosis patients. The linear regression equation for the combined data was ELECTRODE = 0.87[GRAVIMETRIC] + 4.5, and the correlation coefficient was 0.91 [60].

The two techniques gave comparable values, ogives, equivalence points and overlap for both subject groups. This study is claimed to validate the chloride ion-selective electrode test relative to the gravimetric technique. Both techniques can be used as a diagnostic support for cystic fibrosis but the technical skill needed for either test requires meticulous care and constant practice and duplicate runs on any subject are essential to reduce the risk of falst positives and false negatives [60].

## 11. CONCLUSION

Mass CF screening of populations is feasible within the limitations of busy hospital ward routines by personnel prepared to follow simplified, proven procedures for using and maintaining the Orion 417 Sweat Chloride Meter System. Despite some reassuring correlations for the electrode method and the world-wide Gibson and Cooke reference method, there are clearly factors, such as evaporation of sample and/or insufficient sweating, to discourage total confidence in the application of ion-selective electrodes for confirmatory diagnosis.

Shwachman and Mahmoodian behove all clinical laboratory directors to re-examine and standardize their procedures and use the recommendations of the Cystic Fibrosis Foundation. Physicians should be made aware of the current situation and send their patients only to laboratories that perform the test reliably [141].

Acknowledgement - The authors are grateful for the enthusiastic encouragement of Dr.P.T.Bray and to Mrs.P.M.Bevan for her tolerance and patience, particularly with the many Tables to set specifications.

## 12. REFERENCES

1. H.Shwachman, In: *Pulmonary Disorders. Disorders of the Respiratory Tract in Children,* Ed. E.L.Kendig, Saunders, 2nd Edition, Volume 1, (1972). Page 524.

2. L.Nielsen, *Acta Paed. Scand.,* 61, 377 (1972).

3. S.W.Wright and N.F.Marton, *Amer.J.Human Genetics*, 30, 157 (1968).

4. C.R.R.M.Reddy, C.S.Devi, A.M.Anees, D.P.Murtley and G.E.Reddy, *J.Tropical Medicine and Hygiene*, 73, 59 (1970).

5. P.Selander, *Acta Paed.Scand.*, 57, 65 (1962).

6. H.Kollberg, *Cystik Fibrosis i Sverige Speciellt Med Hansyn Till Tidig Diagnos.* (Thesis), Uppsala University (1974).

7. H.Kollberg, *Proc. VIIth International Cystic Fibrosis Congress,* Paris, May 31 to June 3 (1976). Page 144.

8. C.T.Gürson, H.Sertel, M.Gürkan and S.Pala, *Helv.Paed.Acta,* 28, 165 (1973).

9. M.C.Goodchild, J.Insley, D.I.Rushton and H.Gage, *Archs.Disease Childhood*, 49, 739 (1974).

10. R.Prosser, *Proc.VIIth International Cystic Fibrosis Congress,* Paris, May 31 to June 3 (1976). Page 151.

11. P.T.Bray, *Report to Cystic Fibrosis Research Trust*, (London) (1975).

12. R.Prosser, H.Owen, F.Bull, B.Parry, J.Smerkiniche, H.A.Goodwin and J.Dathan, *Archs.Disease Childhood*, 49, 597 (1974).

13. M.S.Honeyman and M.G.Sirken, *Amer.J.Human Genetics*, 17, 461 (1965).

14. M.S.Honeyman and E.Siker, *Connecticut Health Bulletin*, 75, 275 (1961).

15. P.M.Conneally, A.D.Merritt and V.K.Moses, *Program of Amer.Soc. Human Genetics*, Aug. 23rd, p.42, (1965).

16. C.U.Lowe, C.D.May and S.C.Reed, *Amer.J.Diseases Children*, 78, 349 (1949).

17. L.L.Kulczyski, I.B.MacLeod and H.Shwachman, *Amer.J.Diseases Children*, 100, 174 (1960).

18. E.R.Kramm, M.M.Crane, M.G.Sirken and M.L.Brown, *Amer.J.Public Health*, 52, 2041 (1962).

19. H.O.Goodman and S.C.Reed, *Amer.J.Human Genetics*,4, 59 (1952).

20. W.J.Warwick and J.Koroshec, *Minnesota Medicine*, 51, 153 (1968).

21. A.G.Steinberg and D.C.Brown, *Amer.J.Human Genetics*, 13, 416 (1960).

22. A.Bianchi-Bosisic Righetti, M.Migliavacca, L.Prampolini and A.Giunta, *Proc. VIIth International Cystic Fibrosis Congress,* Paris, May 31 to June 3 (1976). Page 153.

23. M.R.Concepcion, *Proc. VIIth International Cystic Fibrosis Congress*, Paris, May 31 to June 3 (1976). Page 160.

24. P.G.Robinson, *Proc. VIIth International Cystic Fibrosis Congress*, Paris, May 31 to June 3 (1976). Page 175.

25. D.A.Stanley, *Abstracts: International Symposium on Selective Ion-sensitive Electrodes*, UWIST, Cardiff, 9-12 April, Paper No.53 (1973)

26. W.J.Warwick, *J.Amer.Med.Ass.*, 198, 59 (1966).

27. W.J.Warwick and S.Monson, *Modern Problems in Paediatrics*, 10, 353 (1967).

28. W.J.Warwick, *Minnesota Medicine*, 50, 1049 (1967).

29. L.W.Matthews, C.F.Doershuk, M.Wise, G.Eddy, H.Nudelman and S.Spector, *J.Pediatrics*, 65, 558 (1964).

30. L.Burton, *The Family Life of Sick Children*, Routledge and Kogem Paul, London (1975).

31. W.Hamilton and C.E.Gray, *Report to Sixth Annual Meeting of European Working Group for Cystic Fibrosis*, Dublin, Ireland. (1975).

32. Cystic Fibrosis Research Trust, "To Conquer Cystic Fibrosis: The First Fifteen Years". CF. Trust, Bromley, Kent. (1980).

33. R.C.Talamo; In Ref.89, Page 8.

34. W.J.Warwick, *Proc. VIIth International Cystic Fibrosis Congress*, Paris, May 31 to June 3 (1976). Page 182.

35. L.Kopito, A.Mahmoodian, R.R.W.Townley, K.T.Khaw and H.Shwachman. *New England J.Med.*, 272, 504 (1965).

36. P.J.Leonard and W.P.Morris, *Archs Disease in Childhood*, 47, 495 (1972).

37. P.G.Robinson and R.B.Elliot, *New Zealand Medical J.*, 79, 1024 (1974).

38. M.N.Green and H.Shwachman, *Pediatrics*, 41, 989 (1965).

39. U.Stephan, E.W.Busch, H.Kollberg and K.Hellsing, *Pediatrics*, 35, 55 (1975).

40. H.Kollberg and K.Hellsing, *Acta Paediatr.Scand.*, 64, 477 (1975).

41. A.Rule, D.Barar and H.Shwachman, *Pediatrics*, 45, 847 (1970).

42. R.Gilly, C.I.Ducos, I.Garcia, D.Periot and A.Revol, *Proc. VIIth International Cystic Fibrosis Congress*, Paris, May 31 to June 3 (1976). Page 139.

43. A.R.R.Cain, A.M.Deall and T.C.Noble, *Archs Disease Childhood*, 37, 131 (1972).

44. D.Kaiser, H.P.Morer and P.Hammershlag, *Proc. VIIth International Cystic Fibrosis Congress*, Paris, May 31 to June 3 (1976). Page 148.

45. H.C.Ryley, L.Neale, T.D.Brogan and P.T.Bray, *Archs Disease Childhood*, 49, 901 (1974).

46. H.Stephan and E.W.Busch, *Pediatrie*, 46, 63 (1974).

47. B.Gottschalk, *Proc. VIIth International Cystic Fibrosis Congress*, Paris, May 31 to June 3 (1976). Page 162.

48. K.Hellsing and H.Kollberg, *Scand.J.Clin.and Lab.Investigation*, 33, 333 (1974).

49. D.Lawson, P.Westcombe and B.Saggers, *Archs Disease Childhood*, 44 715 (1969).

50. U.N.Wiesman, T.F.Boat and P.A.Di Sant' Agnese, *J.Pediatrics*, 76, 444 (1970).

51. W.J.Warwick and L.Hansen, *Pediatrics*, 36, 261 (1965).

52. W.J.Warwick, *J.Amer.Medical Association*, 198, 59 (1966).

53. C.T.Gillespie and A.L.Steeves, *Annals Royal College of Physicians and Surgeons, Canada*, 2, 34 (1969).

54. L.Hansen, M.Buechele, J.Koroshec and W.J.Warwick, *Amer.J.Clin. Path*, 49, 834 (1968).

55. L.Kopito and H.Shwachman, *Pediatrics*, 43, 794 (1969).

56. W.J.Warwick, *Proc. 13th International Congress on Pediatrics, Vienna*, 5, 101 (1971).

57. A.G.Chernian and J.G.Hill, *Clin.Chem*, 17, 652 (1971).

58. M.Green, H.Behendt and G.Libien, *Clin.Chem.*, 18, 497 (1972).

59. J.Steinrud, S.Winkel and E.W.Flensborg, *Danish Medical Bull.*, 21, 251 (1974).

60. W.J.Warwick, L.Hansen, I.Viela and J.Matheson, *Amer.J.Clin.Path.*, 72, 142 (1979).

61. W.J.Warwick and L.G.Hansen, *Clin.Chem.*, 24, 381 (1978).

62. P.T.Bray, G.C.F.Clark, G.J.Moody and J.D.R.Thomas, *Clin.Chim. Acta*, 77, 69 (1977).

63. P.T.Bray, G.C.F.Clark, G.J.Moody and J.D.R.Thomas, *Clin.Chim. Acta*, 80, 333 (1977).

64. P.T.Bray, G.C.F.Clark, G.J.Moody, G.Thomas and J.D.R.Thomas, *Archs Disease Childhood*, 53, 483 (1978).

65. P.R.Finley, J.A.Dye, D.A.Lichti, J.M.Byers and R.J.Williams, *Amer.J.Clin.Path.*, 69, 615 (1978).

66. L.E.Gibson and R.E.Cooke, *Pediatrics*, 23, 545 (1959).

67. R.B.Goldbloom and P.Sekeli, *New England J.Medicine*, 269, 1349 (1963).

68. H.Shwachman, R.Dunham and W.R.Phillips, *Pediatrics*, 32, 85 (1963).

69. H.Shwachman, A.Mahmoodian, L.Kopito and K.T.Khaw, *J.Pediatrics*, 66, 433 (1965).

70. L.E.Fite, *USAEC Contract*, Na-At-(40-1)-3796 (1968).

71. L.E.Fite, R.E.Wainerdi, H.P.Yule, G.M.Harrison, G.Bickers and R.Doggett. *Proc.Int.Conference on Modern Trends in Activation Analysis.* A. and M.College of Texas, Austin, Texas (1968). Page 843.

72. M.T.C.Gandolfo, and A.Strigazzi, *Panminerva Medica*, 11, 582 (1969).

73. S.Stamm, G.L.Woodruff and A.L.Babb, *Cystic Fibrosis Club Abstracts (Atlantic City)*, 10, 22 (1967).

74. G.L.Woodruff, A.L.Babb, W.E.Wilson and Y.Yamamoto, *Nuclear Applications*, 6, 352 (1969).

75. H.Kollberg and O.Landstrom, *Acta Paed.Scand.*, 63, 405 (1974).

76. H.Kollberg and G.Ekbohm, *Acta Paed.Scand.*, 63, 411 (1974).

77. H.Kollberg, *Proc. VIIth International Cystic Fibrosis Congress*, Paris, May 31 to June 3 (1976). Page 171.

78. G.F.Johnsen, M.F.Thompson, S.Fetteroff and A.N.Fasono, *Pediatrics*, 47, 88 (1971).

79. M.A.Nicolaidou, E.Apotalopoulou and V.Samara, *J.Nuclear Medicine*, 7, 153 (1966).

80. M.A.Nicolaidou, A.P.Grimanis and M.Vassilaki-Grimani, *Proc. 13th International Congress on Pediatrics, Vienna*, 5, 509 (1971).

81. S.Wotman, I.D.Mandel, J.Mercadante and C.R.Denning, *Archs.Oral Biol.*, 16, 663 (1971).

82. R.C.Darling, P.A.Di Sant'Agnese, G.A.Perera and D.H.Anderson, *Amer.J.Med.Sciences*, 225, 67 (1953).

83. P.A.Di Sant' Agnese, R.C.Darling, G.A.Perera and E.Shea, *Pediatrics*, 12, 549 (1953).

84. L.Szabo, M.A.Kenny and W.Lee, *Clin.Chem.*, 19, 727 (1973).

85. Orion Research Inc., *Newsletter*, 1, 13 (1969).

86. Orion Research Inc., *Model No.417 Skin Chloride Measuring Instruction Manual*, Cambridge Mass., (1971).

87. P.T.Bray, G.C.F.Clark, G.J.Moody,and J.D.R.Thomas, *Proc. VIIth International Cystic Fibrosis Congress*, Paris, May 31 to June 3, (1976). Page 165.

88. M.Green, A.V.Mascia and H.Behrandt, *J.Lab.Clin.Med.*, 86, 494 (1975).

89. Frankel G.A.P. Conference on *Problems in Sweat Testing*, (Chairman L.F.Gibson), Hilton Head, S.Carolina, February 6-7 (1975).

90. A.L.Tarnoky, V.M.Bayliss, H.J.M.Bowen, R.N.Curnow and J.Sherington, *Clin.Chim.Acta*, 69, 505 (1976).

91. P.A.Di Sant'Agnese: In Ref.89. Page 6.

92. M.S.Sibinga: In Ref.89. Page 8.

93. D.N.Raine, *Lancet*, II, 996 (1974).

94. J.J.Cogswell, R.A.Rivdon and B.Taylor, *Archs Disease Childhood*, 49, 520 (1974).

95. J.K.Sarsfield and J.M.Davies, *Archs Disease Childhood*, 50, 463 (1975).

96. A.D.Griffiths and F.E.Bull, *Archs Disease Childhood*, 47, 132 (1972).

97. P.A.Di Sant'Agnese, *J.Pediatrics*, 81, 196 (1972).

98. H.Shwachman, H.Leubner and P.Catzel, *Advances in Pediatrics*, 7, 249 (1955).

99. K.A.Misch and H.M.Holden, *Archs Disease Childhood*, 33, 179 (1958).

100. A.M.Mauer and C.D.West, *Amer.J.Diseases Children*, 92, 160 (1956).

101. O.Schales and S.S.Schales, *J.Biol.Chem.*, 140, 879 (1941).

102. E.Cotlove, H.V.Trantham and R.L.Bowan, *J.Lab.Clin.Med.*, 50, 358 (1958).

103. D.V.Davies Ed. *Gray's Anatomy*, 34th Edition, Longman, (1972). Page 1350.

104. R.A.Ellis, *J.Cell.Biol.*, 27, 551 (1965).

105. L.E.Gibson and P.A.Di Sant'Agnese, *J.Pediatrics*, 62 855 (1963).

106. L.E.Gibson, *J.Pediatrics*, 81, 193 (1972).

107. W.G.Randell, *Amer.J.Physiol.*, 150, 365 (1947).

108. W.C.Wilson, *Brain*, 57, 422 (1934).

109. I.L.Schwartz and J.H.Thaysen. *J.Clin.Invest*, 35, 114 (1956).

110. D.N.Crozier, *Paediatric Clinics N.America*, 225, 67 (1974).

111. V.Schwarz, C.H.Sutcliffe and P.Style, *Archs Disease Childhood*, 43, 695 (1968).

112. J.Steinrud, *Acta Paed.Scand.*, 61, 378 (1972).

113. P.T.Bray, *Report to Second Annual Meeting of European Working Group for Cystic Fibrosis*, London (1971).

114. F.S.W.Brimblecombe and J.Chamberlain, *Lancet*, II, 1428 (1973).

115. H.Shwachman and A.Mahmoodian, *J.Pediatrics*, 69, 285 (1966).

116. G.H.Bell, J.N.Davidson and H.Scarborough, *Textbook of Physiology and Biochemistry*, 54th Edition. Livingstone, London (1965). Page 643.

117. R.Dobson: In Ref.89. Page 9.

118. F.A.Ibbott, *Clin.Chem.*, 5, 101 (1965).

119. W.Flensborg, and J.Steinrud, *Acta Paed.Scand.*, 61, 378 (1972).

120. P.A.Di Sant'Agnese: In Ref.89. Page 9.

121. Y.Kuno, *Human Perspiration*, Thomas, Illinois, (1956). Page 223.

122. W.Flensborg, *Acta Paed.Scand.*, 61, 377 (1972).

123. C.R.Denning, In Ref.89. Page 11.

124. N.N.Huang, *J.Pediatrics*, 81, 197 (1972).

125. I.H.Blanc, *J.Invest.Dermatol.*, 2, 67 (1939).

126. R.A.Durst, (Ed.) *Ion Selective Electrodes.* Special Publication No.314, NBS, Washington, D.C. (1969).

127. G.J.Moody and J.D.R.Thomas, *Selective Ion-sensitive Electrodes*, Merrow, Watford (1971).

128. G.J.Moody and J.D.R.Thomas In: *Ion-selective Electrodes in Analytical Chemistry*, Ed. H.Freiser, Plenum Press, New York and London, Volume 1. (1978). Page 336.

129. G.Fontaine and J.P.Ferriaux, *La Nouvelle Presse Medicale.* 2, 380 (1973).

130. P.T.Bray, *Report to the Fifth Annual Meeting of the European Working Group for Cystic Fibrosis*, Verona, Italy, 1974.

131. D.Armstrong and J.C.Kramer, *12th Annual Meeting of the Cystic Fibrosis Club*, Atlantic City, April 28th, 1971.

132. Orion Research Inc., *U.K.Patent*, No.1 150 698, April 30th, 1969.

133. A.D.Griffiths and F.E.Bull, *Archs Disease Childhood*, 53, 918 (1978).

134. P.T.Bray, G.J.Moody and J.D.R.Thomas, *Archs Disease Childhood*, 53, 918 (1978).

135. G.J.Moody and J.D.R.Thomas, *Practical Electrophoresis*, Merrow, Watford (1975). Page 9.

136. G.J.Moody and J.D.R.Thomas, *Progress in Medicinal Chemistry*, 14, 51 (1977).

137. R.K.Rhodes and R.P.Buck, *Analytica Chim.Acta*, 113, 67 (1980).

138. NAS/NRC Report, *J.Pediatrics*, 88, 711 (1976).

139. R.A.Nicholson, G.J.Moody and J.D.R.Thomas, Unpublished work.

140. L.Hansen, M.Buechele, J.Korochec and W.J.Warwick, *Minnesota Medicine*, 50, 1191 (1967).

141. H.Shwachman and A.Mahmoodian, *Clin.Chem.*, 25, 161 (1979).

## APPENDIX

Step procedure used with the modified Orion Skin Chloride Measuring System in the Cardiff programme [64].

A *Initial setup.*

1. Fill the reference electrode. Check the battery in the meter.

2. Calibrate the electrode with 0.2M and 0.1M sodium chloride solution.

B *Test procedure.*

1. Wash the volar surface of the right forearm with isopropanol (Mediswab-Pharmax) followed by two washings with gauze soaked in deionised water. Dry the site with a tissue.

2. Place two anode papers and two cathode papers on the iontophoresis block. Add 120 and 100 $mm^3$ respectively of deionised water with a Finnpipette. Leave for 30 s for the water to be absorbed.

3. Place the block on the arm and attach with a Velcro strap. Turn on the current and leave at 1.5 mA for 5 minutes.

4. Remove the iontophoresis block, wash the skin with a gauze soaked in deionised water, and dry with a tissue.

5. Cover the site with a 2.5 cm square of plastic film (Nescofilm) and leave for 3 minutes, or longer if sweating is slow.

6. Remove the film and quickly place the dry electrode on the sweat. Record the meter reading as soon as it settles.

C *Final stage.*

1. Drain the electrode.

2. Wash out the body with deionised water.

3. Store dry.

# THE AUTHORS

Biographical notes on Dr. G.J.Moody and Dr. J.D.R. Thomas appeared in Volume I (page 30) of ION-SELECTIVE ELECTRODE REVIEWS.

*Ion-Selective Electrode Rev.* Vol.2, pp. 117-157. 0191-5371/81/0201-0117$05.00/0


# SURFACE EFFECTS IN RELATION TO THE RESPONSE OF SOLID-STATE ION-SELECTIVE ELECTRODES

Janis Gulens

Atomic Energy of Canada Limited, General Chemistry Branch, Chalk River Nuclear Laboratories, Chalk River, Ontario, K0J 1J0, Canada

CONTENTS

KEYWORDS: ion-selective electrodes; surface effects on selective-electrode response; interference from surface films; surface studies of ion-selective electrodes.

## 1. INTRODUCTION

A tremendous growth has occurred in the past fifteen years in the development and application of ion-selective electrodes; this growth has been reviewed biennially by Buck [1]. The early publications in this field dealt with the compounds and the conditions that could be used to prepare useful electrodes, and established the solution conditions under which the electrodes would respond, the extent of various interferences, etc. As the number of successful applications of ion-selective electrodes to chemical analyses, on-line monitoring and automated analysis increased, attention has focussed more recently on applying ion-selective electrodes to determining increasingly lower concentrations of species and under more varied conditions. As a consequence, publications dealing with the use and study of solid-state ion-selective electrodes began to report effects that had not been observed or reported previously:

(i) that the dynamic response and sensitivity of the electrodes varied with their previous use and conditioning,

(ii) that the limit of Nernstian response and the extent of chemical interferences was not consistent with theoretical predictions or with other published reports, and

(iii) that chemical reactions were occurring at the electrode surface which were affecting not only the visible appearance of the electrode but also its response.

The response of solid-state ion-selective electrodes is governed by ion exchange processes occurring between the electrode membrane (ionic crystal) and the solution [2]. Since the transfer of ions occurs across this membrane/solution interface, it is readily apparent that any changes in the nature and composition of the membrane surface will affect the ion exchange process and hence the response of the ion-selective electrode. This review discusses the reported response of solid-state ion-selective electrodes for copper(II) and copper(I) ions, halide ions, and silver/sulfide ions and attempts to interpret these results in terms of alterations in the nature and composition of the electrode surface arising as a result of redox and/or chemical interactions between the electrode and dissolved species in solution. Areas requiring further attention are identified, both for particular electrodes and for solid-state ion-selective electrodes in general.

## 2. COPPER ION-SELECTIVE ELECTRODES

There are two main types of copper ion-selective electrodes: those based on the sulfide or selenide salts of copper(I), and those based on the corresponding salts of copper(II). In general, the former electrode responds to both free copper(I) and free copper(II) ions, while the latter electrode responds to free copper(II) ions. Under certain solution conditions, however, the response of both electrodes becomes much more complex. There is at present considerable uncertainty and confusion in the literature regarding the response of both types of copper electrodes with regard to:

(i) the lower limit of Nernstian response in non-complexing media,

(ii) the mechanism of chloride ion interference with copper(II) determinations and

(iii) the response to complexing agents such as EDTA and NTA.

Electrode surface effects are of major importance in understanding the electrode response in these areas.

### 2.1 Nernstian response

Despite the low solubility of copper(II) sulfide and copper(I) sulfide ($K_{so}$ = $10^{-36}$ and $10^{-48}$ respectively [3]), many workers have reported that the lower limit of Nernstian response of electrodes made from these salts to free copper ions (in the absence of large amounts of complexed copper) is only $10^{-5}$ or $10^{-6}$ M [4-13]. At copper concentrations less than $10^{-5}$ M, positive deviations from the Nernstian curve were observed indicating that the electrode surface was sensing more copper than had been added. Other workers, however, reported that the lower limit of Nernstian response was $2 \times 10^{-7}$ M [14] or $\sim 10^{-8}$ M [14,15]. There are conflicting claims about this lower limit of Nernstian response because it is affected by

(i) the conditions of the measurement, and/or

(ii) the methods used in preparing the electrode membrane.

2.1.1 Measurement conditions. The physical state of the electrode surface and the composition of the solution in which the measurement occurs both affect the lower limit of Nernstian response. Johansson and Edström [4] were the first to note that the response of an electrode with a $CuS$-$Ag_2S$ membrane was affected by the condition of the electrode surface. An electrode that had been used for two months gave sub-Nernstian slopes, was slow to respond even at high copper concentrations and was pitted at its surface; after the electrode was polished and treated with silicone oil its stability and response improved. The authors concluded that the surface irregularities were producing a mixed potential, that is, when the solution concentration was changed, different areas of the electrode were exposed to different concentrations due to slower mixing within the pits. Polishing the electrode removed the pits and surface irregularities, and the silicone oil treatment filled any remaining depressions, thereby minimizing the mixed potential. Numerous workers have subsequently reported that the electrode response improved after polishing [5-7,14,16].

The lower limit of Nernstian response has been reported to range from $10^{-5}$ to $10^{-8}$ M under a wide range of solution conditions (but in the absence of large amounts of complexed copper). Blaedel and Dinwiddie [5] observed Nernstian response at concentrations greater than $10^{-6}$ M but noted that at lower concentrations, the copper electrode would adsorb copper ions on to its surface. A moderate washing procedure with distilled water would not remove the adsorbed ions but a washing procedure with dilute sulfuric acid overcame this problem. These authors later obtained Nernstian response at $10^{-8}$ M by using a flowthrough microelectrode, but the equilibrium potential at $10^{-8}$ M was attained only after twelve hours [14]. Smith and Manahan [15] also obtained Nernstian response to $2\times10^{-8}$ M, but performed their experiments in a stirred complexing anti-oxidant buffer (CAB) solution consisting of 0.05 M acetic acid-sodium acetate buffer, 0.02 M sodium fluoride and and 0.02 M formaldehyde. The purpose of the formaldehyde was to provide a mild reducing environment and minimize interference from oxidizing agents such as iron(III) [15]. Oglesby, Duer and Millero [17] reported a Nernstian limit of $2\times10^{-7}$ M in distilled water which had not been deaerated and did not contain anti-oxidants. However, Midgely [6] and Moody et al. [12] obtained a limit of Nernstian response of only $10^{-5}$ to $10^{-6}$ M under the same aerated solution conditions.

The presence of oxidants and reductants in solution affects the electrode's response, as could be expected from a consideration of the redox chemistry of sulfide and copper ions. Strong oxidants such as iron(III) ions interfere by oxidizing the electrode surface [15]

$$CuS + 2Fe^{3+} \longrightarrow Cu^{2+} + 2Fe^{2+} + S^{0} \tag{1}$$

and in fact the copper electrode has been used as a sensor for iron(III) ions, based on this reaction [18]. Pungor et al. exposed electrodes made of copper(II) sulfide to strong oxidants such as cerium(IV) or permanganate ions and observed shifts in the electrode's standard potential and a decrease in its Nernstian slope [19]. These observations are consistent with changes in the composition of the electrode membrane as a result of its oxidation. However, exposing the oxidized surface to a reducing environment, such as an ascorbic acid solution, resulted in regeneration of the electrode's response to copper ions [19]. Unfortunately, a mechanism for this regeneration was not presented [19].

Strong reductants, such as ascorbic acid or hydroxylamine hydrochloride, have been reported to interfere with the electrode response at low concentrations by reducing copper(II) ions [15]. However, Pick, Tóth and Pungor [20] reported that a heterogeneous copper(II) sulfide electrode had to be *pretreated* by soaking in a $10^{-2}$ M ascorbic acid solution in order to obtain stable and reproducible potentials, but a more recent publication [19] did not mention this requirement. It would appear that pretreatment of the heterogeneous electrode was required [20] only because of the way in which that electrode had been prepared; other workers have not reported the need for such pretreatment procedures for copper electrodes.

The presence in solution of weaker oxidizing and reducing agents such as dissolved oxygen and formaldehyde respectively, also affect the electrode's response at low copper concentrations. Blaedel and Dinwiddie [5] observed that on exposing a moist electrode to air, copper(II) ions were generated at the electrode surface as a result of surface oxidation

$$2CuS + O_2 + 4H^+ \longrightarrow 2Cu^{2+} + 2H_2O + 2S^0 \qquad (2)$$

In support of this proposal Pungor et al. [19] reported, on the basis of Auger spectroscopic studies of electrode surfaces, that copper(II) sulfide precipitates were readily oxidized in air.

Midgely, in a study of several commercially available copper electrodes, concluded that the positive deviation from Nernstian response observed at copper(II) concentrations less than $10^{-6}$ M was due to the influence of dissolved oxygen [6]. For the CuS-$Ag_2S$ electrodes, the non-Nernstian response was attributed to copper(II) ions that were generated at the electrode surface as a result of the oxidation of the membrane, reaction (2). Additional support for this proposal was provided by the observation that immersion of these electrodes into acid solutions resulted in increased copper(II) ion release, while coating the electrode surface with silicone oil resulted in a decreased rate of copper release both in acidic solutions and in distilled water. Both effects were in agreement with the proposed oxidative dissolution mechanism, reaction (2). Midgely concluded that the $Cu_{1.8}Se$ electrode gave a non-Nernstian response at low copper concentrations because it was measuring mixed potentials [6]. The high electronic conductivity of the $Cu_{1.8}Se$ electrode enabled it to respond directly to the presence of redox species in solution, such as dissolved oxygen [6]. The non-Nernstian response of the Růžička Selectrode (a graphite-Teflon rod impregnated with CuS-$Ag_2S$) was attributed by Midgely to the combined effects of membrane oxidation, reaction (2), and a direct mixed potential interference [6].

Mild reducing agents such as formaldehyde have been added to the solutions to minimize the interference from dissolved oxygen on electrode response [15]. However in other applications, it appears that the formaldehyde also contributes to interference. Hulanicki, Trojanowicz and vel Krawczyk [8] reported a Nernstian response only at copper concentrations greater than $3x10^{-6}$ M for a copper(I) sulfide electrode in the CAB solution, as compared to a limit of Nernstian response of $2x10^{-8}$ M reported for the CuS-$Ag_2S$ electrode in the same CAB solution [15]. In addition, it was noted that a CuS-$Ag_2S$ electrode gave much better accuracy than the copper(I) sulfide electrode in the determination of copper in water [8]. This difference in response between the copper(I) sulfide and the CuS-$Ag_2S$ electrodes may be due to the fact that copper(I) sulfide has very high electrical conductivity [10,21] and thus the copper(I) sulfide electrode may have been responding directly to the presence of formaldehyde, that is, measuring a mixed potential.

Since the CuS-$Ag_2S$ electrode is much less susceptible to direct redox interference than the copper(I) sulfide electrode [6], it gave better response in the CAB solution [8]. However, the presence of formaldehyde (concentration unspecified) in a tris-fluoride buffer solution did not have an effect on the determination of $10^{-3}$ to $10^{-8}$ M copper(II) ions by the copper(I) sulfide electrode [8]. This observation appears surprising in view of the previous discussion but may perhaps be due to the fact that the amount (unspecified) of formaldehyde added to the tris-buffer was not sufficient to generate a mixed potential. Further detailed studies on the factors affecting the response of the copper(I) sulfide electrode at low copper concentrations are required.

Thus, in order to obtain accurate and precise results in the determination of cooper at concentrations less than $10^{-5}$ M, more attention must be given to experimental conditions: the physical state of the electrode surface and the nature of the solution to which it is exposed. Due to the ease by which the electrode surface can be oxidized, the electrode should not be exposed to oxidants. Use of the electrode in air-saturated solutions could leave a residual film of elemental sulfur on its surface and/or generate soluble copper salts at the electrode surface, thereby impairing its response in dilute copper solutions. For the $CuS-Ag_2S$ electrode, the higher solubility of copper(II) sulfide over silver sulfide [3] may result in the outer membrane layer becoming gradually depleted with respect to copper(II) sulfide, also leading to impaired response. Periodic polishing of the electrode surface removes such surface layers and restores the electrode's response. Dissolved oxygen must be removed from the solution for low level copper measurements to minimize interference from the measurement of mixed potentials; however, problems may arise in accomplishing this. The use of reducing agents is not advisable, either because of possible interactions with the copper ions present in solution or because of interference from the measurement of mixed potentials. The solutions can be deaerated with an inert gas but this may not be adequate at exceedingly low copper concentrations.

2.1.2 Electrode preparation. The method by which the electrodes are prepared seriously affects their sensitivity and response. Buck [2,22] has recommended that metal sulfide precipitates used for pressed pellet electrodes be precipitated from an acidified solution of the soluble metal salts using hydrogen sulfide gas. This procedure minimizes the contamination of the precipitate with occluded sulfide salts (since excess sulfide ions are not present) or with coprecipitated hydroxide salts (as would be the case if the precipitation were performed by the addition of an alkaline sulfide solution). A membrane containing these occluded or coprecipitated salts will show deviations from Nernstian response at low concentrations due to interference from the impurities that will be leaching out or dissolving from the membrane.

Heijne, van der Linden and den Boef [13] concluded that in order to obtain $CuS-Ag_2S$ pellets with good mechanical and electrochemical properties, the silver sulfide and copper(II) sulfide salts had to be precipitated by the simultaneous addition of the metal ion solutions to an alkaline sodium sulfide solution. This method of precipitation leads to possible coprecipitation of copper hydroxide and silver hydroxide salts [22] which are much more soluble than the corresponding sulfide salts, but the authors claimed [13] that this fact did not affect the limit of detection of the electrode. However since they observed Nernstian response only at copper concentrations greater than $10^{-5}$ M, their claim does not appear to be justified.

In a subsequent study involving X-ray diffraction analysis, Heijne and van der Linden [23] showed that those pressed pellets which gave a good electrochemical response contained primarily the ternary compound jalpaite, $Ag_{1.5}Cu_{0.5}S$, that is, the membrane contained copper(I) ions. The authors also observed that membranes obtained from precipitates prepared under nitrogen instead of air, showed better response and were less sensitive to subsequent air oxidation effects. A certain amount of readily soluble copper was present in the precipitates and this was identified by X-ray analysis as being copper sulfate. The authors therefore concluded that the precipitates were slowly oxidized by air [23].

In solubility measurements it was observed [23] that the amount of copper(II) which would dissolve from the membrane was significantly higher than that expected on the basis of solubility calculations, and this was attributed to the decomposition of the electrode membrane:

$$Cu_2S \rightleftharpoons CuS + Cu^{2+} + 2e \qquad (3)$$

or

$$Ag_{(2-x)}Cu_xS \rightleftharpoons (1-\tfrac{1}{2}x)Ag_2S + \tfrac{1}{2}xCuS + \tfrac{1}{2}xCu^{2+} + xe \qquad (4)$$

However, the actual oxidizing agent in solution causing this decomposition was not specified. The authors concluded that the formation of ternary compounds such as jalpaite seemed to be essential for obtaining copper electrodes with good mechanical and electrochemical properties. This implied that copper(II) was being reduced to copper(I) in solution with the incorporation of copper(I) ions into the membrane. A mechanism for the reduction of copper(II) ions in solution to copper(I) was not given, but it was suggested that this reduction may be accompanied by the oxidation of sulfide to sulfur [23]. If these electrodes do in fact contain copper(I) ions as suggested, then they should also respond to copper(I) ions in solution [24], but this possibility was not tested [23].

Pungor et al. also have noted the beneficial effects on electrode response of preparing the metal sulfide precipitates in the absence of oxygen [19]. Copper(II) sulfide precipitates were prepared by heating a copper(II) solution with an acidified thiosulfate solution, but electrodes made from precipitates prepared in air did not respond to copper ions while those made from precipitates prepared in a reducing (hydrogen sulfide or nitrogen) atmosphere gave Nernstian response, but only to $10^{-4}$ M [19]. Analysis showed that the latter precipitate consisted mainly of copper(II) sulfide while the composition of the former precipitate was quite different (unspecified). It was also claimed that the heterogeneous silicone rubber copper(II) sulfide electrode was more sensitive than the pressed pellet copper(II) sulfide electrode but the results presented [19] do not confirm this statement.

Koebel [25] discussed the influence on the electrode's response and standard potential of using various metals to make electrical contact to the membrane. Van de Leest [26] showed that electrode response was also affected by the equilibrium properties of this electrical contact. Copper(II) sulfide electrodes were prepared by vapour depositing a layer of copper metal onto a pressed pellet of silver sulfide. Copper(II) sulfide was then produced by heating the coated pellet in the presence of sulfur in a nitrogen atmosphere. Electrical contact to the membranes was made with either graphite or silver metal. For carbon-contacted electrodes good response was observed at copper(II) concentrations greater than $10^{-5}$ M and the experimental value for the standard potential agreed favourably with the calculated value if the electrodes were heated below 200°C. However, when the electrodes were heated above 200°C poor agreement in the standard potential values was observed. For silver contacted electrodes, large differences in the standard potential values were observed at all heating temperatures, indicating that the electrodes were not at thermodynamic equilibrium.

A copper-selective electrode was recently prepared by electroplating copper selenide ($Cu_{1.8}Se$) onto a platinum substrate [27]. The electrodes prepared in this manner gave a rapid linear Nernstian response to $10^{-6}$ M copper(II), and had only a very slight potential drift. The advantages of this electrodeposited electrode over the single-crystal $Cu_{1.8}Se$ electrode were its lack of sensitivity to dissolved oxygen and lack of potential drift. The improved response of the former electrode was attributed to the fact that the copper selenide was prepared by controlled electrodeposition or electrocrystallization, leading to a more uniform deposit [27].

The preparation of precipitates for copper-selective electrodes capable of giving good Nernstian response to low copper concentrations still appears to be a black art. Part of the difficulty lies in not knowing exactly the composition and stoichiometry of the salts in the membranes that do give good response. For example, it still is not clear whether or not copper(I) ions must be present in copper(II) sulfide electrodes to obtain good response. Heijne and van der Linden

[23] did provide evidence from X-ray studies that copper(I) salts were present in those electrodes of good electrochemical response, but Pungor et al. [19] did not report any variations in the copper signal from photo-electron spectroscopic studies of copper(II) sulfide pressed pellets that would indicate the presence of copper(I) ions in the membrane. Moody et al. [12] studied $CuS-Ag_2S$ electrodes using Auger spectroscopy but due to instrumental limitations were unable to resolve the copper peak. There are four known non-stoichiometric compounds of copper sulfide $Cu_{1.92}S$, $Cu_{1.77}S$, $Cu_{1.60}S$ and $Cu_{1.31}S$ [28] but it is not known exactly how the changes in the stoichiometry or how the formation of ternary silver-copper-sulfide complexes [23] in the membrane affect the sensitivity and response of the electrode; the standard potential of the electrode depends on its exact stoichiometry [25,29].

In studying the low-level response of these electrodes, it is important to control and know precisely the purity, composition and stoichiometry of the membrane; thus it is advisable that precipitates for the electrode be prepared in the manner prescribed by Buck [2,22]. Given the ease of air oxidation of sulfide salts it is also advisable to carry out the precipitation reactions in a reducing atmosphere as compared to an oxidizing one [19,23]. When copper sulfide is precipitated in a reducing atmosphere only copper(II) sulfide and copper(I) sulfide (or other non-stoichiometric copper(I) sulfide compounds) will be present, while under oxidizing conditions the precipitate will contain copper(II) sulfide and elemental sulfur (and/or other copper sulfur-oxy salts). Electrodes containing only $CuS-Cu_2S$ should be less susceptible to oxidation effects in solution than those containing $CuS-S^\circ$ and thus the former should give better Nernstian response [30].

A $CuS-Cu_2S$ electrode consists of two phases and three components and according to the phase rule it has three degrees of freedom (temperature, pressure and composition). The $CuS-S^\circ$ electrode, however, consists of two phases but only two components and thus it has only two degrees of freedom (temperature, pressure). When both electrodes are exposed to an oxidizing solution such as one containing dissolved oxygen, the effect of the oxidation reaction on the $CuS-Cu_2S$ electrode will be primarily a change in its composition by the oxidation of copper(I) to copper(II) in the membrane, while the effect on the $CuS-S^\circ$ electrode will be the oxidation of sulfide to elemental sulfur, equation (2) [30].

The $CuS-Cu_2S$ surface cannot undergo the oxidative dissolution described by equation (2) as this would produce another phase on the surface, elemental sulfur, in violation of the phase rule. The sulfide in the $CuS-Cu_2S$ membrane could be oxidized by dissolved oxygen to other soluble sulfur-oxy species and thereby not violate the phase rule, but the oxidation of sulfide ions is more difficult than the oxidation of copper(I) to copper(II) [28]. Changes in the composition of the $CuS-Cu_2S$ electrode as a result of exposure to oxygenated solutions will affect its standard potential [25,29] but should not seriously affect its response to low concentrations of copper(II) ions. The oxidation of the $CuS-S^\circ$ electrode in oxygenated solutions releases copper(II) ions into solution and gives rise to the non-Nernstian response, thus making this electrode more susceptible to oxidation effects than the $CuS-Cu_2S$ electrode. It is apparent, though, that the response properties of non-stoichiometric $CuS-Cu_2S$ electrodes require more detailed study.

## 2.2 Interference from chloride ions

Ross [31] reported that chloride ions would interfere with the response of a $CuS-Ag_2S$ electrode under conditions of suitably high concentrations of both copper(II) and chloride ions due to the formation of a silver chloride layer on the electrode surface

$$Ag_2S + Cu^{2+} + 2Cl^- \rightleftharpoons 2AgCl + CuS \quad (5)$$

The concentration of chloride ions required to effect such an interference, for a given copper concentration, could be calculated from the solubility product values of the silver sulfide, silver chloride and copper(II) sulfide salts [31]. However, numerous authors [7,12,16,17,32-34] have reported that chloride ions will interfere with electrode response at chloride concentrations where the formation of a silver chloride layer on the electrode surface is not possible. A theory involving redox reactions with the electrode surface has been presented recently [34] that provides a basis for understanding the mechanism of the chloride interference.

2.2.1 Evidence of Interference. Crombie, Moody and Thomas [16] observed that on exposing a CuS-$Ag_2S$ electrode to a solution containing both copper(II) nitrate and chloride ions (but at concentrations where silver chloride formation was not possible), the electrode surface became dull and tarnished and the tarnished surface responded to changes in the concentration of chloride ions. By polishing the electrode, the tarnish and hence the interference was removed. The tarnish and the interference was ascribed to the formation of copper(II)-chloro complexes although evidence to support this could not be provided [16].

Jasinski, Trachtenberg and Andrychuk [32] observed that a CuS-$Ag_2S$ electrode gave a cationic type response with a "Nernstian slope" of 52 mV per decade to copper(II) ions in seawater, instead of the expected 29 mV per decade slope. In addition, they observed that various electrodes gave different potential values for a given copper(II) concentration in seawater, and that the "Nernstian slope" of seven electrodes studied ranged from 44 to 52 mV per decade.

Midgely [7] also observed that prolonged immersion of CuS-$Ag_2S$ electrodes in chloride solutions produced a tarnish on the electrode surface with resultant deterioration of the electrode's response, as shown by slower response times and changes in the value of the electrode's standard potential. Midgely concluded that the oxidative dissolution of the electrode surface, equation (2), was being catalyzed by chloride ions and this resulted in electrode tarnishing.

Oglesby, Duer and Millero [17] studied the influence of ionic strength and chloride ion concentration on the response of the CuS-$Ag_2S$ electrode. When the electrode was calibrated in distilled water or in potassium nitrate solutions in the absence of chloride ions, the electrode's response to copper(II) ions was the expected Nernstian slope of 29 mV per decade. However, when the electrode was calibrated in chloride ion solutions, the electrode's response to copper(II) ions increased from a cationic slope of 28 to 49 mV per decade with increasing sodium chloride concentration. The authors concluded that the increase in the slope was due to the increases in chloride concentration and not to changes in the ionic strength of the solution, but they were unable to explain this marked influence of the chloride ions on the electrode's response to copper(II) ions.

In an attempt to determine the mechanism of chloride ion interference, Moody et al. [12] combined a study of the potentiometric response of various CuS-$Ag_2S$ electrodes with a study of their surface composition, after soaking the electrodes in various chloride ion solutions. Using Auger spectroscopy, they observed that chloride ions penetrated some membranes while only a surface contamination of chloride ions was noted for other membranes. For some membranes there was little evidence of sulfide at the electrode surface with the amount of sulfide increasing with depth, while for others the amount of sulfide was constant with depth. One electrode was exposed to a solution which contained both copper(II) nitrate and chloride ions and Auger analysis showed that both chloride and nitrate ions from solution had penetrated this membrane.

The potentiometric studies showed that soaking the electrodes prior to calibration in solutions containing only chloride ions did not affect the electrode's response to copper(II) ions of near-Nernstian cationic slope of ∿29 mV per decade, and the electrodes did not respond to the presence of chloride ions. However, when the electrodes were soaked in chloride ion solutions which also contained copper(II) ions, the electrode's response was altered drastically. After a commercial CuS-$Ag_2S$ electrode had been soaked for ten minutes in a 0.1 M copper(II) nitrate - 0.03 M potassium chloride solution, the electrode did not respond to copper(II) ions (added either as copper(II) nitrate or copper(II) sulfate) but did give a near-Nernstian anionic type response with a slope of 61 mV per decade with respect to chloride ions (added either as potassium chloride or copper(II) chloride). However a CuS-$Ag_2S$ electrode made in the laboratory and soaked for ten minutes in a 0.1 M copper(II) nitrate - 0.1 M potassium chloride solution, did not respond to chloride ions (added either as potassium chloride or as copper(II) chloride) and did not respond to copper(II) ions below $10^{-3}$ M. At concentrations of copper(II) ions above $10^{-3}$ M, the electrode gave a sub-Nernstian cationic type response with a slope of 25 mV per decade. The authors were unable to explain the difference in the response of the two electrodes which had been exposed to a solution containing both copper(II) nitrate and chloride ions but they did conclude that both copper(II) nitrate and chloride ions had to be present in solution to produce the interference. A mechanism for the interference was not proposed.

Recently, Lanza [33] proposed a mechanism to explain the chloride ion interference at CuS-$Ag_2S$ electrodes, and observed reasonably good agreement between his predictions and experimental results. Lanza considered the effect on the electrode's response of the formation of soluble silver chloride complexes, $AgCl_n^{1-n}$, where n is the average coordination number at the particular chloride concentration

$$Ag_2S + Cu^{2+} + 2nCl^- \rightleftharpoons CuS + 2AgCl_n^{1-n} \qquad (6)$$

Three different responses of the electrode could be expected, depending on the concentration of copper(II) and chloride ions in solution:

(i) when the amount of copper(II) ions reacting via equation (6) was very small, the electrode would respond to copper ions with the expected Nernstian slope of 29 mV per decade;

(ii) when all the copper(II) ions added to the solution reacted quantitatively at a suitable chloride concentration to form $AgCl_n^{1-n}$, the apparent electrode response to copper(II) ions would be with a cationic slope of 59 mV per decade due to its response to the $AgCl_n^{1-n}$ complexes formed; the apparent standard potential would be a function of the chloride ion concentration;

(iii) in sufficiently concentrated chloride solutions, the solution would become saturated with silver chloride and the copper(II) concentration would reach a constant maximum value; the calibration slope would be zero.

Experimentally, Lanza observed that when the chloride ion concentration in solution was increased from zero to 1 M, the electrode's response to copper(II) ions increased from a Nernstian slope of 29 mV per decade to a slope of 59 mV per decade as predicted. At even higher chloride concentrations (3 M), the electrode potentials for a given copper concentration increased with time to more positive values; this increase was attributed to the gradual transformation of silver sulfide on the electrode surface to silver chloride, equation (5). Those electrodes which had been exposed to such concentrated chloride and copper solutions that silver chloride formation could occur on the surface, gave a Nernstian response to changes in chloride ion concentration. This response to chloride ions could be eliminated by cleaning the electrode surface with ammonia solution to remove the silver chloride layer.

In a 3 M chloride solution saturated with silver chloride, the electrodes initially did not respond to added copper(II) ions, as predicted by Lanza's theory. However, as the copper(II) concentration was increased above $10^{-4}$ M, large increases in potential were observed. These increases in potential were attributed to the deposition of silver chloride on the electrode surface according to the reaction

$$Ag_2S + 2Cu^{2+} + 4Cl^- \rightleftharpoons 2AgCl + 2CuCl + S \qquad (7)$$

since Lanza observed that elemental sulfur was produced when a 0.1 M copper(II) solution in 3 M potassium chloride was contacted with solid silver sulfide for several days.

Lanza concluded that it was the transformation of silver sulfide to silver chloride on the electrode surface according to reactions (5) and (7) that caused the increases in electrode potential when the electrodes were exposed to concentrated (>1 M) chloride solutions containing copper(II) concentrations greater than $10^{-3}$ M. These surface alteration reactions were also invoked to explain the observation that calibration curves measured in successive runs had constant slopes, but had different apparent standard potentials.

2.2.2 Mechanism of interference. A theory to explain chloride ion interference for copper electrodes has been presented recently by Westall, Morel and Hume [34]. The authors experimentally verified the theory and also applied it to interpret the observations of Jasinski, Trachtenberg and Andrychuk [32] and Oglesby, Duer and Millero [17]. While Lanza's theory [33] was limited to explaining the chloride interference at $CuS$-$Ag_2S$ electrodes, the theory of Westall, Morel and Hume is applicable to all copper ion-selective electrodes.

The Westall-Morel-Hume theory was based on a redox reaction occurring in a thin layer of solution adjacent to the electrode surface between copper(II) ions present in the bulk solution and the electrode surface itself:

$$Cu^{2+} + CuS \rightleftharpoons 2Cu^+ + S^0 \qquad (8)$$

The driving force for this redox reaction was the stabilization of the copper(I) ions produced at the electrode surface by complexation with chloride ions

$$Cu^+ + 2Cl^- \rightleftharpoons CuCl_2^-, \quad K_{CuCl_2^-} \qquad (9)$$

Under the conditions that the bulk copper(I) concentration was zero and that the rate of complexation, equation (9), was fast compared to the rate of the redox reaction, equation (8), the following equations describing the electrode response were derived for three limiting situations in unstirred solutions:

(i) *No chloride interference.* When the concentration of chloride ions was much smaller than the copper(II) ion concentration, very little reduction of copper(II) occurred and the electrode's response was the Nernstian slope of 29 mV per decade.

$$E = E^{01}_{Cu^{2+}/Cu^0} + \frac{RT}{2F} \ln \frac{[Cu(II)]}{a_{Cu^0}} \quad (10)$$

where: [Cu(II)] is the concentration in bulk solution of the added copper(II), $a_{Cu^0}$ is the activity of metallic copper in the membrane phase, and $E^{01}_{Cu^{2+}/Cu^0}$ is the standard potential of the copper(II)-copper metal couple, $E^{0}_{Cu^{2+}/Cu^0}$, corrected for the activity coefficient of copper(II) in the given medium.

(ii) *Full chloride interference.* At high chloride concentrations all the copper(II) at the electrode surface was reduced and complexed to form $CuCl_2^-$; the electrode reponse was

$$E = E^{01}_{Cu^{+}/Cu^0} + \frac{RT}{F} \ln \left\{ \frac{2\,[Cu(II)]}{a_{Cu^0}[Cl^-]^2 . K_{CuCl_2^-}} \right\} \quad (11)$$

that is, a slope of 59 mV per decade change in bulk copper(II) concentration.

(iii) *Critical chloride concentration.* The chloride concentration at which the electrode's response changed from a slope of 29 to 59 mV per decade was designated as the critical chloride concentration, $[Cl^-]^{crit}$. From equations (10) and (11), this was

$$[Cl^-]^{crit} = \left\{ \left[ \exp \frac{F}{RT}\left(E^{01}_{Cu^{+}/Cu^0} - E^{01}_{Cu^{2+}/Cu^0}\right) \right] \frac{2[Cu(II)]^{\frac{1}{2}}}{(a_{Cu^0})^{\frac{1}{2}} . K_{CuCl_2^-}} \right\}^{\frac{1}{2}} \quad (12)$$

Equations (10)-(12), valid for unstirred solutions only, predicted that the electrode's response was also dependent on the activity of elemental copper in the electrode phase, $a_{Cu^0}$. This activity was related to the manner in which the electrode membrane was prepared; for electrodes in contact with metallic silver ($Ag/CuS-Ag_2S$), $a_{Cu^0} = 10^{-6.3}$ M while for those in contact with graphite ($C/CuS-Ag_2S$), $a_{Cu^0} = 10^{-8.6}$ M. Equation (11) also predicted that under conditions where chloride ion interference was present, the electrode would respond to changes in chloride ion concentration with a slope of 118 mV per decade.

Westall, Morel and Hume also included in their discussion [34] the effect of other copper-complexing ligands on electrode response. When ligands capable of forming strong complexes with copper(II) ions were present in addition to chloride ions, the chloride interference was decreased because of the increased difficulty in reducing the complexed copper(II) species. Strong copper(I)-complexing ligands, other than chloride ions, were predicted to have the same effect on electrode response as chloride ions.

The authors also considered the effects on electrode response of the formation of silver chloride complexes $AgCl_n^{1-n}$, as discussed by Lanza [33]; while detailed results were not given, they concluded that consideration of these complexes resulted in very nearly the same electrode potential values since the formation constants for both the silver chloride and copper(I) chloride complexes were very similar. When the electrode was exposed to exceedingly concentrated solutions of both copper(II) and chloride ions, its response was not consistent with equation (11). Westall, Morel and Hume concluded that it was not possible to ascertain whether this was due to blockage of the electrode surface by silver chloride formation (equation (5)), or due to the influence on electrode potential of the creation of concentration gradients of copper(II) sulfide and silver sulfide within the membrane phase arising from reactions (5), (6) and (8).

2.2.3 Significance of the theory of Westall, Morel and Hume. The appearance of the Westall-Morel-Hume theory is most timely as not only has it brought order into the confused situation regarding chloride interference, but also it has significant implications on the use of copper ion-selective electrodes in copper-complexing media.

(i) This theory incorporates the effects of not only complexation reactions but also a redox reaction involving the electrode surface, while Lanza's theory [33] deals only with the effect of silver ion complexation. The Westall-Morel-Hume theory is thus applicable to all copper ion-selective electrodes (that is, $CuS$-$Ag_2S$, copper(I) sulfide and copper(II) sulide) while Lanza's theory is only applicable to those copper ion-selective electrodes containing silver sulfide. To illustrate the importance of the redox reaction relative to silver ion complexation on the interference mechanism, the authors compared the response of a $CuS$-$Ag_2S$ electrode to copper(II) ions in an aqueous chloride solution and in acetonitrile, and observed that the response of the electrode was similar in both solutions. Since both copper(I) and silver(I) ions are complexed to the same extent in aqueous chloride solutions but the complexation of copper(I) ions by acetonitrile is many orders of magnitude stronger than the complexation of silver(I) ions, copper(I) must have been formed at the electrode surface in acetonitrile solutions [34].

(ii) This theory presents a basis for understanding and predicting the response of copper ion-selective electrodes in the presence of not only chloride ions but also other ligands capable of complexing copper(I) and/or copper(II) ions. However, the response in the presence of such ligands must be interpreted with care. When a response slope of 29 mV per decade is obtained in unstirred solutions, the concentration of free uncomplexed copper(II) ions can be measured directly; however, when a response slope of 59 mV per decade is obtained, only the concentration of total copper can be determined. In the transition region between the 29 and 59 mV per decade response slopes, interpretation of the electrode response is complicated; the interpretation becomes even more complex if the solutions are stirred [34].

(iii) The extent of chloride (or other ligand) interference is also dependent on the activity of metallic copper in the membrane phase, $a_{Cu^0}$; since this is a parameter that is governed by the manner in which the electrode is prepared, the extent of interference can be minimized by the appropriate choice of the electrode used. Westall, Morel and Hume showed that in seawater in the absence of other ligands, the silver contacted electrode ($Ag/CuS$-$Ag_2S$) gave a 59 mV per decade response at all copper(II) concentrations while the graphite contacted electrode ($C/CuS$-$Ag_2S$) gave a 59 mV per decade response at copper(II) concentrations less than $4 \times 10^{-5}$ M, but gave a 29 mV per decade response at copper(II) concentrations greater than $4 \times 10^{-5}$ M [34]. Studies of copper(II) speciation in seawater are thus feasible using a $C/CuS$-$Ag_2S$ electrode only when the copper(II) concentration exceeds $4 \times 10^{-5}$ M, but they are not feasible at any concentration level using the $Ag/CuS$-$Ag_2S$ electrode [34].

(iv) Surface studies of electrode membranes could provide direct verification of the proposal that copper(II) is reduced to copper(I) at the membrane surface. Moody et al. [12] used Auger spectroscopy to study electrodes that had been stored in chloride solutions; unfortunately due to instrumental limitations they were unable to resolve the copper peak in the spectrum and thus it was not possible to determine if any copper(I) had been produced at the surface of the electrode which had been stored in the solution containing both copper(II) and chloride ions.

(v) More attention should be given to the effect on electrode response of using internal filling solutions that contain high concentrations of copper(II) and chloride ions. Reactions (5), (8) and (9) will also occur at the inner membrane/solution interface, thereby altering the nature and composition of this surface. If this inner surface is covered with a layer of elemental sulfur or silver chloride and/or develops concentration gradients of copper(II) sulfide or silver sulfide within the membrane, problems may arise in attaining proper Nernstian response, as has already been reported [12], in addition to problems in attaining rapid electrode response or steady potential values. The use of solid electrode contacts [25,34,43] appears advisable.

(vi) The response of the copper ion-selective electrodes to changes in the chloride ion (or other ligand) concentration should be studied. Under the conditions where the full chloride interference is observed, the electrode should respond to changes in the chloride ion concentration in unstirred solutions with a slope of 118 mV per decade, equation (11). However, various workers have reported that $CuS$-$Ag_2S$ electrodes which had been previously exposed to concentrated copper(II) solutions containing the appropriate amount of chloride ions, gave a "Nernstian" anionic type response of 59 mV per decade to changes in the chloride ion concentration, in stirred solutions [12,16,33]. This 59 mV per decade response to chloride ions may indicate that the electrode surface had already been appreciably covered* by a layer of silver chloride, equation (5), or may reflect the fact that the response was obtained in stirred solutions, that is, conditions where equation (11) is no longer applicable [34].

(vii) The theory of Westall, Morel and Hume can be applied to the experimental results reported by other workers, in addition to those already discussed [34]. The apparent standard potential of copper ion-selective electrodes has been observed to increase to more positive values with increased length of exposure to chloride solutions containing copper(II) ions [12,16,17,33], yet cleaning or polishing the surface restored the potential to its initial value. These observations are consistent with the mechanism wherein copper(II) sulfide and perhaps silver sulfide are removed from the membrane surface while elemental sulfur and perhaps silver chloride are formed on the surface. This results in a sulfur-rich membrane with a possible change in the copper(II) sulfide stoichiometry and thus a more positive standard potential value. Polishing the surface removes the non-stoichiometric films and restores the electrode to its original state.

Hulanicki and Lewenstam [10] observed that a copper(I) sulfide electrode responded to copper(I) ions in a potassium chloride solution with a 59 mV per decade slope, and responded to copper(II) ions in a potassium nitrate solution with a 29 mV per decade slope. The response to copper(II) ions was attributed to an exchange reaction at the electrode surface

$$Cu^{2+} + CuS \rightleftharpoons 2Cu^{+} + CuS, \quad K \tag{13}$$

* Note by Editor: Hardly possible in Refs. 12 & 16, for salts of copper(II) other than nitrate do not promote chloride interference.

and the electrode response was governed by the equation

$$E = E^{o} + \frac{RT}{F} \ln \left\{ [Cu^{+}] + K \cdot [Cu^{2+}]^{\frac{1}{2}} \right\} \qquad (14)$$

The exchange constant, K, for reaction (13) was experimentally determined to be $6.3 \times 10^{-7}$ as compared to a calculated value of $4 \times 10^{-13}$. When the electrode was exposed to a 1 M potassium chloride solution containing equimolar amounts of copper(I) and copper(II) ions, it responded with a 55 mV per decade slope and was influenced by the concentration of both copper(I) and copper(II) ions. Hulanicki and Lewenstam [10] attributed this response to the strong complexation of copper(I) by the chloride ions, thus lowering the concentration of free copper(I) ions in solution to such a low level that the electrode could also respond to the copper(II) ions present. However the response can also be explained by applying the mechanism of Westall, Morel and Hume wherein the copper(II) in the chloride solution is converted to copper(I) at the electrode surface

$$2Cu^{2+} + Cu_2S + 8Cl^- \rightleftharpoons 4CuCl_2^- + S^o \qquad (15)$$

(viii) The copper(I) sulfide (or copper(II) sulfide) electrode could be used to verify the Westall-Morel-Hume theory that the redox reactions at the membrane surface are of primary importance relative to the mechanism involving silver-ion complexation as proposed by Lanza [33] since the copper(I) sulfide electrode does not contain any silver sulfide. Some possible experiments would be the following:

(a) the copper(I) sulfide electrode should give two distinct regions of linear response with slopes of 59 and 29 mV per decade to changes in copper(II) ion concentration in a solution of fixed chloride concentration; the point at which the response will change from a slope of 59 to 29 mV per decade will be a function of both the copper(II) and chloride ion concentrations, equation (12);

(b) a response similar to that in (a) should also be observed when the copper(I) sulfide electrode is calibrated in copper(II) solutions containing other ligands which strongly complex copper(I) ions; and

(c) the response of the copper(I) sulfide electrode to both copper(I) and copper(II) should be studied in a solution that does not contain copper(I)-complexing ligands (but under anaerobic conditions where copper(I) is stable and does not become oxidized to copper(II)); under these conditions, the electrode should only respond to changes in the concentration of copper(I) ions and not to changes in the copper(II) concentration if the exchange constant for reaction (13) is less than or equal to $6.3 \times 10^{-7}$.

## 2.3 Response to complexing agents

The copper ion-selective electrode responds to the presence of copper(II)-complexing ligands, L, in solution as these affect the concentration of free copper(II) ions present at the electrode surface

$$CuS + L \rightleftharpoons CuL^{2+} + S^{2-} \quad (16)$$

As a result, the copper electrode can be used as an indicator electrode for the direct or indirect potentiometric determination of metal ions by complexometric titrations. However, the response of this electrode to certain ligands indicates that surface reactions, in addition to solubility effects, are again affecting the electrode's response.

Good agreement has generally been observed between the response of the copper electrode to various ligands and the theoretical predictions of the response, except when polyaminocarboxylic acid ligands such as EDTA, NTA or EGTA have been used [9,35-40]. When the electrode's response to changes in pH was measured in solutions containing only ligands such as EDTA, NTA or EGTA without any added copper(II) ions, the potential-pH response was identical to that observed in solutions that contained copper(II) ions in the presence of an excess of EDTA, NTA or EGTA [35,39]. In addition, the apparent concentration of free copper(II) ions present at the electrode surface as calculated from the measured potential was many orders of magnitude larger than that calculated on the basis of complexation equilibria [9,35,39,40].

A fully satisfactory explanation for these observations has not yet been proposed. However these results have been interpreted as indicating that the response of the electrode depends only on the bulk ligand concentration, not the bulk copper(II) concentration [35], and that these ligands are inducing the dissolution of the membrane surface thereby leading to artificially high surface concentrations of copper(II) ions [35]. Olson et al. [36] compared the response of the silver sulfide electrode to the response of the $CuS\text{-}Ag_2S$ electrode towards these ligands, and concluded that the response of the $CuS\text{-}Ag_2S$ electrode was governed by interactions between copper(II) ions and the ligand, not by interactions between the silver ions and the ligand. Further support for the proposal that the artifically high surface concentration of copper(II) ions was due to ligands interacting with the membrane [35] was provided by the observation that changes in the stirring rate altered the electrode potential in an EDTA solution [36,41] and that the potentials measured in moderately alkaline and acidic solutions were not readily reproducible [36]. In addition, prolonged exposure of the electrodes to these ligands resulted in electrode tarnishing and deterioration of their response [37,38]; similar tarnishing effects were observed in the presence of humic and fulvic acids [38].

Interference from EDTA and NTA on the response of copper ion-selective electrodes can be minimized by the use of a ligand interference preventive buffer (LIPB) [41]. This solution consists of 0.2 M nitric acid-0.2 M potassium nitrate-0.2 M triethylenetetramine (trien). The trien forms a very strong complex with copper(II) ions, displacing any other ligands that may be complexed to copper(II) [41]. When the LIPB solution was added to a solution containing EDTA or NTA, the electrode potential stabilized and was independent of the stirring rate [41]. Evidently, when the copper(II) ions dissolved from the membrane surface become strongly complexed by the trien, interference from EDTA or NTA is minimized.

Further work is required in order to understand the interactions between the electrode surface and these ligands.

## 3. HALIDE ION-SELECTIVE ELECTRODES

The most common halide ion-selective electrodes are based on homogeneous pressed pellets ($AgX\text{-}Ag_2S$ or $Hg_2X_2\text{-}HgS$) and heterogeneous membranes ($AgX$ or $AgX\text{-}Ag_2S$ dispersed in silicon rubber or graphite). Significant deviations from the predicted Nernstian response have been reported for halide-selective electrodes in solutions containing not only low concentrations but also high concentrations of the halide ions. The methods used to prepare the electrodes also have had a strong influence on the Nernstian response. Due to the high solubility of the silver halide salts, many ions interfere with the response of electrodes based on the silver halides and significant deviations between the predicted and observed values of the selectivity coefficients for these electrodes have been reported. Electrode surface effects again play a major role in understanding the response of these electrodes under a wide range of conditions.

## 3.1 Nernstian response

3.1.1 Mechanism of response. The general mechanism of the response of solid-state silver halide membrane electrodes is based on rapid reversible interfacial ion exchange, and has been thoroughly discussed [2,42]. The potential of a silver halide membrane electrode depends on the nature of the internal electrical contact. This potential is the same as the potential of a classical second-order electrode (Ag/AgX) if either an internal ionic (silver ion) contact or a silver metal contact is used in the membrane electrode [43]. The defect structure or nonstoichiometry of the silver halide layer used in a second-order electrode does not affect its response, but these parameters do affect the potential of the silver halide membrane electrode [43]. Halide-selective membrane electrodes based on mercury(I) salts, $Hg_2X_2$, appear to function like second-order electrodes, a fact that influences their method of preparation and use.

Lechner and Sekerka (44,45) were the first to report that halide-selective membranes could be prepared by pressing at elevated temperatures a mixture of mercury(II) sulfide and mercury(I) halide salts; an internal metallic mercury pool was used for the electrical contact. The quality of the pellet produced was a function of the length of time and temperature of pressing. It was reported both by Lechner and Sekerka [44,45] and by other authors [46,47,90] that the $Hg_2X_2$-HgS electrodes had to be preconditioned by soaking in dilute (pH 3.5) nitric acid [45] or in distilled water [44,46,47] for two or three days before use, but the reasons for this preconditioning were not presented. The response of these electrodes was presumed to be governed by the solubility of the mercury(I) halide salt in the membrane [44-48] similar to the silver halide electrodes, but the latter do not require a preconditioning procedure. The $Hg_2Cl_2$-HgS electrode was shown to respond directly to the presence of sulfite ions and it was proposed [47] that the response was due to the effect of the reaction

$$Hg_2Cl_2 + 2SO_3^{2-} \rightleftharpoons Hg + Hg(SO_3)_2^{2-} + 2Cl^- \qquad (17)$$

on the equilibrium

$$Hg_2Cl_2 \rightleftharpoons Hg_2^{2+} + 2Cl^- \qquad (18)$$

Hulanicki, Lewandowski and Lewenstam [48] investigated the response of the $Hg_2Br_2$-HgS electrode and provided an explanation not only for the response mechanism but also for the need to precondition the electrodes. It was observed that when membrane pellets containing $Hg_2Br_2$-HgS-$Ag_2S$ were heated, sublimation occurred and a crystalline substance identified as mercury(II) bromide was deposited on the walls of the heating chamber. Analysis by X-ray diffraction of the membrane constituents before and after pressing and heating, showed that mercury(I) bromide was being lost from the membrane, and that mercury(II) bromide was being formed in it. In addition, it was observed that during the slight pressing needed to affix the silver metal contact to the membrane, droplets of metallic mercury appeared on the walls of the die. The authors concluded that disproportionation was occurring during the heating and pressing stages of the membrane preparation

$$Hg_2Br_2 \rightleftharpoons Hg + HgBr_2 \quad (19)$$

and that this disproportionation reaction provided a basis for understanding the electrode's response:

(i) The electrodes required preconditioning such as soaking in dilute acid or distilled water in order to leach out from the surface layer the mercury(II) halide salts which are more soluble than the corresponding mercury(I) salts [48]. This conclusion was further supported by the observation that those membranes that already contained a small amount of metallic mercury prior to pressing, did not require to be heated and did not require to be preconditioned in order to give good response [48]. The presence of the metallic mercury in the mixture during pressing hindered the disproportionation, reaction (18). (A similar disproportionation has been reported during the pressing of mercury(I) sulfide precipitates [49].)

(ii) These electrodes were second-order electrodes, with the response governed by the reaction

$$Hg_2Br_2 + 2e \rightleftharpoons 2Hg + 2Br^- \quad (20)$$

The $Hg/Hg_2Br_2$ couple was dispersed in an electron-conducting matrix mercury(II) sulfide (and silver sulfide) and the electrode potential was given by

$$E = E^o_{Hg_2^{2+}/Hg} + \frac{RT}{2F} \ln K_{so} - \frac{RT}{F} \ln [Br^-] \quad (21)$$

where $K_{so}$ is the solubility product for $Hg_2Br_2$.

Additional evidence for the conclusion that the electrode was functioning as a second-order electrode was provided by the observation that its response to soluble mercury(I) and mercury(II) ions was identical to the response given by a mercury metal electrode. The fact that this electrode functions as a second-order electrode is significant in that the membrane must contain metallic mercury in order to respond. Since the amount of mercury in the membrane as a result of disproportionation is a function of the temperature and applied pressure, the quality of the electrode produced will also be influenced by these variables, as observed by Lechner and Sekerka [44]. By ensuring that metallic mercury is already present in the membrane during the pressing stage, the heating and preconditioning of the pellet can be eliminated [48]. It is also expected that because of the presence of metallic mercury, these electrodes will be more sensitive to redox couples in solution than membrane electrodes which respond primarily on the basis of ion exchange reactions [43].

3.1.2 Response in concentrated solutions. Bagg and Rechnitz [50] observed that on exposing $AgX-Ag_2S$ ($X = Cl^-, Br^-, I^-$) electrodes for a prolonged time to concentrated solutions of their anions, the electrodes showed evidence of surface damage. After immersion for more than one hour in solutions more concentrated than 1 molal, the surface of the chloride and bromide electrodes became matte and the electrode response in dilute solutions was no longer Nernstian. For the iodide electrode, surface deterioration was observed to begin at concentrations greater than 0.2 molal. Polishing the surface restored the sensitivity of the electrodes, provided that the exposure to the concentrated solutions had not been for too long.

In a subsequent study, the surfaces of AgI-$Ag_2S$ electrodes were examined with a scanning electron microscope after exposure for varying periods of time to solutions of varying iodide concentration [51]. Exposing an electrode to a $10^{-3}$ M or $10^{-1}$ M iodide solution for up to 17 h resulted in only a slight deterioration in the appearance of the electrode surface, although the surface was more pitted after exposure to the $10^{-1}$ M solution than to the $10^{-3}$ M solution, and the electrode continued to give Nernstian response. On exposing the electrode to a 0.6 M iodide solution the surface became considerably more pitted and after 17 h immersion, it gave very slow and unsteady response with sub-Nernstian slopes [51]. The authors concluded that the surface deterioration was due to the leaching of soluble silver iodide complexes, $AgI_{x+1}^{x-}$, from the membrane surface

$$AgI + xI^- \rightleftharpoons AgI_{x+1}^{x-} \tag{22}$$

The silver sulfide remaining in the membrane was unable to be as sensitive to changes in the iodide concentration as the original surface, giving rise to the sub-Nernstian response; the resultant membrane was also rough and porous, and the dynamic response of the electrode was slow and unsteady.

Malissa et al. [52] also used a scanning electron microscope to study the effect of exposing a silver iodide silicone rubber membrane electrode to concentrated iodide solutions. After soaking the electrodes in potassium iodide solutions more concentrated than 0.6 M for over 24 h, the electrodes did not respond at all to iodide ions. However, the electrodes could be regenerated by soaking in distilled water or 0.1 M potassium chloride or 0.1 M lithium chloride solutions for at least 5 h. On this basis, the authors concluded that the mechanism of iodide interference as proposed by Sorrentino and Rechnitz [51] was incomplete. Malissa et al. observed that surface dissolution due to the formation of the $AgI_3^{2-}$ complex was occurring, in agreement with the proposal of Sorrentino and Rechnitz [51], and that the dissolution continued until the silver iodide was completely dissolved, but they also observed that potassium and iodide ions were chemisorbed onto the active surface of the electrode. The authors concluded that an electrochemically inactive iodide-containing layer was being produced on the electrode surface, causing the electrode to cease responding to iodide ions in solution. However, the nature of the inactive layer and the mechanism for its removal by distilled water or chloride solutions, were not identified [52].

Baucke [53] observed that the response of AgI-$Ag_2S$ electrodes deteriorated after exposing the electrodes to hydroiodic acid at elevated temperatures; the response time increased and the electrodes acquired a yellow coating which was shown to be elemental sulfur. Baucke concluded that traces of iodine, formed in the solution due to the ingress of air, were oxidizing the electrode surface:

$$6I^- + O_2 + 4H^+ \rightleftharpoons 2I_3^- + 2H_2O \tag{23}$$

$$I_3^- + Ag_2S \rightleftharpoons S^0 + 2AgI + I^- \tag{24}$$

The electrode surface was thus covered with a non-conducting layer of elemental sulfur on which more silver iodide was precipitated; the deterioration of the electrode's response was due to the elemental sulfur layer. Baucke concluded that reactions (23) and (24) may also have occurred in the experiments reported by

Sorrentino and Rechnitz [51], except that the dissolution of the silver iodide substrate according to reaction (22) would prevent the build-up of a sulfur layer on the electrode's surface.

The reactions occurring at the electrode surfaces when exposed to concentrated iodide solutions appear to be complex, and the influence of dissolved oxygen on the possible reactions must also be considered as emphasized by Baucke [53]. The conclusion by Malissa et al. [52] that an electrochemically inactive chemisorbed iodide-containing layer is formed on the electrode surface is difficult to reconcile with their observations [52] that:

(i) during prolonged exposure to the solution, the entire silver iodide layer will dissolve from the membrane. One would expect that if the chemisorbed layer is electrochemically inactive, it would block further dissolution of the silver iodide; and

(ii) the concentration of apparent free iodide at the surface of the coated electrode is $10^{-1}$ M as calculated from the measured potential. Such a high apparent solubility is not consistent with such a long (2 h) soaking period required to regenerate electrode response and is also not consistent with the conclusion that the chemisorbed layer is electrochemically inactive. Much more positive values of the electrode potential would be expected if the chemisorbed layer is electrochemically inactive.

Further studies on the reactions occurring at the surfaces of these electrodes are needed.

3.1.3 <u>Response in dilute solutions.</u> At low halide concentrations, the solubility of the membrane itself must be considered when predicting the response of the halide ion-selective electrodes. The electrode response under these conditions for silver halide based electrodes is [42]

$$E = E^0_{AgX} - \frac{RT}{F} \ell n \left\{ \frac{[X^-] + ([X^-]^2 + 4K_{so})^{\frac{1}{2}}}{2} \right\} \qquad (25)$$

where $K_{so}$ is the solubility product for AgX. On the basis of the solubility product values, deviations from the Nernstian slope of 59 mV per decade should begin at halide ion concentrations of approximately $10^{-5}$, $10^{-6}$ and $10^{-8}$ M for the silver chloride, bromide and iodide electrodes respectively. However deviations from Nernstian response have been observed, particularly for the silver iodide electrode, at considerably higher concentrations than predicted by equation (25).

Morf, Kahr and Simon [54] proposed a theory wherein the detection limit of silver compound electrodes was governed not only by the solubility of the membrane material but also by the activity of the silver defects in the membrane surface. They stated that the activity of silver defects at the membrane surface was roughly constant for a given set of experimental parameters, but could be changed by a different conditioning of the membrane. The defect activity for the silver halide electrodes was given as $10^{-8.8}$, $10^{-6.3}$, $10^{-6.0}$ M for silver chloride, bromide and iodide respectively. Using electrodes that they themselves had made, they presented experimental evidence that appeared to confirm their theory. For a silver iodide membrane electrode, they observed that the electrode potential changed by approximately 200 mV as the iodide activity was decreased from $10^{-5}$ to $10^{-7}$ M, before levelling off as predicted by equation (25). However, they also stated that continuous use of silver iodide membrane electrodes in iodide solutions resulted in a reduction in the activity of the silver defects, leading to response curves that appeared to follow equation (25) more closely. These workers also claimed that this theory was verified by the response of their home-made silver chloride and silver sulfide electrodes.

Several groups have convincingly shown that this defect theory is not valid and have presented alternative explanations for the experimental results presented in support of the defect theory [54]. Kontoyannakos, Moody and Thomas [55] showed that, at iodide concentrations less than $10^{-5}$ M, oxidation reactions involving dissolved oxygen could significantly decrease the concentration of added iodide ions

$$4I^- + O_2 + 4H^+ \longrightarrow 2I_2 + 2H_2O \tag{26}$$

The linear Nernstian response of $AgI-Ag_2S$ electrodes was extended to almost $10^{-7}$ M when the dissolved oxygen was removed from solution by saturating with nitrogen or by adding an antioxidant such as ascorbic acid, as compared the limit of $10^{-5}$ M as reported by Morf, Kahr and Simon [54]. When iodide ions were lost from solution as a result of the oxidation reaction the electrode potential shifted to more positive values, thus offering one explanation for the large shifts in potential observed at concentrations less than $10^{-5}$ M by Morf, Kahr and Simon [54].

The results reported by Morf, Kahr and Simon [54] can also be explained by the leaching of soluble silver ion impurities from the membrane [2,22]. The presence of these silver ions at the electrode surface would only be noticeable at low iodide (or other halide) concentrations and would result in large changes in the potential to more positive values. As was discussed in Section 2.1.2, the silver iodide precipitate may become contaminated with soluble silver complexes through coprecipitation or occlusion reactions if the precipitates are improperly prepared [2,22]. The observation that the response of the silver iodide electrodes improved with continued use [54] can thus be attributed to the decrease in concentration of the silver ion impurities in the membrane, and not to changes in the defect concentration.

Hulanicki, Lewenstam and Majzurawska [56] also studied the response of silver iodide electrodes and obtained results that confirmed the conclusion that the activity of silver defects did not govern electrode response at low concentrations. By working in solutions containing ascorbic acid, Nernstian response could be attained at iodide concentrations as low as $2x10^{-7}$ M. However, significant adsorption of iodide ions onto the electrode surface was observed at iodide concentrations of $10^{-7}$ M. A typical adsorption isotherm was obtained and the saturation level of the electrode surface corresponded to a concentration of $1.5x10^{-7}$ M at 20°C. In addition, the extent of adsorption was strongly affected by the composition and state of the electrode surface and the adsorption increased with an increase in the mole fraction of silver iodide in the $AgI-Ag_2S$ membrane. Freshly polished membranes showed decreased adsorption and the adsorption was significantly decreased when the silver iodide was dispersed in a plastic material (heterogeneous membrane electrode).

Hulanicki, Lewenstam and Majzurawska [56] also noted that the concentration of ascorbic acid in solution affected the response of the electrode at iodide concentrations below $10^{-5}$ M. The electrode potentials in a 0.1 per cent ascorbic acid solution were considerably more positive than those measured in a 4 per cent ascorbic acid solution, with the latter potentials corresponding to those based on equation (25). The positive deviations in potential in the 0.1 per cent ascorbic acid solution were ascribed to the presence of an excess of silver ions at the electrode surface, arising from the leaching of soluble silver salts from the membrane; the soluble silver salts were produced by the air oxidation of iodide ions during the preparation and washing of the silver iodide precipitate.

When the ascorbic acid concentration in solution was increased to 4 per cent, the decrease in concentration of excess silver ions at the electrode surface was attributed to the reduction of silver ions by ascorbic acid [56]. An alternative and equivalent explanation for this influence of ascorbic acid concentration on electrode potential is that the 0.1 per cent solution was not sufficiently concentrated to reduce at pH 7 all the dissolved oxygen present. Iodide ions would be lost from solution due to oxidation by the residual dissolved oxygen, equation (26), resulting in the potential shifting to more positive values, and increasing the ascorbic acid concentration to 4 per cent ensured the absence of dissolved oxygen and eliminated the loss of iodide. An empirical equation, a modified form of equation (25) was proposed [56] to take into account the effect of ion adsorption and surface reduction processes. Since the parameters in this equation vary from one membrane to the next and with the composition, roughness etc. of a given membrane, the quantitative value of the equation is limited. Nevertheless, it is useful in that it does provide a feasible explanation of the factors that may affect the response of these electrodes at low concentrations.

Van de Leest and Geven [57] also reported that silver and iodide ions were adsorbed at the surface of silver-silver iodide electrodes, resulting in a deterioration of the electrode's response. The interference occurred after the electrode had been exposed to a concentrated (greater than 1 M) silver ion or a concentrated (unspecified concentration) iodide ion solution. The interference could be eliminated by converting the adsorbed ions into silver iodide by either chemical or electrochemical treatment. It was concluded that the interfering ions were adsorbed at the grain boundaries [57].

Van de Leest and Geven [57] also observed that pressed pellet electrodes of $Ag_3SI$ and $Ag_3SBr$ gave poor response and attributed this to the fact that silver and bromide ions had been adsorbed onto the $Ag_3SI$ and $Ag_3SBr$, respectively, during the precipitation reactions; chemical treatment of the membranes was again claimed to have removed the interference. While the effect of the chemical or electrochemical treatment of the electrodes was undoubtedly to improve their response, it is not clear that all the interference can be ascribed to adsorbed ions. The Ag/AgI electrode used by van de Leest and Geven had a limit of Nernstian response to silver ions of only $10^{-5}$ M, both in the untreated state and after the excess adsorbed silver ions had been removed by either the chemical or electrochemical method; the deviations from Nernstian response occur at much higher concentrations than predicted [42] or observed by others [55,56]. While details were not given about the method of preparation of the Ag/AgI electrode [57], it would appear that the untreated electrode already contained a large excess of soluble silver ions, thereby accounting for the poor sensitivity to silver ions. The dissolution of the electrode membrane, equation (22), must also be considered [50-52] in addition to possible adsorption reactions, when the effects of storing the iodide electrodes in concentrated iodide solutions are studied.

Iodide-selective electrodes have also been used successfully for the indirect potentiometric determination of various reducing agents [58,59]; the determination is based on the reaction

$$I_2 + Red \rightleftharpoons 2I^- + Ox \tag{27}$$

where Red and Ox represent the reduced and oxidized forms of the reducing agent, respectively. The ethanolic iodine solution used for the oxidation undergoes hydrolysis:

$$I_2 + 2H_2O \rightleftharpoons HIO + I^- + H^+ \tag{28}$$

and it has been observed [58] that when the iodide-selective electrode was immersed for a long period of time in solutions containing up to $10^{-4}$ M HIO, the electrode sensitivity decreased by two orders of magnitude, but polishing the surface restored the initial sensitivity of the electrode. The decreased sensitivity on prolonged exposure to HIO was presumably due to oxidation of the silver-sulfide at the electrode surface [53].

3.1.4 Electrode preparation. As previously discussed, the method of preparing precipitates for pressed pellet electrodes affects their sensitivity and response because of possible contamination of the precipitate by more soluble occluded or coprecipitated salts [2]. In addition to these problems, Vesely [60] observed that the properties of iodide ion-selective electrodes containing only silver iodide, depended on the ratio of the β-AgI and γ-AgI phases in the membrane. The β-AgI phase was produced when the precipitation was performed in the presence of an excess of iodide ions while the γ-AgI phase was produced when the precipitation was carried out in the presence of an excess of nitrate ions (that is, excess of silver ions). However the amount of each phase actually present in the electrode membrane depended not only on the method of precipitation of the silver iodide but also on the pressure and length of time of pressing of the precipitate pellets. Since at ambient temperatures and pressures the γ-AgI in the membrane was unstable and was gradually converted to the β-AgI phase, and since the properties of these phases (volume, adsorption properties, solubility and perhaps standard potential) could be different, Vesely concluded that it was not advisable to prepare iodide-selective electrodes consisting only of silver iodide. Iodide electrodes made from AgI-$Ag_2S$ mixtures, however, did not change their properties with time and it was presumed that this was due to the formation of the ternary compound, $Ag_3SI$ [60].

Thin films of silver halides have been prepared by alternative methods [61,62] and have been applied as halide ion-selective electrodes [61]. Van de Leest deposited thin silver halide films onto solid ionic conductors, such as $Ag_2S$, $Ag_3SBr$ or $Ag_3SI$ by either chemical or electrochemical methods. In the chemical method, the solid ionic conductor was pressed into a pellet and then heated in a halogen atmosphere

$$Ag_2S + X_2 \rightleftharpoons 2\ AgX + S \tag{29}$$

while in the electrochemical method an anodic current was passed through a pellet of the ionic conductor, which was contacted with silver metal, while immersed in a solution containing the appropriate halide ion. Silver ions generated within the membrane reacted with the halide ions in solution at the membrane/solution interface to form a thin layer of the silver halide. Electrodes responsive to chloride, bromide and iodide ions were prepared by the chemical method and X-ray diffraction data showed that the substrates were entirely covered with layers of pure silver halide, ranging in thickness from 1-5 μm. The response of the electrodes was fast, and a AgI-$Ag_2S$ electrode gave Nernstian response to iodide ions at $10^{-6}$ M. The response of the thin-layer AgCl-$Ag_2S$ electrode was claimed to be faster than that of a conventional pressed mixture of AgCl-$Ag_2S$; this was explained by the fact that silver chloride crystals prepared by solid-state reactions were superior, having a smaller solubility product value, than those prepared by wet chemical methods [61]. Similar conclusions have also been made for copper ion-selective electrodes which were prepared by electrodeposition [27].

While van de Leest did not report whether the silver iodide layer formed on the substrate surface by either the chemical or electrochemical means was present as the β or γ-AgI phase [61], a recent study has shown that silver iodide films, prepared by exposing a vacuum deposited silver layer to iodine vapour, consisted entirely of γ-AgI [62]. A silver iodide precipitate, prepared in the presence of an excess of iodide ions, consisted of both the β and γ phases, and it was observed that the precipitated silver iodide particles showed a considerable increase in crystal size after they had been aged in a $10^{-5}$ M iodide solution for at least 6 h at 80°C; however, the deposited silver iodide films did not show ageing effects. Properties of silver iodide films such as conductivity, crystal structure and crystal size are strongly affected by the conditions used in preparing the films [62]. Since these properties may also affect the solubility of silver iodide films, thereby influencing the response and sensitivity of ion-selective electrodes incorporating these films, further detailed investigations are required. The influence of film thickness, ranging from several micrometres to several millimetres, on these properties may be important when comparing the response of electrodes which have been prepared by widely differing techniques.

## 3.2 Selectivity Coefficients

The Nicolsky equation

$$E_A = E_A^o + \frac{RT}{F} \ln \left\{ [A] + k_{A,B}^{pot} [B] \right\} \quad (30)$$

expresses empirically the potential response for an ion-selective electrode to its primary univalent ion A in the presence of an univalent interfering ion B. $k_{A,B}^{pot}$ is defined as the potentiometric selectivity coefficient and it expresses the degree of preference of the electrode to ion A relative to ion B. This equation assumes that the ion-exchange reaction involving ions A and B and the electrode surface, is rapid and reversible. For a solid-state silver halide electrode, AgX, immersed in a solution containing not only $X^-$ but also another ion $Y^-$ with which it forms a sparingly soluble salt, the ion-exchange reaction is

$$AgX + Y^- \rightleftharpoons AgY + X^- \quad (31)$$

Assuming that this exchange reaction is rapid and reversible, and that the mixed-phase AgX-AgY formed at the electrode surface is in thermodynamic equilibrium with both halides in solution, it has been presumed that the electrode response can be given by the Nicolsky equation [42,63]

$$E_{X,Y} = E_X^o - \frac{RT}{F} \ln \left\{ [X^-] + k_{X,Y}^{pot} [Y^-] \right\} \quad (32)$$

where the potentiometric selectivity coefficient is now related to the equilibrium constant, K, for reaction (31) and is given by the ratio of the solubility products, $K_{so}$, of the salts [42,63]

$$k_{X,Y}^{pot} = K = \frac{K_{so}(AgX)}{K_{so}(AgY)} \quad (33)$$

There are two procedures for experimentally determining selectivity coefficients: the separate solution, and the mixed solution method. In the first procedure, Method I, the electrode potential is measured in separate solutions containing $X^-$ and $Y^-$ only

$$E_X = E_X^o - \frac{RT}{F} \ln [X^-] \quad (34)$$

$$E_Y = E_X^o - \frac{RT}{F} \ln k_{X,Y}^{pot}[Y^-] \quad (35)$$

and when the concentrations of $X^-$ and $Y^-$ are equal

$$k^{pot}_{X,Y} = \exp\left\{\frac{E_X - E_Y}{(RT/F)}\right\} \tag{36}$$

In the mixed solution method, Method II, the potential is again measured in a solution containing $X^-$ only, equation (34), and then compared to the potential measured in a solution that contains both $X^-$ and $Y^-$, equation (32). When the concentrations of $X^-$ and $Y^-$ are equal

$$k^{pot}_{X,Y} + 1 = \exp\left\{\frac{E_X - E_{X,Y}}{(RT/F)}\right\} \tag{37}$$

Serious discrepancies have been reported between the values of the selectivity coefficients calculated on the basis of the ratio of the solubility products, equation (32), and the values determined experimentally by Method I or Method II [64,65], but some workers have reported good agreement [63]. These discrepancies have arisen because some of the experiments were performed under conditions where the assumption that the mixed-phase layer was in equilibrium with the solution, was not valid.

Hulanicki and Lewenstam [65] developed a diffusion-layer model to interpret this variability in the values of the selectivity coefficients for various solid-state electrodes. Their model was based on the assumption that a mixed phase was produced at the electrode surface, and is valid only for this condition. The model outlined various factors which could affect the measured values of the selectivity coefficient and the influence of some of these factors has been verified experimentally [66,67].

3.2.1 Diffusion-layer model. In the diffusion-layer model, the response of a halide ion-selective electrode to a solution containing both $X^-$ and $Y^-$ is governed by the concentration of $X^-$ at the electrode surface, $[X^-]_0$. This surface concentration is related to the bulk concentration of $X^-$ and $Y^-$, their diffusion coefficients $D_X$ and $D_Y$ respectively, and the equilibrium constant, K, for the surface exchange reaction, equation (31)

$$[X^-]_o = \frac{KD_X}{KD_X + D_Y}[X^-] + \frac{KD_Y}{KD_X + D_Y}[Y^-] \tag{38}$$

Thus, the electrode response to a solution containing both $X^-$ and $Y^-$ is

$$E_{X,Y} = E^o_X - \frac{RT}{F}\ln[X^-]_o$$

$$= E^o_X - \frac{RT}{F}\ln\left\{\frac{KD_X}{KD_X + D_Y}[X^-] + \frac{KD_Y}{KD_X + D_Y}[Y^-]\right\} \tag{39}$$

By using the concepts of the diffusion-layer model to calculate the selectivity coefficients, the values calculated depend not only on the value of K but also on whether Method I or Method II is used for the calculation. Equation (39) was used [65] to represent the response of the electrode when exposed to solutions containing $Y^-$ only, or containing $X^-$ and $Y^-$. Hulanicki and Lewenstam [65] showed that the calculated value of the selectivity coefficient could range in value from the ratio $D_X/D_Y$ to the value of the equilibrium constant, K, but of more importance, it was also shown that these values were related to physical processes occurring at the electrode surface.

For a silver/silver chloride electrode exposed to a solution containing equimolar concentrations of bromide and chloride ions, the selectivity coefficient calculated on the basis of the diffusion-layer model, equation (39), was 1.02 while the value calculated on the basis of the solubility product ratio, equation (33), was 340. Experimentally Hulanicki and Lewenstam [65] determined that when the chloride and bromide concentration was $10^{-3}$ M, the selectivity coefficient determined by Method I and Method II was 2.1 and 2.7 respectively, but when the concentration of both halide ions was increased to $10^{-1}$ M, the experimental selectivity coefficients determined by Method I and II were 330 and 310. Hulanicki and Lewenstam also observed that on transferring the Ag/AgCl electrode from a $10^{-3}$ M solution of chloride and bromide ions to a $10^{-2}$ M chloride-bromide solution, the selectivity coefficient calculated from the initial change in potential was 2.0, in agreement with the value determined in the $10^{-3}$ M solution; however as the electrode was exposed to the $10^{-2}$ M solution for longer periods of time, the potential changed with time until it eventually acquired a stable value, corresponding to a selectivity coefficient of 330.

These changes in the value of the selectivity coefficient as a function of solution concentration and/or length of exposure were interpreted [65] as being caused by changes in the composition of the electrode surface. In the $10^{-3}$ M halide solution, the electrode surface was partially covered by a mixed phase of AgCl-AgBr, and the selectivity coefficient was thus given by the diffusion-layer model. In the $10^{-1}$ M halide solution, the bromide ion concentration was sufficiently large that the electrode surface was covered rapidly and completely by a silver bromide layer and thus the diffusion-layer model was no longer valid, and under these conditions the selectivity coefficient was given by the ratio of the solubility products. In the experiment where the electrode was transferred from the $10^{-3}$ to $10^{-2}$ M solution, the electrode was initially covered by a mixed phase layer of AgCl-AgBr and the initial potential response due to the changes in the chloride and bromide ion concentration was consistent with that predicted on the basis of the diffusion-layer model. Upon continued immersion of the electrode in the $10^{-2}$ M solution for a longer time, the electrode potential changed as the silver bromide layer continued to grow on the electrode surface, until the surface was covered and the potential was constant. The selectivity coefficient associated with this new potential was equal to that observed in the $10^{-1}$ M halide solution, and was determined by the solubility product ratio.

The experiments of Hulanicki and Lewenstam [65] demonstrate that selectivity coefficients vary according to the physical processes occurring at the electrode surface; thus, only under certain circumstances are they governed by equilibrium thermodynamic relationships. Since different workers have used different conditions and approaches in measuring selectivity coefficients, it is not surprising that widely differing results have been reported. Hulanicki and Lewenstam [65] concluded that the determination and interpretation of selectivity coefficients requires the knowledge of various non-thermodynamic parameters such as the degree of surface coverage of the electrode by the interfering compound, the rate of stirring (mass transport) and the length of time that the electrode is exposed to the interferent solution. These parameters are in turn affected by factors such as solution composition and concentration, electrode surface area and electrode pretreatment (the initial state of the surface) [65].

The importance of the physical processes occurring at the electrode surface on the determination of selectivity coefficients was illustrated by a scanning electron microscope (SEM) study of the iodide interference on silver chloride electrodes [67]. When a Philips chloride electrode was used to determine the selectivity coefficient for iodide ions, the separate solution procedure

(Method I) in $10^{-3}$ M halide solutions gave a value of 14, while the mixed solution procedure (Method II) in $10^{-4}$ M solutions gave values of 1.4 and 5 for stationary and stirred solutions respectively. These values are many orders of magnitude lower than the value of $10^6$ calulated from the solubility product ratio. The surface appearance of the electrodes was examined by SEM after exposure to solutions with different iodide concentrations for varying periods of time, in order to correlate the potentiometric response and electrode appearance.

The potentiometric response to chloride ions of electrodes that had been stored for 90 h in a $10^{-4}$ M iodide solution, or for 24 h in a $10^{-2}$ M chloride-$10^{-4}$ M iodide solution was unchanged. However, electrodes that had been exposed for 30 min to a $10^{-2}$ M iodide solution, or for 45 min to a $10^{-1}$ M iodide solution, gave almost no response to chloride ions. SEM examination of the electrode surfaces showed that those electrodes which had been stored in the $10^{-4}$ M iodide solution were covered by a layer of loosely packed crystals of silver iodide that were much larger than the silver chloride crystals in the membrane. Because of the large spacings between the loosely deposited silver iodide crystals, the silver chloride membrane was still in free contact with the sample solution and was able to respond to changes in the chloride concentration.

The silver iodide layers grown from a $10^{-2}$ M iodide solution consisted of small crystals with small holes between them. The layers grown from the $10^{-1}$ M iodide solution contained large flat crystals which blocked the membrane surface completely from the sample solution preventing the electrode from responding to the presence of chloride ions (ageing of silver iodide films has also been reported to affect their crystal size [62]).

The selectivity coefficients measured in the dilute iodide solutions were thus obtained with an electrode containing a mixed-phase layer, while for the value measured in the concentrated iodide solution the surface was covered with a layer of silver iodide; thus, the measured values of the selectivity coefficient agreed with the predictions of Hulanicki and Lewenstam [65]. In the dilute iodide solutions, stirring the solution altered the value of the selectivity coefficient as was predicted [65], due to the change in the rate of mass transport of iodide ions to the surface affecting the rate of growth of silver iodide crystals.

3.2.2 Verification of diffusion-layer concepts. Recently, Rhodes and Buck [66] have experimentally verified the conclusions of Hulanicki and Lewenstam [65] that the selectivity coefficients vary as a function of electrode surface coverage and exposure time. In modelling the observed response, Rhodes and Buck have also provided valuable information on the mechanism of the interference.

Rhodes and Buck [66] studied the interference of bromide ions at rotating anodized Ag/AgCl electrodes as a function of bromide-chloride concentration, thickness of the silver chloride layer and exposure time. When electrodes covered with thin silver chloride films were exposed to mixed bromide-chloride solutions, the potential-time response consisted of four distinct regions; at short times there was a potential overshoot (region 1) followed by a potential plateau (region 2). The potential then decreased to more negative values and began to level off to the breakdown point (region 3), where the potential fell sharply to those negative values expected for a silver bromide electrode (region 4). The length of the plateau in region 2, as well as the abruptness of the transition from regions 2 to 3, and regions 2 to 4, depended on the bromide concentration in solution and on the thickness of the silver chloride coating.

Using independent chloride and bromide-selective electrodes, Rhodes and Buck [66] simultaneously monitored the decrease in bromide ion concentration and the increase in chloride ion concentration in solution, as functions of the time of exposure of the anodized Ag/AgCl electrode to the bromide-containing solution. By assuming that the decrease in the bromide concentration in the bulk solution was due to the incorporation of bromide ions into the surface together with a release of an equivalent amount of chloride ions into solution, the rate of silver chloride depletion and silver bromide formation was modelled. The silver chloride layer was converted to silver bromide via the diffusion of bromide ions into, and chloride ions out of the electrode interior.

The extent of electrode conversion increased with increasing bromide concentration, but was independent of the initial chloride concentration in solution. By analysis of the potential-time data, Rhodes and Buck [66] also showed that the coverage of the electrode surface by silver bromide increased with time, and that the value of the selectivity coefficient also varied with time. Thus, the selectivity coefficient ranged in value from 4 to 32 in the plateau region, from 10 to 74 in the breakdown region and attained a limiting value of 450 in the final region of constant potential values, when the electrode surface was completely converted from silver chloride to silver bromide. The potential-time response observed by Rhodes and Buck was qualitatively the same as that observed by Hulanicki and Lewenstam [65] on transferring the Ag/AgCl membrane electrode directly from a $10^{-3}$ M to a $10^{-2}$ M bromide solution and the interpretation of the data was qualitatively similar as well.

In calculating the surface coverage from the potential data, Rhodes and Buck [66] proposed two different models for the electrode response. In the first model, the measured electrode potential was assumed to be equal at all points on the electrode surface and was determined by the average surface silver ion concentration, $[\overline{Ag^+}]$

$$E_{meas} = E^{o}_{Ag+/Ag} + \frac{RT}{F} \ln [\overline{Ag^+}] \tag{40}$$

This latter value was defined in terms of the solubility products of the salts, the fractional surface coverage, S, and the bulk solution concentrations

$$[\overline{Ag^+}] = (1-S)\, K_{so(AgCl)}/[Cl^-] + (S)\, K_{so(AgBr)}/[Br^-] \tag{41}$$

In the second model, the measured potential was assumed to be a mixed potential, due to the average potential of the two salts

$$E_{meas} = (1-S)\, E_{AgCl/Cl^-} + (S)\, E_{AgBr/Br^-} \tag{42}$$

Using these two models, Rhodes and Buck [66] determined from the potential-time data the rate constant for the ion-exchange reaction at the electrode surface, equation (31). They concluded that the second model, that is, the assumption that the measured potential was a mixed potential, provided the most consistent, solution-independent rate constant and was thus the better model for these processes.

Rhodes and Buck [66] also noted that the thickness of the silver chloride layer affected the determination of the selectivity coefficient. When a standard Orion electrode ($Ag/AgCl-Ag_2S$, about 4 mm thick) was exposed to low bromide concentrations, the value of the selectivity coefficient was small, 5-10, and remained relatively constant implying that the surface coverage by silver bromide was also remaining constant. It was concluded [66] that the silver bromide which was originally formed at the chloride electrode surface was diffusing into the thick silver chloride membrane; the latter functioned as a sink for bromide ions and could scavenge sufficiently low concentrations of bromide ions to prevent them from interfering with the electrode's response [65].

The observations and conclusions of Rhodes and Buck [66] have significant implications on the use and construction of ion-selective electrodes. When solid-state electrodes are exposed to ions which form a less soluble compound with one of the membrane components, the thickness of the membrane becomes an important parameter. If the electrode contains only a thin film of the electroactive material, prolonged exposure of that electrode to such an interference may result in the conversion of a substantial fraction of the electroactive material. Due to the thinness of the electroactive layer, it may not be possible to restore the original electrode response by polishing.

For thicker membranes, additional problems may also arise because the inward diffusion of the interfering ion, $Y^-$, into the membrane will cause a non-uniform variation with depth of the electrode's composition (AgX-AgY) as a function of the concentration and time of exposure to the interference. Consequently, when the electrode is polished or otherwise treated to remove the outermost interference layer (AgY), the non-uniform mixed-phase layer AgX-AgY will be exposed to the solution. Since the resulting measured potential will be a mixed potential, dependent on the relative amounts of AgX and AgY in the membrane, the electrode response will be variable and unpredictable. Thus, it is not advisable to use thin membranes (usually prepared by solid-state or electrochemical reactions) in the presence of interferences, or to expose thick membranes for long periods of time to concentrated solutions of interferences; the interference may be obvious (change in electrode potential due to complete coverage of the electrode surface) or much more subtle (measurement of mixed potentials due to changes in the composition of the membrane in its interior as well as at its surface).

On the basis of the work of Hulanicki and Lewenstam [65] and Rhodes and Buck [66], it is debatable whether or not the concept of "a potentiometric selectivity constant" is really valid for silver halide selective electrodes. The usefulness of the concept of a selectivity constant is that it predicts, on the basis of equilibrium thermodynamic principles, those ions and their concentration that will interfere with the response of the electrode. However, the extent of the interference by a given ion at a fixed concentration depends on a time-dependent variable, the fractional surface coverage; this parameter is, in turn, a complex function of non-thermodynamic parameters such as electrode surface area and thickness, solution concentration and length of exposure time to it, past history of the electrode, etc. It is more useful to regard the measured potential as being a mixed potential with a concentration-dependent surface coverage.

## 4. SILVER SULFIDE ELECTRODES

Silver sulfide is widely used in making ion-selective electrodes; the silver sulfide can be used alone as an electrode of the first kind ($Ag_2S$, responsive to silver ions), alone or mixed with other silver salts as electrodes of the second kind ($Ag_2S$, responsive to sulfide ions; AgX-$Ag_2S$, responsive to halide ions) and mixed with metal salts as electrodes of the third kind (MS-$Ag_2S$, responsive to metal ions where M = $Cu^{2+}$, $Pb^{2+}$, $Cd^{2+}$).

Chemical reactions involving the surface of the mixed salt electrodes were discussed in the previous sections; the majority of the reactions involved only the silver halide or copper(II) sulfide component of the membrane since silver sulfide is more inert (less soluble, more difficult to oxidize and/or reduce) than these components. However, recent studies of silver sulfide electrodes have shown that chemical reactions can also alter the surface of the silver sulfide electrode, thereby affecting its response to low levels of sulfide ions.

## 4.1 Low-level sulfide response

The silver sulfide electrode is a silver ion conductor and it responds to changes in the concentration of sulfide ions in solution since they alter the concentration of silver ions present at the electrode surface as a result of the solubility of the silver sulfide membrane. However, the testing of the electrode's response to sulfide ions is complicated by the fact that sulfide ions are extensively hydrolyzed in aqueous solutions and also are very susceptible to air oxidation. Since hydrogen sulfide is a weak acid

$$H_2S \rightleftharpoons H^+ + HS^-, \quad K_{a1} = 10^{-7} \qquad (43)$$

$$HS^- \rightleftharpoons H^+ + S^{2-}, \quad K_{a2} \sim 10^{-13} \qquad (44)$$

the sulfide ion concentration is related to the total dissolved sulfide concentration $[S^{2-}]_{TOT}$ through the equations

$$[S^{2-}]_{TOT} = [H_2S] + [HS^-] + [S^{2-}] \qquad (45)$$

$$[S^{2-}] = \frac{[S^{2-}]_{TOT} \cdot K_{a1} \cdot K_{a2}}{[H^+]^2 + K_{a1}[H^+] + K_{a1}K_{a2}} \qquad (46)$$

Since the silver sulfide electrode responds directly to changes in the concentration of the sulfide ion, it is evident from equation (46) that pH changes as well as changes in $[S^{2-}]_{TOT}$ will affect the electrode's response.

### 4.1.1 Nernstian and super-Nernstian response.

Hseu and Rechnitz [68] showed that the silver sulfide electrode would give Nernstian response to sulfide ion concentrations as low as $10^{-7}$ M. However, this response was attained by varying the pH of solutions that were well-buffered with respect to the total dissolved sulfide, that is, contained $3x10^{-5}$ M total dissolved sulfide. Baumann [69] attempted to measure directly levels of total dissolved sulfide as low as $3x10^{-7}$ M by working in highly alkaline solutions, and observed significant deviations from the expected Nernstian response. The response above $1.5x10^{-6}$ M total dissolved sulfide was nearly Nernstian, but at lower concentrations the calibration curve showed negative deviations from Nernstian response. This super-Nernstian response corresponded to increased curvature of the calibration curve towards more positive potential values indicating apparent sulfide ion concentrations lower than those added. These measurements were performed in alkaline solutions containing ascorbic acid which was added to remove the dissolved oxygen from solution and thus prevent the loss of dissolved sulfide through air oxidation reactions. Baumann also observed that the response time of the sulfide electrode varied with its history of use and with the solution concentration.

Crombie, Moody and Thomas [70] showed that the silver sulfide electrode would give Nernstian response to concentrations of total dissolved sulfide as low as $10^{-6}$ M in a sodium hydroxide (1 M) solution which was deaerated with nitrogen. If, however, ascorbic acid was added to the alkaline solution to remove more effectively the low levels of dissolved oxygen present, the Nernstian response could be extended an order of magnitude lower to $2x10^{-7}$ M total dissolved sulfide. Super-Nernstian response was observed when the calibration curve was extended to concentrations of total dissolved sulfide lower than $10^{-7}$ M even in the presence of ascorbic acid. The authors concluded that this response was due to the loss of sulfide ions from solution as a result of possible oxidation and/or adsorption reactions [70].

These results reported by Crombie, Moody and Thomas demonstrated convincingly that the defect theory of Morf, Kahr and Simon [54], to the effect that the concentration ($3x10^{-6}$ M) of silver ion defects in the silver sulfide membrane determined the limit of detection of the electrode, was in error as has already been discussed in greater detail in Sections 2.1 and 3.1.

A silver sulfide electrode has been used in an on-line monitor for continuous $H_2S$-in-water measurements, wherein the sample was mixed with an alkaline solution containing a reducing agent such as ascorbic acid or hydrazine [71,72]. It was reported that when the electrode was exposed to solutions containing 1 per cent hydrazine for several weeks, it gradually lost its sensitivity, had a super-Nernstian response and the membrane became dull and tarnished in appearance [71]. Polishing the membrane removed the tarnished layer and restored the electrode's sensitivity and Nernstian response at the $3x10^{-6}$ M level. One electrode which had been in continuous use in the $H_2S$-in-water monitor for three weeks failed to give a Nernstian response at the $10^{-5}$ M level, even after polishing [72]. Subsequent tests showed that the super-Nernstian response was due to interference from deposits of metallic silver at the electrode surface.

4.1.2 <u>Interference from deposits of metallic silver.</u> After further polishing of the "failed" electrode [72], this electrode and a new silver sulfide electrode were calibrated simultaneously in the same alkaline-ascorbic acid solution. The new electrode gave Nernstian response to $10^{-7}$ M total sulfide while the failed electrode gave super-Nernstian response at concentrations lower than $3x10^{-6}$ M [73]. Since both electrodes were exposed simultaneously to the same solution, the difference in their response had to be related to differences in the electrodes themselves. After further use in the laboratory it was noted that the new electrode also began to deviate from Nernstian response at $10^{-6}$ M total sulfide, while the failed electrode began to deviate at $10^{-5}$ M. Similar observations were noted for other silver sulfide electrodes; the concentration of total dissolved sulfide at which super-Nernstian response was obtained varied from one electrode to another but it was generally in the range $10^{-5}$-$10^{-6}$ M. The response time of a given electrode at a given concentration also increased as the electrode response at that concentration deteriorated from Nernstian to super-Nernstrian [73].

Films would rapidly appear on the surface of the failed electrode during its exposure to alkaline-ascorbic acid solutions. SEM examination showed that there were various deposits on the highly pitted or porous surface and that the most prevalent of these appeared as platelets; X-ray spectra showed that these platelets were rich in silver but depleted in sulfur and it was presumed that they could be deposits of either silver oxide ($Ag_2O$) or metallic silver [73]. On the basis of these SEM results it was concluded that the super-Nernstian response was due to the measurement of mixed potentials which arose because of the heterogeneity of the electrode surface. The silver-rich platelet deposits were assumed to be metallic silver and it was proposed that the presence of the metallic silver at the membrane/solution interface gave rise to mixed potentials because of the response of the silver metal to redox couples (such as ascorbic acid) present in solution. It was proposed [73] that the silver metal accumulated at the membrane/solution interface either by the rapid diffusion of silver through the membrane from the solid silver contact at the back of the membrane, or by redox reactions between reducing agents (ascorbic acid, hydrazine) and the membrane surface.

Subsequent studies by X-ray diffractometry (XRD) confirmed the presence of metallic silver at the silver sulfide membrane/solution interface [74].

The XRD studies revealed that varying amounts of metallic silver were present at the surfaces of the electrodes examined, but those electrodes which had been exposed for a prolonged period to an alkaline-ascorbic acid solution contained the most metallic silver. The silver layer could be removed by polishing and then a new layer of metallic silver could be re-deposited onto the electrode surface by re-exposure to the alkaline-ascorbic acid solution. Attempts to observe the diffusion of silver metal from the silver metal contact to the outer membrane surface by heating an electrode were unsuccessful. Thus, it was concluded [74] that the primary source of the silver metal observed at the electrode surface was the reduction of the silver sulfide at the membrane surface by the ascorbic acid present in the solution.

Voltammetric studies showed that these silver sulfide electrodes responded to the presence of redox couples in solution, and that the voltammetric response was qualitatively similar to that observed at a silver metal electrode, both in the absence and presence of electroactive species in solution [74]. Well-defined voltammetric oxidation peaks were observed for the oxidation of ascorbic acid or iron(II) ions at the silver sulfide electrode; the oxidation of the iron(II) ions was diffusion-controlled over a limited range of concentration and voltage scan rate, while the oxidation of the ascorbic acid involved a pre-adsorption step. These voltammetric results were in contrast to a previous report that there was a large resistance to electron-transfer across the silver sulfide/solution interface [75]. When graphite-contacted silver sulfide electrodes were used, there was no apparent voltammetric response to the redox couples in solution; this was attributed [74] to the much lower electronic conductivity of graphite-contacted silver sulfide electrodes as compared to silver-contacted silver sulfide electrodes [75].

To explain the observed response of the silver-contacted silver sulfide electrodes to the redox couples in solution, it was proposed [74] that the electrode membrane contained continuous "threads" of metallic silver extending from the back of the membrane through to the electrode/solution interface. These threads formed as a result of the reduction of the silver sulfide to metallic silver by the ascorbic acid and the subsequent inward diffusion of the metallic silver to the back of the membrane, meeting the metallic silver diffusing outwards from the silver contact at the back of the membrane. The resulting threads could therefore function as the sites for the electron-transfer reactions involving the redox species both in the voltammetric and potentiometric studies. In the potentiometric studies, the electrode would thus be measuring mixed potentials, giving rise to the super-Nernstian response [74]. However, it was not possible to differentiate between electrodes that gave good or poor Nernstian response to sulfide ions on the basis of their voltammetric response to iron(II) ions, since all the silver contacted electrodes gave qualitatively the same voltammetric response. Further studies are needed in this area.

4.1.3 <u>Interference from other reactions.</u> The response of the silver sulfide electrode to low levels of sulfide may also be affected by surface oxidation processes as well as by surface reduction reactions. As was discussed in Section 2.1, various workers reported that metal sulfide electrodes became oxidized on exposure to oxygen, iron(II) ions and iodine, and similar oxidations can be expected to occur at the silver sulfide electrode; such reactions have not yet been reported or studied for electrodes containing only silver sulfide. Presumably this is because of the experimental problems associated with making reliable measurements at total dissolved sulfide concentrations less than $10^{-6}$ M, and because of the difficulty in differentiating between the effects of oxidation of the electrode surface and oxidation of the added sulfide ions. However, it was reported recently [76] on the basis of ESCA studies of a silver sulfide-tris(thiourea) copper(I) monohydrogenphosphate electrode that the silver sulfide in the membrane became oxidized on exposure to aerated distilled water. Measurements showed that after exposing the electrode for one week to deaerated and aerated distilled water, the apparent silver ion solubility [77] was $6x10^{-6}$ M and $8x10^{-5}$ M respectively. This increase in silver ion solubility in the oxygenated solution was attributed to the oxidation of the silver sulfide to the more soluble silver sulfate [76]. Surface oxidation effects of this kind can also be expected to affect the response of the silver sulfide electrodes at low sulfide concentrations.

It has been claimed [78] that in order to obtain Nernstian response below $10^{-6}$ M total dissolved sulfide in alkaline solutions, the alkaline-ascorbic acid solution used in the measurements had to be "pre-titrated" with a sulfide ion solution to remove heavy metal impurities; use of such "pre-titrated" solutions was claimed to give near-Nernstian response to total dissolved sulfide levels approaching $10^{-8}$ M [78]. While the presence of heavy metal impurities can undoubtedly affect the response and sensitivity of the electrode to low levels of sulfide [69], the practice of "pre-titrating" the solution does not appear to be totally justified or advisable. Nernstian response as low as $10^{-7}$ M has been attained in alkaline-ascorbic acid solutions without pre-titration of the reagents [70,73] and there is no significant difference in response between using reagent grade and extra-pure chemicals [79]. It seems probable that the reported improvement in sensitivity was due to "pre-titration" of the electrode surface. The response of the silver sulfide electrode improves with successive use during the day [69,79] and hence the effect of the pre-titration may have been to overcome interference from soluble silver ions present as a result of oxidation of the electrode surface, or to overcome interference from the mixed potential arising from the metallic silver that may have been formed at the electrode surface as a result of storing and conditioning the electrode in an alkaline-ascorbic acid solution.

4.1.4 Minimizing interference. To minimize interference in the response of silver-contacted silver sulfide electrodes, it is necessary to eliminate the reducing agents from solution. This minimizes the interference from mixed potentials as it not only prevents the reduction of the silver sulfide to metallic silver but also eliminates the source of the redox couples in solution which are sensed by the metallic silver. Elimination of the reducing agents, however, poses problems in measuring levels of total dissolved sulfide less than $10^{-6}$ M in alkaline solutions, because of the loss of sulfide ions through oxidation by traces of dissolved oxygen present in solution [70]. Graphite-contacted silver sulfide electrodes could be used instead of the silver-contacted electrodes, as the former appear to be less sensitive to redox couples in solution [74]. However, prolonged or continual use of these graphite-contacted electrodes in alkaline-ascorbic acid solutions is not advisable since reduction of the electrode surface to metallic silver will result in changes of the silver sulfide stoichiometry, resulting in changes of the electrical conductivity and standard potential of the electrode [25].

The best way to accurately measure low levels of dissolved sulfide and eliminate the problems associated with the use of reducing agents, is to work in an acidic (pH 5) environment. Under these acidic conditions, reducing agents need not be added to the acid reagent which is mixed with the sample stream because of the very slow rate at which hydrogen sulfide is oxidized by dissolved oxygen in this pH range [80]. In addition, since the amount of acid required to bring the sample to a pH less than 5 is generally very small, the amount of dissolved oxygen introduced into the sample by the reagent is also very small. The sensitivity of the electrode is not compromised by working in these acidic conditions, as Nernstian response to total dissolved sulfide levels as low as $1x10^{-7}$ M in a pH 5 solution can be obtained [81]. A silver sulfide electrode has been operating for twelve months in a continuous $H_2S$-in-water monitor wherein the sample is acidified prior to measurement, and deterioration of the electrode's response or sensitivity has not been observed [81].

4.2 Other electrodes based on silver sulfide

Apart from the copper and halide ion-selective electrodes discussed previously, lead and cadmium ion-selective electrodes have been prepared using $PbS-Ag_2S$ and $CdS-Ag_2S$ precipitates [49,82,83], and other chalcogenides [84].

The conditions for precipitation of the mixed salts and for pressing of the resultant membranes were investigated thoroughly to obtain electrodes with the best electrochemical response [49,83]. As in the case of the $CuS-Ag_2S$ electrodes [22], the best results were obtained when the precipitation was performed by adding a solution containing both the silver and metal ion to an excess of sulfide ions. However in this case, deoxygenation of the solutions did not affect the response of the resultant electrodes [83]. The precipitate containing the lead sulfide was dried under a stream of nitrogen prior to heating and pressing [83], but apparently the presence of an inert atmosphere was not required for the cadmium electrode [49]. X-ray diffractometry showed that the electrodes contained only the metal sulfide salts, and that ternary compounds were not present.

Ageing effects in the use of these electrodes have been reported [49,82,83,85,86] for the electrode response became sluggish and was accompanied by drifting potentials and poor reproducibility. Good response may be restored by polishing the membrane and the ageing effects have been attributed to surface oxidation effects [82,83]. Because of the higher solubility of the lead or cadmium sulfide as compared to the silver sulfide, the surface reaction presumably involves the oxidation of the metal sulfide salt to elemental sulfur or to sulfate

$$2MS + O_2 + 4H^+ \rightleftharpoons 2M^{2+} + 2S^0 + 2H_2O \qquad (47)$$

$$MS + 2O_2 \rightleftharpoons M^{2+} + SO_4^{\ 2-} \qquad (48)$$

Interference from strong oxidizing agents such as iron(III) and dichromate ions has also been reported [87] and can again be attributed to membrane oxidation effects. A more subtle interference was observed when the lead electrode was used for the indirect determination of sulfate in organic solvents; the lead electrode was "poisoned" by minute amounts of peroxide present in the organic solvents such as p-dioxane [88,89]. The poisoning was presumably due to oxidation of the sulfide by the peroxide.

A study of the lead and cadmium electrodes in the presence of complexing agents showed that their response was very similar to that of the copper ion-selective electrode. The lead and cadmium electrodes responded correctly to the lead and cadmium ion activities, respectively, in bulk solution in the presence of the ligands, trien and tetren. In the case of EDTA, both electrodes appeared to respond only to the bulk concentration of EDTA as their response was independent of the presence of the corresponding metal-EDTA complexes [49,83]. Similar anomalous behaviour of the copper electrode to EDTA was discussed in Section 2.3, and an explanation for the response of all three electrodes to the presence of EDTA is needed.

## 5. AREAS REQUIRING FURTHER ATTENTION

Specific topics requiring further investigation have been identified when individual electrodes were discussed in the previous sections. However, there are also more general areas requiring further consideration so that the response of the solid-state electrodes can be better understood and/or predicted. Some of these considerations are significant from a fundamental-theoretical point of view and may also have some importance from a commercial-production point of view.

## 5.1 Methods of electrode preparation

The methods used to prepare the membranes for solid-state electrodes affect significantly the response of these electrodes. However the preparation of properly functioning homogeneous and heterogeneous electrodes from precipitated metal salts is still a black art, as there appear to be problems in combining good electrochemical properties with good mechanical properties of the membrane.

The electrochemical properties of the membrane (sensitivity, response time, potential stability) may be affected by factors such as purity of the precipitate (due to the method and conditions used in precipitating the metal salts), the conditions used in pressing the pellet, the mole ratio of the metal salts, the phase or modification of the salts, the crystal size and the ionic and electronic conductivities of the metal salts. The mechanical properties of the membranes (the strength and durability of the pellets) can be affected not only by the conditions used to prepare the precipitates but also by the conditions used in pressing the pellets (temperature, pressure and time). The mole ratio of the salts in the pellet can also affect its mechanical properties. There is an apparent interaction between the parameters controlling the mechanical and electrochemical properties, but the details of this interaction are not known.

The influence of the purity of the metal salt precipitates on the electrochemical response can be readily understood. However, the influence of phase changes (e.g. $\gamma$-AgI vs $\beta$-AgI), the formation of ternary compounds ($Ag_{1.5}$-$Cu_{0.5}S$), changes in stoichiometry (e.g. $Cu_{1.76}S$ vs CuS), changes in crystal size (e.g. due to ageing effects in solution) on the electrode's response and sensitivity are not at all well understood. In addition, not enough is known about the influence of temperature, pressure and length of time of the pellet pressing process on the above variables and on the mechanical properties of the pellets themselves. Increased use should be made of surface analysis techniques such as X-ray diffractometry, ESCA and Auger Spectrometry to study the pressed pellets; results from these surface studies could then be correlated with the results of careful studies of electrode response in an attempt to understand the controlling variables. Some studies of this type have already been reported, but more are required.

It appears that some benefits can be gained from preparing electrodes by solid-state vapor-phase interactions at elevated temperatures, or by electrochemical synthesis. Much purer and better-defined salts should be produced by these procedures since the conditions for these syntheses can be controlled much better than those for the wet chemical precipitation reactions. However, as the electrodes are cooled, phase changes and decomposition reactions may occur and the influence of such reactions should be considered.

The thickness of the membrane used should also be of some concern. If a thin silver-salt based electrode containing a solid silver contact is used in a reducing environment, metallic silver may quickly saturate the membrane and give rise to interference by the measurement of mixed potentials. Similarly if thin electrodes are used in solutions containing interferences or resulting in oxidation of the membrane surface, the life-time of the electrode may be limited due to the rapid removal of the thin electroactive layer by repeated polishing.

## 5.2 Studies of electrode response

The influence of membrane properties on the response of the electrode is generally not evident until the electrode is exposed to concentrations lower than $10^{-5}$ M. Thus, it hardly seems appropriate to conclude that an electrode gives "good response" if it is not calibrated or tested at concentrations where such surface effects can be detected, despite the analytical difficulties encountered in preparing reliable and stable dilute solutions of air-sensitive compounds, or problems associated with the loss of ions through adsorption or complexation reactions. Attention must be directed not only at the solution conditions but also at the condition of the electrode surface during rigorous and thorough calibration of electrodes in low concentration regions to permit an assessment of the variables controlling electrode response.

The mixed solution method, Method II, appears to be more appropriate for determining selectivity coefficients for interfering ions. In this method the electrode is exposed to both ions in solution and thus the electrode can indicate much more effectively and accurately the extent to which the reactions at its surface have attained equilibrium. However, variables such as electrode surface area and thickness, the previous history of the electrode's use, the solution concentration and length of time of exposure to the solution must also be considered when interpreting this response.

## 5.3 Electrical contact to the membrane

Hitherto, attention has been directed at the influence on electrode response of reactions occurring at the membrane's exterior surface/solution interface, but the ion-exchange reactions occurring at the membrane's inner surface are equally important in governing the electrode's response [42,43]. While experimentally it may be most convenient to make electrical contact with the membrane by using an internal liquid contact, there can be practical problems in using such contacts under conditions of very high or very low temperatures or where rapid and large oscillations in temperature are encountered. Solid electrical contacts can be used instead and these have various advantages over the liquid contacts; the requirements of the solid conductors which can be used for making contact to different membranes have been discussed [25,42,43]. When different solid contacts are used, changes in the stoichiometry of the membrane phase occur with resultant changes in the standard potential and conductivity of the electrode [25,26,75].

A prime requirement of solid electrical contacts to the membrane is that rapid, reversible exchange of the potential determining ion must occur at the solid/membrane interface [25,42]. In many cases, however, this solid/solid junction is an electrical contact between a conductor and a semi-conductor and little is known about the electrode processes at a solid conductor/semi-conductor interface [92]. The response of an ion-selective electrode may be affected, particularly in solutions of low concentration where the exchange current densities at the outer membrane/solution interface will be small, because of the presence of a solid/solid electrical contact in the electrode since the procedure used in making the solid/solid contact can seriously affect the exchange reactions at this interface. For example, the exchange reactions at the $Ag/Ag_2S$ interface can be slow due to the presence of a high ohmic contact resistance [25,26,75,91,93] unless special precautions are taken in preparing this interface [25,75]. Even then, problems may arise from the fact that the $Ag/Ag_2S$ interface may not be at equilibrium with the result that the contact resistance may increase with time [75,91].

The problems associated with making the solid/solid electrical contact may perhaps be alleviated if the electroactive membranes are prepared by solid-state or electrochemical reactions on appropriate conducting supports. This may prove to be a significant advantage for the preparation of electrodes via this route as compared to the precipitation-pressed pellet method. The solid/solid contact at a conductor/semi-conductor interface is an area that requires and is currently receiving more attention, both on a theoretical and experimental basis.

## 6. ACKNOWLEDGEMENT

The author thanks Dr. Andrzej Lewenstam for valuable discussions during the latter's visit to Canada.

## 7. REFERENCES

1. R.P. Buck, Anal. Chem., 50, 17R (1978).

2. R.P. Buck, Theory and principles of membrane electrodes. In: Ion-Selective Electrodes In Analytical Chemistry, Vol. 1, Ed. H. Freiser, Plenum, New York (1978).

3. R.M. Smith and A.E. Martell, Critical Stability Constants, Vol. 4, Plenum, New York (1976).

4. G. Johansson and K. Edström, Talanta, 19, 1623 (1972).

5. W.J. Blaedel and D.E. Dinwiddie, Anal. Chem., 46, 873 (1974).

6. D. Midgely, Analytica Chim. Acta, 87, 7 (1976).

7. D. Midgely, Analytica Chim. Acta, 87, 19 (1976).

8. A. Hulanicki, M. Trojanowicz and T. vel Krawczyk, Water. Res., 11, 627 (1977).

9. A. Hulanicki, M. Trojanowicz and M. Cichy, Talanta, 23, 47 (1976).

10. A. Hulanicki and A. Lewenstam, Talanta, 23, 661 (1976).

11. H. Hirata, K. Higashiyama and K. Date, Analytica Chim. Acta, 51, 209 (1970).

12. G.J. Moody, N.S. Nassory, J.D.R. Thomas, D. Betteridge, P. Szepesvary and B.J. Wright, Analyst, 104, 348 (1979).

13. G.J.M. Heijne, W.E. van der Linden and G. den Boef, Analytica Chim. Acta, 89, 287 (1977).

14. W.J. Blaedel and D.E. Dinwiddie, Anal. Chem., 47, 1070 (1975).

15. M.J. Smith and S.E. Manahan, Anal. Chem., 45, 836 (1973).

16. D.J. Crombie, G.J. Moody and J.D.R. Thomas, Talanta, 21, 1094 (1974).

17. G.B. Oglesby, W.C. Duer and F.J. Millero, Anal. Chem., 49, 877 (1977).

18. Y.S. Fung and K.W. Fung, Anal. Chem., 49, 497 (1977).

19. E. Pungor, K. Tóth, M.K. Pápay, L. Pólos, H. Malissa, M. Grasserbauer, E. Hoke, M.F. Ebel and K. Persy, Analytica Chim. Acta, 109, 279 (1979).

20. J. Pick, K. Tóth and E. Pungor, Analytica Chim. Acta, 65, 240 (1973).

21. K. Okamoto and S. Kawai, Japan J. Appl. Phys., 12, 1130 (1973).

22. R.P. Buck, Anal. Chem., 48, 23R (1976).

23. G.J.M. Heijne and W.E. van der Linden, Analytica Chim. Acta, 93, 99 (1977).

24. L.F. Heerman and G.A. Rechnitz, Anal. Chem., 44, 1655 (1972).

25. M. Koebel, Anal. Chem., 46, 1559 (1974).

26. R.E. van de Leest, Analyst, 102, 509 (1977).

27. M. Neshkova and H. Sheytanov, J. Electroanal. Chem., 102, 189 (1979).

28. M. Lamache and D. Bauer, Anal. Chem., 51, 1320 (1979).

29. T. Hepel, M. Hepel and M. Leszko, Analyst, 102, 132 (1977).

30. A. Lewenstam, private communication (1979).

31. J.W. Ross, Solid-state and liquid membrane ion-selective electrode. In: Ion-Selective Electrodes, Ed. R.A. Durst, Natl. Bur. Stand. (U.S.) Spec. Publ., 314, 82 (1969).

32. R. Jasinski, I. Trachtenberg and D. Andrychuk, Anal. Chem., 46, 364 (1974).

33. P. Lanza, Analytica Chim. Acta, 105, 53 (1979).

34. J.C. Westall, F.M.M. Morel and D.N. Hume, Anal. Chem., 51, 1792 (1979).

35. G.J.M. Heijne and W.E. van der Linden, Analytica Chim. Acta, 96, 13 (1978).

36. V.K. Olson, J.D. Carr, R.D. Hargens and R.K. Forcé, Anal. Chem., 48, 1228 (1976).

37. M.F. El-Taras, E. Pungor and G. Nagy, Analytica Chim. Acta, 82, 285 (1976).

38. I. Sekerka and J.F. Lechner, Anal. Lett., A 11, 415 (1978).

39. G. Nakagawa, H. Wada and T. Hayakawa, Bull. Chem. Soc. Japan., 48, 424 (1975).

40. Y. Umezawa, Y. Imanishi, K. Sawatari and S. Fujiwara, Bull. Chem. Soc. Japan., 52, 945 (1979).

41. A. Jyo, T. Hashizume and N. Ishibashi, Anal. Chem., 49, 1868 (1977).

42. R.P. Buck, Anal. Chem., 40, 1432 (1968).

43. R.P. Buck and V.R. Shepard, Anal. Chem., 46, 2097 (1974).

44. J.F. Lechner and I. Sekerka, J. Electroanal. Chem., 57, 317 (1974).

45. I. Sekerka and J.F. Lechner, J. Electroanal. Chem., 69, 339 (1976).

46. P.K.C. Tseng and W.F. Gutknecht, Anal. Lett., 9, 795 (1976).

47. P.K.C. Tseng and W.F. Gutknecht, Anal. Chem., 48, 1996 (1976).

48. A. Hulanicki, R. Lewandowski and A. Lewenstam, Analytica Chim. Acta, 110, 197 (1979).

49. W.E. van der Linden and R. Oostervink, Analytica Chim. Acta, 108, 169 (1979).

50. J. Bagg and G.A. Rechnitz, Anal. Chem., 45, 271 (1973).

51. M.H. Sorrentino and G.A. Rechnitz, Anal. Chem., 46, 943 (1974).

52. H. Malissa, M. Grasserbauer, E. Pungor, K. Tóth, M.K. Pápay and L. Pólos, Analytica Chim. Acta, 80, 223 (1975).

53. F.G.K. Baucke, Fresenius Z. Anal. Chem., 282, 105 (1976).

54. W.E. Morf, G. Kahr and W. Simon, Anal. Chem., 46, 1538 (1974).

55. J. Kontoyannakos, G.J. Moody and J.D.R. Thomas, Analytica Chim. Acta, 85, 47 (1976).

56. A. Hulanicki, A. Lewenstam and M. Majzurawska, Analytica Chim. Acta, 107, 121 (1979).

57. R.E. van de Leest and A. Geven, J. Electroanal. Chem., 90, 97 (1978).

58. R. Christova, M. Ivanova and M. Novkirishka, Analytica Chim. Acta, 85, 301 (1976).

59. S. Ikeda and J. Motonaka, Analytica Chim. Acta, 90, 257 (1977).

60. J. Vesely, Coll. Czech. Chem. Comm., 39, 710 (1974).

61. R.E. van de Leest, Analyst, 101, 433 (1976).

62. K.J. Peverelli and H.P. van Leeuwen, J. Electroanal. Chem., 99, 157 (1979).

63. E. Pungor and K. Tóth, Analytica Chim. Acta, 47, 291 (1969).

64. G.A. Rechnitz and M.R. Kresz, Anal. Chem., 38, 1786 (1966).

65. A. Hulanicki and A. Lewenstam, Talanta, 24, 171 (1977).

66. R.K. Rhodes and R.P. Buck, Analytica Chim. Acta, 113, 67 (1980).

67. H.A. Klasens and J. Goossen, Analytica Chim. Acta, 88, 41 (1977).

68. T.M. Hseu and G.A. Rechnitz, Anal. Chem., 40, 1054 (1968).

69. E.W. Baumann, Anal. Chem., 46, 1345 (1974).

70. D.J. Crombie, G.J. Moody and J.D.R. Thomas, Analytica Chim. Acta, 80, 1 (1975).

71. J. Gulens and B. Labbatte, A. Energy Can. Ltd., AECL-Rep., AECL-5542 (1976).

72. J. Gulens, K. Jessome and C.K. MacNeill, Analytica Chim. Acta, 96, 23 (1978).

73. J. Gulens and B. Ikeda, Anal. Chem., 50, 782 (1978).

74. J. Gulens and D.W. Shoesmith, submitted for publication (1980).

75. M. Koebel, N. Ibl and A.M. Frei, Electrochim. Acta, 19, 287 (1974).

76. A.C. Wilson and K.H. Pool, Analytica Chim. Acta, 109, 149 (1979).

77. A. Ringbom, Complexation in Analytical Chemistry, Interscience, New York (1963).

78. I. Sekerka and J.F. Lechner, Analytica Chim. Acta, 93, 139 (1977).

79. J. Gulens and A. Geerts, unpublished results.

80. K.Y. Chen and S.C. Morris, Environ. Sci. Technol., 6, 529 (1972).

81. J. Gulens, unpublished results.

82. M. Mascini and A. Liberti, Analytica Chim. Acta, 60, 405 (1972).

83. G.J.M. Heijne, W.E. van der Linden and G. den Boef, Analytica Chim. Acta, 100, 193 (1978).

84. H. Hirata and K. Higashiyama, Talanta, 19, 391 (1972).

85. P. Kivalo, R. Virtanen, K. Wickström, M. Wilson, E. Pungor, K. Tóth and G. Sundholm, Analytica Chim. Acta, 87, 387 (1976).

86. P. Kivalo, R. Virtanen, K, Wickström, M. Wilson, E. Pungor, G. Horvai and K. Tóth, Analytica Chim. Acta, 87, 401 (1976).

87. M.J.D. Brand, J.J. Militello and G.A. Rechnitz, Anal. Lett., 2, 523 (1969).

88. W. Selig, Mikrochim. Acta (Wien), 168 (1970).

89. W. Selig and A. Salomon, Mikrochim. Acta (Wien), 663 (1974).

90. G.B. Marshall and D. Midgely, Analyst, 103, 438 (1978).

91. R.K. Rhodes and R.P. Buck, Analytica Chim. Acta, 110, 185 (1979).

92. M. Kleitz and J. Dupuy, Editors, Electrode Processes in Solid State Ionics, D. Reidel, Boston (1976).

93. K. Camman and G.A. Rechnitz, Anal. Chem., 48, 856 (1976).

# THE AUTHOR

Dr. Janis Gulens

Janis Gulens was born in Germany in 1944 of Latvian parents. He emigrated to Canada in 1950, and received his B.Sc. in 1967 from the University of Toronto. He obtained his Ph.D. in 1971 from Queen's University (Kingston, Ontario) working with Professor John A.Page on "Electrochemical Studies of Osmium Ammine Complexes". The following 18 months were spent with Professor Fred C.Anson at California Institute of Technology as a Research Fellow, on a National Research Council of Canada Postdoctoral Fellowship. He joined Atomic Energy of Canada Limited in 1973 and since that time has been working in the Chemistry and Materials Division at the Chalk River Nuclear Laboratories.

His research interests are in electroanalytical and environmental chemistry, with current emphasis on ion-selective electrodes and their application to environmental monitoring and analysis of surface and ground waters.

Dr. Gulens is a member of the Chemical Institute of Canada, the American Chemical Society and the Electrochemical Society.

*Ion-Selective Electrode Rev.* Vol.2, pp. 159-218. 0191-5371/81/0201-0159$05.00/0


# ANALYTICAL CONTROL OF DRUG-TYPE SUBSTANCES WITH MEMBRANE ELECTRODES

Vasile V. Cosofret

Institute of Chemical and Pharmaceutical Research, 74351, Sos. Vitan, 112 Bucharest-4, Romania

CONTENTS

KEYWORDS: Analysis of drugs; direct potentiometry; drugs - inorganic and organic; ion-selective electrodes for drugs and drug-type substances.

## 1. INTRODUCTION

Developments with ion-selective electrodes undoubtedly represents the greatest progress in the field of analytical electrochemistry for the last 10-15 years. The publishing of a journal (Ion-Selective Electrode Reviews) specially dedicated to reviewing progress supports various other activities, such as, the first book on the theoretical and practical aspects on the ion-selective membrane electrodes by Moody and Thomas (1) and the reviews periodically published by Buck (2-6) in "Analytical Chemistry" and by Koryta (7,8) in "Analytica Chimica Acta".

During the last ten years, several symposia and conferences on ion-selective electrodes have been held and their proceedings have been published (9-13). Some recent books have brought together the principal theoretical and practical advances of ion-selective electrodes (14-24) while other books contain comprehensive chapters on the subject (25-29).

Control of the quality of drugs is one of the analytical chemistry branches with broad impact on public health and hence the finding of new sound, quick and accurate methods for determination of the purity of the active ingredients is welcomed.

In the last few years the application of ion-selective membrane electrodes to drug-quality control has been shown to be viable, but unfortunately none of the modern pharmacopoeias has so far introduced the ion-selective electrode assay technique. Nevertheless, this review demonstrates that ion-selective electrodes have a definite role in the assay of drug-type substances.

## 2. INORGANIC COMPOUNDS

Several inorganic compounds are used in medical practice and many of them are as yet irreplaceable from the therapeutical point of view. The determination of inorganic drug-substances with ion-selective electrodes has been performed from the emmergence of these electrochemical sensors. In most cases, however, reference has not been made to the therapeutical action of inorganic drug-substances determined by ion-selective electrodes. For the wide range of inorganic materials that have been determined with ion-selective electrodes, the following review has selected only the principal aspects concerning the utilization of these devices for the determination of the best known and most used drug-substances.

### 2.1 Aluminium

Aluminium is used as a therapeutic only for its local effects. Its soluble salts (aluminium acetate, aluminium chloride, aluminium and potassium sulphate, aluminium sulphate) have an astringent and antiseptic action. Insoluble compounds, e.g., aluminium hydroxide, are strong adsorbants, and are used as antacids.

Aluminium in solution can be determined by means of direct or indirect complexometric titration at concentrations of $10^{-2}$ and $10^{-5}$ M, respectively (30). $Al^{3+}$ ions in solution show a marked tendency to hydrolyse with the formation of soluble polynuclear hydroxo- and aquo-complexes and a precipitate of aluminium hydroxide. The complexometric determination of aluminium at low concentrations is difficult because both aluminium hydroxide and the soluble polynuclear complexes are rather unreactive while aluminium ions react relatively slowly with EDTA (31). Also, the back titration of excess of EDTA with a colorimetric indicator is difficult because the colour change at the end-point is not easy to detect.

An aluminium ion-selective membrane electrode is not available, but $Al^{3+}$ determination is possible by using the $LaF_3$ crystal membrane electrode whereby aluminium is titrated with fluoride in ethanol (32). In the presence of sodium ions an insoluble cryolite ($Na_3AlF_6$) or a similar compound is formed:

$$Al^{3+} + 6NaF \longrightarrow Na_3AlF_6 + 3\,Na^+ \quad (1)$$

The ratio of six fluorides to one aluminium favours the determination of small quantities of aluminium. Samples may be titrated potentiometrically with 0.1M NaF at pH 5.0 in _ca_. 70% (m/v) ethanol.

A simpler method for aluminium determination is also described in the literature. This is based on the titration to a fixed potential of a fluoride-selective electrode, at a constant pH, when the fixed potential represents a constant level of free fluoride (33,34).

Chang and Cheng (35) proposed a new method for the determination of aluminium indirectly with a divalent ion-selective electrode (Orion, Model 92-32). The method is also suitable for other tri- and tetravalent cations. The procedure is based on the differentiation between stability constants of Al(III)-EDTA and Mg-EDTA complexes. If both Al(III) and Mg-EDTA are present, the amount of Al(III) can be determined by measuring the amount of magnesium displaced. In the measurement of Al(III), the displacement reaction occurs below pH 5.0 at a temperature of 80°C; however, a turbidity occurs when the Al(III) concentration is not strong enough to prevent the formation of aluminium hydroxide. The proposed method for determining $Al^{3+}$ ions is based on two steps; the first is the displacement reaction and the second is the potential measurements of free $Mg^{2+}$ ions with a divalent electrode. The optimum pH range for determining $Al^{3+}$ is 6.5 - 6.8. In Mg-EDTA solution, $Mg^{2+}$ forms a stable complex with EDTA above pH 7.0, but at pH below 7.0, the Mg-EDTA complex tends to dissociate to give an increase in free $Mg^{2+}$ ions which results in an increase in the potential with interference in the measurement. Above pH 10, magnesium forms $MgOH^+$ and $Mg(OH)_2$, which are not measured by the divalent electrode. Therefore, the optimum pH range for the two steps lies between 7 and 10. The slope of the calibration curve for $Al^{3+}$ ions in the displacement reaction of $Al^{3+}$ with Mg-EDTA was 29.3 mV decade $^{-1}$ (because the slope is very near to the theoretical value, displacement can be regarded as complete). The precision in this method was found to be the same as that for the measurement of the divalent ions, being ± 2% in the range $2 \times 10^{-3}$ - $2 \times 10^{-4}$M.

2.2 Ammonium

The ammonium ion has an important role in the living organism, being a part of the renal regulation of acid-base balance. Ammonium chloride is a systemic acidifying substance which is rapidly absorbed from the gastro-intestinal tract.

A common electrode, the ammonia electrode, can be used both for ammonia solution (a local irritant) and for other ammonium salts (carbonate, bicarbonate, etc.). This electrode is available from manufacturers: Orion (Model 95-10), Philips (IS-561-$NH_4$), Beckman (No.39565), EIL (8002-8), etc.

Commercially available ammonia electrodes have a hydrophobic gas-permeable membrane to separate the alkaline test solution from the internal solution (0.1 M $NH_4Cl$). A glass pH-electrode and a Ag/AgCl reference electrode are immersed in the internal solution. Since the internal ammonium ion concentration is very large, the equilibrium equation for the hydrolysis of ammonia, namely

$$[NH_4^+] \; [OH^-] \; / \; [NH_3] \quad = \quad K \sim 2 \times 10^{-5} \qquad (2)$$

can be approximated by

$$[OH^-] \; / \; [NH_3] \quad = K' \qquad (3)$$

for small values of $[NH_3]$.

Thus, when the electrode is immersed in an alkaline test solution, ammonia diffuses through the membrane until the partial pressure of ammonia is the same on the both sides of the membrane. The $NH_3$ concentration in the filling solution is thus altered and causes a pH change which is monitored by the glass electrode. The potential of the electrode sensing element with respect to the internal reference element, varies in a Nernstian manner with changes in the hydroxide level:

$$E = E_o - S \log [OH^-] \qquad (4)$$

where S is the numerical value of the slope. Since the hydroxide concentration is proportional to ammonia concentration (eq.3) the electrode response to ammonia is also Nernstian:

$$E = E_o' - S \log [NH_3] \qquad (5)$$

A thorough treatment of the theory of the operation of ammonia electrodes, as well as for other sensing elements from the same category ($SO_2$, $NO_2$, $H_2S$, HCN) which are commercially available has been given by Ross et al. (36).

The need for determining ammonium ion in its various compounds has promoted investigations resulting in $NH_4^+$ - selective membrane electrodes. As a result of the development of such a cation selective glass-electrode (37), ammonium ion determination is possible both in aqueous solutions and in non-aqueous media. Modified $NH_4^+$-sensitive glass electrodes (38, 39) are also known and have been obtained by covering the cationic ($H^+$, $Na^+$ or $K^+$)- selective glass electrode by an ammonia permeable membrane. Similarly, ammonium ion-selective electrodes have been developed, based on ion-exchange resins (40) or neutral carrier agents (41-44).

## 2.3 Calcium

Calcium is essential for the good functioning of tissues in normals and for maintaining functional intergrity of central and peripheral nervous system. It is also necessary for normal functioning of the heart, being one of the factors influencing blood coagulation. Although calcium ion activities play an important role in biological systems, reliable measurement of the free ions has become possible only recently (45).

In 1967, Ross (46) obtained the first electrode selective to $Ca^{2+}$. This was a really valuable electrode, containing a liquid membrane based on the calcium salt of didecylphosphoric acid (0.1 M) dissolved in di-n-octylphenylphosphonate. The fair selectivity of this sensor towards typical components of blood serum has led to numerous applications in the biomedical sciences and eventually also to its use in analytical, inorganic, organic and physical chemistry. This type of electrode was manufactured by Orion (Model 92-20 and 93-20), soon followed by Beckman (No.39608), Corning (EE-Ca), EDT Research, Philips (IS-560-Ca and ID-561-Ca), Radiometer (F2110 Ca and F2112 Ca), etc.

These electrodes are subject to pH interference to an extent depending on the calcium levels in the samples. The chemical properties of the calcium alkylphosphate-based membranes also create certain problems during the use of these electrodes for indication during complexometric determination of calcium in samples, such as serum or sea water (47), where the sodium level is so high that it impairs electrode performance. Various attempts have been made to overcome these drawbacks.

Different electroactive materials and devices have been suggested (48-57) with various measures of success. Also, the Orion electrode has been modified. A significant advance was the electrode of Thomas et al. (52,53) in which the original ion-exchanger was incorporated in a PVC matrix. Use of such plastics matrix membranes results in substantially prolonged electrode life time, and allows the reservoir of ion-exchanger within the electrode body to be discarded. Sensors of calcium bis di(4-alkylphenol)phosphates (alkyl = hexyl, octyl and 1,1,3,3-tetramethylbutyl) in conjunction with dioctylphenylphosphonate solvent mediator in PVC matrices give good calcium ion-selective electrodes with near-Nernstian slopes and detection limits of $1.9 \times 10^{-6}$ to $2.7 \times 10^{-6}$ M $Ca^{2+}$ in the absence of ion buffers. Electrodes with the octyl-based sensor, when calibrated at 25°C with calcium-ion buffers, gave calcium-ion detection limits of ca. $10^{-8}$ M (54).

The e.m.f. response of the electrodes in $10^{-3}$ M $CaCl_2$ in 0.1M NaCl was stable within a pH range of from below 5 to about 9. All the above electrodes showed good selectivity towards $Ca^{2+}$ over $Na^+$, $K^+$ and $Mg^{2+}$.

Some other very interesting works covering the obtaining, characterization and applications of new calcium-selective electrodes have been developed by Thomas and co-workers (55-65). In addition, the neutral carrier sensor of Simon and co-workers is a significant development (49,66). The general progress in designing calcium ion-selective electrodes has recently been reviewed (67, 68).

Růžička and Tjell (69) have followed their experience with the Orion electrode for calcium measurements in a flowing stream by the introduction of other electrode designs (90) including the di(octylphenyl)phosphate calcium ion sensor (71). The membranes were prepared by having a mixture of dioctylphenylphosphonate and the calcium salt of the new calcium ion-sensor, di-n-octylphenyl-phosphoric acid, in the mass ratio 10:1, and combined with PVC dissolved in tetrahydrofuran. The membrane was cast in a 40 mm diameter glass ring by the method of Griffiths et al.(53).

Pranjić-Anušić (72) have determined the calcium contents of some pharmaceutical products by potentiometric digital titration with EDTA using a Philips calcium ion-selective electrode (Model, IS-561-Ca). For example, for the potentiometric titration of calcium gluconate, the sample was first acidified with hydrochloric acid solution, followed by boiling for 5 min. The solution was then buffered to pH 9.2 (borate buffer) and the titration performed with 0.05 M EDTA solution. (In digital titrations of samples from Vital effervescent tablets and Vital granules, containing calcium gluconate, relative standard deviations of ± 0.7% and ± 0.5%, respectively, were obtained). The total potential change between the beginning of titration and 100% excess of titrant was about 100 mV.

## 2.4 Iron

Many drugs have an influence on blood and haematopotic organs, either directly or indirectly. Among them, iron(II or III) has an important role and it is accepted that the iron ions from inorganic salts are absorbed, assimilated and incorporated by haemoglobin.

An electrode selective to $Fe^{2+}$ or $Fe^{3+}$ ions is not yet commercially available, but determination of the ions using membrane-electrodes is not difficult with either tetrachloroferrate(III) selective electrodes, or indirectly with other ion-selective electrodes, e.g., $Cu^{2+}$, $Hg^{2+}$, etc.

Cattrall and Chin-Poh Pui (73) have determined iron(III) by using a coated-wire ion-selective electrode sensitive to tetrachloroferrate(III) ions. The electroactive membrane is composed of the tetrachloroferrate(III) salt of Aliquat 336 S combined with PVC. A linear near-Nernstain response for this electrode is obtained over the total iron(III) concentration range of $10^{-1}$ - $10^{-4}$ M for solutions containing a controlled total chloride ion concentration of 6M. The electrode was used in potentiometric titration with EDTA (74).

A liquid-membrane electrode for iron(III) determination has been described by Hopirtean (75). The electro-active component was triphenylpyrilium tetrachlorferrate ($TPh\pi \cdot FeCl_4$) and the liquid membrane electrode was obtained by soaking the pores of a $G_4$ glass frit made hydrophobic and attached to the end of a glass tube with $TPh\pi \cdot FeCl_4$ solution in tetrachloroethane. The organic solution was also used as internal solution with a saturated calomel electrode as inner reference electrode. The electrode response is Nernstian towards $FeCl_4^-$ ions over $10^{-1}$ - $10^{-4}$ M concentration range (slope 58.5 mV decade $^{-1}$).

For the complexometric titration of iron(III), Hopirtean et al. (76) used a $Hg^{2+}$ -sensitive membrane electrode. The determination was based on the substitution reaction of the Hg-EDTA complex. The $Hg^{2+}$ liberated is titrated at pH4 (adjusted with hexamine) with EDTA solution in the presence of the $Hg^{2+}$ -sensitive membrane electrode ($K_{FeY}$ = 1.26 x $10^{25}$ and $K_{HgY^{2-}}$ = 6.31 x $10^{21}$). A blank correction was needed because the $HgY^{2-}$ used contained a small excess of $Hg^{2+}$.

An interesting procedure for iron(III) determination has recently been described by Lazarou and Hadjiioannou (77) using a perbromate-selective membrane electrode, containing the water-insoluble Crystal Violet-perbromate dissolved in chlorobenzene as the electro-active material. The electrode exhibits rapid and near Nernstian response (slope 57-61 mV decade$^{-1}$) to perbromate activity from $10^{-5}$ to $10^{-2}$ M in the 2 to 10 pH range. The determination of iron(II) is based on the reaction:

$$2Fe^{2+} + BrO_4^- + 2H^+ \longrightarrow 2Fe^{3+} + BrO_3^- + H_2O \qquad (6)$$

The relative standard deviation for the determination of a 4 x $10^{-4}$M Fe(II) sample was 1.2% (n=5).

An automatic reaction rate method for the determination of iron(II), especially in the presence of 1,10-phenanthroline acting as an activator has recently been described (78). The procedure is based on iodide reduction of perbromate to bromate in alkaline or neutral solutions:

$$BrO_4^- + 3I^- + 2H^+ \longrightarrow BrO_3^- + I_3^- + H_2O \qquad (7)$$

and to bromide in acidic solutions. The reaction proceeds very slowly under all conditions of acidity and alkalinity. However, when iron(II) is added, the iron is oxidized by perbromate (see eq.6) and this induces the oxidation of iodide. The kinetics of the iron(II)-induced perbromate-iodide reaction was studied with an iodide-selective electrode (Orion, Model 94-53) to monitor the course of the reaction. The time required for the reaction to consume a fixed amount of iodide, and thus for the potential to increase by a preselected amount (10.0 mV) is measured automatically and related directly to the iron(II).

## 2.5 Magnesium

Magnesium, like calcium, is an ion of the utmost physiological and pharmacological importance, being needed for the functional integrity of the neuromuscular system. At higher concentrations, it is an inhibitor for the neuro-muscular system and for the central nervous system.

Up to date, there is no selective electrode for magnesium, but there are divalent ion electrodes. The Orion divalent cation electrode (Model 93-32) consists of an electrode body and a replaceable pre-tested sensing module. The sensing module contains a gelled internal filling solution - a liquid ion - exchanger of an organic amine which can form complexes with all divalent metal ions, and a membrane saturated with liquid ion-exchanger. The membrane separates samples from electrode filling solution and is selective for divalent cations, e.g., $Ca^{2+}$, $Mg^{2+}$, etc. If a complexing agent is used to mask all divalent metal ions except magnesium, the divalent ion electrode will be selective for $Mg^{2+}$ ions in the presence of the complexing agent. For example, ethyleneglycol-bis (2-aminoethyl-ether)tetraacetic acid (EGTA) makes it simple to distinguish $Mg^{2+}$ from $Ca^{2+}$ (79).

Cheng and Cheng (80) report the results of an attempt to adapt a divalent ion electrode (Orion, Model 93-32) as a selective electrode for $Mg^{2+}$ ions with EGTA. When the interfering polyvalent cations are absent, addition of EGTA is not necessary. The calibration curve is linear to $Mg^{2+}$ in the concentration range 1M - 6 x $10^{-6}$M, with a slope of about 25 mV decade $^{-1}$. A pH of 7.0 $\pm$ 0.2 was selected for measuring the activity of magnesium. The electrode exhibited good response times (99% response in 1 min or less) for concentrations above 2 x $10^{-5}$M. Magnesium was determined by direct potentiometry in solutions of between $10^{-5}$ and $10^{-1}$M (80).

It is also possible to determine magnesium by potentiometric titration at pH 9.7 (adjusted with glycine-NaOH buffer) with EDTA solution with a calcium ion-selective electrode (81). The electrode response to changes in magnesium activity is slower than for calcium. Magnesium in the 0.7-5 mg range was determined semi-automatically with average errors of about 0.3%. The titration technique yielded more accurate results than those obtained by the direct method for the titration curve is reproducible and the equivalence point is easily located.

## 2.6 Sodium and potassium

Sodium and potassium ions play an important part in cellular metabolism, being unreplaceable from this point of view.

Most of sodium and potassium salts for therapeutical purposes are used mainly for the action of their anions rather than of the cations (see Table 1). When the pharmaceutical product is used for its cation action ($Na^{+}$ or $K^{+}$), a supplimentary assay is recommended, and a $Na^{+}$- or $K^{+}$- selective membrane electrode may be used as indicator electrode.

Several manufacturers include a sodium-selective glass electrode among their range of products, the principal ones being the Orion 94-11, Corning 47621000, Beckman 39278, Philips G15Na, Radiometer G502 and Schott 9601 electrodes. A neutral carrier membrane electrode is now available (Philips IS-561-Na) and this is less subject to interference by hydrogen ions than the glass electrodes, but its selectivity with respect to other ions is generally poorer and it has a higher limit of detection ($10^{-5}$M, in the 7-10 pH range) (23).

For potassium there are three different types of commercially available electrodes, whose characteristics and properties vary

TABLE 1. Sodium and potassium drug substances

| Name and Formula | Function | Name and Formula | Function |
|---|---|---|---|
| Potassium acetate $CH_3COOK$ | Electrolyte replenisher | Sodium carbonate $Na_2CO_3 \cdot 10H_2O$ | Systemic alkalizer |
| Potassium bicarbonate $KHCO_3$ | Treatment of gastric hyper-acitidy | Sodium chloride NaCl | Electrolyte replenisher |
| Potassium bromide KBr | Sedative and hypnotic | Sodium fluoride NaF | Dental carriers prophylactic |
| Potassium chloride KCl | Electrolyte replenisher | Sodium lactate $C_3H_5\ NaO_3$ | Fluid and electrolyte replenisher |
| Potassium citrate $C_6H_5K_3O_7 \cdot H_2O$ | Electrolyte replenisher | Sodium nitrite $NaNO_2$ | Antidote to cyanide poisoning |
| | | Sodium phosphate $Na_2HPO_4 \cdot 7H_2O$ | Saline purgative |
| Potassium iodide Kl | Antifugal; expectorant; source of iodine | Sodium salicylate $C_7H_5NaO_3$ | Antipyretic and analgesic |
| Sodium amino-salicylate $C_7H_6NaO_3N \cdot 2H_2O$ | Antibacterial (tuberculo-static) | Sodium and potassium tartrate $C_4H_4KNaO_6 \cdot 4H_2O$ | Saline purgative |
| Sodium biphosphate $NaH_2PO_4 \cdot H_2O$ | Cathartic | | |
| Sodium bicarbonate $NaHCO_3$ | Systemic alkalizer | Sodium thiosulphate $Na_2S_2O_3 \cdot 5H_2O$ | Antidote to cyanide poisoning |
| | | Sodium sulphate $Na_2SO_4 \cdot 10H_2O$ | Saline purgative |

especially from the point of view of selectivity and life times.

Potassium ion-selective glass electrodes are durable and have stable standard potentials and sensitivities, but poor selectivity for $K^+$ over $Na^+$ and $NH_4^+$. These electrodes are discussed in detail by Lakshiminarayanaiah (18) and by Moody and Thomas (1). Among such electrodes are the EIL-GKN 33, Beckman 39137 and Corning 47622000 types.

Liquid ion-exchanger electrodes of the neutral carrier type have relatively high selectivities for potassium but are generally less stable than the glass membrane type.

The neutral carrier electrodes are usually based on valinomycin (Orion 93-19, Philips IS-561-K and IS-560-K, Metrohm EA 301K, Radiometer F2312K, Beckman 39622, Radelkis OP-K-711D, Etc.). This antibiotic, produced by the Streptomyces fulvissimus, forms complexes with alkali metal ions (82), the stabilities of which decrease in the order $Rb^+ > K^+ \gg Na^+ > Li^+$. The remarkable difference in the stabilities of the potassium and sodium complexes is not only of importance in biomedical sciences (83) but is also the reason for the high selectivity of the potassium electrode ($k^{pot}_{K,Na} \sim 10^4$) based on this compound.

The third type of potassium ion-selective electrodes are based on liquid ion-exchangers containing derivitives of potassium tetraphenylborate, but these are not of the high quality of the valinomycin systems.

The following sub-sections describe possible determinations of the principal salts of $Na^+$ and $K^+$ based on the estimation of the respective inorganic anion. Salts of organic anions will be discussed with the organic compounds of the second part of this work.

2.6.1 Bicarbonate, carbonate, carbon dioxide. Electrodes for the measurement of the partial pressure of carbon dioxide in solution or in the gaseous phase are now manufactured by Orion (Model 95-02) and Radiometer (Type E 5036). Since the introduction of the $pCO_2$ electrode by Stow et al. (84) and Severinghaus and Bradley (85) over 20 years ago, the basic construction and principles of operation of potentiometric gas-sensing electrodes have remained essentially unchanged. The Severinghaus type $pCO_2$ electrode consists of an internal pH electrode and reference electrode contained within an outer electrode body and gas permeable membrane. The pH sensor is positioned within the electrode so that a thin film of the internal electrolyte, which contains bicarbonate, exists between the surface of the pH electrode and the gas permeable membrane.

Jensen and Rechnitz (86) found that the electrode response to $H_2CO_3$ is given by

$$E = E' + \frac{RT}{F} \ln (K_1/N) + \frac{RT}{F} \ln [H_2CO_3] \qquad (8)$$

where E' is a function of the internal reference electrode and the concentration of the reference ion; N is the concentration of sodium bicarbonate in the internal electrolyte and $K_1$ is the first equilibrium constant of carbonic acid.

Theoretical aspects concerning the functioning of carbon dioxide sensing electrode have been discussed by other authors in the last few years (36, 87-90).

More recently, it has been recognized that the glass membrane pH electrode can be replaced by electrodes sensitive to other ions, thereby providing potentiometric sensors selective to a wider range of different gases (87). Also, Růžička and Hansen (91) have shown that it is possible to dispense with the gas-permeable membrane in the so-called air-gap electrode. The air gap separates the electrolyte layer from the sample solution. The advantages of air-gap electrodes in comparison with classical membrane electrodes have been well advocated (92-96).

Herman and Rechnitz (97, 98) have reported a liquid membrane electrode which is highly selective to carbonate over $Cl^-$, $SO_4^{2-}$ and $HCO_3^-$. The electrode was constructed with an Orion 92-liquid membrane electrode body and a Millipore VC cellulose acetate support membrane. Commercially available hydrophobic Orion calcium membranes appeared to give identical response. The active liquid phase consisted of Aliquat 336S, dissolved in trifluoroacetyl-4-butylbenzene (1%, v/v). The salt was used in chloride form or after conversion to the hydrogen carbonate form. A mixture of sodium chloride and sodium hydrogen-carbonate, both 0.1 M, was used as an inner reference solution. The response of the electrode is linear from $10^{-7}$ to $10^{-2}$M carbonate, with a Nerstian slope.

2.6.2 Bromide, chloride, iodide. Membrane electrodes selective to halide ions contain either heterogeneous or homogeneous membranes. In heterogeneous solid-state membrane electrodes the silver halide precipitate is held in a polymer matrix (paraffin wax, silicone rubber, etc.). The functional quality of the membrane depends on the physical nature and quality of the precipitate incorporated into the membrane.

Homogeneous solid-state electrodes for $Cl^-$, $Br^-$ and $I^-$ are prepared and supplied by manufacturers. They give stable and reproducible responses for a large range of solution conditions and concentrations. The disadvantages of certain membranes (silver chloride and silver bromide) are their resistance and the need to work in constant light because of the development of photoelectric potentials, but the membrane materials are now mixed with silver sulphide. Silver iodide is a good conductor, but this material in pure form is difficult to form into a pellet and is also mixed with silver sulphide.

Růžička et al. (99,100) prepared and studied electrodes sensitive to halides by activating the surface of graphite electrodes with electroactive precipitates of the silver halides. The approach of using only a superficial layer of the electroactive compound adsorbed on the surface of an electrochemically inert, but conducting material has the advantage that the surface can easily be replaced by either the same compound or by another electroactive material depending on whether the electrode should be renewed or be made sensitive to another ion.

2.6.3 Fluoride. The discovery of the fluoride electrode with a membrane of lanthanum fluoride crystal doped with europium (101) can be considered as the most important discovery in the field of the membrane electrodes. The importance of the fluoride in various fields makes for an extensive use of this electrode. Hundreds of papers have been published since 1966 with respect to the construction, characterization and applications of lanthanum fluoride. Doping with europium increases the membrane conductance. The electrodes are manufactured by Orion (94-09), Beckman (39600), Coleman (3-803), Metrohm (EA-306F), Tacussel (PG 4), Radiometer (F1052), Corning (EE-F), Philips (IS-550-F),

Radelkis (OP-F-711), Leeds & Northrup (117405), EDT Research, etc.

The fluoride electrode responds in both neutral and moderately acidic media for the fluoride ion concentration range of 1 to $10^{-5}$ or $10^{-6}$M, and in pure fluoride solutions, the response even reaches the limit of $10^{-7}$M (102).

Lingane (103, 104) employed the lanthanum fluoride electrode for titrations of fluoride with thorium, lanthanum and calcium ions in order to define the optimum conditions of these titrations. He (104) found that the lanthanum ion was the best of the titrant metal ions studied and the optimum pH conditions are a neutral, unbuffered solution. As shown in Fig.1, the titration curve is further improved very considerably by the addition of 60 to 70% (v/v) ethanol since this decreases the solubility of lanthanum fluoride and enhances the rate of potential change at the equivalence point.

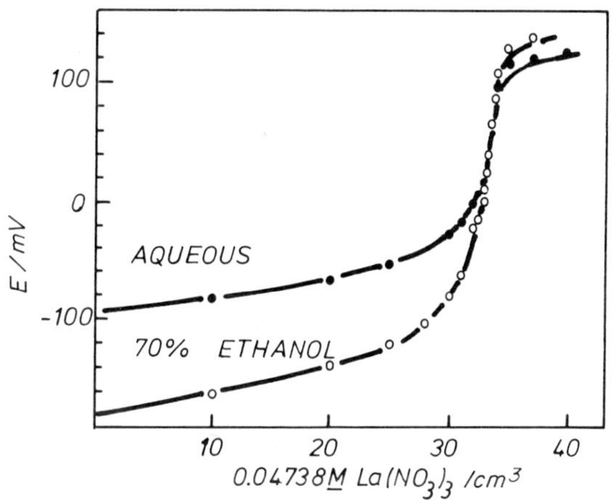


Fig.1. Titration curve of fluoride (200 mg in 100 cm$^3$) by lanthanum nitrate in water and 70% ethanol (104).

2.6.4 Phosphate. Commercial selective electrochemical sensors to phosphate ion are not available, although there are numerous reports of attempts to construct phosphate-sensitive electrodes.

Pungor et al. (105,106) described the preparation of a phosphate ion-selective membrane electrode based on the incorporation of insoluble inorganic phosphates in a silicone-rubber matrix.

Guilbault and co-workers (107-110) proposed various electroactive materials for phosphate sensitivity, namely, systems of silver, thiourea and glutaraldehyde with dibasic phosphate (110), 12-molybdophosphoric - and 12-tungstophosphoric acids (107), quaternary ammonium, phosphonium and arsonium salts (109) and immobilized alkaline phosphatase and glucose oxidase (108). Novozamsky and Riemsdijk (110) reported a selective $PO_4^{3-}$ electrode for the system proposed by Růžička et al. (100) using a silver phosphate sensing element, while Tacussel and Fombon (112) have constructed an electrode based on a simple crystalline sensing element in pellet form which may be obtained by pressing mixtures of silver phosphate and chalcogenides.

Selig (113) determined phosphate with a lead-selective electrode (Orion, Model 94-82). This follows from the fact that orthophosphate forms one of the most insoluble lead salts ($-\log P_s = 42.1$ at 25°C) and can therefore be titrated with lead perchlorate in aqueous solution. Since insoluble phophates are readily soluble in acids and basic lead salts precipitate in very basic solutions (pH $>$10), pH control by buffering is necessary. An ammonium acetate buffer adjusted to pH 8.70 to 9.15 with ammonia has been found satisfactory to ensure an initial pH of 8.25 to 8.75 (113). At this pH the normal orthophosphate, $Pb_3(PO_4)_2$, precipitates. At lower pH, the calculated titre increases, perhaps indicating formation of such species as $Pb(H_2PO_4)_2$ and $PbHPO_4$.

The last method is recommended here for the determination of phosphate ions having regard to the fact that many companies are now manufacturing good lead ion-selective electrodes.

2.6.5 Sulphate. Efforts to prepare useful sulphate ion-selective membrane electrodes have been largely unsuccessful. Data for $BaSO_4$-cellophane membranes (114) and precipitate impregnated silicone-rubber membranes (115) show that potentiometric response to sulphate can be achieved, but it is extremely difficult to obtain selectivity. However, Rechnitz et al. (116, 117) reported an electrode with good response and selectivity for sulphate ions, optimum results being obtained for a membrane consisting of 32 mole % $Ag_2S$, 31 mole % PbS, 32 mole % $PbSO_4$ and 5 mole % CuS. The electrode had a slope of 29 mV decade$^{-1}$ over the $10^{-1}$ to $10^{-4}$M sulphate range.

Although not of the potentiometric type, an enzyme electrode has been constructed for the assay of sulphate ion (118) and is based on inhibition of the reaction:

$$\text{4-Nitrocatechol sulphate} \xrightarrow[\text{sulphatase}]{\text{aryl-}} \text{4-Nitrocatechol} + SO_4^{2-} \quad (9)$$

The steady-state current arising from oxidation of the 4-nitrocatechol product is measured at +0.8 V vs. S.C.E. The competitive inhibition of this reaction by added sulphate ion causes a decrease in this steady-state current in a linear relationship to $pSO_4^{2-}$ in the 2-4 range. The arylsulphate sulphohydrolase enzyme is chemically immobilised in a layer on a platinum electrode.

Ross and Frant (119) found that the lead ion-selective electrode (Orion, Model 94-82) permits the potentiometric determination of end-points in direct titrations of sulphate with lead salt solutions. The relatively high solubility of lead sulphate in aqueous media ($P_S = 1.6 \times 10^{-8}$) limits the magnitude of the potential change at the end-point and makes it impractical to titrate dilute solutions. The addition of dioxane gives greater decrease in lead sulphate solubility than similar amounts of ethanol. The optimum dioxane concentration in the solution to be titrated was about 50% by volume (119). Sulphate may also be titrated with barium chloride or perchlorate to an end point determined by a barium ion-selective electrode (120,121) which may be obtained from EDT Research.

## 2.7 Zinc

In therapy, zinc is used exclusively for its local actions, some of its compounds (zinc oxide, zinc sulphate, zinc undecenoate, etc.) being valuable antiseptics and astringents.

Even if a zinc(II) ion-selective electrode is not commercially available, zinc determination may be based on alternative methods or by the use of non-commercial zinc(II) ion-selective electrodes which are readily made.

For example, Fogg et al. (122) described a liquid-membrane electrode based on Brilliant Green-tetrathiocyanatozincate, which had near-Nernstian response to $Zn^{2+}$ ions in the presence of thiocyanate, while Rachman'ko et al. (123) used the tributyloctadecylammonium salt of $Zn(SCN)_4^{2-}$ in the electro-active membrane. In the presence of 0.2M $SCN^-$, the electrode displays a Nernstian behaviour down to $10^{-5}$M zinc.

Cattrall and Chin-Poh Pui have described coated wire ion-selective electrodes sensitive to halocomplexes of iron(II) (73,74), mercury(II) (124) and copper(II) (125) but a similar approach can be used for zinc (126) using the chlorozincate(II) salt of Aliquat 336S in PVC (30%, w/w) to coat a bead formed on the tip of a platinum wire. The electrodes were conditioned initially for 3h in a 0.1M Zn(II) solution containing an excess of chloride. The electrode prepared with the Aliquat salt extracted from 1M total chloride solution, gave the largest change in potential per decade change in zinc concentration. At a very high concentration of free chloride (8M), the electrode responds in a Nerstian manner down to $10^{-5}$M total zinc with a slope of 28-29 mV decade$^{-1}$. The response time of the electrode was relatively short with about 5 min being required to reach a stable potential ($\pm$ 0.1 mV) in a $10^{-5}$M zinc(II) solution containing 3M chloride; the response was instantaneous in 0.1M zinc(II) solution.

The determination of zinc(II) (e.g., from zinc oxide) may be carried out by titration with EDTA solutions (in 3M sodium chloride solution) at pH 5.4 (adjusted with sodium hydroxide and solid sodium acetate).

## 2.8 Other inorganic salts

Some other inorganic compounds having therapeutical activity, are used in medical practice (see Table II).

There are ion-selective membrane electrodes for all cations of the salts listed in Table II as well as simple and rapid methods for their determination.

2.8.1 Barium. Barium from barium sulphate, following dissociation under conditions described by the U.S.Pharmacopoeia (127), can be determined using either commercial barium ion-selective electrodes (e.g. EDT, Philips, IS-561-Ba, linear response from $10^{-1}$ - $10^{-5}$M $Ba^{2+}$; Corning, EE-Ba, linear response from $10^0$ - $10^{-5}$M $Ba^{2+}$) or by indirect titration (R-titration, after Orion Res.) using CuCDTA or CdEDTA as titrant with copper (Orion Model 94-29) and cadmium ion-selective (Orion Model 94-48) electrodes, respectively as indicator electrodes.

There are described in the literature some other possibili-

TABLE II. Other inorganic salts used as drug-substances

| Denomination and formula | Category | Denomination and formula | Category |
|---|---|---|---|
| Barium sulphate $BaSO_4$ | Radiopaque medium | Lead acetate $(CH_3COO)_2Pb.3H_2O$ | For treatment of bruises & sprains |
| Copper sulphate $CuSO_4.5H_2O$ | Astringent | Silver nitrate $AgNO_3$ | Caustic, astringent |

ties for barium determination with ion-selective electrodes. Thus, liquid membranes based on tetraphenylborate salts of barium complexes of poly-(ethylene glycol) derivatives dissolved in 4-nitroethylbenzene give ion-selective electrodes with high selectivity for $Ba^{2+}$ over $Ca^{2+}$ $Mg^{2+}$ and other cations (128,129). These salts can similarly be used in PVC matrices with 2-nitrophenyl phenyl ether as solvent mediator (120).

A barium ion-selective liquid electrode based on a neutral carrier, N,N,N',N' - tetraphenyl-3,6,9-trioxaundecane diamide, in a PVC matrix has recently been described (130). The membranes were about 0.2 mm thick, and were incorporated into a Philips IS-560 electrode body. The response of the neutral carrier electrode is linear in the $10^{-1}$ - $8x10^{-6}M$ range $Ba^{2+}$, with a slope of 32.5± 1.0 mV for the linear regression. In unbuffered $Ba^{2+}$ systems, the detection limit is well below $10^{-6}M$.

2.8.2 Copper. Several different types of copper ion-selective electrodes are available, all having solid-state membranes with the commonest consisting of a mixture of $Ag_2S$ and CuS (Orion 94-29, Radiometer F3002, Leeds & Northrup 117403). The Radiometer F 1112 Cu electrode has a single-crystal of $Cu_{1.8}Se$. Other electrodes include the Beckman 39612, Philips IS-550-Cu, Tacussel PCU 2, etc.

Since Frant and Ross (131) introduced the use of metal sulphide membranes for ion-selective electrodes as sensors for $Cu^{2+}$, $Cd^{2+}$ and $Pb^{2+}$, many authors have published methods for the preparation of copper(II) ion-selective electrodes (132-139, etc.). It is generally assumed that the response of these electrodes is related to the dissolution process of the membrane material, and that the lower limit of detection of the electrode should correspond to the solubility of this material (137,138,140). All these electrodes showed linear response ranges only for copper(II) concentrations between $10^{-1}$ and $10^{-7}M$ when calibrated with Cu(II) solutions.

Liquid membranes have also been used for constructing copper(II) ion-selective membrane electrodes (141-146) especially for use in potentiometric titration where $Cu^{2+}$ ions are implicated.

For the copper(II) determination, the most appropriate

procedure seems to be titrations with EDTA as titrant (in ammonia or acetate buffer).

2.8.3 Lead. Since the development of the solid-state lead(II) ion-selective electrode by Frant and Ross (131) several similar electrodes have been described (140,147-151). Membranes can be prepared from the sulphide mixture by a variety of methods: for example, homogeneous membranes have been prepared by pressing (134,140), pressing and sintering (132,148) or melting (147). Heterogeneous membranes have been prepared by incorporating the sulphide mixture into a silicone-rubber (152) or polyethylene matrix (151).

Monocrystal membranes have been prepared from synthetic lead chalcogenide (153,154). Instead of sulphides, tellurides or selenides (150) can be used as membrane materials. Precipitate based-electrodes can also be prepared by applying the precipitate to the end of a hydrophobised Teflon rod (155). Such an electrode exhibits an almost Nernstian slope within the pPb range of 2-11, and its sensitivity in lead(II)-buffered solutions is close to the theoretical limit imposed by the conditional solubility product of lead sulphide.

Lead membrane electrodes can be used for direct potentiometric determination of $Pb^{2+}$ ions but the electrodes have been used most frequently as indicating electrodes in potentiometric titrations. Titration with EDTA or NTA at pH 4.7 are recommended although the potential change is smaller then for copper(II). With TETREN and TRIEN, titrations have to be carried out at higher pH, but problems then occur because of protolysis of $Pb^{2+}$ and the titrations can give erratic results. Titrations with EDTA in TRIS buffer (pH 8.0) have also been successful (156).

Many companies supply lead(II) ion-selective membrane electrodes, e.g., Orion 94-82, Radiometer F3004, Leeds & Northrup 117407, Tacussel PPB1 and Activion 00315016 (with a neutral carrier incorporated in a PVC membrane). The general characteristics are much improved over earlier electrodes.

2.8.4 Silver. Most manufacturers supply silver(I) ion-selective electrodes with silver sulphide in the membrane: Orion 94-16, Philips IS-550-S, Beckman 39610, Metrohm EA 306 S/Ag, Radiometer F1212S and F1712S, Coleman 3-805, Tacussel PAG 2, etc.

The crystal membrane electrodes suffer less interference from strongly oxidixed solutions, but no other advantages have been clearly demonstrated and most frequently they are costlier than simple silver metal electrodes. The silver(I) ion-selective electrodes are sensitive both to $Ag^{+}$ and to sulphide as well as certain other ions (18,157,158).

Some recently published works discuss the behaviour of new silver(I) ion-selective electrodes based on liquid membranes (146,160-164). Sensitivity, reproducibility, response time and other characteristics are inferior compared with those of solid-membrane electrodes.

## 3. ORGANIC COMPOUNDS

Many organic compounds are used as drugs, some of these have already been described in relation to assay by ion-selective electrodes in the book of Baiulescu and Cosofret (21) or in the review papers of Selig (165,166). Some of the drug-type compounds are discussed here is the context of their determination with ion-selective electrodes.

## 3.1 Halogenated compounds

The titrimetric determination of halogens have been made by both residual and instrumentation methods for detecting the titration end point. The conventional methods (Mohr, Volhard, Fajans) are no longer adequate for rapid work and there are also some limitations regarding their applications since the methods often need large quantities of sample material.

With the advent of ion-selective membrane electrodes, there was the possibility of developing simple, rapid and precise methods for the determination of halogens as halides both separately and possibly in the presence of other halogens by direct potentiometry or in potentiometric titrations.

For the determination of halogenated organic compounds there are new analytical methods, which in principle, do not greatly differ from the conventional ones. Most of these reprsent variants of the Kolbe method, after Schöniger (167).

Walisch and Jaenicke (168) have pointed out that potentiometric indication of the end-point in argentimetric titrations has three advantages; one is related to the rather high solubility of some silver salts in water, the second to the slow attainment of equilibrium and hence to difficulty in automation, and the third to the slow indication of the end-point. These disadvantages do not arise in the case of non-aqueous solvents which have previously been saturated with silver halide (160,170).

In order to determine 0.3 - 10 mM of halogens from 10-500 $\mu g$ of organic substances, for example, benzene hexachloride and pediculicide, Krijgsman et al. (171) developed a simple, precise and rapid method permitting automatic detection of the end-point. After combustion of the sample in a stream of oxygen, in a quartz tube, at 900-1000$^{o}$C, the halide is titrated in 80% acetic acid with silver nitrate solution, the indicator electrode being a $Ag_2S$-membrane electrode (Orion, Model 94-16). The results obtained for 40-70 $\mu g$ samples of several organic substances containing 17-75% of halogen gave relative standard deviation between 0.5 and 1.5% and the means had no bias.

The values obtained for chlorine and bromine in volatile organic compounds ($CHCl_3$, $CHBr_3$, etc. - anaesthetics) by the Schöniger method may be lower than theoretical, because of incomplete combustion, but in the method of Potman and Dahmen (172) the sample is introduced into the combustion system by injection after combustion in a stream of oxygen at 1000$^{o}$C. The gases produced are absorbed in 80% acetic acid containing nitric acid, hydrogen peroxide and mercury(II) chloride or bromide. The titrant is 0.1N mercury(II) and ion-selective electrodes with $Ag_2S$ and AgI membranes, respectively, are used as indicator electrodes. In the Floret method (173) sodium metabisulphite is used for absorbtion of the combustion gases and 0.01M silver nitrate used as titrant.

Hassan (174) used solid-state chloride and bromide electrodes (Orion, Model 94-17 and Model 94-35, respectively) for the micro-determination of chlorine and bromine in organic compounds. After combustion in an oxygen-filled flask, potentiometric titration with silver nitrate solution at pH 5-7 in presence of 50% dioxan was carried out. The absolute error was ± 0.3% for both elements.

## 3.2 Hydroxy derivatives

3.2.1 Alcohols. Selective spectrophotometric enzymatic methods utilising alcohol dehydrogenase (175, 177) or alcohol oxidase (178) for the determination of alcohols are described in the literature but these cannot be completed with an electroanalytical device such as an ion-selective electrode.

Alcohol oxidase catalyses the oxidation of primary aliphatic alcohols having a small number of carbon atoms according to the reaction:

$$RCH_2OH + O_2 \xrightarrow[\text{oxidase}]{\text{Alcohol}} RCHO + H_2O_2 \qquad (10)$$

Guilbault and Lubrano (179) used alcohol oxidase for the determination of alcohol (ethanol-anaesthetic) and the hydrogen peroxide formed in the reaction was determined amperometrically.

Nanjo and Guilbault (180) have described an alcohol enzyme electrode, useful for assaying blood ethanol in concentrations as low as 10 mg%. The immobilized alcohol oxidase was mounted on the surface of a platinum electrode (Beckman, No.39273) and secured with nylon cloth and O-rings; the electrode was stored in buffer solution (0.1M phosphate buffer, pH 8.2) at room temperature. The enzyme electrode and a reference electrode were placed in a stirred buffer solution at an applied potential of -0.6 V *vs.* S.C.E. When the current was constant the sample solution was pipetted into the buffer solution, and the initial rate of change in the dissolved oxygen limiting current and the final steady-state current were recorded. When the ethanol is added to the buffer solution it diffuses into the layer of chemically bound alcohol oxidase where it is oxidized at the expense of dissolved oxygen and converted into hydrogen peroxide and acetaldehyde. The amount of substrate present was calculated from a calibration curve of initial rate or steady-state current *vs.* substrate concentration.

Mascini and Palleschi (181) used a tubular iodide membrane electrode designed for flow analysis and alcohol oxidase for the determination of ethanol in the $10^{-2}$ - $10^{-4}$M range. The indicator reaction (catalyzed by molybdate, rather than the more expensive peroxidase) is:

$$3I^- + H_2O_2 + 2H^+ \xrightarrow{MoO_4^{2-}} I_3^- + 2H_2O \qquad (11)$$

The change in iodide concentration is measured by an iodide-selective electrode, and is related to the original ethanol concentration. For the enzymatic reaction, a buffer of pH 7.5, iodide concentration of $10^{-4}$M and enzyme concentration of 0.5g $dm^{-3}$ were found to be optimal.

The variation of the electrode potential as a function of substrate concentration follow an empirical relationship:

$$E = 58 \log [I_o] / ( [I_o] - K [S] ) \qquad (12)$$

where $[I_o]$ is the initial iodide concentration, K is a constant and [S] is the concentration of substrate. The precision of the determination of ethanol was 5 - 10%.

3.2.2 Phenols. Phenols (phenol, cresols, pyrogallol, picric acid, etc.) are used in medical practice for their strong antiseptic action. For their quantitative determination with membrane electrodes it is possible to recommend the general methods described by Macholán and Schaněl (182) and Gur'ev et al. (183), as well as the particular methods referring to the determination of picric acid (184,185), but the methods are not necessarily based on potentiometric priciples.

A biochemical sensor with insolubilized poly-phenol oxidase (o-diphenol oxidase) has been used for the rapid and technically simple determination of phenlic substrates (phenol, p-cresol, pyrocatechol and pyrogallol) in quantities of 20 -200 mmol per 3 $cm^3$ of reaction mixture, by measurement of the uptake of oxygen as co-substrate of the enzymic reaction (182). The active membrane was prepared by insolubilization of the enzyme by glutaraldehyde on polyamide netting to give good mechanical stability.

Before use, the enzyme membrane was stretched over the outer surface of the hydrophobic membrane of the oxygen electrode and fixed by a rubber-ring. For the determination of phenolic substrates, the oxygen electrode with the attached enzyme membrane was placed in a thermostated (30$^oC$) reaction vessel containing 3.0$cm^3$ of 0.1M potassium phosphate buffer (pH 7.0) saturated with air. After the electrode current became constant, increasing volumes (5-25 $mm^3$) of 10 mM substrate solution were added at 1 min intervals (magnetic stirring); the unknown sample was added between the measurements. Each addition of substrate manifested itself as a recorder wave indicating the decrease of the electrode current. The concentration of the unknown sample was determined with the aid of a calibrating curve expressing the dependence of the height of electrode response (in mm) on the quantity of the phenolic standard used for the individual addition.

The dependence of electrode response on the pH of the reaction mixture was examined with seven substrates in 0.1M potassium phosphate and acetate buffers. The effect of hydrogen ion concentration on the magnitude of electrode response was small for most substrates in the commonly used pH-range of 5.0 - 8.5 but a pH-optimum of 7.0 was found for pyrocatechol. Measurements were therefore carried out at pH 6.5 - 7.0 in most cases.

For most readily oxidized substrates the calibration curves were linear in the concentration range 6.6 - 66 $\mu$M for membrane bound polyphenol oxidase from mushroom and potato sources (Fig.2).

A picrate-selective electrode with an active membrane of tetrapentylammonium picrate in 2-nitrotoluene (0.01M solution) has been reported (186). The electrode was constructed by using the body of an Orion 92 electrode equipped with Orion 92 membranes (e.g., 92-81-04 perchlorate membrane).

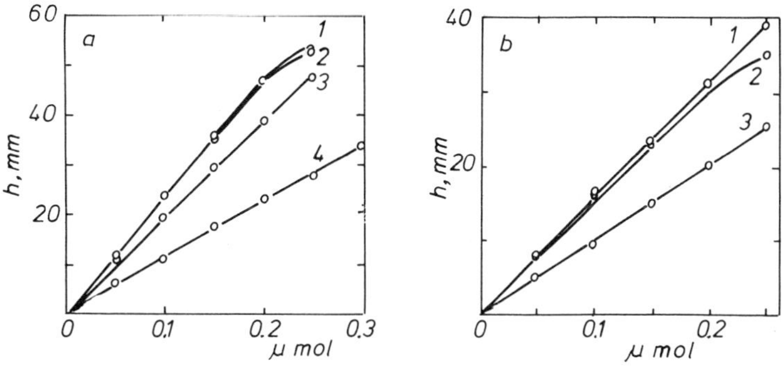


Fig.2. Calibration curves for various polyphenol oxidase substrates with initial $pO_2$ 150 mm Hg (182). The diagram shows a plot of wave height of electrode response at steady state (h) vs. quantities of substrate added to $3 cm^3$ of 0.1M potassium phosphate buffer at pH = 7.0 - 7.1 at $30^oC$
(a) Membrane with mushroom polyphenol oxidase (4U): 1, p-cresol; 2, phenol, 3, pyrocatechol; 4, pyrogallol.
(b) Membrane with potato polyphenol oxidase (3.6 U): 1, pyrocatechol; 2, p-cresol; 3, pyrogallol (pyrogallol was determined at pH 6.5).

The internal reference solution was 0.01M sodium picrate plus 0.1M sodium chloride.

By using a measuring cell of type (I)

| | | | | | |
|---|---|---|---|---|---|
| Ag/AgCl | Reference solution | Membrane $(C_5H_{11})_4NC_6H_2N_3O_7$ | Test solution | $KNO_3$ salt bridge | AgCl/Ag (I) |

a linear reponse was obtained over the $10^{-2}$ - $10^{-5}$M picrate range of slope 58 mV $decade^{-1}$. Between pH 3 and 10, the e.m.f. was practically independent of pH but at higher pH the e.m.f. increases slowly, probably because of complex formation between picrate and $OH^-$ ions. The response times were quite short (less than 4s including the mixing and recorder time) but dependent on the concentration change, the stirring rate and efficiency of solution mixing. It should therefore be possible to employ the picrate-selective electrode in titrations with continuous addition of titrant and for continuous monitoring. The electrode can also be used for direct potentiometry.

3.2.3 Glucose. Many enzyme electrodes have been developed for the in vitro measurement of the glucose concentration (fluid nutrient and electrolyte replenisher) in various body fluids. The specificity for glucose is obtained with glucose oxidase which catalyzes the reaction between glucose and molecular oxygen to produce gluconic acid and hydrogen peroxide:

$$\text{Glucose} + O_2 + H_2O \underset{\text{oxidase}}{\overset{\text{glucose}}{\rightleftharpoons}} \text{gluconic acid} + H_2O_2 \qquad (13)$$

Since the reaction rate depends on glucose concentration, the change in concentration of the reactants or products can be monitored and used to obtain an output signal proportional to glucose concentration. One approach utilizes oxygen electrodes to measure the amount or rate of oxygen consumption (186-190), while the other approaches are based on the direct or indirect determination of the amount or rate of hydrogen peroxide formation. Direct polarographic measurement (191,192) or indirect reaction with another agent to produce a more readily quantified ion or compound (193-195) are used for peroxide detection.

In the first case, the enzyme electrodes are constructed by placing a thin immobilised layer of glucose oxidase around a platinum or special thick-film electrode. When the electrode is placed in a glucose solution, glucose diffuses into the gel layer where it undergoes reaction by equation 13. The hydrogen peroxide produced then diffuses towards both the sensing surface and the solution, the rate of production being followed amperometrically by the electrode. In the second case, iodide ion is used as indicator by reacting with hydrogen peroxide produced according to equation 13 (see also equation 11). The decrease in iodide activity may be monitored with an iodide selective electrode.

Thévenot et al. (196,197) have described a trace glucose analyzer using electrochemical sensors. The differential device includes:

(a) a glucose sensor consisting of a modified gas electrode in which the pH detector was replaced by a platinum disc and the porous film by a collagen membrane on which $\beta$-D-glucose oxidase has been covalently bound after an acyl-azide activation process;

(b) a compensating electrode mounted with a nonenzymatic collagen membrane.

After injection of a glucose-containing sample into the reaction vessel fitted with the probes, an anodic current is detected at the enzyme working electrode. Current outputs of both electrodes are subtracted and twice differentiated; a steady state is reached and the stationary and dynamic responses are recorded. Both reponses are proportional to glucose concentration in the 0.1 $\mu$M - 2 mM range and the reproducibility was found to be better than 2% using these conditions.

Another good method for glucose determination is based on the use of a solid-state copper(II) ion-selective electrode (Orion, Model 94-29) by a standard addition technique (198). Following sample treatment with a standard amount of Stanley-Benedict reagent (0.1M $CuSO_4$ and 1M in sodium carbonate and also with citric acid), the amount of unreduced copper(II) is monitored with the copper(II) ion-selective electrode. For this, a known small volume of a concentrated copper(II) solution is added and the amount of unreduced copper(II) for each glucose standard or sample calculated by using the standard addition technique. The unreduced copper(II) is directly related to the amount of glucose in the sample under specific conditions and the amount of glucose can be found from a calibration graph prepared with standard glucose solutions. Amounts of glucose in the range 0.25 – 2 mg were determined with an average error of about 2%.

## 3.3 Carbonyl compounds

Representative members of this class are formaldehyde (a disinfectant) and hexamine (an antibacterial agent).

An argentimetric potentiometric method for the precise determination of formaldehyde using an iodide ion-selective electrode (Orion, Model 94-53) (190), the sample is treated with 0.1M iodine-iodide solution (in 2M alkaline medium) such that the excess of iodine is at least 100%. The iodide ion is determined by potentiometric titration with standard silver nitrate solution at pH $<2$ (adjusted with 2M nitric acid):

$$HCHO + I_2 + 2KOH \longrightarrow 2KI + HCOOH + H_2O \qquad (14)$$

$$KI + AgNO_3 \longrightarrow AgI + KNO_3 \qquad (15)$$

Blank runs are made on identical amounts of iodine-iodide solution.

Hadjiioannou and co-workers (200) have recently reported a kinetic method for determining formaldehyde and hexamethylenetetramine (hexamine) with a cyanide ion-selective electrode (Orion, Model 94-06). The method is based on the well-known addition reaction of hydrogen cyanide to carbonyl compounds whereby the hydrogen cyanide takes part in a nucleophillic attack of the positively polarized carbon atom to form a cyanohydrin:

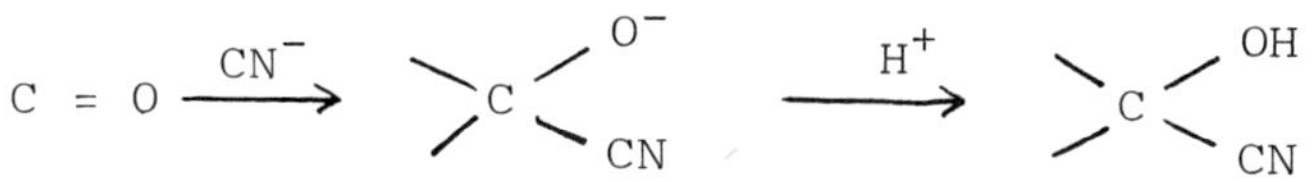

(16)

In practice, a very dilute cyanide solution (<u>ca</u>. $1.6 \times 10^{-5}$M) is treated with a large excess of unknown or standard formaldehyde solution, and the reaction monitored with the cyanide ion-selective electrode. The time required for the potential to change by a preselected amount (8.0 mV) is measured automatically and related directly to the formaldehyde concentration. The average error in the determination of 60-300 $\mu$g of formaldehyde in a sample volume of 1.0cm$^3$ was about 1.3% and measurement times were about 18-80s.

The kinetic method (200) for the determination of formaldehyde has also been applied to the determination of hexamine which hydrolyses on heating to 60°C for 30 min to formaldehyde:

$$(CH_2)_6N_4 + 6H_2O + 4H^+ \longrightarrow 4NH_4^+ + 6HCHO \qquad (17)$$

50-250 $\mu$g of hexamine in 0.5cm$^3$ could be determined with an average error of about 1.6%. The accuracy was further tested by comparing values for pharmaceutical tablets containing hexamethylenetetramine mandelate with those obtained by the U.S.Pharmacopoeia titrimetric method based on modified Nessler reagent. There was satisfactory agreement between the two methods.

## 3.4 Carboxylic acids and derivatives

3.4.1 Acetic acid. The importance of acetic acid in the pharmaceutical organic industry (for its expectorant action and as a pharmaceutical acidifying agent) or in fermentation processes has resulted in interest for developing acetate electrochemical sensors (180, 201-203). It is of interest that a wide variety of solvent extraction (ion association, as well as chelating) systems may be used in making ion-selective electrodes for carboxylic acids. The acetate of the methyltricaprylylammonium ion (Aliquat 336S) dissolved in 1-decanol functions effectively as organic phase component in liquid-liquid membrane electrodes for the determination of acetate within the $10^{-1}$ to $10^{-5}$M concentration range with a slope of 53.0 mV $decade^{-1}$ (201, 202). The electrode assembly consisted of an Orion calcium electrode barell in which the liquid phase was placed and an Orion 92-20 membrane was used to separate the organic phase from the test solution. The organic phase was a 10% (v/v) solution of Aliquat 336S in 1-decanol, converted to acetate form by repeated shaking with 0.1M sodium acetate. Similar electrodes were obtained for p-toluene-sulphonate, formate, propionate, oxalate, benzoate, salicylate and other organic acids or inorganic anions. The response time of the electrodes varied, but a constant potential reading was usually obtained within 1 min. The concentration of the acetate ion was determinable to about 0.7% within the useful concentration range of the electrode.

A microbial sensor (Fig.3) consisting of immobilized yeasts, a gas-permeable Teflon membrane and an oxygen electrode has recently been described for the determination of acetic acid (203).

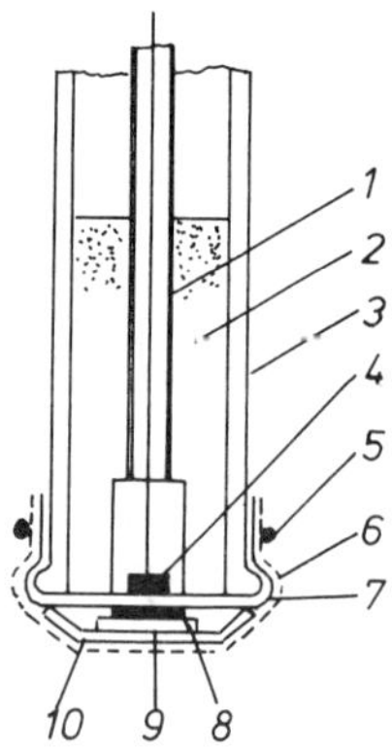


Fig.3. A microbial sensor for acetic acid: 1, aluminium anode; 2, electrolyte; 3, insulator; 4, platinum cathode; 5, rubber ring; 6, nylon net; 7, eflon membrane; 8, micro-organisms; 9, acetylcellulose membrane; 10, porous Teflon membrane (203).

The oxygen electrode consisted of a Teflon membrane (50 μm thick), a platinum cathode, an aluminium anode and a saturated potassium chloride electrolyte. The porous membrane (1.4 cm diameter circle) with immobilized micro-organisms was fixed on the surface of the Teflon membrane. This was covered with a gas-permeable Teflon membrane (Millipore, type FH, 0.5 μm pore size) or a silicone rubber membrane (Radiometer, type D 606), the whole arrangement being held together with nylon net.

The electrode is suitable for the continuous determination of acetic acid in fermentation broths. When an acetic acid solution is pumped through the flow system, the current decreases to a steady state with a reponse time of 8 min. The relationship between the current decrease and the acetic acid concentration was linear up to 54 mg $dm^{-3}$, with a relative standard deviation of about 6% at higher concentrations.

3.4.2 Benzoic acid. The determination of benzoic acid (an anti-bacterial and antifungal agent) using a liquid membrane electrode has recently been described (204). The benzoate liquid-membrane electrode consists of a 2 mm thick liquid membrane composed of a $10^{-2}$M solution of trimethylcetylammonium benzoate (TMCAB) in 1-decanol held between Millipore cellulose acetate discs (diameter 13 mm, thickness 0.1 mm, pore diameter 0.3 μm). The electrode gives rapid ( $<$30s ) and reproducible response and was employed for the direct determination of sodium benzoate in commercial expectorating medicinal syrups of varying composition (substituted diamines, citric acid, sugar, alcohols, menthol, ammonium chloride, etc.). In some cases, pre-extraction was necessary, always with pH-adjustment.

3.4.3 Citric acid. There are just two methods for the determination of citric acid or citrate anion by ion-selective electrodes. Lazarou and Hadjiioannou (205) used a perbromate ion-selective electrode for the kinetic potentiometric determination of citric acid. Citric acid and perbromate do no react, but when iron(II) is added, the iron is oxidized by perbromate and the reaction induces the oxidation of citric acid by perbromate. Under controlled conditions there is a linear relation between the increase in the electrode potential within a fixed period of time (6 min ) and the amount of the citric acid present. Citric acid in the $10^{-4}$ - $10^{-3}$M range can be determined with relative errors of about 1%. The relative standard deviation for the determination of 4 x $10^{-4}$M citric acid sample was 1.0% (n = 5).

El Tara et al. (200), in a study of the behaviour of the silicone rubber-based copper(II) ion-selective electrode (133) in the presence of various complexing agents, found that the calibration curve for citrate was linear in the $10^{-2}$ - $10^{-6}$M with a slope of 29 mV $decade^{-1}$. Since the electrode potential in citrate solutions was found to be very stable even in the absence of copper(II), the concentration of citrate ions can be determined by direct potentiometry with this electrode. The response of copper(II) ion-selective electrodes to the concentration of citrate and other complexing agents in solutions free of $Cu^{2+}$ ions may be explained on the same basis as the response of silver halide membrane electrodes to complexing agents such as cyanide. Although the theoretical understanding of the behaviour of ion-selective electrodes in the presense of complexing agents is far from complete, the interpretation is based on the dissolution of the precipitate incorporated in the sensing membrane of the electrode. Thus, the potential response of the copper electrode observed in copper-free solutions is the result of the steady-state of a continuous dissolution process of copper(II) sulphide precipitate from the sensing surface.

The concentrations of solutions of sodium citrate and citric acid may also be determined by potentiometric titration with copper sulphate solution as titrant, and 0.1M sodium hydrogen-carbonate for buffering the pH change during the titration. Several commercially available pharmaceutical preparations containing citrate ion were analyzed by this method with an error of $<4\%$.

3.4.4 Salicylic acid. Anion-responsive electrodes based on ion-association extraction systems, some of which are sensitive to salicylate have been discussed in sections 2.4.1 and 2.4.2. The organic phase consists of methyltricaprylylammonium salicylate dissolved in 1-decanol, and salicylate can be determined in the $10^{-1}$ - $10^{-3}$M range at a slope of 56 mV decade$^{-1}$ (201,202). The concentration of the salicylate anion can be determined to within 0.7% ($\pm$ 0.2 mV) over the useful concentration range of the electrode.

Two salicylate-selective electrodes have been constructed by Haynes and Wagenknecht (207). The two organic phases were prepared by dissolving tetraheptylammonium iodide in 1-decanol (1.0M and 0.2M respectively) and exchanging the iodide for salicylate by repeated shaking with an aqueous solution of sodium salicylate (1.0M and 0.2M respectively). An Orion liquid membrane electrode barell was used with and Orion 92-19 porous membrane separating the organic phase from the test solutions. The aqueous reference phase was 0.1M in both sodium chloride and sodium salicylate. In the concentration range of $10^{-1}$ - $10^{-3}$M, both electrodes showed a near-Nernstian slope of 58 mV decade$^{-1}$. The response of these electrodes was almost instantaneous and the optimum pH was 8.5.

3.4.5 EDTA (disodium salt). There is no known electrode sensitive to EDTA which is used in medicine as a metal complexing agent, but many papers mention the possibility of its determination by its capacity to complex cations in the ion-selective membranes of solid-state membrane electrodes.

The response of copper(II) ion-selective electrodes in solutions containing complexing agents, such as EDTA has been discussed in several papers (206,208-214). To verify the response of the copper(II) ion-selective electrode (Orion, Model 94-29) to EDTA a $10^{-3}$M $Ca^{2+}$ solution has been titrated with $10^{-2}$M EDTA solution at pH 12 and the titration curve allowed easy location of the equivalence point (214) The electrode did not respond to $Ca^{2+}$ ions, showing that the change of potential is related to changing EDTA activities (214). The electrode response in copper-free solutions was measured at different pH values. The calibration curves for EDTA showed that near-Nernstian responses were obtained at pH 4 and 7.6, whereas the response at pH 11 was suppressed by the interfering effect of $OH^-$ ions. At pH 2, the formation constants of copper complexones decrease to levels which do not appreciably affect electrode potential. Experimentally useful electrode responses to complexones can be achieved in the 4-8 pH range. Response times at pH 4 and 7.6 were about 5 min at $10^{-3}$M complexone concentration and increased to 15 min at $10^{-6}$M. The electrode potentials were not stable at pH 11 and the electrode was poisoned but proper function could be restored by repolishing the sensing membrane. Similar deterioration of electrode function occurred by prolonged exposure of the electrode to high concentrations ($>10^{-4}$M) of complexones. The limits of detection with the electrode, at pH 7.5 were $\sim10^{-7}$M for EDTA.

Theoretical and experimental curves for the titration of EDTA with copper(II) are presented in Fig.4 for different concentrations of the ligand and there is good agreement between theory and experiment.

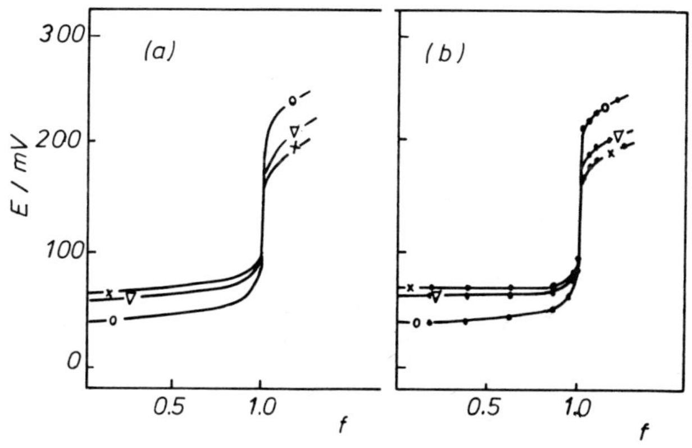


Fig.4. Titration curves for EDTA with copper(II): (—o—), $10^{-3}$M; (—▽—), $10^{-4}$M; (—x—), 5 x $10^{-5}$M. (a) Theoretical; (b) experimental (213).

Lead ion-selective electrodes (Orion, Model 94-82) and cadmium ion-selective electrodes (Orion, Model 94-48) also respond to EDTA in the range $10^{-3}$ - $10^{-7}$M (214).

A perchlorate ion-selective electrode, which also responds to periodate has been evaluated as an end-point sensor in catalytic titrations involving periodate indicator reactions (215). In catalytic titrations, the catalyst serves as a titrant for solutions containing a specied (inhibitor) which reacts stoichiometrically and very quickly with it. In addition, there is a certain reagent mixture for the indicator reaction which is also catalyzed by the titrant and the rate of which is monitored to locate the titration end-point.

Hadjiioannou et al. (215) thus developed a semi-automatic potentiometric method for the direct catalytic titration of EDTA with standard manganese(II) solution using the periodate-diethylaniline reaction as the indicator reaction, and a perchlorate ion-selective electrode to monitor the indicator reaction.

The perchlorate ion-selective electrode (Orion, Model 92-81) responds rapidly to changes in periodate concentration and it can follow the indicator reaction with continuous addition of titrant from a constant rate burette. During the titration, the potential remains practically constant until the equivalence point is reached; the excess of manganese(II) then catalyzes the indicator reaction and a large increase in the potential occurs because of the periodate reduction. The sharpness of the potential break increases with increasing titrant concentration. The end-point is read from the recorded potentiometric titration curve. The determinations were carried out in a citrate-phosphate buffer (pH 7.0) using 0.05M sodium metaperiodate and 0.05M diethylaniline solutions.

The periodate-diethylaniline reaction is ideal as an

indicator reaction in catalytic titrations with EDTA because it is strongly catalyzed by traces of Mn(II) ($10^{-7}$M). Furthermore, the Mn-EDTA reaction is very rapid and the Mn-EDTA complex exhibits a very weak catalytic effect on the indicator reaction. EDTA in the 7.5 x $10^{-6}$ - $10^{-3}$M range can be determined with an average error and precision of 1.6% and 1.3% respectively.

3.5 Basic nitrogen compounds.

3.5.1 Choline and its esters. Choline and its esters which act as parasympathicomimetics have been determined by means of ion-selective electrodes (216-221) and both liquid and solid-state membrane electrodes have been used.

The liquid membrane electrode used by Baum (216) had a high selectivity for acetylcholine relative to choline (15:1) and inorganic cations. A linear response was obtained in the concentration range of $10^{-1}$ - $10^{-5}$M for choline and $10^{-1}$ - $10^{-6}$M for acetylcholine; the slope was almost Nernstian in the first case and Nernstian in the second, while response times were rapid even at low organic cation concentration.

The solid-state electrode described by Baum et al. (217) consists of an electroactive PVC membrane prepared from a solution of acetylcholine tetra-4-chlorophenylborate in a phthalate ester which also serves as plasticizer for PVC. The disc-shaped membrane was placed in the bottom cap of a Corning liquid membrane assembly, and a saturated solution of potassium chloride and silver chloride was used as internal reference solution.

Comparative data for the phthalate ester-plasticized PVC electrodes towards choline and choline esters in single electrolyte solution are presented in Table III and along with those for a liquid membrane electrode having 3-nitro-2-xylene as solvent (217). The responses were generally linear from about $10^{-1}$M to an activity of about $10^{-3}$ - $10^{-4}$M.

TABLE III. Comparative performance of various electrodes with respect to choline and choline esters (218).

| | Selectivity ratios [a] | | |
|---|---|---|---|
| | Liquid membrane | Plasticized dibutylphthalate | PVC dioctyl-phthalate |
| Choline | 1 (45) | 1 (54) | 1 (58) |
| Acetylcholine | 37.6 (53) | 6.87 (56) | 5.43 (59) |
| Methacholine | 121 (59) | 19.1 (60) | 11.5 (55) |
| Butyrylcholine | 407 (56) | 50.0 (58) | 41.7 (53) |

[a] Figures in parentheses are electrode slopes in the $10^{-1}$ - $10^{-3}$M range.

Kina et al. (221) developed a very sensitive methacholine electrode of response range down to $10^{-6}$M with dipicrylamine anion in nitrobenzene as the liquid ion-exchanger. The electrode reponded rapidly with the equilibrium potential being reached within one second, although the actual response time depends on the efficiency of solution mixing. Such a rapid response is useful for potentiometric titration and the methacholine-selective electrode was so used in titrations of methacholine chloride with sodium tetraphenylborate solution as titrant. The end-point data agreed well with those of argentimetric titrations of the chloride ion.

An interesting electrode, selective to acetylcholine, was recently constructed by using immobilized acetylcholinesterase on the active surface of a glass pH electrode (220). Acetylcholinesterase (AChE) catalyses the hydroysis of the acetylcholine according to:

$$(H_3C)_3\overset{+}{N}CH_2CH_2OCOCH_3 + HOH \xrightarrow{AChE} CH_3COO^- + (H_3C)_3\overset{+}{N}CH_2CH_2OH + H^+ \qquad (18)$$

The acetic acid released is detected by the pH glass electrode. The electrode has a linear response to acetylcholine in the $10^{-2}$ - $10^{-4}$M range in phosphate buffer (pH 8.3 - 8.6).

3.5.2 <u>Chlorpheniramine</u>. A PVC matrix membrane electrode (222) sensitive to the antihistamine chlorpheniramine [3-(p-chlorphenyl)-3-(2-pyridyl)-propyl-dimethylamine] may be prepared by taking a tetrahydrofuran solution of PVC (12.5% PVC) a 1,2-dichloroethane solution of the tetraphenylborate of chlorpheniramine (18.75%) and dioctylphthalate in the mass ratio of 8:4:2. The mixture is spread on a glass plate and left for 48 hrs for the solvent to evaporate. A 5-10 mm diameter membrane disc is attached to the end of a PVC tube to act as electrode membrane.

Chlorpheniramine measurements were carried out with the following electrochemical cell:

$$\underbrace{\text{Ag/AgCl} \;\Big|\; \text{reference solution} \;\Big|\; \text{PVC membrane}}_{\text{chlorpheniramine electrode}} \;\Big|\; \text{sample solution} \;\Big|\; \text{S.C.E.} \qquad (II)$$

and the response was Nernstian to the chlorpheniramine cation over a $10^{-1.5}$ - $10^{-4.5}$M range. There was no pH interference in the 4.5 to 8.0 range and the interference of sodium, potassium, ammonium and calcium ions were extremely low. Ephedrine, methylephedrine, caffeine and antipyrine which are usually contained in common drugs did not interfere between pH 5 and 7, but there was interference from diphenhydramine and tripelenamine ions.

The chlorpheniramine electrode was useful as an indicator electrode for determinations by potentiometric titration with sodium tetraphenylborate titrant.

3.5.3 <u>Clobutinol</u>. A clobutinol-selective membrane electrode may be based on a PVC matrix membrane containing clobutinol-tetraphenylborate salt and dioctylphthalate (223) in a similar manner to that described above for chlorpheniramine (section 3.4.2).

The electrode was Nernstian between $10^{-1.5}$ and $10^{-4.5}$M for the clobutinol cation [1-(4-chlorophenyl)-4-dimethylamine-2,3-dimethyl-butan-2-ol: an antitussive]. The response was independent of pH from 3.0 to 8.0 while the selectivity coefficients $k^{pot}_{clobutinol,B}$, calculated by the separate solution methol (1) showed that only diphenhydramine, tripelenamine and chloropheniramine interfere ($k^{pot}_{clobutinol,B}$ were 1.25, 1.41 and 1.0, respectively). The response times of the electrode were about 2 min in $10^{-1.5}$ and $10^{-4.5}$M solutions and almost instantaneous in between $10^{-2}$- $10^{-4}$M.

3.5.4 Ephedrine and methylephedrine. Ion-selective electrodes with a liquid membrane or PVC matrix membrane sensitive to the bronchodilatant ephedrine and the antitussive and antihistaminic di-methylephedrine, have been developed by Fukamaki et al. (214). An organic solvent solution of the tetraphenylborate salt of the particular cation was used as the liquid membrane in an Orion electrode body (Model 92). The PVC matrix membrane was prepared by taking a PVC solution (20% m/m tetrahydrofuran solution), dioctylphthalate and the tetraphenylborate salt of the cation concerned in the ratio of 25:10:2. The mixture was spread on a glass plate and left for more than 48 hours for the solvent to evaporate. A 5-10 mm diameter membrane disc was attached to a PVC tube or to an Orion electrode body.

The methylephedrine (MEP)-selective electrode of the liquid membrane and PVC matrix membrane types exhibited Nernstian responses to the MEP cation down to $10^{-4}$M but the ephedrine (EP)-selective electrode was Nerstian down to just $10^{-3}$M (Fig.5).

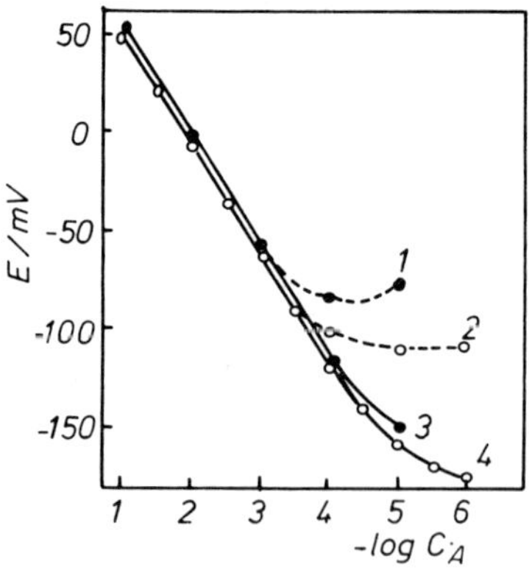


Fig.5 Responses (E) of electrodes to concentrations ($C_A$) of methylephedrine and ephedrine (224). Curve 1, liquid membrane and curve 2, PVC matrix membrane electrodes for 1- and 2-ephedrine. Curve 3, liquid-membrane and curve 4, PVC matrix membrane electrodes for 3- and 4-methylephedrine. (Solvent in membranes is nitrobenzene).

The membrane potentials of both electrodes were independent of pH from 1.5 to 8.0 while the interferences of sodium, potassium, ammonium and calcium ions were extremely low for the determination of methylephedrine and ephedrine by both electrodes. The presence of caffeine, antipyrine, vitamin C, etc., contained in the usual drugs caused no disturbance to the response of the methylephedrine electrode.

Besides its use for the direct potentiometry of ephedrine or methylephedrine, the PVC matrix membrane electrode was useful as an indicator electrode in the potentiometric titration of methylephedrine by precipitation with $10^{-2}$M sodium tetraphenylborate titrant.

An electrode sensitive to ephedrine was constructed by Luca et al. (225) and was based on a liquid membrane of the tetraphenylborate of ephedrine in nitrobenzene ($10^{-3}$M). The electrode responds in the $10^{-2}$ - $10^{-5}$M range with a slope of about 59 mV $decade^{-1}$.

3.5.5 Methadone. The use of methadone [6-(dimethulamino)-4,4-diphenyl-3-heptanones narcotic and analgesic] in opiate-addiction treatment programmes has created a need for rapid, simple, yet accurate analytical methods for monitoring drug levels in biological fluids, in order to evaluate patient compliance.

Srianujata et al. (226) have described the design of a modified version of a coated-wire ion-selective electrode originally described by Freiser et al. (227,228). The new design is smaller, less fragile, more stable and provides faster response times and was used to monitor urinary levels of methadone, a hydrophobic drug which is made cationic at low pH by protonation of the amine.

The electrode was constructed by coating a Ag/AgCl electrode with a sensitive polymer membrane, prepared as a 1:1 mixture of PVC and dioctylphthalate in 1,2-dichloroethane maintained at 60°C. The Ag/AgCl electrode tip was dipped into the polymer solution a total of 4 to 6 times and allowed to dry after each treatment. The electrode was finally equilibrated with 0.1M KCl solution for 12 hours before use.

A typical cell for measuring the potentials developed was:

$$\text{Ag/AgCl} \;\Big|\; 0.1\text{M KCl} \;\Big|\; \begin{matrix}\text{plastic}\\ \text{membrane}\end{matrix} \;\Big\|\; \begin{matrix}R_4N^+\\ \text{test solution}\end{matrix} \;\Big|\; \text{S.C.E.} \qquad \text{(III)}$$

The electrodes gave good, apparently Nernstian responses over wide concentration ranges to tetrabutylammonium bromide (TBA) and tetrapentylammonium bromide (TPA). The sensitivity limit of the electrode was approximately $10^{-6}$M for TBA and $10^{-8}$M for TPA. The response of the coated wire towards varying concentrations of methadone hydrochloride in water was identical to that for TBA. A linear response was observed in the $10^{-6}$ to $10^{-2}$M range with a slope of 59 mV $decade^{-1}$. Measurements for methadone were made in acidic solution to ensure protonation of the tertiary amine group.

The usefulness of the coated-wire plastic electrode is still somewhat unsatisfactory for direct potentiometric measurements because of the need for frequent standardization. However, for some applications, where the highest accuracy is not required, it might be possible to estimate the methadone concentration from a single potentiometric measurement.

As is known, tertiary ammonium salts can be titrated with sodium tetraphenylborate to give stable precipitates. The specificity and sensitivity of the method is dependent, in part, on the dissociation constant of the complex and the constants for potential interferents. Tetraphenylborate was used to titrate the protonated form of methadone and the decrease in analyte activity was monitored by the plastic electrode. Methadone can be determined titrimetrically to levels of 5 x $10^{-5}$M with an accuracy of ± 5% and precision of ca.1%.

3.5.6 General. The assaying of amine drugs in formulations by precipitation with excess tetraphenylborate (229) is well known. The above shows that titrations can be carried out directly and this can result in considerable savings in time. The method has also been applied to the assay of procyclidine hydrochloride, cyclizine hydrochloride and diethylcarbonazine citrate in tablets using nitrobenzene and/or bis(2-ethylhexyl)phthalate in PVC on a graphite support as indicator electrodes (230).

## 3.6 Heterocyclic nitrogen compounds

3.6.1 Barbital and phenobarbital. An electrode sensitive to barbital (sodium 5,5'-diethylbarbiturate; a hypnotic and sedative) has been obtained by using cetylpyridinium bromide solution (0.05%) in nitrobenzene as active membrane, and supported in the pores of a $G_4$ glass frit attached to the extremity of a glass tube (231). The organic solution was also used as reference solution. This electrode responds to barbital in the $10^{-1}$M up to 5.6 x $10^{-4}$M range with a slope of 50 mV $decade^{-1}$.

Carmack and Freiser (239) used a coated-wire phenobarbital electrode based on ion-pair complex between phenobarbital and the quaternary ammonium cation, tricaprylylmethylammonium for analyses of phenobarbital (sodium 5-ethyl-5-phenyl-barbiturate; a hypnotic and sedative) solutions. The coated-wire electrodes gave a linear response (slope 55 ± 2 mV $decade^{-1}$) from $10^{-1}$M to approximately $10^{-4}$M.

High precision (relative standard deviation of ± 1.3%) quality control-type analyses are possible with this electrode because the approximate phenobarbital content of a tablet is known beforehand. A standard solution containing the approximate concentration can be prepared and measurements can be performed repeatedly on it and on the sample solution until a reproducible (± 0.1 mV) potential difference is obtained between the two solutions. Usually, the difference in potential was within 1-2 mV.

3.6.2 Fluorouracil and other fluorine-drug substances.

The literature describes many techniques for determining organo-fluorine compounds using a fluoride ion-selective electrode (233-237). Most of the pre-treatment methods involve decomposition of the organic substance by combustion in an oxygen atmosphere. This has some disadvantages, such as incomplete combustion, with poor release of fluorine leading to low accuracy for those compounds having a small fluorine content. The method is not generally convenient for the routine determination of the fluorine bound to various organic pharmaceutical substances because of such practical difficulties. Among the other decomposition procedures, reductive cleavage by means of a stabilized, highly reactive reagent, e.g., ether-sodium aromatic hydrocarbon complexes, seems to be suitable owing to the rapidity, ease and precision with which the organically bound fluorine is released. Such complexes have been generally used for the rapid decomposition of organic compounds containing other halogens as well as for fluorine. The most effective complex in terms of stability, reactivity at room temperature and active sodium concentration is the dimethoxyethane-sodium biphenyl complex (sodium biphenyl reagent).

Jones et al. (236) have described conditions for the quantitative ionization of the fluorine bound in drug-substances by use of the sodium biphenyl reagent. The fluorine released is determined by direct potentiometry with a fluoride ion-selective electrode (Orion, Model 94-09). The average relative standard deviation was ± 0.38% for 12 fluorinated organic compounds containing from 4 to 21% organically bound fluorine. The average absolute error, assuming 100% purity was less than 0.17% for all the compounds tested.

Rogers and Wilson (235) introduced a method for the estimation of two alkali-labile trifluoromethyl compounds present in the same solution (5-trifluoromethyl-2'-deoxyuridine and its metabolite, 5-trifluoromethyluracil). The method is based on the differential rates of hydrolysis of the carbon-fluorine bonds of the two compounds and the measurement of the released fluoride by means of an Orion fluoride ion-selective electrode.

Mertens et al. (237) have determined the fluorine in a series of drug-substances (fluanisone, haloperidol, fluorometholane, trifluoperidol, dexamethasone, triamcinolone acetonide, etc.) as well as in various tablets, creams and ampoules, by an automatic potentiometric method with a fluoride ion-selective indicator electrode and a reference electrode connected into a Technicon module (Fig.6).

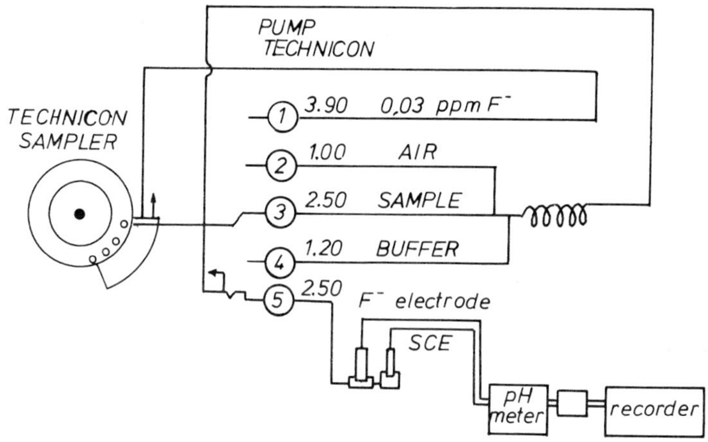


Fig.6. Assembly for the automatic potentiometric determination of fluoride (237).

With regard to the procedure, TISAB buffer was added after introduction of the sample and the solution homogenized. Two methods were used for the destruction of the organic compounds. One was the Schöniger method, and the other use of sodium peroxide in a Parr microbomb. For substances containing the $-CF_3$ group (e.g., trifluoperidol) addition of potassium nitrate and lactose was necessary for complete transformation into hydrogen fluoride. Recoveries for several drugs containing from 4 to 17% fluorine ranged from 96.4% to 100.1% for the Schöniger method of decomposition and from 97.4% to 101.5% for the Parr bomb method.

3.6.3 Inactin and thiopental. Cosofret and Bunaciu (238) have investigated the potentiometric response of the silver suphide crystal membrane electrode (Orion, Model 94-16) to thiobarbiturates. In the presence of thiobarbiturate ion, S,S'-disubstituted thiobarbiturates exist in two predominant forms in alkaline solution (239) while the reaction between silver ions and thiobarbiturate itself is given by

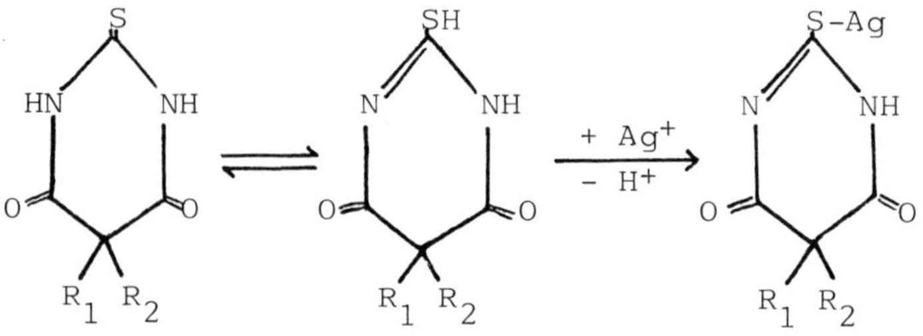

(19)

with electrode response to thiobarbiturate being according to

$$E = E_0' + S \log a_{thiob.} \quad (20)$$

where S is the slope of the electrode which in this case should be -59.1 mV decade$^{-1}$ according to the stoichiometry, and $a_{thiob.}$ represents the activity of the thiobarbiturate ion.

Fig.7 shows plots of electrode response for thiopental and inactin in 0.1M sodium hydroxide. The linear response range in both cases is about $10^{-3}$ to $10^{-5}$M thiobarbiturate with slopes agreeing with the 1:1 silver; thiobarbiturate stoichiometry (51 mV for inactin and 54 mV for thiopental, respectively).

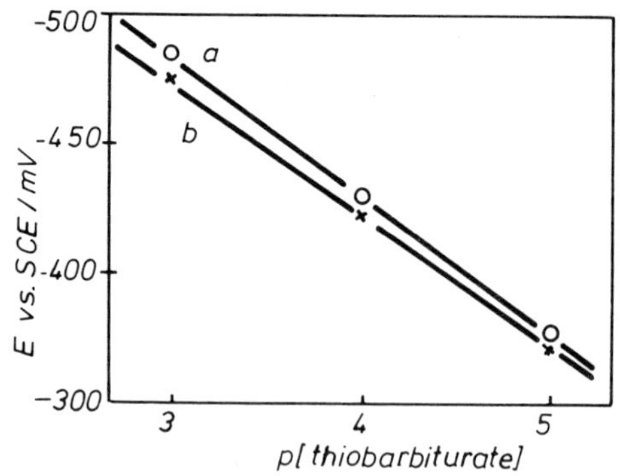


Fig.7. Response of the silver sulphide crystal membrane electrode to thiopental (a) and inactin (b) (238).

In dilute solutions, the response times of the electrode are long, about 15 min for a normal measurement. Nevertheless, potentiometric titrations, with good results up to 5 x $10^{-4}$M thiobarbiturate were recommended. (238).

3.6.4 Isoniazid. An indirect potentiometric method for determining isoniazid (hydrazide of 4-pyridine carboxylic acid; an antibacterial and tuberculostatic agent) based on the quantitative oxidation with an excess of chloramine-T (CAT), followed by measurement of the unconsumed CAT with a chloamine-T selective electrode has recently been described (240). At pH 5-6 two moles of CAT are consumed per mole of isoniazid:

$$C_5H_4NCONHNH_2 + 2CH_3C_6H_4SO_2NCl + HOH \longrightarrow C_5H_4NCOOH + N_2 + 2CH_3C_6H_4SO_2NH_2 + 2\ Cl^- \quad (21)$$

Amounts of isoniazid in the 0.28-14 mg range can be determined with a relative standard deviation of about 1-2%. This method has been used for the determination of isoniazid in pharmaceutical preparations and the values obtained in injection solutions and tablets by this potentiometric method were comparable with those obtained by the USP standard iodometric method (241).

### 3.6.5 as-Triazines

Some asymmetrical triazines (inhibitors in the enzymatic catalysis of the proteic biosynthesis) have been determined by potentiometric titration with silver nitrate solution using as indicator electrode a silver (I) ion-selective electrode whose liquid membrane consists of the silver(I) chelate of 1(2',3',5'-tri-O-benzoyl-β-D-ribofuranozyl)-4-thioxo-5-methyl-thio-6-azauracil in chloroform in conjunction with a sulphide ion-selective electrode (Radelkis, type OP-S-711 D) (242). The determinations need to be carried out in 1:10 methanol-water (or acetate buffer solution of pH 6.0).

One mole of as-T-1 or as-T-4 (see Table 1V) reacts with one mole of silver nitrate and only one break appears in the titration curve. In the case of as-T-2 and as-T-3 one mole of the as-triazine reacts with two moles of silver nitrate to give two breaks, the first is not very obvious but corresponds to a stoichiometry of 1:1 mole of reactant to reagent. The potential data may be correlated with IR spectra.

TABLE IV. as-Triazines analyzed potentiometrically (242).

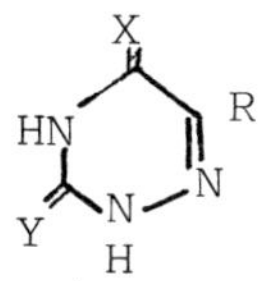


| X | Y | R | Short notation |
|---|---|---|---|
| S | O | Br | as-T-1 |
| S | S | $OCOC_2H_5$ | as-T-2 |
| S | S | $SCH_3$ | as-T-3 |
| O | O | SNa | as-T-4 |

The absolute error for the determination of as-triazines with the silver (I) ion-selective electrode does not exceed ± 1% (242).

### 3.7 Nitrogen derivatives of carbonic acid (biguanides)

The interest in guanidine derivatives (Table V) arises from the antimalarian properties of proguanyl. Some hypoglicaemic substances of medical interest have been found among the mono-or di-substituted derivatives of the biguanide type (Table V) and they have been adopted in therapy. The hydrochlorides are used in the oral treatment of diabetes.

TABLE V. Some biguanide drug substances.

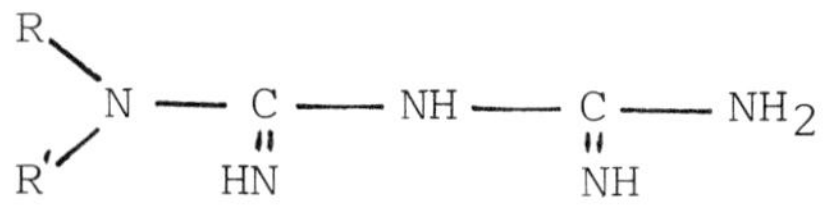


| R | R' | Denomination |
|---|---|---|
| $CH_3$ | $CH_3$ | Metformin |
| H | $n\text{-}C_4H_9$ | Buformin |
| H | $-CH_2-CH_2C_6H_5$ | Fenformin |
| H | $-p\text{-}Cl-C_6H_4-$ | Proguanil |

A method for determining alkyl-1-biguanides is based on the use of a liquid copper(II) ion-selective membrane (243). It is based on the formation of complexes of the type $[Cu(Big)_2]X_2$ by the reaction between copper(II) ammine complexes and biguanides. There are several ways in which this method can be used for determining biguanides:

(a) using an excess of copper(II) ammine in order to precipitate $[Cu(Big)_2]X_2$ and subsequent determination of excess copper(II) in solution by potentiometric titration in the presence of a copper(II) ion-selective electrode:

(b) transformation of the biguanide salt into the free base which is then determined by direct potentiometric titration with an aqueous solution of copper sulphate in the presence of a copper(II) ion-selective electrode; and

(c) separation of the water-insoluble complex and dissolution of a precisely known quantity in a suitable solvent and subsequent determination of $Cu^{2+}$ ions in the solution by a convenient method, such as with a copper(II) ion-selective electrode.

Methods (b) and (c) are lengthier because quantitative conversion of the biguanide salt into free base is needed, also the filtration and drying stage are time-consuming and susceptible to errors. For method (a), the $[Cu(Big)_2]SO_4$ complex is formed by adding an excess of ammoniacal copper sulphate (with known titre) to an aqueous solution of biguanide hydrochloride. The excess of $Cu^{2+}$ ions is determined by potentiometric titration with 0.1M EDTA, using a copper(II) ion-selective electrode as indicator.

There are (Fig.8) two potential jumps in the titration curve corresponding to excess copper(II) sulphate and complexed copper, respectively. The error of the method does not exceed ±0.8%.

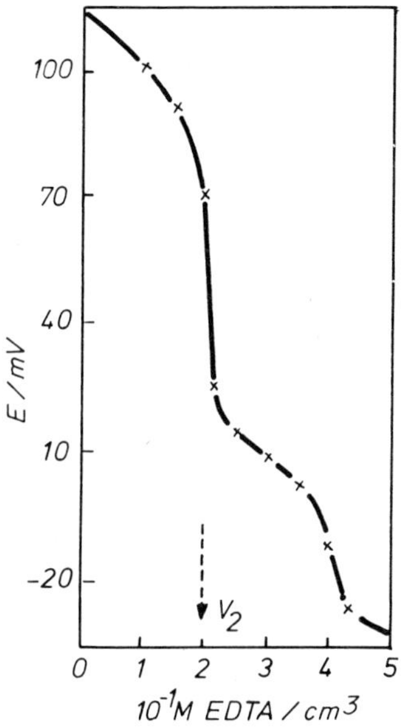


Fig.8. Potentiometric titration curve of an ammoniacal mixture of $[Cu(Big)_2]\ SO_4$ and $CuSO_4$ according to method (a). (243).

## 3.8 Alkaloids

The alkaloids, especially their halide salts, are extensively used drugs, and their determination (in the presence of other components) by conventional analytical methods is laborious and time-consuming.

Kalman et al. (244) have determined several of halide salts by potentiometric titration, using $10^{-2}$M silver nitrate as titrant and chloride or bromide ion-selective electrodes (Radelkis, type OP-711) as indicator electrodes. The results showed differences ranging from -0.6% to +2.3% between the amounts taken and found; these are rather large compared with those of other determinations using the same type of electrodes and may be attributed to poorer identification of the end points.

The possibility of determining various alkaloid halide salts, separately or in mixtures with other substances by direct potentiometry with ion-selective electrodes has also been studied (245), but the errors were much larger than those obtained by potentiometric titration. Measurements made in 0.1M potassium nitrate media gave the best results but even then the interfering effects could not be avoided. The errors were mainly negative and this was probably due to the uncertainty in activity coefficients.

Cosofret et al. (164) used a silver(I) ion-selective membrane electrode for the determination of scopolamine hydrobromide and N-butylscopolammonium bromide in pharmaceutical preparations. The results obtained are comparable to those obtained by using a commercial bromide-selective electrode (Radelkis type OP-Br-7111-D).

The precipitation of univalent cations of alkaloids with sodium tetraphenylborate may be potentiometrically monitored with a potassium ion-selective electrode (Crytur, type 19-15) (246) or platinum wire coated by a plastic membrane of poly(vinyl chloride) with dibutylphthalate and potassium tetraphenylborate (247). The titration curves of $10^{-2}$M cinchonine, brucine, codeine, quinine and strychnine with 5 x $10^{-2}$M sodium tetraphenylborate are shown in Fig.9. The break at the equivalence point was graphically determined with a ±2% error. The samples titrated in 100 $cm^3$ volume correspond to 13-16 μg alkaloid $cm^{-3}$, showing a high sensitivity of the potassium ion-selective electrode towards these alkaloids.

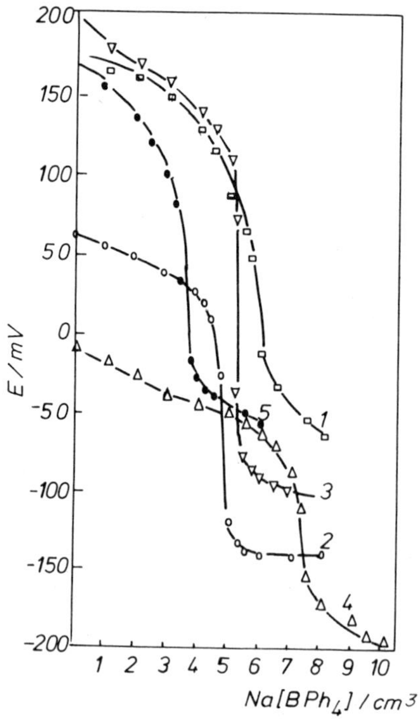


Fig.9. Titration curves of cinchonine (1), brucine (2), quinine (3), codeine (4) and strychnine (5). (247).

The coated-wire potassium ion-selective electrode obtained by Hopirtean and Stefănigă (247) exhibits a near-Nernstian response to quinine (range $10^{-1}$ - $10^{-7}$M, slope 55 mV decade$^{-1}$) brucine ($10^{-1}$ - 3 x $10^{-6}$M, slope 54 mV decade$^{-1}$) and cinchonine ($10^{-1}$ - $10^{-5}$M, slope 61 mV decade$^{-1}$).

A new electrode sensitive and reasonably selective to strychnine has recently been described (248). It is based on a strychnine-picrolonate ion-pair complex in nitrobenzene in a liquid membrane used in an Orion (Model 92) barrel with an Orion 92-81-04 porous membrane to separate the organic phase from the test solution. the cell used is represented by:

| | | | | |
|---|---|---|---|---|
| Ag/AgCl<br>0.01M KCl and<br>0.01M strychnine sulphate | $10^{-3}$M strychnine-picrolonate in nitrobenzene | ‖ | test solution | S.C.E. (IV) |

The electrode gave a near-Nernstian response with an average cationic slope of 57 mV decade$^{-1}$ over the 5 x $10^{-2}$M to 5 x $10^{-5}$M range, and 52 mV decade$^{-1}$ over 5 x $10^{-2}$M to 7 x $10^{-7}$M, in potassium sulphate background and pure water, respectively.

Direct potentiometric determination of samples down to 5 ppm showed an average recovery of 101.3% and a mean standard deviation of 2.4%. Pontentiometric titration with either picrolonic acid or sodium tetraphenylborate gave an average recovery of 99.1% and a mean standard deviation of 1.2%.

### 3.9 Sulphur compounds

3.9.1 Chloramine-T. A chloramine-T (CAT)-selective electrode with a liquid membrane of nickel-bathophenanthroline-chloramine-T dissolved in 2-nitro-4-cymene has been described (249). The sensor is Orion 92-07-01 nitrate liquid ion-exchanger converted to the chloramine-T form. The body of an Orion series 92 electrode equipped with an Orion polycellulose acetate membrane, for example, a 92-07-04 porous nitrate membrane was used for setting up the membrane electrode, the measuring cell being:

| | | | | | | |
|---|---|---|---|---|---|---|
| Ag/AgCl | internal solution<br>0.1M CAT+<br>0.1M NaCl | Membrane saturated with CAT-exchanger | Test solution | $KNO_3$ salt bridge | KCl | AgCl/Ag (V) |

The CAT-electrode was conditioned by soaking in a stirred 0.1M CAT solution for at least 12 hours after which the electrode presents a linear response to chloramine-T (used as a bactericide) at pH 7.0 (phosphate buffer) in the $10^{-1}$ - $10^{-4}$M range. During the operative life of the electrode (about 1 month) the slope was 59-62 mV decade$^{-1}$ at 25$^{o}$C but after 1 month the slope decreased gradually (249). The electrode can be used practically for the determination of chloramine-T either by direct potentiometry or by potentiometric titration.

3.9.2 Cistaphos. Ionescu et al. (250) described conditions for the quantitative determination of Cistaphos (sodium S-(2-amino-ethyl) thiophosphate; used as a radioprotecting agent) by means of various ion-selective electrodes.

The method is based on the reaction:

$$H_2NCH_2CH_2SPO_3Na + HOH \xrightarrow{H+} H_2NCH_2CH_2SH + NaH_2PO_4 \qquad (22)$$

which, since the S-P bond is very labile in acidic medium, takes place quantitatively at pH 2 to 3 (adjusted with 1M $HClO_4$ solution) in less than 1 min at 100°C. The products are stable for 15 to 20 min.

For the quantitative determination of the thiol product of equation (22) the authors (250) have used the responses of ion-selective electrodes. However, potentiometric titration of cysteamine with silver nitrate solution using a silver sulphide crystal membrane electrode (Orion, Model 94-16) was not satisfactory in either acidic or basic media owing to small changes of the e.m.f. and to the slow response around the equivalence point.

Since the silver sulphide membrane electrode also responds to $Hg^{2+}$ ions, and since mercury(II) salts are preferred to silver salts as titrants for thiols (the mercury(II) mercaptides being considerably more stable than the silver mercaptides), 0.1 or 0.01M mercury(II) nitrate solutions were used as titrant for the thiol product of equation (22). Fig.10 shows the potentiometric titration curves for 3.08 mg of cysteamine under different conditions. In all cases, the curves are well defined and the localization of the titrant volume may easily be made. The titration curves correspond to a 1:2 stoichiometry.

The possibility of using a silver sulphide or copper(II) ion-selective membrane electrode by direct potentiometry was also considered for the determination of cysteamine. The response of the silver sulphide electrode to cysteamine was linear within the $10^{-2}$ - $10^{-4}$M range with a slope of -87 mV decade$^{-1}$, which is consistent with the value of -90 mV decade$^{-1}$, corresponding to the formation of a 2:3 Ag-cysteamine complex:

$$2Ag^+ + 3H_2NCH_2CH_2S^- \longrightarrow Ag_2(H_2NCH_2CH_2S-)_3^- \qquad (23)$$

The copper(II) ion-selective electrode (Orion, Model 94-29) responds well to cysteamine and the electrode function is stable and reproducible over a wide concentration range ($10^{-2}$ - $10^{-6}$M) and the response times allow use of the electrode for direct potentiometry. In this case the slope is -59 mV decade$^{-1}$, in good agreement with the theoretical value of -59 mV decade$^{-1}$ corresponding to a redox interaction at the membrane-solution interface:

$$2Cu^{2+} + 4H_2N(CH_2)_2SH \longrightarrow 2H_2N(CH_2)_2S^- Cu^+$$

$$+ H_2N(CH_2)_2SS(CH_2)_2NH_2 + 4H^+ \qquad (24)$$

whereby the $Cu^{2+}$ ion oxidizes cysteamine to cystamine and resulting in the formation of the $Cu^+H_2NCH_2CH_2S^-$ electroactive species.

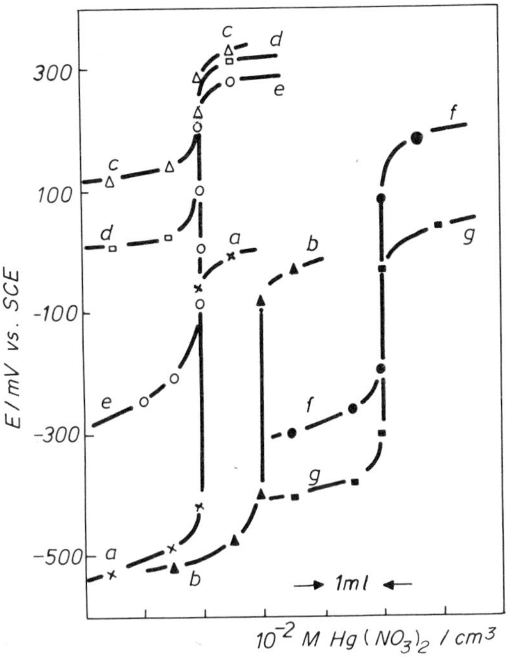


Fig.10. Potentiometric titration of 3.08 mg cysteamine under different conditions:

(a) in $10^{-1}$M NaOH; (e) at pH 7.5;

(b) in $10^{-2}$M NaOH; (f) at pH 6.0 (adjusted with hexamine)

(c) at pH 2.5; (g) at pH 8.0 (adjusted with 2 cm$^3$ pyridine)

(d) at pH 4.0;

(Curves b, f and g have been displaced horizontally for clarity) (250).

3.9.3 Sulphonamides. An electrode sensitive to sulphonamides depends on ion-association pairs between iron(II)-bathophenanthroline chelate and sulphonamides. These are extracted into nitrobenzene (5 x $10^{-5}$M) to form the active membrane material (251). The sample and the reference solutions are separated by having the liquid membrane in the bottom of a U-shaped glass tube. Electrodes for sulphamerazine (4-methyl-2-sulphanilamidopyrimidine) and sulphisomidine (2,4-dimethyl-6-sulphanilamidopyrimidine) were used to exemplify sulphonamides. The response was linear in the $10^{-1}$ - $10^{-3}$M range with a slope of 57 mV decade$^{-1}$, at pH 9.0 (adjusted with tromethamine and sulphuric acid buffer). The presence of a tenfold molar excess of urea, glycine, aminopyrine, or 4-aminobenzoic acid, agents known to interfere in the usual colorimetric analysis of sulpha drugs did not interfere with the observed potential.

3.9.4 Thiourea and derivatives. In studies on the potentiometric determination of thiourea and its derivatives(used as thyroid inhibitors), Pungor and co-workers (252 - 255) used a sulphide ion-selective electrode while Cosofret and Baiulescu (164,165,256) have used silver ion-selective electrodes having liquid membranes. In both cases, an aqueous silver nitrate solution was used as titrant. Thiourea could be titrated in the $10^{-1}$ - $10^{-3}$M concentration range in the presence of 1M or 0.1M sodium hydroxide. The reactions that take place during the titration were easily followed with these ion-selective electrodes.

The titration curves of thiourea with silver nitrate solution show two potential breaks and both may be used for the determination of thiourea in the sample. The volume of titrant corresponding to the first potential break gives the quantity of thiourea in the sample, and the volume corresponding to the second break is equivalent to twice the thiourea content. The following reactions take place (253):

$$(H_2N)_2C{=}S + 2\ AgNO_3 \longrightarrow Ag_2S + H_2N{-}C{\equiv}N + 2HNO_3 \qquad (25)$$

$$H_2N{-}C{\equiv}N + 2AgNO_3 \longrightarrow Ag_2N{-}C{\equiv}N + 2HNO_3 \qquad (26)$$

Equation (25) corresponds to the first break in the titration curves when silver sulphide precipitates. The resulting cyanamide subsequently reacts with silver nitrate to form a precipitate of silver cyanamide corresponding to the second break.

Hassan (257) has described the reaction conditions of iodine with thiourea and its derivatives (and other thiocarbonyl and thiol compounds) under which mg to g amounts of these compounds can be determined using the iodide (Orion, Model 94-53) or sulphide ion-selective electrodes (Orion, Model 94-16). It was found that a variety of thiocarbonyl compounds react with 0.5M iodine in carbon tetrachloride in the presence of pyridine with a stoichiometric liberation of 4 $\pm$ 0.04 equivalents of hydriodic acid per group. The presence of pyridine enhances reproducibility and precision of the results. Pyridine has influence on the electron affinity of iodine and possibly modifies its redox potential.

The reaction of thiourea with iodine may be represented thus:

$$(H_2N)_2C{=}S + 2C_5H_5NI_2 + 2H_2O \longrightarrow (H_2N)_2C{=}SO_2 + 4HI + 2C_5H_5N \qquad (27)$$

The iodide ions in an aqueous extract were potentiometrically titrated with 0.005M silver nitrate using the solid-state silver sulphide electrode. The results obtained with samples down to 70 $\mu$g corresponded to an average recovery of 98.6% and there were no interferences by unsaturated materials. Measurements of the iodide ion using an iodide ion-selective electrode and the known addition technique has given results with an absolute error of 2%.

The fact that potentials measured directly with a sulphide ion-selective electrode (Radelkis, type OP-S-711-D) in 0.1M solutions of 2-thiouracil, 6-methyl-2-thiouracil and 2,4-dithiouracil, in 0.1M sodium hydroxide, were -120, -80 and -470 mV, respectively, versus a saturated calomel electrode demonstrates the definite sulphide function of these compounds (255).

In solution buffered with acetate buffer at pH 5.6, only one break occurred in the potentiometric titration curves for 2-thiouracil with silver nitrate as titrant. Under the same conditions, one mole of 2,4-dithiouracil reacted with two moles of silver nitrate and breaks appeared in the titration curve.

The direct potentiometric titration of methimazol (1-methyl-2-mercapto-imidazole) with copper(II) sulphate solution at pH 5.6 gave one sharp potential change (ca. 140 mV) at the equivalence point for a copper(II) ion-selective indicator electrode at a drug to $Cu^{2+}$ ion molar ratio of 2:1. A similar procedure for methylthiouracil at pH 11.1 gave two potential breaks corresponding to 2:1 and 1:1 molar ratio, respectively (258). This method was satisfactory for the determination of methimazol and methylthiouracil in pharmaceutical preparations with relative standard deviation of 1.87% and 1.32%, respectively.

### 3.10 Organic mercury compounds

Many organic compounds of mercury (merbromine mersalyl, nitromersal, etc.) as well as some inorganic ones (mercury(II) chloride, ammoniated mercury, etc.) have a well-established reputation as powerful diuretic drugs. To date there are no commercially-available electrodes for their sensitivity to mercury(II), but there are numerous instances in the literature of mercury(II) responsive electrodes.

Růžička and Lamm (259) mentioned the preparation of a mercury(II) ion-selective electrode by activating the surface of a graphite rod with mercury(II) sulphide or its admixture with silver sulphide. No details were given about the preparation of the sulphides. Anfält and Jagner (260) prepared a mercury(II) ion-selective electrode by coating a silver rod with silver sulphide by anodic oxidation in a sulphide medium followed by precipitation of mercury(II) sulphide on the surface. This was accomplished by immersing the electrode in a solution of sodium sulphide and then adding a solution of mercury(II) nitrate until the free $Hg^{2+}$ ion concentration was about $10^{-2}$M. A similar procedure was proposed by van de Leest (261). None of these electrodes behaved in a manner suitable for the direct determination of mercury.

Kahr (262) found that it was difficult to press good pellets from black cubic mercury(II) sulphide (metacinnabar), whereas the red hexagonal sulphide (cinnabar) could not be pelleted at all. Mixed precipitates containing mercury(II) sulphide and bismuth sulphide were easily pelleted and membranes prepared in this way showed a better response. Popescu et al.(263) proposed a silver iodide-silver sulphide mixture for mercury(II) determination and such mixture may be obtained by precipitating sulphide and iodide from a solution of sodium sulphide and potassium iodide with silver nitrate. More recently, van der Linden and Oostervink (264) have examined the electrochemical behaviour of membranes pressed from silver sulphide and mercury(II) sulphide, but it was very difficult to obtain electrodes that behaved reproducibly.

A coated-wire ion-selective electrode for determining mercury(II) in chloride and iodide solutions has been described by Cattrall and Pui (124). In this case, the electroactive membranes consist of chloromercurate(II) and iodomercurate(II) salts of Aliquat 336S, combined with PVC. The response of both electrodes was rapid, requiring about 5 min to reach a stable potential in the solutions of the lowest concentrations and instantaneously at high concentration. The chloromercurate electrode was evaluated for the potentiometric titration of mercury(II) with EDTA.

Determination of mercury(II) in aqueous solutions by potentiometric titration with EDTA at pH 6 (adjusted with hexamine) have been carried out by others with the same liquid-state mercury(II) ion-sensitive electrodes based on Aliquat 336S salts (145,162,265-267). There were large potential breaks at the equivalence point and a maximum error of 0.5% for the titration of 18-46 mg mercury(II) with 0.1M EDTA.

Overman (268) used an iodide-selective electrode (Orion, Model 94-53) for detecting the end-point in the potentiometric titration of mercury(II) with sodium iodide solutions and the relative standard deviation for titrating 1 mg was $\pm$ 0.4%.

For the determination of mercury in organic compounds, the combustion products of the Schöniger oxidation are absorbed in either nitric acid or in an aqueous solution of bromine. Mercury(II) may then be determined gravimetrically or volumetrically, but the methods require a large amount of substance. This inconvenience was eliminated in a microassay procedure for mercury in organic substances (269). In this method an iodide ion-selective electrode (Orion Model 94-53) was used as indicator electrode, but experiments for the determination of mercury(II) by direct potentiometry did not give satisfactory results, especially in the concentration range of $10^{-3}$ to $10^{-4}$M.

Campiglio (269) used Overman's method (268) for determinating mercury(II) with 5 x $10^{-3}$M sodium iodide titrant. Thus, the sample (3-5 mg) was burnt in an oxygen flask and the products absorbed in 4 $cm^3$ concentrated nitric acid. The absorbtion solution is boiled for 6 min, during which complete oxidation to mercury(II) occurs. The optimum pH for the subsequent potentiometric titrations is in the range 1 to 2.

The results obtained by Campiglio (269) were within the usual $\pm$ 0.3% limit of error. Nevertheless, chlorine and bromine, frequently found in organic compounds of mercury negatively influence the mercury(II) titration by lowering the potential change around the equivalence point.

## 3.11 Vitamins

3.11.1 Ascorbic acid (Vitamin C). For the determination of ascorbic acid and other reductants which react stoichiometrically with iodine, Christova et al. (270) proposed a method based on the quantitative oxidation in solutions of pH 2.0 with 0.1M virtually iodide free ethanolic solution of iodine, followed by measurement of the activity of the iodide ions formed with an iodide ion-selective electrode (Radiometer F 1032 or Crytur iodide electrode). E.m.f. measurements were first made on a blank, and then for solutions of different concentrations of ascorbic acid, at intervals of 15s, beginning with the lowest. The electrodes had slopes of 59$\pm$ 1 mV decade$^{-1}$ at ascorbic acid levels between $10^{-3}$ and $10^{-5}$M.

Ascorbic acid was determined by direct potentiometry in injection solutions or in tablets with a relative standard deviation of about 1 to 2% for 0.7 to 35 mg of ascorbic acid, using a chloramine -T selective electrode (244). Analysis of aqueous ascorbic acid solutions of known concentrations by potentiometric titration with chloramine-T, in phosphate buffer, showed that ascorbic acid in the range 4 to 40 mg might be determined with relative errors and relative standard deviation of about 1%.

3.11.2 Cyanocobalamin (Vitamin $B_{12}$). Two methods are known for the determination of cyanocobalamin with ion-selective electrodes (271, 272). Both consist of the quantitative liberation of hydrogen cyanide from cyanocobalamin in pure form as well as in pharmaceutical preparations. In both methods a cyanide ion-selective electrode (Radelkis type OP-CN-7111) was used.

The cyano group is released by reduction with reducing agents such as ascorbic acid, tin(II) chloride in HCl, calcium hypophosphite in hydrochloric or sulphuric acid under reflux or by exposure to a strong source of visible light at room temperature (two 500W lamps for 30 min). The determinations were carried out by direct potentiometry or by potentiometric titration.

Both the reduction and illumination methods in a combination with a cyanide ion-selective electrode have proved accurate and reproducible as well as being simple and rapid for routine use.

The decomposition of cyanocobalamin and some analogs (cobinamide and hydroxocobalamin) in a stream of nitrogen at a suitable temperature has been investigated by Goldstein and Duca (272). By heating at 120°C (for cobinamide) or at 140-145°C (for cyanocobalamin) in sodium hydroxide solution (pH 12-13) while bubbling nitrogen through the mixture, the cyano group can be converted into cyanide and quantitatively determined by a cyanide ion-selective electrode. The error does not exceed $\pm$ 3%.

3.11.3 Nicotinic acid and derivatives

The determination of nicotinic acid is an analytical problem of great interest because of its biological activity and that of its derivatives, particularly the amide. This activity is connected with liver functions in the presence of PP vitamin in the human body. Studies (273-275) of the determination of the materials with membrane electrodes have been made with heterogeneous solid membrane electrodes (copper nicotinate as sensor, and dipped into different media: polyethene silicone rubber or paraffin) and liquid membrane electrodes (an ester of nicotinic acid as sensor).

The solid electrodes are mechanically advantageous for they are compact and do not need an internal reference electrode and are sufficiently fast in response. However, the liquid membrane cetyl nicotinate electrode, was found to be more selective than the solid-membrane electrode. The electrodes are convenient for titrating of nicotinate with sulphuric acid or hydrochloric acid solutions.

The determination of nicotinamide (and of N-methyl-nicotinamide) with the nicotinate liquid membrane electrode is based on the transformation of nicotinamide into nicotinate:

$$\text{nicotinamide} + \text{HOH} \rightleftharpoons \text{nicotinate}^- + NH_4^+ \qquad (28)$$

The nicotinate ion is then titrated potentiometrically with 0.05M sulphuric acid solution (274).

3.11.4 Vitamin $B_1$ and vitamin $B_6$. Ion-selective membrane electrodes sensitive to vitamin $B_1$, (thiamine hydrochloride) and vitamin $B_6$ (pyridoxine hydrochloride) have been prepared from ion-association extraction systems (223,277). The vitamins are easily extracted into organic solvents, such as 1,2-dichloroethane and nitrobenzene with tetraphenylborate or dipicrylamine anion from acidic aqueous media (pH3 5) and the organic solutions can be used as the corresponding vitamin-sensitive membranes.

Concentrations of the sensor within the membranes were $1 \times 10^{-4}$M. Acidic aqueous solution of vitamin $B_6$ is stable but the aqueous solution of vitamin $B_1$, is comparatively stable only in the pH 2 to 4 range. Therefore, the vitamin $B_1$ solution was prepared immediately before the measurement of membrane potential. The electrodes exhibit typical responses to divalent (vitamin $B_1$) and univalent (vitamin $B_6$) cations. The main characteristics of these vitamin-sensitive electrodes are presented in Table VI from which it can be seen that the electrodes have the high selectivities for vitamin ions over $NH_4^+$, $Na^+$ and $K^+$ ions.

TABLE VI. Performance of the vitamin-sensitive electrodes (277)

| Electrode | Solvent | Exchange site | Slope /mV decade$^{-1}$ | Useful range | Selectivity coefficients |
|---|---|---|---|---|---|
| Vitamin $B_1$ | 1,2-di-chloro-ethane | TPB | -30 | $10^{-1}$-$10^{-5}$ (pH 3-5) | $NH_4^+=10^{-4}$; $Na^+$& $K^+= 10^{-4}$; Vit.$B_6$ = 70 |
| Vitamin $B_6$ | nitro-benzene | dipicryl-amine | -57 | $10^{-1}$-$10^{-5}$ (pH 3-5) | $NH_4^+=8\text{x}10^{-3}$; $Na^+=6\text{x}10^{-4}$; $K^+=2.5\text{x}10^{-2}$; Vit.$B_1=10^{-1}$ |

## 3.12 Antibiotics

3.12.1 Cephalosporins. *Citrobacter freundii* releases cephalosporinase which catalyses the following reaction of cephalosporin with the release of hydrogen ions:

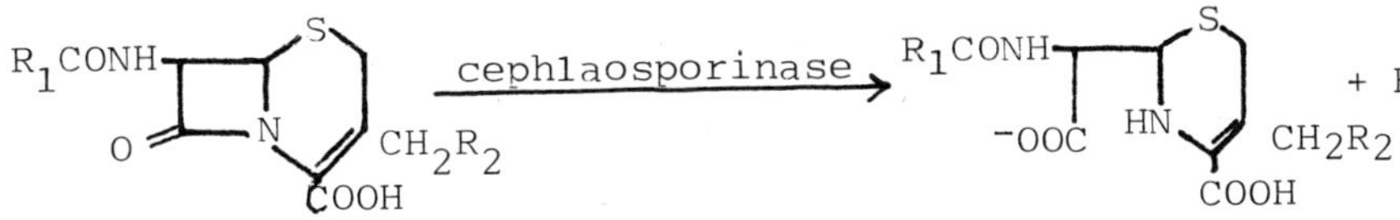


(29)

Cephalosporin may therefore be determined from the $H^+$ concentration generated in a medium by using immobilized cephalosporinase (278). Immobilization of cephalosporinase was difficult because the molecular mass of the enzyme is only 30 000 and the enzyme is unstable. Therefore, whole cells of Citrobacter freundii were immobilized in a collagen membrane. Such a microbial sensor composed of a bacteria-collagen membrane reactor and a combined glass electrode can be used for the determination of cephalosporins.

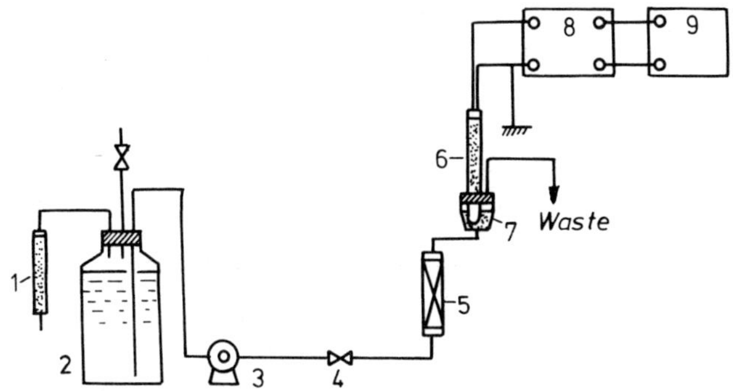


Fig.11. Immobilised whole cell-based flow-type sensor for cephalosporins:
1, soda lime; 2, buffer reservoir; 3, peristaltic pump;
4, sample inlet; 5, immobilized whole cell reactor;
6, combination glass electrode; 7, sensing chamber;
8, amplifier; 9, recorder (278).

The system used for the continuous determination of cephalosporins is illustrated in Fig.11. The reactor was a biocatalytic type with a spacer located in the centre (279). The inner volume of the reactor was 4.1 $cm^3$. The bacteria-collagen membrane was rolled into a plastic net and inserted into the reactor. The pH of the sample solution was measured by a combination glass electrode and displayed on a recorder. Phosphate buffer (pH 7.2) was transferred continuously to the reactor and sensing chamber. Sample solutions (10 $cm^3$) containing various amounts of cephalosporins were transferred to the reactor, at 2 $cm^3$ $min^{-1}$ by a peristaltic pump and the hydrogen ion concentration in the sensing chamber determined continuously. The electrode potential difference in the sensing chamber increased with time until a maximum was reached. The time required for the maximum to be reached (the response time) depends on the flow-rate and the activity of the bacteria-collagen membrane. A linear relationship was obtained between the logarithm of the cephalosporin concentration and the potential difference. This sensor was used for the determination of 7-phenylacetylamidodesacetoxysporanic acid (phenylacetyl-7-ADCA), cephaloridine, cephalothin and cephalosporin C.

Reproducibility of the system was determined with phenyl-acetyl-7 ADCA solution (125 $\mu g\ cm^{-3}$), the relative standard deviation being 10% (2 mV) for 10 experiments.

3.12.2 Chloramphenicol. Determination of chloramphenicol (D-threo-2-dichloroacetamido-1-p-nitrophenylpropane-1,3-diol) and its esters has been, and is still,the subject of detailed investigations since these broad-spectrum antibiotics have wide therapeutic uses throughout the world.

In a recent investigation Hassan and Eldesouki (280) have carried out an easy reduction procedure with cadmium metal of chlor-amphenicol and its esters in various pharmaceutical preparations, without prior extraction, and followed by a rapid instrumental measure-ment of the released cadmium ions and amine by atomic-absorbtion spectrometry, spectrophotometry and potentiometry with a cadmium ion-selective electrode (Orion Model 94-48).

Six equivalents of cadmium ion are quantitatively released per mole of chloramphenicol after 15 min (in 0.05M hydrochloric acid). However, quantitative reductions of chloramphenicol stearate, palmit-ate and succinate esters under the same conditions are not possible, probably because these esters have low aqueous solubility (280). Prior hydrolysis, by treatment with 1M alcholic potassium hydroxide at room temperature, leads to complete reduction. The free carboxyl-ate ions released by hydrolysis do not interfere in the subsequent measurement procedures.

For the prescribed conditions, chloramphenicol has no available sites other than the $-NO_2$ group for reduction with cadmium metal, and the reaction proceeds to the corresponding amine:

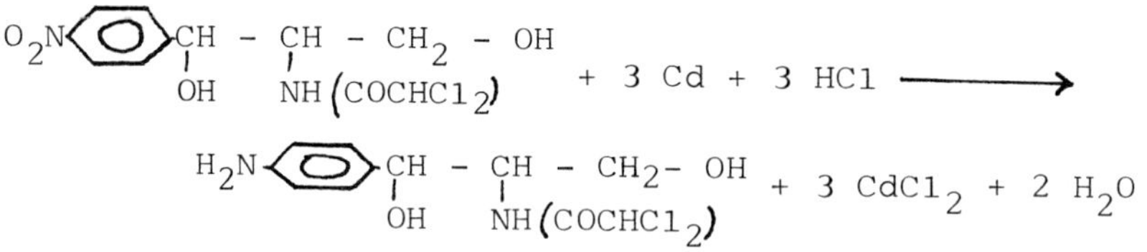

(30)

Released cadmium ions can be measured by potentiometric titrations with EDTA using a cadmium ion-selective electrode as end-point detector. The pH is adjusted to 10 with ammonia. The results showed an average recovery of 99.3% and a mean standard deviation of ± 0.5%.

3.12.3 Penicillins. Papariello et al. (281,282) have developed an electrode which is specific for the analysis of penicillins. This electrode is based on the concept of an enzyme electrode and is pre-pared by immobilizing penicillinase in a thin membrane of polyacryl-amide gel moulded around and in intimate contact with a pH glass electrode. When the electrode is exposed to an aqueous solution of penicillin adjusted to a pH of 6.4, the immobilized enzyme hydrolyses the penicillin to produce the corresponding penicillic acid:

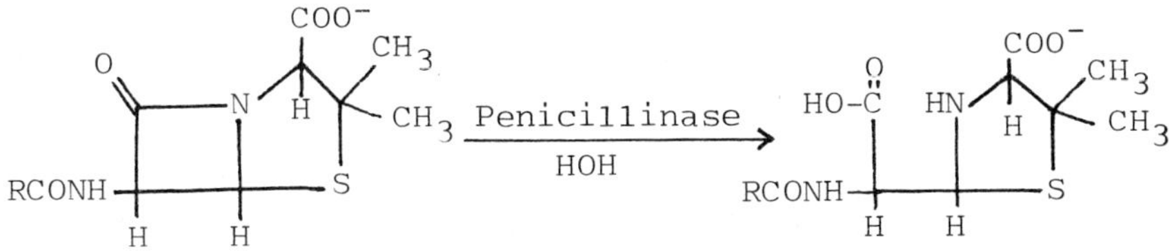


(31)

The increase in hydrogen ion concentration from the penicillic acid is sensed by the glass electrode. It is clear from the reaction that the electrode potential can be related to the penicillin concentration in the following manner with a slope of 59 mV decade$^{-1}$.

$$E = E_o + \frac{2.303RT}{F} \log a_{H^+} \qquad (32)$$

$$E = E_o + \frac{2.303RT}{F} \log [\text{penicillin}] \qquad (33)$$

However, electrode response is not completely Nernstian for many penicillins (Sodium Ampicilin-52, Sodium Nafcillin monohydrate-42, Potassium Penicillin G-44, Potassium Penicillin V-40, etc.). The slope is affected by age and frequency of use of the electrode. Never-the-less the electrode is analytically useful in the penicillin concentration range of $10^{-4}$ to $5 \times 10^{-2}$M and exhibits a relatively quick response (15 to 30 s) to most penicillins.

A new simple, rapid and accurate method, free from many of the defects usually encountered in other methods of the analysis of total penicillins has been developed by Hassan et al. (283). The method is based on disulphurization with potassium EDTA at pH 4.5 (acetate buffer) by using a lead ion-selective electrode (Orion Model 94-82).

Some pure penicillin samples were analysed by procedure using a lead ion-selective indicator electrode and the results compared with those obtained by applying the standard iodometric procedure of the British Pharmacopoeia (284). The results of the two methods agreed within $\pm$ 0.5%. Analysis of a penicillin sample takes less than 15 min whereas the iodometric procedure requires 60 min.

## 4. REFERENCES

1. G.J.Moody and J.D.R.Thomas, Selective Ion Sensitive Electrodes, Merrow, Watford (1971).
2. E.C.Toren Jr. and R.P.Buck, Anal.Chem., 42, 284R (1970).
3. R.P.Buck, Anal.Chem., 44, 270R (1972).
4. R.P.Buck, Anal.Chem., 46, 28R (1974).
5. R.P.Buck, Anal.Chem., 48, 23R (1976).
6. R.P.Buck, Anal.Chem., 50, 17R (1978).

7. J.Koryta, Analytica Chim.Acta, 61, 329 (1972).

8. J.Koryta, Analytica Chim.Acta, 91, 1 (1977).

9. R.A.Durst (Ed.), Ion Selective Electrodes, NBS, Spec.Publ.314, Washington (1969).

10. E.Pungor (Ed.), Ion-Selective Electrodes,Symposium held at Matrafüred (Hungary) 1972, Akademiai Kiadó, Budapest (1973).

11. E.Pungor (Ed.), Ion-Selective Electrodes, 2nd Symposium held at Matrafüred (Hungary) 1976, Akademiai Kiadó, Budapest (1977).

12. E.Pungor (Ed.), Ion-Selective Electrodes, Conference held at Budapest (Hungary) 1977, Akademiai Kiadó, Budapest (1978).

13. IUPAC International Symposium on Ion-Selective Electrodes, 1973, UWIST, Cardiff, (1973).

14. H.J.Berman and N.C.Hebert (Eds.), Ion-Selective Microelectrodes, Plenum Press, New York, (1974).

15. M.Kessler, L.C.Clarc, D.W.Lubers, I.A.Silver and W.Simon (Eds.) Ion and Enzyme Electrodes in Biology and Medicine, Urban and Schwarzenberg, München, (1976).

16. Theory and Applications of Ion-Selective Electrodes in Physiology and Medicine, Extended Abstracts of Symposium held at Dortmund (West Germany), 1977 , published in Arzneim,-Forsch./Drug Res., 28, (I), 705 and 866 (1977).

17. K.Cammann, Das Arbeiten mit Ionenselektiven Elektroden, Springer Verlag, Berlin, (1973) and (1977).

18. N.Lakshiminarayanaiah, Membrane Electrodes, Academic Press, London (1976).

19. J.Koryta, Ion-Selective Electrodes, Cambridge University Press, Cambridge (1975).

20. P.L.Bailey, Analysis with Ion Selective Electrodes, Heyden, London (1976).

21. G.E.Baiulescu and V.V.Cosofret, Applications of Ion Selective Membrane Electrodes in Organic Analysis, Wiley, Chichester (1977).

22. H.Freiser (Ed.), Ion Selective Electrodes in Analytical Chemistry, Vol.1, Penum Press, New York (1978).

23. D.Midgley and K.Torrance, Potentiometric Water Analysis, Wiley, Chichester (1978).

24. J.Vesely, D.Weiss and K.Stulik, Analysis with Ion Selective Electrodes, Wiley, Chichester (1978).

25. J.Tenygl, Ion-Selective Electrode Analysis, MTP Series, Physical Chemistry, Vol.12, Analytical Chemistry-Part 1 (T.S.West, Ed.) Butterworths, London (1972).

26. C.D.Ferris, Introduction to Bioelectrodes, Plenum Press, New York (1974).

27. J.I.Clerc, H.J.Degenhart and E.Pretsch, Ion Selective Electrodes Clinical Biochemistry, Vol.1 (H.Ch.Curtius and M.Roth, Eds.), W.de Gryther, Berlin (1974).

28. G.Svehla (Ed.),Comprehensive Analyical Chemistry, Vo.8, Elsevier, Amsterdam (1977).

29. J.C.Johnson, Immobilized Enzymes Preparation and Engineering, N.D.C., Park Ridge (1979).

30. J.Kragten, Analyst, 99, 43 (1974).

31. N.Radić, Analyst, 101, 657 (1967).

32. M.S.Frant and J.W.Ross, Science, 154, 1553 (1966).

33. B.Jaselkis and M.K.Bandemer, Anal.Chem., 41, 855 (1969).

34. A.Hamola and R.O.James, Anal.Chem., 48, 776 (1976).

35. F.C.Chang and K.L.Cheng, Analytica Chim.Acta, 76, 177 (1975).

36. J.W.Ross, J.H.Riseman and J.A.Krueger, Pure and Appl.Chem., 36, 743 (1973).

37. G.Eisenman, D.O.Rudin and J.U.Casby, Science, 126, 831 (1957).

38. J.G.Montalvo and G.G.Guilbault, Anal.Chem., 41, 1897 (1969).

39. J.G.Montalvo, Analytica Chim.Acta, 65, 189 (1973).

40. M.Adhikari, M.Paul and D.Gangopadhyay, Indian J.Chem., 8, 79 (1970).

41. R.P.Scholer and W.Simon, Chimia, 24, 372 (1970).

42. R.E.Cosgrove, C.A.Mask and I.H.Krull, Anal.Letters, 3, 457 (1970).

43. G.A.Rechnitz and E.Eyal, Anal.Chem., 44, 370 (1972).

44. O.K.Stefanova and I.V.Rusina, Elektrokhimiya, 14, 882 (1978).

45. W.Simon, D.Ammann, M.Oehme and W.E.Morf, Ann.N.Y.Acad.Sci., 307, 52 (1978).

46. J.W.Ross, Science, 156, 1378 (1967).

47. M.Mascini and A.Liberti, Analytica Chim.Acta, 53, 202 (1971).

48. A.Ansaldi and S.I.Epstein, Anal.Chem., 45, 595 (1973).

49. D.Ammann, M.Gueggi, E.Pretsch and W.Simon, Anal.Letters, 8, 709 (1975).

50. R.W.Cattrall, D.M.Drew and J.C.Hamilton, Analytica Chim.Acta, 76, 269 (1975).

51. M.Ghosh, M.R.Dhaneshwar, R.G.Dhaneshwar and B.Gosh, Analyst, 103, 768 (1978).

52. G.J.Moody, R.B.Oke and J.D.R.Thomas, Analyst, 95, 910 (1970).

53. G.H.Griffiths, G.J.Moody and J.D.R.Thomas, Analyst 97, 420 (1972).

54. A.Craggs, G.J.Moody and J.D.R.Thomas, Analyst, 104, 412 (1979).

55. L.Keil, G.J.Moody and J.D.R.Thomas, Analytica Chim. Acta, 96, 171 (1978).

56. G.J.Moody, N.S.Nassoury and J.D.R.Thomas, Analyst, 103, 68 (1978).

57. A.Craggs, G.J.Moody and J.D.R.Thomas, J.Chem.Educ., 51, 541 (1974).

58. S.Kamata, G.J.Moody and J.D.R.Thomas, Analytica Chim.Acta, 108, 385 (1979).

59. L.Keil, G.J.Moody and J.D.R.Thomas, Analyst, 102, 274 (1977).

60. A.M.Y.Jaber, G.J.Moody and J.D.R.Thomas, Analyst, 102, 943 (1977).

61. A.Craggs, G.J.Moody and J.D.R.Thomas, Analyst, 104, 961 (1979).

62. G.J.Moody, N.S.Nassoury and J.D.R.Thomas, Talanta, 26, 872 (1979).

63. A.M.Y.Jaber, G.J.Moody and J.D.R.Thomas, Polypropylene Glycol Adducts as Sensors for Ion-Selective Electrodes. In: Ion-Selective Electrodes, Ed. E.Pungor, Akademiai Kiadó, Budapest (1978), p.411.

64. A.Craggs, L.Keil, G.J.Moody and J.D.R.Thomas, Talanta, 22, 907 (1975).

65. A.M.Y.Jaber, G.J.Moody, J.D.R.Thomas and A.Willcox, Talanta, 24, 655 (1977).

66. D.Ammann, R.Bissig, M.Güggi,E.Pretsch, W.Simon, I.J.Borowitz and L.Weiss, Helv.Chim.Acta, 58, 1535 (1975).

67. J.D.R.Thomas, Lab.Practice, 27, 857 (1978).

68. G.J.Moody and J.D.R.Thomas, Ion-Selective Electrode Reviews, 1, 3 (1979).

69. J.Růžička and J.Ch.Tjell, Analytica Chim.Acta, 47, 475 (1969).

70. J.Růžička, C.G.Lamm and J.Ch.Tjell, Analytica Chim.Acta, 62, 15 (1972).

71. J.Růžička, E.H.Hansen and J.Ch.Tjell, Analytica Chim.Acta, 67, 155 (1973).

72. Ž.Pranjić-Anušić, Acta Pharm.Jugoslav., 29, 29 (1979).

73. R.W.Cattrall and Chin-Poh Pui, Anal.Chem., 47, 93 (1975).

74. R.W.Cattrall and Chin-Poh Pui, Analytica Chim.Acta, 78, 463 (1975)

75. E.Hopirtean, Rev.Roum.Chim., 22, 1385 (1977).

76. E.Hopirtean, C.Liteanu and R.Vlad, Talanta, 22, 912 (1975).

77. L.A.Lazarou and T.P.Hadjiioannou, Anal.Letters, 11, 779 (1978).

78. L.A.Lazarou and T.P.Hadjiioannou, Anal.Chem., 51, 790 (1979).

79. M.Marhol and K.L.Cheng, Anal.Chem., 42, 652 (1970).

80. K.L.Cheng and K.Cheng, Mikrochim.Acta, 1974 p.385.

81. T.P.Hadjiioannou and D.S.Papastathopoulos, Talanta, 17, 399 (1970).

82. L.A.R.Pioda, V.Stankova and W.Simon, Anal.Letters, 2, 665 (1969).

83. P.Laüger, Science, 178, 24 (1972).

84. R.W.Stow, R.F.Baer and B.F.Randall, Arch.Phys.Med., 38, 646 (1957).

85. J.W.Severinghaus and A.F.Bradley, J.Appl.Physiol., 13, 515 (1958).

86. M.A.Jensen and G.A.Rechnitz, Anal.Chem., 51, 1972 (1979).

87. P.L.Bailey and M.Riley, Analyst, 100, 145 (1975).

88. P.L.Bailey and M.Riley, Analyst, 102, 213 (1977).

89. M.Mascini and C.Cremisini, Analytica Chim.Acta., 97, 237 (1978).

90. K.Livansky, Coll.Czech.Chem.Commun., 43, 3042 (1978).

91. J.Růžička and E.Hansen, Analytica Chim.Acta, 69, 129 (1974).

92. E.H.Hansen, H.B.Filho and J.Růžička, Analytica Chim.Acta, 71, 225 (1974).

93. J.Růžička, E.H.Hansen, P.Bisgaard and E.Reymann, Analytica Chim. Acta, 72, 215 (1974).

94. E.H.Hansen and J.Růžička, Analytica Chim.Acta, 72, 353 (1974).

95. U.Fiedler, E.H.Hansen and J.Růžička, Analytica Chim.Acta, 74, 423 (1975).

96. J.Fligier and Z.Gregorowicz, Analytica Chim.Acta, 90, 263 (1977).

97. H.B.Herman and G.A.Rechnitz, Science, 184, 1074 (1974).

98. H.B.Herman and G.A.Rechnitz, Analytica Chim.Acta, 76, 155 (1975).

99. J.Růžička and C.G.Lamm, Analytica Chim.Acta, 54, 1 (1971).

100. J.Růžička, C.G.Lamm and J.Ch.Tjell, Analytica Chim.Acta, 62, 15 (1972).

101. M.S.Frant and J.W.Ross, Science 154, 1553 (1966).

102. W.E.Bazzelle, Analytica Chim.Acta, 54, 29 (1971).

103. J.J.Lingane, Anal.Chem., 39, 881 (1967).

104. J.J.Lingane, Anal.Chem., 40, 935 (1968).

105. E.Pungor and K.Tóth, Mikrochim.Acta, 2, 656 (1964).

106. E.Pungor, K.Tóth and J.Havas, Mikrochim.Acta, 4, 689 (1966).

107. G.G.Guilbault and P.J.Brignac, Analytica Chim.Acta, 56, 139 (1971).

108. G.G.Guilbault and M.Nanjo, Analytica Chim.Acta, 78, 69 (1975).

109. M.Nanjo, T.J.Rohm and G.G.Guilbault, Analytica Chim.Acta, 77, 19 (1975).

110. F.R.Shu and G.G.Guilbault, Anal.Letters, 5, 559 (1972).

111. I.Novozamsky and W.H.Riemsdijk, Analytica Chim.Acta, 85, 41 (1976).

112. J.Tacussel and J.J.Fombon, New Solid State Ion-Selective Electrodes for $PO_4^{3-}$ and $Cl^-$ Activity Measurements. In: Ion Selective Electrodes, Ed. E.Pungor, Akademiai Kiadó, Budapest (1978), p.567.

113. W.Selig, Mikrochim.Acta, 564 (1970).

114. P.Hirsch-Ayalon, Electrochim.Acta, 10, 773 (1965).

115. E.Pungor and J.Havas, Acta Chim.Acad.Sci.Hung., 50, 77 (1966).

116. G.A.Rechnitz, G.H.Fricke and M.S.Mohan, Anal.Chem., 44, 1098 (1972).

117. G.A.Rechnitz and M.S.Mohan, Anal.Chem., 45, 1323 (1973).

118. T.Cserfalvi and G.G.Guilbault, Analytica Chim.Acta, 84, 259 (1976).

119. J.W.Ross and M.S.Frant, Anal.Chem., 41, 957 (1969).

120. A.M.Y.Jaber, G.J.Moody and J.D.R.Thomas, Analyst, 101, 179 (1976).

121. Dilys L.Jones, G.J.Moody, J.D.R.Thomas and M.Hangos, Analyst, 104, 973 (1979).

122. A.G.Fogg, M.Duzinkewycz and A.S.Pathan, Anal.Letters, 6, 1101 (1973).

123. E.M.Rachman'Ko, G.L.Starobinets, V.L.Lamako and A.Beisis, Vests. Akad.Nauk BSSR, Ser.Khim.Navuk., 6, 68 (1978).

124. R.W.Cattrall and Chin-Poh Pui, Anal.Chem., 48, 552 (1976).

125. R.W.Cattrall and Chin-Poh Pui, Analytica Chim.Acta, 83, 355 (1976).

126. R.W.Cattrall and Chin-Poh Pui, Analytica Chim.Acta, 87, 419 (1976).

127. U.S.Pharmacopoeia, XIX, p.45.

128. R.J.Levins, Anal.Chem., 43, 1045 (1971).

129. R.J.Levins, Anal.Chem., 44, 1544 (1972).

130. M.Güggi, E.Pretsch and W.Simon, Analytica Chim.Acta, 91, 107 (1977).

131. M.S.Frant and J.W.Ross, Ger.Offen.1942397 (12 Mar.1970).

132. H.Hirata and K.Higashiyama, Talanta, 19, 391 (1972).

133. J.Pick,K.Tóth, E.Pungor, Analytica Chim.Acta, 61, 169 (1972).

134. J.D.Czaban and G.A.Rechnitz, Anal.Chem., 45, 471 (1973).

135. A.Hulanicki, M.Trojanowicz and M.Cichy, Talanta, 23, 47 (1976).

136. T.Hepel, M.Hepel and M.Leszko, Analyst 102, 132 (1977).

137. J.G.M.Heijne, W.E.van der Linden and G.den Boef, Analytica Chim. Acta, 89, 287 (1977).

138. G.J.M.Heijne and W.E.van der Linden, Analytica Chim.Acta, 93, 99 (1977).

139. A.Hulanicki and A.Lewenstam, Talanta, 23, 661 (1976).

140. M.Koebel, Anal.Chem. 46, 1559 (1974).

141. V.V.Cosofret, Rev.Chim.(Bucharest),27, 240 (1976).

142. V.V.Cosofret and P.G.Zugravescu, Rev.Chim.(Bucharest), 28, 785 (1977).

143. G.E.Baiulescu and V.V.Cosofret, Rev.Chim.(Bucharest), 27, 158 (1976).

144. G.E.Baiulescu and V.V.Cosofret, Rev.Chim.(Bucharest), 26, 1051 (1975).

145. G.E.Baiulescu, V.V.Cosofret and M.Blasnic, Potentiometric Determination of Alklxanthates with Ion-Selective Electrode In: Ion-Selective Electrodes, Ed. E.Pungor, Akademiai Kiadó, Budapest (1978) p.207.

146. J.Růžička and J.Ch.Tjell, Analytica Chim.Acta, 51, 1 (1970).

147. A.V.Gordievski, V.S.Shterman, A.Y.Syrchenkov, N.I.Savvin, A.F.Zhukov and Y.I.Urasov, Zh.Anal.Khim., 27, 2170 (1972).

148. H.Hirata, K.Higashiyama and K.Date, Analytica Chim.Acta, 54, 415 (1971).

149. H.Hirata and K.Kigashiyama, Bull,Chem.Soc.Jap., 44, 2420 (1971).

150. H.Hirata and K.Kigashiyama, Analytica Chim. Acta, 57, 476 (1971).

151. M.Mascini and A.Liberti, Analytica Chim.Acta, 60, 405 (1972).

152. H.Hirata and K.Date, Anal.Chem., 43, 279 (1971).

153. V.Majer, J.Vesely and K.Stulik, Anal.Letters, 6, 577 (1973).

154. D.Cörmös, I.Haiduc and P.Stetin, Rev.Roum.Chim., 20, 259 (1975).

155. E.H.Hansen and J.Růžička, Analytica Chim.Acta, 72, 365 (197 ).

156. G.J.M.Heijne, W.E.van der Linden and G.den Boef, Analytica Chim. Acta, 100, 193 (1978).

157. C.Liteanu, I.C.Poescu and H.Naşcu, Rev.Roum.Chim., 17, 1651 (1972).

158. J.Vesely, O.J.Jensen and B.Nicolaisen, Analytica Chim.Acta, 62, 1 (1972).

159. D.J.Crombie, G.J.Moody and J.D.R.Thomas, Analytica Chim.Acta, 80, 1 (1975).

160. V.V.Cosofret, Rev.Chim.(Bucharest), 25, 836 (1974).

161. G.E.Baiulescu, V.V.Cosofret and C.Cristescu, Rev.Chim.(Bucharest), 26, 429 (1975).

162. V.V.Cosofret, Doctoral Theses, Polytechnic Institute of Bucharest, Bucharest (1975).

163. V.V.Cosofret, Rev.Roum.Chim., 23, 1489 (1978).

164. V.V.Cosofret, C.Ştefănescu and A.A.Bunaciu, Talant , 26, 1035 (1979).

165. W.Selig, Ion-Selective Electrodes in Organic Elemental and Functional Group Analysis. A Review, Lawrence Livermore Laboratory, REPT, UCRL, 52393 (1977).

166. W.Selig, Ion-Selective Electrodes in Organic Elemental and Functional Group Analysis. A Review, Lawrence livermore Laboratory, REPT, UCRL, 52393 (1978).

167. W.Schöniger, Mikrochim.Acta, 123 (1955).

168. W.Walisch and O.Jaenicke, Mikrochim.Acta, 1147 (1967).

169. R.A.Hofstader, Microchem,J., 11, (1966).

170. W.Krijgsman, F.F.Mansveld and B.Griepink, Z.Anal.Chem., 249, 368 (1970).

171. W.Krijgsman, B.Griepink, J.F.Mansveld and W.J.Oort, Mikrochim. Acta, 1970, p.793.

172. W.Potman and E.A.M.F.Dahmen, Mikrochim.Acta, 1977, p.303.

173. A.Floret, Bul.Soc.Chim.France, 11, 2350 (1974).

174. S.S.M.Hassan, Z.Anal.Chem., 266, 272 (1973).

175. R.K.Bonnichsen and H.Theorell, Scand.J.Clin.Lab.Invest., 3, 58 (1951).

176. R.K.Bonnichsen and C.T.Lundgren, Acta Pharmacol.Toxicol., 13, 256 (1957).

177. M.Brower and J.E.Woodbridge, Clin.Chem., 16, 539 (1970).

178. G.G.Guildbault and S.H.Sadar, Anal.Letters, 2, 41 (1969).

179. G.G.Guilbault and G.J.Lubrano, Analytica Chim.Acta, 69, 189 (1974).

180. M.Nanjo and G.G.Guilbault, Analytica Chim.Acta, 75, 169 (1975).

181. M.Mascini and G.Palleschi, Analytica Chim.Acta, 100, 215 (1978).

182. L.Macholán and L.Schanel, CollCzech.Chem.Commun., 42, 3667 (1977).

183. I.A.Gur'ev, E.A.Gushchina and E.N.Mitina, Zh.Anal.Khim., 34, 1184 (1979).

184. S.G.Back, Anal.Letters, 4, 793 (1971).

185. T.P.Hadjiioannou and E.P.Diamandis, Analytica Chim.Acta, 94, 443 (1977).

186. S.J.Updike and G.P.Hicks, Nature, 214, 986 (1967).

187. A.H.Kadish, R.L.Litle and J.C.Sternberg, Clin.Chem., 14, 116 (1968).

188. S.P.Bessman and R.D.Schultz, Trans.Am.Soc.Artif.Intern.Organs., 19, 361 (1973).

189. M.K.Weibel, W.Dritschilo, H.J.Bright and A.E.Humphrey, Anal. Biochem., 52, 402 (1973).

190. H.J.Kunz and M.Stasny, Clin.Chem., 20, 1018 (1974).

191. M.Nanjo and G.G.Guilbault, Analytica Chim.Acta, 73, 367 (1974).

192. D.R.Thévenot, R.Sternberg, P.R.Coulet and D.C.Gautheron, Bioelectrochem, Bionerg., 5, 548 (1978).

193. R.A.Llenado and G.A.Rechnitz, Anal.Chem., 45, 826 (1973).

194. R.A.Llenado and G.A.Rechnitz, Anal.Chem., 45, 2165 (1973).

195. G.Nagy, L.H.von Storp and G.G.Guilbault, Analytica Chim.Acta, 66, 443 (1973).

196. D.R.Thévenot, P.R.Coulet, R.Sternberg and D.C.Gautheron, Bioelectrochem, Boienerg., 5, 541 (1978).

197. D.R.Thevenot, R.Sternberg, P.R.Coulet, J.Laurent and D.C.Gautheron, Anal.Chem., 51, 96 (1979).

198. D.S.Papastathopoulos, D.P.Nikolelis and T.P.Hadjiiaonnou, Analyst, 102, 852 (1977).

199. S.Ikeda, Anal.Letters, 7, 343 (1974).

200. M.A.Koupparis, C.E.Efstathiou and T.P.Hadjiioannou, Analytica Chim.Acta, 107, 91 (1979).

201. C.J.Coetzee and H.Freiser, Anal.Chem., 40, 2071 (1968).

202. C.J.Coetzee and H.Freiser, Anal.Chem., 41, 1128 (1969).

203. M.Hikamu, T.Kubo and T.Yasuda, Analytica Chim.Acta, 109, 33 (1979).

204. M.T.Benignetti, L.Campanella and T.Ferri, Z.Anal. Chem., 296, 412 (1979).

205. L.A.Lazarou and T.P.Hadjiioannou, Analytica Chim.Acta, 108, 375 (1979).

206. M.F.El-Taras, E.Pungor and G.Nagy, Analytica Chim.Acta 82, 285 (1976).

207. W.M.Haynes and J.H.Wagenknecht, Anal.Letters, 4, 491 (1971).

208. G.Nakagawa, H.Wada and T.Hayakawa, Bull.Chem. Soc.Jap 48, 424 (1975).

209. A.Hulanicki, M.Trojanowicz and M.Cichy, Talanta, 23, 47, (1976).

210. V.K.Olson, J.D.Carr, R.D.Hargens and R.K.Force, Anal.Chem., 48, 1228 (1976).

211. W.E.van der Linden and G.J.M.Heijne, The Behaviour of Solid-state Ion-Selective Electrodes in the Presence of Complexing Agents. In: Ion-Selective Electrodes, Ed. EPungor, Akadmiai Kiadó, Budapest (1978), p.445.

212. G.J.M.Heijne and W.E.van der Linden, Analytica Chim.Acta, 96, 13 (1978).

213. J.M.van der Meer, G.den Boef and W.E.van der Linden, Analytica Chim.Acta, 79, 27 (1975).

214. I.K.Sekerka and J.F.Lechner, Anal.Letters, 11, 415 (1978).

215. T.P.Hadjiioannou, M.A.Koupparis and C.E.Efstathiou, Analytica Chim.Acta, 88, 281 (1977).

216. G.Baum, Anal.Letters, 3, 105 (1970).

217. G.Baum, F.B.Ward and S.Yaverbaum, Clinica Chim.Acta, 36, 405 (1972).

218. G.Baum, M.Lynn and F.B.Ward, Analytica Chim.Acta, 65, 385 (1973).

219. E.Hopirtean and M.Miklos, Rev.Chim.(Bucharest), 29, 1178 (1978).

220. C.T.Minh, R.Guyonnet and T.Beaux, C.R.Acad.Sci.Paris, Ser.C., 286, 115 (1978).

221. K.Kina, N.Maekawa and N.Ishibashi, Bull.Chem.Soc.Japan, 46, 2772 (1973).

222. K.Fukamachi and N.Ishibashi, Bunseki Kagaku, 27, 152 (1978).

223. K.Fukamachi and N.Ishibashi, Yakukaku Zasshi, 99, 126 (1979).

224. K.Fukamachi, R.Nakagawa, M.Morimoto and N. Ishibashi, Bunseki Kagaku, 24, 428 (1975).

225. C.Luca, C.Baloescu, G.Semenescu, T.Ţolea and E.Semenescu, Rev. Chim.(Bucharest), 30, 72 (1979).

226. S.Srianujata, W.R.White, T.Higuchi and L.A.Sternson, Anal.Chem., 50, 232 (1978).

227. R.W.Cattrall and H.Freiser, Anal.Chem., 43, 1905 (1971).

228. R.W.Cattrall, S.Tribuzio and H.Freiser, Anal.Chem., 46, 2223 (1974).

229. C.A.Johnson and R.E.King, J.Phar,Pharmacol., 14, 77T (1962).

230. J.M.Campbell, B.Demetriou and R.Jones, Analyst, 105, 605 (1980).

231. E.Hopirtean and E.Veress, Rev.Roum.Chim., 23, 273 (1978).

232. G.D.Carmack and H.Freiser, Anal.Chem., 49, 1577 (1977).

233. S.S.M.Hassan, Mikrochim.Acta, 889 (1974).

234. W.Selig, Mikrochim.Acta, 229 (1970).

235. W.J.Rogers and J.A.Wilson, Anal.Biochem., 32, 31 (1969).

236. B.C.Jones, J.E.Heveran and B.Z.Senkowsky, J.Pharm.Sci., 60, 1036 (1971).

237. J.Mertens, P.van den Winkel, A.Henrion-Boeckstijns and D.L.Massart, J.Pharm.Belg., 29, 181 (1974).

238. V.V.Cosofret and A.A.Bunaciu, Anal.Letters, 12, 617 (1979).

239. W.F.Smyth, G.Svehla and P.Zuman, AnalyticaChim.Acta, 51, 463 (1970).

240. M.A.Koupparis and T.P.Hadjiioannou, Talanta, 25, 477 (1978).

241. U.S.Pharmacopoeia, XIX, p.273.

242. V.V.Cosofret, C.Cristescu and P.Zugrăvescu, Determination of Some as-Triazines with a New Silver Ion-Selective Membrane Electrode In: Ion-Selective Electrodes, Ed. E.Pungor, Akademiai Kiadó, Budapest (1978), p.325.

243. G.E.Baiulescu, V.V.Cosofret and F.G.Cocu, Talanta, 23, 329 (1976).

244. J.Kálmán, K.Tóth and D.Küttel, Acta Pharm.Hungary, 41, 267 (1971).

245. J.Kálmán, K.Tóth and D.Küttel, Acta Pharm.Hungary, 42, 152 (1972).

246. K.Vytra and V.Říha, Cesk.Farm., 26, 9 (1977).

247. E.Hopirtean and E.Stefănigă, Rev.Roum.Chim., 21, 305 (1976).

248. S.S.M.Hassan and M.B.Elsayes, Anal.Chem., 51, 1651 (1979).

249. M.A.Koupparis and T.P.Hadjiioanou, Analytica Chim.Acta, 94, 367 (1977).

250. M.Ionescu, V.V.Cosofret, T.Panaitescu and M.Costescu, Anal. Letters, 13, A(9) (1980).

251. N.Hazemoto, N.Kamo, Y.Kobatake, J.Pharm.Sci., 65, 435 (1976).

252. M.K.Papay, K.Tóth and E.Pungor, Analytica Chim.Acta, 56, 291 (1971).

253. E.Pungor, K.Tóth and M.K.Pápay, Chem.Anal.(Warsaw), 17, 947 (1972).

254. M.K.Pápay, V.P.Izvekov, K.Tóth and E.Pungor, Analytica Chim.Acta, 69, 173 (1974).

255. M.T.Neshkova, V.P.Izvekov, M.K.Pápay, K.Tóth and E.Pungor, Analytica Chim.Acta, 75, 439 (1975).

256. G.E.Baiulescu and V.V.Cosofret, Ion-Sensitive Electrodes in Organic Analysis. Lecture presented at the Univ. of Sassari (italy), 20 May 1976.

257. S.S.M.Hassan, Mikrochim.Acta, 405 (1977).

258. L.W.Przyborowski, Cupric Ion-Selective Electrode for the Pontentiometric Determination of Drugs. In: Ion-Selective Electrodes. Ed. E.Pungor, Akademiai Kiado, Budapest (1978) p.519.

259. J.Růžička and C.G.Lamm, Analytica Chim.Acta, 53, 206 (1971).

260. T.Anfält and D.Jagner, Analytica Chim.Acta, 55, 477 (1971).

261. R.E.van der Leest, Analyst, 102, 502 (1977).

262. G.Kahr, Diss.ETH-Zürich, No.4927 (1972).

263. I.C.Popescu, E.Hopirtean, L.Savici and R.Vlad, Rev.Roum.Chim., 20, 993 (1975).

264. W.E.van der Linden and R.Oostervink, Analytica Chim.Acta, 108, 169 (1979).

265. V.V.Cosofret, Rev.Chim.(Bucharest), 27, 240 (1976).

266. G.E.Baiulescu and V.V.Cosofret, Talanta, 23, 677 (1976).

267. V.V.Cosofret, P.G.Zugravescu and G.E.Baiulescu, Talanta, 24, 461 (1977).

268. R.F.Overman, Anal.Chem., 43, 616 (1971).

269. A.Campiglio, Mikrochim.Acta, 71 (1977).

270. R.Christova, M.Ivanova and M.Novkirishka, Analytica Chim.Acta, 85, 301 (1976).

271. M.Dessouky and E.Pungor, Analyst, 96, 442 (1971).

272. S.Goldstein and A.Duca, J.Pharm.Sci., 65, 1831 (1976).

273. L.Campanella and D.Gozzi, Anal.Chim.(Rome), 67, 345 (1977).

274. L.Campanella, G.De Angelis, T.Ferri and D.Gozzi, Analyst, 102, 723 (1977).

275. L.Campanella, T.Ferri, D.Gozzi and G.Scorcelletti, Dynamic Behaviour of a Liquid Membrane Electrode Reversible to Nicotinate Ion. In: Ion-Selective Electrodes. Ed.E.Pungor, Akademiai Kiadó, Budapest (1978). p.307.

276. L.Campanella, T.Ferri and D.Gozzi, Rev.Roum.Chim. 23, 281 (1978).

277. N.Ishibashi, K.Kina and N.Maekawa, Chem.Letters, 119 (1973).

278. K.Matsumoto, H.Seijo, T.Watanabe, I.Karube, I.Satoh, and S.Suzuki, Analytica Chim.Acta, 105, 429 (1979).

279. I.Satoh, I.Karube and S.Suzuki, J.Solid-Phase Biochem., 2, 1 (1977).

280. S.S.M.Hassan and M.H.Eldesouki, Talanta, 26, 531 (1979).

281. G.J.Papriello, A.K.Mukherji and C.M.Shearer, Anal.Chem., 45, 790 (1973).

282. L.F.Cullen, J.F.Rusling, A.Scheifer and G.J.Papariello, Anal. Chem., 46, 1955 (1974).

283. S.S.M.Hassan, M.T.M.Zaki and M.H.Eldesouki, Talanta, 26, 91 (1979).

284. British Pharmacopaeia, 1973.

# THE AUTHOR

Dr. Vasile V. Cosofret

Vasile V. Cosofret is a Main Researcher in the Institute of Chemical and Pharmaceutical Research, Bucharest and Associate Assistant in the Department of Analytical Chemistry, Polytechnic Institute of Bucharest. He reveived the B.S. and Ph.D. degrees in analytical chemistry from the University of Bucharest and Polytechnic Institute of Bucharest, respectively in 1970 and 1975. He is the author of several papers in electroanalytical chemistry, especially on the role of ion-selective electrodes in organic and pharmaceutical analysis.

Dr. Cosofret is co-author, with Professor G.E.Baiulescu of "Applications of Ion-Selective Membrane Electrodes in Organic Analysis" (Wiley, 1977). His research interests include thin-layer chromatography, uv and visible spectroscopy, polarography and ion-selective electrodes.

*Ion-Selective Electrode Rev.* Vol.2, pp. 219-256. 0191-5371/81/0201-0219$05.00/0


# RECENT TITLES—II

G. J. Moody and J. D. R. Thomas

Chemistry Department, Redwood Building, U.W.I.S.T., Cardiff CF1 3NU, Wales

CONTENTS

1. ELECTRODE DEVELOPMENT AND NEW ELECTRODES, INCLUDING ASPECTS OF SELECTIVITY, RESPONSE TIMES, ETC.

1.1 Solid-state devices, including non-pH glass electrodes

Lanthanum fluoride single crystals with or without europium fluoride, and their response to fluoride ions. T. Aomi, Denki Kagaku oyobi Kogyo Butsuri Kagaku, 47, 733 (1979).

Buffers for fluoride calibration in the low concentration range. M. Trojanowicz, Talanta, 26, 985 (1979).

Ion-sensitive capillary electrode ($LaF_3$ or AgCl). H. Marsoner and C. Ritter. Ger. Offen., 2,854,444 (Cl. G01N27/30) 05 July 1979.

Preparation, analytical characteristics and application of 733-type iodide ISEs. Tsinhai Inst. of Salt Lake Research, Fen Hsi Hua Hsueh, 6, 6 (1978).

Response time of solid-state membrane electrodes (iodide, chloride) under fast changes of concentration in streaming solutions. A. Dencks and R. Neeb, Fresenius' Z. Anal. Chem., 297, 121 (1979).

Carbonate membrane selective electrode. N.V. Garbuzova, A.L. Grekovich, L.I. Ishutkina, V.S. Keravan and E.A. Materova, Ion. Obmen i Ionometriya, (2), 156 (1979).

Anion (phosphate) selective electrode (from $Ca_4P_2O_9$,$Ag_3PO_4$ and uranyl or thorium phosphate in PVC). K-H. Houng, Bull. Inst Chem., Acad. Sin., 26, 31 (1979).

Solid-state ISEs for potentiometric determination of phosphate. D. Midgley, Talanta, 26, 261 (1979).

Studies with parchment supported cobalt(II) phosphate membrane for poetntial use for a phosphate ISE. M.N. Beg, F.A. Siddiqi, M. Arshad, R. Shyam and M.M. Beg, Fresenius' Z. Anal. Chem., 298, 157 (1979).

Experience with mixed-salt ion-selective sulphate electrodes. K. Nagy and T.A. Fjeldly, Talanta, 26, 811 (1979).

Chemically treated tin oxide electrode response to pH and sulphide. H.A. Laitinen and T.M. Hseu, Anal. Chem., 51, 1550 (1979).

A new ammonium ion-sensing heterogeneous membrane electrode (of ammonium phosphomolybdate + methacrylate powder). P. Longhi, T. Mussini, F.M. Nardi and S. Rondinini, Nouv. J. Chim., 3, 649 (1979).

Inexpensive, miniature potassium ISE. H. Rosenberg, Anal. Biochem., 96, 90 (1979).

Selection of glass for the manufacture of ISEs for measuring sodium content in the working environment of thermal power plants. A. Zielinski, Energetyka, 33, 151 (1979).

Electrochemical properties of sodium and potassium glass micro-electrodes. C.O. Lee, Biophys. J., 27, 209 (1979).

Thallium(I)-selective solid-membrane electrode. A.K.Jain, S.Agrawal and R.P.Singh, Anal.Letters, Part A 12, 995 (1979).

Sudy and development of two types of specific electrodes with a polycrystalline sensitive element for the determination of thallium(I) ions. J.J.Fombon, Y.Oddon, J.Tacussel and A.Tranquard. Analusis, 7 494 (1979).

Investigation and development of selective membrane electrodes. XIV. Determination of mercury(II) and thallium(III). A.V.Kopytin, A.F.Zhukov, Yu.I.Urusov, L.A.Kopytina and A.V.Gordievskii. Zh.Anal. Khim., 34, 465 (1979).

Super-high sensitivity of a copper(II) selenide-silver sulphide solid-membrane copper(II) ion-selective electrode in several metal buffer solutions. Y.Umezawa, Y.Imanishi, K.Sawatari and S.Fujiwara, Bull. Chem.Soc.Japan, 52, 945 (1979).

Membrane (CuS + conductive acetylene black in PTFE) for copper(II) ion-sensitive ISEs. H.Strzelecki, W.Chrzanawski. Polish Patent, 105,827, (Cl.C 08J5/22) 15 February 1980.

ISEs with sensors of electrolytically plated chalcogenide coatings. Part I. Copper ISE based on copper selenide. M.Neshkova and H.Sheytanov, J.Electroanal. Chem.Interfac.Chem., 102, 189 (1979)

The behaviour of Copper(II) selective electrode in chloride containing solutions. P.Lanza. Anal.Chim.Acta, 105, 53 (1979).

Chloride interference in copper(II) ISE measurements. J.C.Westall, M.M.Morel, D.N.Hume. Anal.Chem., 51, 1792 (1979).

Formation and properties of mixed cadmium sulphide-silver sulphide and mixed mercury sulphide-silver sulphide membranes for electrodes selective to cadmium(II) and mercury(II). W.E.Van der Linden and R.Oostervink, Analytica Chim.Acta, 108, 169 (1979).

Lead-selective electrodes based on lead and silver sulphides. Yu.G.Vlasov, Yu.E.Ermolenko and O.A.Iskhakova. Zh.Anal.Khim., 34, 1522 (1979).

Composition of a membrane of an ISE for determining lead ions (whereby adding 5-10% graphite to 65-75% PbS and 15-25% polystyrene increased accuracy). T.P.Ushangishvili, V.A.Dolidze, A.N.Khutsishvili, A.G.Rodichev and V.G.Krungak. U.S.S.R. Patent 676,918 (Cl.GolN27/30) 30 July 1979.

Preparation of a rare earth ISE (with $CeO_2$ membrane). Y.Takasaka and Y.Suzuki. Bull.Chem.Soc.Japan, 52, 3455 (1979).

Use of crystalline solid electrolytes as membranes in ISEs. M.Voinov. In (Eds: P.Hagenmuller and W.van Gool). "Solid Electrolytes". Academic Press, New York. (1976) p.527.

Use of tin oxide ($SnO_x$) semi-conductors as selective electrode indicators. L.Arsov and P.Kirkov. Prilozi-Makedon.Akad.Nauk.Umet., od. Prir.-Mat.Nauki, 8, 13 (1976). Chem.Abst.219283, 91, (1979).

Ceramic materials for ion-selective membrane electrodes. H.Hirata, Natl.Tech.Rep. (Matsushita Elect.Ind.Co.), 25, 985 (1978).

Solid-state ISEs with intergrated electronics. T.A.Fjeldly, K.Nagy and J.S.Johannessen. J.Electrochem.Soc., 126, 792 (1979).

Crystalline and pressed-powder, solid-membrane electrodes. R.P.Buck, in (Ed. A.K.Covington) "Ion-Sel.Electrode Methodol", I, 175 (1979).

## 1.2 Liquid ion-exchanger devices, including those trapped in PVC and other polymer matrix membranes

### 1.2.1 General.

Heterogeneous-membrane, carbon-support and coated-wire ISEs. R.W. Cattrall, in (Ed: A.K.Covington) "Ion-Sel.Electrode Methodology", I, 131 (1979).

Response characteristics of (coated-wire and conventional ISEs based on dinonylnaphthalene - sulphonic acid. C.R.Martin and H.Freiser, Anal.Chem., 52, 562 (1980).

Ion-selective coated (with PVC and sensor for $Na^+$, $K^+$, $NH_4^+$ or $Ca^{2+}$ and plasticiser) glass electrodes. J.G.Schindler, M.V.Guelich, W.Schael, G.Stork, W.Schmid,K.D.Karaschinski, H.E.Braun and H.Maier. Fresenius' Z. Anal.Chem., 299, 115 (1979).

Effect of the solvent nature on electrode properties of liquid and film ion-selective membranes (experimental data). A.L.Grekovich, O.A.Concharuk and K.N.Mikhel'son. Ion.Obmen i Ionometriya, (2),125 (1979).

Effect of nature of solvent on electrode properties of ion-selective liquid and film membranes. A.L.Grekovich, E.A.Materova and K.B.Mikhel'son, Ion.Obmen i Ionmetriya (Leningrad), (2), 111 (1979).

Research and development of selective membrane electrodes. Communication 15. Effect of selectivity of ion-exchange and solvent extraction of acids on electrode functions of liquid ion-exchangers. A.V. Go dievskii, Yu.I.Urosov, V.V.Sergievskii, A.F.Zhukov and A.V.Kopytin, Zh.Anal.Khim., 34, 1252 (1979).

Applications of liquid ion-exchangers in ISE technology. J.D.R.Thomas, Proc.Ion-Exch.Symp. (Ed: G.T.Gadre) CSMCRI, Bhavnagar, India. 313 (1978).

ISE assembly (for potassium). F.F. Spaziani and J.E.Fowler, British Patent, 2,025,629 (Cl. G01N27/40). 23 January 1980.

Ion-selective multidisk electrodes (for calcium, potassium and anions). J.G.Schindler, A.Moennich, W.Riemann and H.E.Braun. Biomed.Tech., 22, 209 (1977).

Potassium-selective micro electrode (in hypodermic needle). E.Niki and K.Uchida. Japan Kokai Tokkyo Koho 79,141,697 (Cl G01N27/30) 5 November 1979.

Poly(vinyl chloride) matrix membrane ion-selective electrodes. G.J.Moody and J.D.R.Thomas, in (Ed: A.K.Covington), Ion-Sel.Electrode Methodol. 1, 111 (1979).

### 1.2.2 Cation electrodes.

Manufacture of PVC calcium electrode (with calcium didecylphosphate plug and methylphosphonic acid methyl octyl ester solvent). T.Lu, T-L.Lin, T-F.Liu, F-Y.Li and Yu-Y.Chang, Fen.Hsi Hua Hsueh, 6, 485 (1978).

Calcium ISE without internal reference solution. Sh.K.Norov and O.G.Vartanova, Zh.Anal.Khim., 34, 1500 (1979).

A poly(vinyl chloride)-matrix membrane calcium-selective electrode based on calcium bis 2,6-dinitro-4-(1,1,3,3-tetramethylbutyl)phenoxide S.Kamata, G.J.Moody and J.D.R.Thomas, Analytica Chim.Acta, 108, 385 (1979).

Divalent (water hardness) ISEs(based on calcium bis(didecylphosphate) in) poly(vinyl chloride) and poly(methyl acrylate) matrix membranes. S.K.A.G. Hassan, G.J. Moody and J.D.R. Thomas, Analyst, 105, 147 (1980).

Calcium ISE measurements in the presence of complexing ligands. A.Craggs, G.J.Moody and J.D.R.Thomas, Analyst, 104, 961 (1979).

Preparation of a calcium ISE in a PVC matrix, and determination of calcium in water. S.W.Sun, X.Wang and Q.E.Ge, Hua Hsueh Tung Pao, 32 (1979).

Progress in designing calcium ISEs. G.J.Moody and J.D.R.Thomas, Ion-Selective Electrode Rev., 1, 3 (1979).

Interference studies on a poly(vinyl chloride)-matrix membrane calcium ISE based on a calcium bis-[4-(1,1,3,3-tetramethylbutyl)phenyl-phosphate] as sensor and tripentyl phosphate as solvent mediator. G.J.Moody, N.S.Nassory and J.D.R.Thomas, Talanta, 26, 873 (1979).

PVC ISEs based on calcium bis-[dialkyl- and di(4-alkylphenyl)phosphates] and mixed solvent mediators. A.Craggs, B.Doyle, S.K.A.G.Hassan, G.J.Moody and J.D.R.Thomas, Talanta, 27, 277 (1980).

Calcium-selective film electrode based on 2-ethylhexyl phosphate. O.G.Vartanova, S.Khuzhaev, E.S.Gureev and Sh.K.Norov, Zh.Anal.Khim., 34, 982 (1979).

Ion-selective polymeric-membrane electrodes with immobilised ion-exchange sites. Part I. Development of a calcium electrode. L.Ebdon, A.T.Ellis and G.C.Corfield, Analyst, 104, 730 (1979).

(Liquid cation-exchangers in microcapillaries) for single cell measurements with ISEs. J.L.Walker in "Medical and Biological Applications of Electrochemical Devices". Ed. J.Koryta. J.Wiley, Chichester. (1980).

Electrodes based on salts of dialkyldithiophosphoric acid. E.A.Materova, V.V.Muklovikov and M.G.Grigor'eva, Ion.Obmen i Ionometriya, (2), 142 (1979).

Film membrane electrodes based on alkyl sulphoxides which are selective to uranium(VI). E.A.Materova, N.S.Lebedeva and T.Ya Bart. Ion. Obmen i Ionometriya, (2), 134 (1979).

Calcium ISE potassium zinc ferrocyanide in a PVC matrix. W.Dollislager and L.Heerman, J.Electrochem.Soc., 126, 347 (1979).

Study of electrode properties of ammonium-selective membranes in solutions containing foreign cations. O.K.Stefanova, N.V.Rozhdestvenskaya and I.V.Rusina. Ion.Obmen i Ionometriya, (2), 166 (1979).

(A $Cu(py)_2$ $(SCN)_2$ liquid membrane) ISE for determining nickel(II) ion. A.Suteu, I.Simcsak and I.Pintyak. Lucr.Stiint.- Inst.Invatamint Super.Orandea, Ser.A, 101 (1977).

Study of the properties of a copper-selective BMSA electrode with a chelating active substance (LlX-64N chelating liquid ion-exchanger based on 2-hydroxybenzophenone derivatives). K.Sykut, J.Dumkiewicz, R.Dumkiewicz and Urszula Loboda. Buil.Lubel.Tow.Nauk., Mat.Fiz-Chem., 21,23 (1979).

Development of polymeric membranes for zinc ISEs. U.Fiedler-Linnersund and K.M.Bhatti, Analytica Chim.Acta, 111, 57 (1979).

Liquid ISE based on zinc tetrathiocyanate anion . E.M.Rakhman'ko, G.L.Starobinets, V.L.Lomalo and A.Beisis, Vesti Akad.Navuk BSSR, Ser.Khim.Navuk, 6, 68 (1978).

1.2.3 Anion electrodes.

Limits of detection and selectivity coefficients of a PVC-based anion-selective electrode (using capriquat). N.Kamo, Y.Kobatake and K.Tsuda, Talanta, 27, 205 (1980)

Construction of a liquid-membrane-type histamine ISE. M.Kudoh, M.Kataoka and T.Kambara, Bunseki Kagaku, 28, 705 (1979).

(PVC) nitrate-ion-selective electrode based on the copper(I) - neocuproine [2,9-dimethyl-1,10-phenanthroline] complex. M.Trojanowicz. Fresenius' Z.Anal.Chem., 297, 414 (1979).

Nitrate ISE (of basic dye in nitrobenzene). J.Senkyr and J.Petr. Chem.Listy, 73, 1097 (1979).

Evaluation of a nitrate ISE. N.A.Fayad and J.F.Tyson, Int.Lab., 9, 49 (1979).

Ion-selective picrate electrode with liquid membrane. D.Homolka, Collect.Czech.Commun., 44, 3644 (1979).

Liquid membrane electrode based on picrate ion. A.L.Gulevich, E.M.Rakhman'ko and G.L.Starobinets, Vesti Akad.Navuk BSSR, Ser.Khim. Navuk, (4), 77 (1979).

Ion-selective membrane electrodes for organic analysis (III). Picrate-selective liquid membrane electrode. E.Hopirtean and E.Radulescu. Stud.Univ. Babes-Bolyai, (Ser.) Chem., 24, 23 (1979).

Acetate ion-selective membrane electrodes based on liquid anion-exchange systems with a series of solvents containing a trifluoroacetyl group. E.A.Materova, S.A.Ovchinnikova, V.S.Karavan and L.O.Ishutkina. Elecktro Khimiya, 15, 1185 (1979).

Benzoate liquid-membrane ISE. M.T.Benignetti, L.Campanella and T.Ferri, Fresenius' Z.Anal.Chem., 296, 412 (1979).

Effect of foreign anions on the potential of a liquid bromide electrode. R.Del Toro Denis, E.M.Rakhman'ko and G.L.Starobinets. Vesti Akad.Navuk BSSR, Ver.Khim.Navuk, (3), 29 (1979).

Liquid-membrane-type tetraphenylborate ISE. M.Kataoka, M.Kudoh and T.Kambara, Denki Kagaku Oyobi Butsuri Kagaku, 46, 548 (1978).

A hydrogen chromate ISE of poly(vinyl chloride) matrix membrane type. T.Iimori, M.Sugawara and T.Kambara, Denki Kagaku, 47, 549 (1979).

Investigation of PVC matrix based selective chromate and perchlorate sensitive liquid membrane electrode. F.Jasmin, Iraqi J.Sci., 20, 430 (1979).

Perchlorate-selective electrodes with Urushi as the membrane matrix. K.Hiiro, A.Kawahara and T.Tanaka, Analytica Chim.Acta, 110, 321 (1979).

1.3 Carrier-complex and related systems

Liquid membrane pH electrode based on a synthetic proton carrier. D.Erne, D.Ammann and W.Simon, Chimia, 33, 88 (1979).

Design, properties and applications of neutral ionophores. W.Simon and E.Carafoli. Methods Enymol., 56, (Biomembranes, Part G), 439 (1979).

Ionophores of extremely high electrophilicity as selective components for liquid membrane electrodes. U.Oesch, D.Ammann, E.Pretsch and W.Simon, Helv.Chim.Acta, 62, 2073 (1979).

Lifetime of neutral carrier based ion-selective liquid in membrane electrodes. U.Oesch and W.Simon, Anal.Chem., 52, 692 (1980).

Potassium film electrode based on the macrocyclic polyether dibenzo-24-crown-8. S.L.K.Norov, E.S.Gureev, A.K.Tashmukhomedova, O.G.Vartanova and N.Zh.Saifullina, Elekrokhimiya, 15, 943 (1979).

PVC membrane potassium ISE with 4,4′-di-tert-butyldibenzo-30-crown-10 as active agent. Lanchow Chemical Physics Laboratory, Fen.Hsi Hua Hsueh, 7, 20 (1979).

Electrochemical properties of ion-selective membranes based on valinomycin in solution containing picrate ions. V.E.Yurinskaya, O.K.Stefanova, E.A.Materova and O.A.Yukhno. Elektrokhimiya, 16, 320 (1980).

Modified potassium electrode based on valinomycin. K.N.Mikhel'son and A.L.Grekovich, Ion.Obmen i Ionometrya, (2), 171 (1979).

Electrode properties of potassium-selective membranes with different valinomycin content. O.K.Stefanov and E.D.Suglobova, Elektrokhimiya, 15, 1710 (1979).

Low impedance coaxial potassium ion-selective microelectrodes. E.Ujei, O.Keller, J.Machek and V.Pavlik, Pfluegers Arch., 382, 189 (1979).

Neutral-carrier sodium ion-selective micro-electrode for intracellular studies. R.A.Steiner, M.Oehme, D.Ammann and W.Simon, Anal.Chem., 51, 351 (1979).

Caesium-selective PVC membrane electrodes based on bis(crown ethers). K.Kimura, H.Tamura and T.Sono, J.Electroanal.Chem. Interfacial Electrochem., 105, 335 (1979).

12,18-Dimethyl-5,15,25-trioxa-12,18-diazanonacosane-6,13,17,24-tetrone as a liquid selective for calcium ions. V.M.Dziomko, I.S.Markovich, A.M.Kapustin, G.M.Sorokina, L.T.Blokina, N.A.Bolotina, N.V.Kruglova and Yu.S.Ryabokobylko. U.S.S.R.Patent. 694.499 (Cl. C07C103/34), 30 October 1979.

Interferences of a barium ISE used for the potentiometric titration of sulphate. D.L.Jones, G.J.Moody, J.D.R.Thomas and M.Hangos, Analyst, 104, 973 (1979).

Calcium-selective electrode (with cyclic $(Me_2 SiO)_{12}$ sensor). C.J.Olliff and G.R.Pickering, British Patent, 1,558, 553 (Cl G01N27/30) 3 January 1980.

Calcium-selective electrodes. W.Simon, D.Ammann, M.Oehme and W.E.Morf, Ann.N.Y.Acad.Sci., 307, 52 (1978).

## 1.4 Enzyme and bacteria-based sensors

Enzyme electrodes. T.Ya.Bart and E.A.Materova, Ion.Obmen i Ionometriya, (2), 73 (1979).

On a new enzyme transducer combination; the enzyme transistor. B.Danielsson, L.Lundstrom, K.Mosbach and L.Stiblert, Anal.Lett., Part B, 12, 1189 (1979).

Design and response characteristics of an enzyme electrode for measurement of penicillin in fermentation broth. S.O.Enfors and H.Nilsson, Enzyme Microb.Technol., 1, 260 (1979).

Asymmetrical coupling of enzymic systems on collagen membranes. Application to multi-enzyme electrodes. P.R.Coulet and C.Bertrand, Anal.Letters, Part B. 12, 581 (1979).

Enzyme collagen membrane for electrochemical determination of glucose. D.R.Thevenot, R. Sternberg, P.R.Coulet, J.Laurent and D.C.Gautheron, Anal.Chem., 57, 97 (1979).

Development of an enzymic electrode (as an) oil in water sensor. A.M.Cundell and E.Findl. Report, (Bio Res.Inc., Farmingdale, N.Y. U.S.A.), ER-013179, 29 pages (1979).

Measuring cell containing enzyme membrane electrodes for micro determinations. J.L.Romette, G.Gellf and D.Thomas. Fr.Demande, 2,391,467, (Cl.G01N27/28), 15 December 1978.

Chemically modified electrodes bearing grafted enzymes. C.Bourdillon, J.P.Bourgeois and D.Thomas. Biotechnol. Bioeng., 21, 1877 (1979).

(Polarographic) enzyme electrodes and (cellulose acetate) membrane containing an immobilised enzyme. K.Yoda, R.Urakabe and T.Tsuchida. Ger.Offen., 2,903,216 (Cl. G01N27/48) 2 August 1979.

Enzyme-specific electrode of glucose based on the amperometric detection of hexacyanoferrate(II). J.Mahenc, H.Aussaresses. C.R. Hebd.Seances Acad.Sci.,Ser C. 289, 357 (1979).

Novel-type (glucose-sensitive amperometric) molecule-selective sensors (of immobilized glucose oxidase). Preliminary communication. J.Havas. Magy.Kem.Foly., 85, 329 (1979).

Glucose-oxygen-selective electrodes (with immobilised glucose oxidase). H Takahara. Japan Kokai Tokkyo Koho, 79, 41,191 (Cl. G01N27/30) 02 April 1979.

Enzyme electrodes(with immobilized redox enzyme and a membrane with a pinhole). H.Takahara. Japan Kokai Tokkyo Koho, 79 43,796 (Cl.G01N27/30) 06 April 1979.

Enzyme electrodes (with additional enzyme layer to destroy interfering reaction products). H.Takahara. Japan Kokai Tokkyo Koho, 79 81,177 (Cl. C25B11/00), 28 June 1979.

Films and biochemically-active coatings (of enzymes, antibodies agglutimines, enzyme inhibitors, antigens, haptens, metabolic effectors or combinations thereof for substrate specific electrodes). J.C.Quennesson and D.Thomas, Fr.Demande. 2,391,254, (Cl. C09D3/04) 15 December 1978.

Enzyme "sequence" electrode for D-gluconate. M.A.Jensen and G.A.Rechnitz. J.Membrane Sci., 5, 117 (1979).

Cellulase production and ammonia metabolism in trichoderma reesei on high levels of cellulose. D.Sternberg and S.Dorval, Biotechnol.and Bioeng., 21, 181 (1979).

Studies on quantitative physiology of trichoderma reesei with two-stage continuous culture for cellulase production. D.Ryu, R.Andreotti, M.Mandels, B.Gallo and E.T.Reese, Biotechnol.and Bioeng., 21, 1887 (1979).

Effect of initial substrate concentration of the rate of nitrification in a batch experiment. W.K.Shieh and E.J.La Motta, Biotechnol. and Bioeng., 21, 201 (1979).

Biochemical cell for the determination of lactate. J.J.Kulys and G.-J.S.Svirmickas, Anal Chim.Acta, 109, 55 (1979).

Potentiometric enzyme electrode for lactate. T.Shinbo, M.Surgiwra and N.Kamo, Anal.Chem., 51, 100 1979).

Neutral lipid sensors based on immobilized lipase and a flow-through pH electrode. I.Satch, I.Karube, S.Suzuki and K.Aikawa, Anal.Chim. Acta, 106, 369 (1979).

Fundamental studies on (an enzyme) bio-electrochemical sensor. II. Amine sensor. S.Suzuki, I.Karube and I.Satoh, Asahi.Garasu Kogyo Gijutsu Shoreikai Kenkyu Hokaku, 32, 355 (1978).

Flow-through enzyme electrode based on dehydrogenases with regenerated coenzyme (nicotin-amide-adenine dinucleotide). J Kulis and A.Malinauskas, Zh.Anal.Khim., 34, 876 (1979).

Electrochemical bio-sensors in assay of antibiotics. M.Thompson. P.J.Holik and E.A.Stubley. Analytica Chim.Acta, 104, 195 (1979).

Bacterial membrane electrode for determination of nitrate. R.K.Kobos, D.J.Rice and D.S.Flournoy, Anal Chem., 51, 1122 (1979).

Release of weak and strong copper-complexing agents by algea. D.M.McKnight and F.M.M.Morch. Limnol. and Oceanog., 24, 823 (1979).

## 1.5 Field effect transistor (FET) and related systems

ISFET with heterogeneous membranes. B.Shiramizu, J.Janata and S.P.Moss. Analytica Chim.Acta, 108, 161 (1979).

The ion-sensitive field effect transistor in rapid acid-base titrations. M.Bos, P.Bergveld and A.M.W.van Veen-Blaauw. Analytica Chim.Acta, 109, 145 (1979).

Ion-selective field-effect transistors (ISFET) - a new type of electrode for chemical analysis and biomedical research. Y.G.Vlasov, J.App.Chem. U.S.S.R. (Eng.Trans.) 52, 1 (1979).

Chalcogenide glasses as ion-selective (for copperII) and lead(II) ions) materials for solid-state electrochemical sensors. A.E.Owen. J.Non-cryst.Solids, 35-36, 999 (1980).

(A review of) ion-sensitive field effect transistors. J.Janata and R.J.Huber. Ion-Sel.Electrode Rev., 1, 31 (1979).

Ion sensor (with sensor in Urushi lacquer adhered to gate). K.Koshiishi. Japan Kokai Tokkyo Koho, 79,141,698 (Cl. G01N27/00) 5 November 1979.

ISFETS using inorganic gate thin films. H.Abe, M.Esashi and T.Matsuo, IEEE Trans.Electron.Devices, ED-26, 1939 (1979).

Field-effect transistor-type chemical sensor element. K.Koshiishi and H.Aoki. Japan Kokai Tokkyo Koho, 79,141,699 (Cl.G01N27/00) 5 November 1979.

Basic properties of the electrolyte-silicon dioxide-silicon systems: physical and theoretical aspects (to improve understanding of ISFETs). W.M.Sin abd R.S.C.Cobbold. IEEE Trans.Electron Devices, ED-26, 1805 (1979).

ISEs based on field transistors with a silver chloride - silver bromide membrane sensitive to silver, bromide and chloride ions. R.P.Bak, Yu.G.Vlasov and D.E.Khekleman. Zh.Prikl.Khim., 52, 2601 (1979).

Study of coated-wire potassium-valinomycin and sodium-monensin ion-sensing systems by use of a conventional field-effect transistor. M.Thompson, U.J.Krull and P.J.Worsfold. Talanta, 26 , 1015 (1979).

Fabrication of a silver-, chloride- and bromide-responsive ion-selective field-effect potentiometric sensor. Yu. G. Vlasov, D.E. Hackleman and R.P. Buck. Anal. Chem., 51, 1570 (1979).

## 1.6 Miscellaneous

Ligand structure and complexation. L. Ion-selective crown ether dyes. J.P.Dix and F.Voegtle. Chem.Ber., 113, 457 (1980).

Ion-sensitive capillary electrode modules (for liquid membrane, capillary-shaped glass and capillary-shaped solid body electrodes). H.Marsoner and C.Ritter. U.K.Patent 2,014,734 (Cl.G01N27/26) 30 August 1979.

Carbon paste for calcium-selective electrode. G.A.Qureshi, Libyan J.Sci., 8A, 37 (1978).

Recommendations for publishing manuscripts on ion-selective electrodes. Ion-Selective Electrode Rev., 1, 139 (1979).

Voltammetric ISE for determination of nitrate. J.A.Cox and G.R.Litwinski. Anal.Chem., 51, 554 (1979).

## 2. MECHANISTIC AND THEORETICAL ASPECTS

Theoretical principles of ISEs. K.Cammann. Gewaesserschutz, Wasser, Abwasser, 39, 1 (1979).

pX standards. A.K.Covington in (Ed: A.K.Covington). "Ion-Sel. Electrode Methodol", 1, 67 (1979).

A unified approach to glass electrode theory. W.E.Morf, Talanta, 26, 719 (1979).

Contribution to the function of the internal metallic lead of ion-selective disc electrodes. J.G.Schindler, G.Stork, H.J.Struh, W.Schmid and K.D.Karaschinski. Fres.Z.Anal.Chem., 295, 248 (1979).

Chloride interference in copper(II) ISE measurements. J.C.Westall, F.M.M.Morel and D.N.Hume. Anal.Chem., 51, 1792 (1979).

Surface morphology of ion-selective membrane electrodes. II. Studies on the copper(II)-selective electrode. E.Pungor, K.Tóth, M.K.Papay, L.Polos, H.Malissa, M.Graaserbauer, E.Hoke, M.F.Ebel and K.Persy. Analytica Chim.Acta, 109, 279 (1979).

Behaviour of iodide-selective electrodes at low concentrations of iodide. A.Hulanicki, A.Lewenstam and M.Maj-Zurawska. Analytica Chim.Acta, 107, 121 (1979).

Competitive ion-exchange evaluation of the bromide interference on anodized silver /silver chloride electrodes. R.K.Rhodes and R.P.Buck. Analytica Chim.Acta, 113, 67 (1980).

Response time of solid state membrane electrodes (iodide and chloride) under fast charges of concentration in streaming solutions. A.Dencks and R.Neeb. Fres.Z.Anal.Chem., 297, 121 (1979).

Physicochemical study of solid electrolytes to be used as membranes of ion-selective electrodes. I. Ionic and electronic conductivity of silver chloride-silver sulphide and silver bromide - silver sulphide solid electrolytes. Yu.G.Vlasov and S.B.Kochergin. Ion.Obmen i Ionometriya, (2), 243 (1979).

Interfacial properties of precipitate-based ISEs. Rotating disc impedance measurements of the silver sulphide/silver (aqueous) interface. R.K.Rhodes and R.P.Buck. Analytica Chim.Acta, 110, 185 (1979).

The mechanism of the potential response of bromide-selective electrodes based on mercury salts. A.Hulanicki, R.Lewandowski and A.Lewenstam. Analytica Chim.Acta, 110, 197 (1979).

Response time curves of ion-selective solid-state membrane electrodes under fast concentration changes in streaming solutions. Application of model theories to an experimental example. A.Dencks and R.Neeb. Fresenius' Z.Anal.Chem., 298, 131 (1979).

(Collection of) selectivity coefficients of ISEs. IUPAC Analytical Chemistry Division. Pure Appl.Chem., 51, 1913 (1979).

Comparison of time constants for liquid ion-exchange membrane electrode responses determined by an impedance method and an activity step method. F.S.Stover, T.R.Brumieve and R.P.Buck. Analytica Chim. Acta, 109, 259 (1979).

Ionic permeability of adsorbed membrane protein monolayers. M.Blank, L.Soo and R.E.Abbott. J.Electrochem.Soc., 126, 1471 (1979).

Dynamic response relationship for an ion-exchange liquid membrane electrode. D.Gozzi, L.Campanella and T.Ferri. Electrochim.Acta, 24, 817 (1979).

A study of the mechanism of response of liquid ion-exchanger calcium selective electrodes. Part II. Conductimetric investigation of the ionic dissociation in liquid ion-exchangers. N.D.Van Man and C.Gavach. J.Electroanal.Chem., Interfac.Electrochem., 97, 163 (1979).

Site concentration determination in liquid ion exchanger membranes. Part III. Experimental. R.P.Buck, F.S.Stover and D.E.Mathis. J.Electroanal Chem., Interfac.Electrochem., 100, 63 (1979).

Investigation and development of selective membrane electrodes. XV. Effect of selectivity of ion exchange and solvent extraction of acids on the electrode functions of liquid ion exchangers. A.V.Gordievskii, Yu I.Urisov, V.V.Sergievskii, A.F.Zhukov and A.V.Kopytin. Z.Analytica Khim., 34, 1252 (1979).

Studies of concentration potentials across heavy-metal-soap liquid ion-exchange membranes. M.A.Beg and A.Nabi. J.Electroanal.Chem., Interfac.Electrochem., 103, 137 (1979).

Bipolar pulse conductance measurements with a calcium ISE. C.R.Powley, R.F.Geiger and T.A.Nieman. Anal.Chem., 52, 701 (1980).

Selectivity rating of calcium ion-selective electrodes. G.J.Moody, N.S.Nassory and J.D.R.Thomas. Proc.Anal.Div.Chem.Soc., 16, 32 (1979).

Kinetic study of the distribution of electrically central ionspheres between a solvent polymeric membrane and an aqueous phase. U.Oesch and W.Simon. Helv.Chim.Acta, 62, 754 (1979).

Ion transport in free and supported nitrobenzene Aliquat nitrate liquid membrane ISEs. III. Potentiometric limits of detection and selectivity. D.E.Mathis, R.M.Freeman, S.T.Clark and R.P.Buck. J.Membrane Sci., 5, 103 (1979).

Matrix of coefficients of specificity of ISEs with liquid ion-exchangers. V.A.Mikhailov, V.V.Osipov and E.N.Graf. Elektrokhimiya, 15, 1042 (1979).

## 3. SOLUTION CHEMISTRY AND ISES

Comparison between measured and calculated concentrations of calcium ions in buffers. O.Scharff. Analytica Chim.Acta, 109, 291 (1979).

Buffers for fluoride electrode calibration in the low concentration range. M.Trojanowicz. Talanta, 26, 985 (1979).

Multi component two-phase buffer systems. T.J.Janjic, E.B.Milosavljevic and M.K.Srdanovic. Analytica Chim.Acta, 107, 359 (1979).

Single ion activities in aqueous solutions analagous to sea water. R.A.Robinson and R.G.Bates. Marine Chem., 7, 281 (1979).

A simple system for the measurement of ion activities with solvent polymeric membrane electrodes. H.Affolter and E.Seigel. Anal.Biochem., 97, 315 (1979).

The mean activity and the activities of the separate ions of sodium decanoate above and below the CMC determined by a surfactant selective Ag/Ag decanoate electrode. E.Vikinstad. J.Colloid.Interfac.Science, 72, 68 (1979).

Ion association: ISE experiment. M.M.Emara, N.A.Farid and C.T.Lin. J.Chem.Educ., 56, 620 (1979).

Variation of the critical micelle concentration of sodium dodecyl sulphate with ionic strength measured by selective-ion membrane electrodes. J.E.Newbery. Colloid and Polymer Science, 257, 773 (1979).

A kinetic approach to ion-selectivity in aqueous electrolyte solutions of alkali ions. (Review). S.V.Talekar. Biophys.Membr. Transp., (1), 197 (1979).

## 4. STANDARD ADDITION METHODS

Comparative study on the precision of potentiometric techniques applied with ISEs. G.Horvai and E.Pungor. Analytica Chim.Acta, Part I. Direct techniques, 113, 287 (1980). Part 2. Multiple standard addition, 113, 295 (1980).

Standard addition method used in ISE analysis applications. K-L.Wu, Fen Hsi Hua Hsueh, 6, 266 (1978).

Determination of the equivalence point in potentiometric titrations with Gran's first method used to test the electrode response. E.Still. Analytica Chim.Acta, 107, 337 (1979).

Generalised standard addition method. B.E.H.Saxberg and B.R.Kowalski. Anal.Chem., 51, 1031 (1979).

Injection techniques in dynamic flow-through analysis with electro-analytical sensors. E.Pungor, Z.Feher, G.Nagy, K.Tóth, G.Horvai and M.Gratzi. Analytica Chim.Acta, 109, 1 (1979).

Microprocessor-based millivoltmeter (with known addition facility) in application of ISEs. G.J.Moody and J.D.R.Thomas. Lab.Practice, 28, 125 (1979).

## 5. POTENTIOMETRIC TITRATIONS

Determination of the equivalence point in potentiometric titrations with Gran's first method used to test the (copper) electrode response. E.Still. Analytica Chim.Acta, 107, 377 (1979).

Linearised titration curves, titration to a given pH value, novel considerations in potentiometry. A.Ivaska. Kem.Kozlemenyek, 51, 129 (1979).

Two-phase potentiometric titration of organic bases with use of a picrate-selective electrode. I.A.Gur'ev and T.S.Vyatchanina. Zh. Anal.Khim., 34, 978 (1979).

Two-phase potentiometric titration of phenols with an ISE. I.A.Gur'ev, E.A.Gushchina and E.N.Mitna. Zh.Anal.Khim., 34, 1184 (1979).

Use of copper and silver ion-selective electrodes in potentiometric titrations. C.J.Coetzee and S.R.Grobler. Tydskr.Natuurwet., 19, 12 (1979).

Comparative study on the precision of potentiometric techniques applied with ISEs. Part 3. Potentiometric titrations. G.Horvai and E.Pungor. Analytica Chim.Acta, 116, 87 (1980).

A semi-automated single-cell gradient-titration system, using ISEs as end-point sensors. G.P.Bound and B.Fleet. Talanta, 27, 257 (1980).

Linear titration plots with ISEs. A.Ivaska. Talanta, 27, 161 (1980).

Titration of microgram amounts of sulphide with a sulphide-selective electrode. T.M.Florence and Yvonne J.Farrar. Analytica Chim.Acta, 116, 175 (1980).

Microtitration of various anions with quaternary ammonium halides using solid-state electrodes. W.Selig. Microchem.,J. 25, 200 (1980).

New titrant for perchlorate, hexadecyltrimethylammonium bromide. W.Selig. Talanta, 26, 1061 (1979).

Determination of cobalt with a new ISE based on the benzalkonium tetrathiocyanantocobaltate (II) ion-pair complex. K.Burger and G.Petho. Magy.Kem.Foly., 85, 424 (1979).

Potentiometric titration of fluoride, sulphate, chromate, molybdate, tungstate, oxalate, phosphate, pyrophosphate and hexacyanoferrate(II) ions with lead(II) solutions and a fluoride-selective electrode. C.E.Efstathiou and T.P.Hadjiioannou. Analytica Chim.Acta, 109, 319 (1979).

Single-point titrations IV. Determination of acids and bases with flow-injection analysis. O.Åstrom. Analytica Chim.Acta, 105, 67 (1979).

Use of a copper(II) ISE as end-point detector in precipitation titration of oxalate. M.Taddia. Fresenius' Z.Anal.Chem., 296, 149 (1979).

Potentiometric titrations of halide mixtures with an iodide-selective electrode. J.Motonaka, I.Junko, Sanae and N.Tanaka. Analytica Chim. Acta. 105, 417 (1979).

The use of iodate and periodate with iodine in the potentiometric titration of arsenite, sulfide and sulfite with mercury(II). H.Khalifa, N.T.Abdel-Ghani and M.S.Risk. Microchem. J., 24, 310 (1979).

Kinetic potentiometric determination of citric acid with a perbromate selective electrode. L.A.Lazarou and T.P.Hadjiioannou. Analytica Chim.Acta, 108, 375 (1979).

Titrimetric determination of some analytically significant organic reagents using a cadmium(II) selective electrode Crytur. K.Vytras, and M.Novotna-Horcicova. Coll.Czech.Chem.Commun., 44, 1126 (1979).

## 6. AUTOMATIC ANALYSIS AND CONTINUOUS MONITORING

Computor automation of potentiometric analysis with ISEs. J.Slanina, F.Bakker, J.J.Mols, J.E.Ordelman and A.G.M.Bruyn-Hes. Analytica Chim.Acta, 112, (1979). Comput.Tech.Optimiz., 3, 45.

Microcomputor-controlled potentiometric analysis systems. C.R.Martin and H.Freiser. Anal.Chem., 51, 803 (1979).

Automated potentiometric analysis with selective electrodes. L.P.Rigdon, C.L.Pomernacki, D.J.Balaban and J.W.Frazer. Analytica Chim.Acta, 112, (1979) Comput.Tech.Optimiz., 3, 397.

Automatic and microcomputor controlled potentiometric titrimetry. L.M.Doane, T.J.Stock and J.D.Stuart. J.Chem.Educ., 56, 415 (1979).

A microcomputor based data acquisition and processing system for electrochemical analysis. T.E.Edmonds. Analytica Chim.Acta, 108, 155 (1979).

Apparatus for (automatically) measuring (sodium and potassium) ion activity. T.Yamamoto, H.Takekawa, T.Banno and K.Koshiishi. U.S. Patent, 4,172,777 (Cl.20X-195R; G01N27/46) 30 October 1979.

(Automatic response measuring) biological microanalyser for potassium and sodium ions. J.Daroczy, J.Erdelyi, J.Havas, G.Jozan, L.Kecskes and K.Nyiro. Hung.Sci.Instrum., 45, 47 (1979).

Continuous and automatic detection of the concentration of gaseous inorganic fluorine compounds and chlorine compounds using ISEs. G.Blazevic, M.Boehner and E.Scheubeck. Fresenius' Z.Anal.Chem., 298, 12 (1979).

Automatic titration by stepwise additions of equal volumes of titrant. Part IV. General purpose programme for evaluating potentiometric acid-base titrations. A.Johansson and S.Johansson. Analyst, 104, 601 (1979).

Automatic sampling and monitoring of potentiometric electrodes. Steady state response by first and second derivative techniques. D.T.Skogberg, T.Richardson and T.Blasczyk. Anal.Chem., 51, 2054 (1979).

Injection techniques in dynamic flow-through analysis with electro-analytical sensors. E.Pungor, Z.Feher, G.Nagy, K.Tóth, G.Horvai and M.Gratzl. Analytica Chim.Acta, 109, 1 (1979).

Calibration of ISEs in a flow-through system. G.Horvai, K.Tóth and E.Pungor. Magy.Kem.Foly., 85, 382 (1979).

Multiple measurement system for the electrochemical analysis of flowing liquids and gases. J.G.Schindler. Biomed.Tech., 22, 235 (1977).

A new method for continuous in vivo measurements with ion-selective disk electrodes; the course of potassium and calcium ion concentrations in the blood of alert rats after bolus injections. O.Aziz, J.G.Schindler and R.Deunhardt. Biomed.Tech., 23, 194 (1979).

ISEs: application in flowing systems. K.Tóth, G.Nagy, Zs.Feher and E.Pungor. Euroanal 3, Rev.Anal.Chem., (Conf,), 3rd.1978. (Editor: D.M.Carroll) App.Sci.,Barking. (1979) page 173.

ISEs for continuous measurement of chloride and cyanide ions. H.Galster. Gewaesserschutz, Wasser, Abwasser, 39, 143 (1979).

## 7. GENERAL ASPECTS OF METHODOLOGY

Problems associated with use of ISEs in analysis. Ta.S.Khvorostin, M.S.Shikanova, I.G.Raskina, T.A.Roberts and E.A.Ivanova. Zh.Anal. Khim., 34, 1694 (1979).

Use of simple digital voltmeters in studying environmental pollutants by means of ion-sensitive measuring devices. W.Maurer. GIT Fachz. Lab., 23, 806 (1979).

Buffers for fluoride-electrode calibration in the low concentration range. M.Trojanowicz. Talanta, 26, 985 (1979).

The ion-sensitive field-effect transistor ISFET in rapid acid-based titrations. M.Bos, P.Bergveld and V-B.Blaauw. Analytica Chim.Acta, 109, 145 (1979).

Testing ISEs in continuous measuring systems. J.G.Connery, R.D.Jurenko and E.W.Shaffer. U.S.Patent, 4,189,367 (Cl.204-195G; G01N27/46) 19 February 1980.

Analytical methods involving ISEs (including flow methods). K.Tóth, G.Nagy and E.Pungor. in (Ed: A.K.Covington) "Ion-Sel.Electrode Methodol." 2, 65 (1979).

Practical techniques for ISEs. R.J.Simpson. in (Ed: A.K.Covington) "Ion-Sel.Electrode Methodol." 1, 43 (1979).

Instrumentation for ISEs. P.R.Burton in (Ed: A.K.Covington) "Ion-Sel. Electrode Methodol." 1, 21 (1979).

Microanalytical techniques with inverted solid-state ISEs. I. Nanolitre volumes. G.L.Vogel, L.C.Chow and W.E.Brown. Anal.Chem., 52, 375 (1980). II. Microlitre volumes. G.L.Vogel and W.E.Brown. Anal. Chem., 52, 377 (1980).

Determination of cyanide present in solution. R.Nakamwa and K.Takahara. Japan Kokai Tokkyo Koho, 79 52,592 (Cl.G01N31/32), 25 April 1979.

Thermal stabilization or temperature compensation in concentration measurement with ISEs. G.Kasseker. Gewaesserschutz, Wasser, Abwasser, 39, 159 (1979).

Processing of numerical data from measurements with ISEs. M.Trojanowicz. Chem.Anal.(Warsaw), 24, 659 (1979).

Some problems of using ISEs (for 2- and multicomponent solutions) in analytical practice. Ya S.Khvorastin, M.S.Shikanova, I.G.Raskina, T.A.Roberts and E.A.Ivanova. Zh.Anal.Khim., 34, 1694 (1979).

Response of solid-state membrane electrodes (for iodide or chloride) to rapid changes in concentration in flowing solutions. A.Dencks and R.Need. Fresenius' Z.Anal.Chem., 297, 121 (1979).

Limit of detection in analysis with ISEs. D.Midgley. Analyst, 104, 248 (1979).

Automatic sampling and monitoring of potentiometric electrodes; steady-state response by first and second derivative techniques. D.Skogberg, T.Richardson and T.Blasczyk. Anal.Chem., 51, 2054 (1979).

ISE measurements. J.L.Walker. Methods Enzymol., 56, (Biomembranes, Part G) 359 (1979).

Cleaning electrodes. P.Luethi and W.Steinmetz. British U.K.Patent Appl., 2,004,970 (Cl.B083/02) 11 April 1979.

Method and apparatus for measuring ion activity. N.Suzuki and S.Yoshinari. Ger.Offen., 2,908,719 (Cl. G01N27/46) 13 Sept.1979.

A simple system for the measurement of ion activities with solvent polymeric membrane electrodes. H.Affolter and E.Sigel. Anal.Biochem., 97, 315 (1979).

## 8. APPLICATION OF ISEs

### 8.1 Coordination complexes and reaction kinetics

Comparison between measured and calculated concentration of calcium ions in (ion) buffers. O.Scharff. Analytica Chim.Acta, 109, 291 (1979).

The disproportionation of hypoiodous acid. T.R.Thomas, D.T.Spence and R.A.Hasty. J.Inorg.Nucl.Chem., 42, 183 (1980).

The reaction of the alkaline earth metal oxides with iodine in the presence of water as part of a thermochemical cycle. G.F.V.Mason, J.D.Farr and M.G.Bowman. J.Inorg.Nucl.Chem., 42, 799 (1980).

Complex formation of trace elements in geochemical systems. I. Potentiometric study of fluorocomplexes of rare earth elements in fluorite bearing model systems. B.A.Bilal, F.Hurmann and W.Fleischer. J.Inorg.Nucl.Chem., 41, 347 (1979).

Complex formation of trace elements in geochemical systems. II. Stability of rare earth fluorocomplexes in fluorite-bearing model systems at various ionic strengths. B.A.Bilal and P.Becker. J.Inorg.Nucl.Chem., 41, 1607 (1979).

Complex formation of trace elements in geochemical systems. III. Studies on the dissociation of fluorocomplexes of rare earth elements in fluorite-bearing model systems. B.A.Bilal and V.Ko J.Inorg.Nucl.Chem., 42, 629 (1980).

The estimation of acid dissociation constants in sea water from potentiometric titrations with strong base. I. The ionic product of water - $K_W$. A.G.Dickson and J.P.Riley. Marine Chem., 7, 89 (1979).

Stability of the calcium sulphate ion pair at the ionic strength of sea water by potentiometry. B.Elgquist and M.Wedborg. Marine Chem., 7, 273 (1979).

Kinetic potentiometric determination of citric acid with a perbromate-selective electrode. L.A.Lazarou and T.P.Hadjiioanou. Analytica Chim.Acta, 108, 375 (1979).

Characteristics of the use of a bromide-selective electrode for studying self-oscillating chemical systems. A.S.Kovalenko and L.P.Tikhonova. Teor.Eksp.Khim., 14, 558 (1978).

Kinetic determination of formaldehyde and hexamethylenetetramine (hexamine) with a cyanide-selective electrode. M.A.Koupparis, C.E.Efstathiou and T.P.Hadjiioanou. Analytica Chim.Acta, 107, 91 (1979).

### 8.2 Mineralised tissue and dental materials

The antimicrobial effect of fluoride (acidulated phosphate, sodium and tin(II)) on *Actinomyces viscosus*. N.A.Yoon and C.W.Berry. J.Dent.Res., 58, 1824 (1979).

The influence of fluoride on the seeded growth of apatite from stable supersaturated solutions at pH 7.4. E.D.Eanes. J.Dent.Res., 59, 144 (1980).

Effect of sodium fluoride on viability growth of *Streptococcus mutans*. L.R.Brown, S.F.Handler, I.S.Horton, J.L.Streckfuss and S.Dreizen. J.Dent.Res., 59, 159 (1980).

The effects of water conditioning on fluoride concentration. K.W.Groman, M.M.Nazif and T.Zullo. J.Dent.Child., 47, 21 (1980).

Ionic transport through bis-glycidylmethacrylate films. B.Williams and J.A.Fraunhoffer. J.Oral Rehabilitation, 7, 131 (1980).

Fluoride concentrations in fissure and cervical enamel of unerupted teeth. I.Gedalia, Mr.Soldinger, B.Azaz and W.A.Soskoline. J.Oral Rehabilitation 7, 73 (1980).

Association between fluoride concentration in successive layers of human enamel and individual dental caries experience. R.Schamschula, H.Agus, G.Charlton, J.L.Duppenthaler and P.Un. Archs oral Biol., 24, 847 (1980).

Absorption of fluoride from fluoride dentifrices. J.Ekstrand and M.Ehrnebo. Caries Res., 14, 96 (1980).

Penetration of fluoride from sodium monofluorophosphate into artificially produced incipient enamel lesions. J.R.Mellberg. Caries Res., 14, 115 (1980).

Rate of fluorine uptake by surface and subsurface sound enamel from sodium monofluorophosphate. J.R.Mellberg. Caries Res., 14, 50 (1980).

Bone fluoride concentration associated with fluoridated drinking water. J.Charen, D.R.Taves, J.W.Stamm and F.M.Parking. Calcif. Tiss.Int., 27, 95 (1979).

Characterisation of glass ionomer cements. 6. A study of erosion and water absorption in both neutral and acidic media. S.Crisp, B.G.Lewis and A.D.Wilson. J. of Dent., 8, 68 (1980).

Influence by patterns of water intake on the occurence of incisor fluorosis and on the fluoride uptake by bones and incisor teeth in the rats. Y.Iizuka, T.Yasaki, K.Hara, K.Ishida and K.Inami. Bull. Kanagawa Dent.Coll., 6, 45 (1978).

In vitro fluoride uptake, distribution and retention by human enamel after 1 - and 24 hr application of various topical agents. D.H.Retief, P.G.Sorvas, E.L.Bradley, R.E.Taylor and A.R.Walker. J.Dent.Res., 59, 573 (1980).

In vivo remineralisation using a sustained topical fluoride delivery system. L.J.Abrahams, M.Yonese, W.Higuchi, J.L.Fox and G.T.Charbeneau. J.Dent.Res., 59, 583 (1980).

The effect of sodium fluoride in acid etching solution on sealant bond and fluoride uptake. T.Takahashi, Y.Arakawa, T.Matsukuko and M.Takeuchi. J.Dent.Res., 59, 625 (1980).

Effect of fluoride on ion exchange, remineralisation and acid resistance of surface enamel. L.A.Aponte-Merced and F.F.Feagin. J.Oral.Path., 8, 333 (1979).

Use of high performance liquid chromatography techniques to study the protection of hydroxy/apatite by fluoride and glycerophosphate against demineralization in vitro. T.H.Grenby and J.M.Bull. Caries Res., 14, 221 (1980).

Ultrastructural study and comparative analysis of fluorine content of enameloid in sea-water and fresh-water sharks. G.Daculsi and L.M.Kerebel. Archs Oral Biol., 25, 145 (1980).

Estimated fluoride intake of average two-year-old children in four dietary regions of the United States. R.H.Ophaug, L.Singer and B.F.Harland. J.Dent.Res., 59, 777 (1980).

Systemic absorption of fluoride from alginate impression material in humans. G.M.Whitford and J.Ekstrand. J.Dent.Res., 59, 782 (1980).

The effect of fluoride and mono fluorophosphate treatment on the electrochemical properties of bovine tooth enamel. J.W.E.Van Dijk, J.M.P.M.Borggreven and F.C.M.Driessens. Arch.Oral.Biol., 24, 753 (1979).

## 8.3 Biomedical fluids

### 8.3.1 Blood fluids.

Determination of inorganic sulphide and cyanide in blood using ISEs: application to investigation of hydrogen sulphide and cyanide poisoning. B.H.McAnalley, W.T.Lowry, R.D.Oliver and J.C.Garriott, J.Anal. Toxicol., 3, 111 (1979).

Limitations of the cyanide electrode for use in plasma and whole blood. J.R.Kistner, D.E.Longnecker, E.D.Miller and A.D.Lescanic. Anesthesiol.Analg., 58, 457 (1979).

Detection of fluorinated anesthetic metabolites by sodium fusion test. J.J.Soltis and A.J.Gandolfi. Anesthesiol. Analg., 59, 61 (1980).

Metabolism of Synthane: Comparison with in vivo and in vitro defluorination of other halogenated hydrocarbon anaesthetics. R.I.Mazze, W.J.Beppu and B.A.Hitt. Brit.J.Anaesthesiol., 51, 839 (1979).

Serum inorganic fluoride levels in obese patients during and after enflurane anesthesia. J.B.Bentley, R.W.Vaughan, M.S.Miller, J.M. Calkins and A.J.Gandolfi. Anesthesiol Analg., 58, 409 (1979).

Sevoflurane anaesthesia and biotransformation in man. D.A.Holaday and F.R.Smith. Anesthesiol., 51, S27 (1979).

Enzymatic defluorination of a halothane metabolite. D.D.Denson, J.A.Thurman and B.H.Hitt. Anesthesiol., 51, S240 (1979).

Plasma fluoride ion and urinary fluoride excretion in children following application of the fluoride-containing varnish Duraphat®. J.Ekstrand, G.Koch and L.G.Petersson. Caries Res.,14, 185 (1980).

Isoniazid metabolites and anaesthetic metabolism. M.P.Fish and S.A.Rice. Anesthesiol, 51, S256 (1979).

Deuterated methoxyflurane ($a_4$ - MOF) anesthesia. J.M.Baden, S.A.Rice, D.D.Denson and R.I.Mazze. Anesthesiol., 51, S263 (1979).

Accelerated fall in serum bromide level after administration of perchlorate to man. S.Seyfert. Eur.J.Clin.Pharm., 16, 351 (1979).

Succinylcholine and serum calcium. D.Glomaud. Anesthesiol., 51, 567 (1979).

Accute ammonia toxity and ammonia excretion in rainbow trout (Salmo gairdneri). B.A.Hillaby and D.J.Randall. J.Fish Res.Board Can., 36, 621 (1979).

More on $Na^+$ determination by ISE vs flame photometry. R.L.Coleman. Clin.Chem., 25, 1865 (1979).

Clinical comparison of ISEs and flame photometry for the determination of serum sodium and potassium ions. C.J.Preuss and C.Fuchs. Clin. Chem.Clin.Biochem., 17, 639 (1979).

Evaluation of an instrument (Nova-1) for direct potentiometric analysis of sodium and potassium in blood and their indirect potentiometric determination in urine. J.H.Ladenson. Clin.Chem.,25,757 (1979).

Serum potassium levels following transfusion of frozen erythrocytes. T.L.K.Rao, M.Maltric, M.R.Salem and A.A.El-Etr. Anesthesiol, 52, 170 (1980).

Continuous intraoperative measurements of sodium, potassium and calcium with carrier membrane disc electrodes. J.G.Schindler, G.Stork, R.Dennhardt, W.Schal, H.E.Braun, K.D.Karaschinski, W.Schmid and E.Fresenius. J.Clin Chem.Clin.Biochem., 17, 573 (1979).

New ISEs and their clinical and biological application (for $Na^+$, $K^+$, $Ca^{2+}$, $Cl^-$ and $H^+$ in blood serum and whole blood). D.Ammann, H.B.Jenny, P.C.Meier and W.Simon. Anal.Chem.Symp.Ser. 1979 (Published 1980). 2 (Electroanal.Hyg.Environ.Clin.Pharm.Chem.,),page 3.

Study of blood serum electrolyte levels using ISEs. A.K.Merzon, V.V.Kolomiets, E.A.Materova and S.E.Didina. Ionnyi Obmen Ionometriya, 2, 191 (1979).

A new method for continuous in vivo measurements with ion-selective disk electrodes: the course of potassium and calcium ion concentrations in the blood of alert rats after bolus injections. O.Aziz, J.G.Shindler, R.Dennhardt. Biomed.Tech., 23, 194 (1978).

Methods for determination of total calcium and calcium fractions in serum. K.Paschen. Aerztl.Lab., 25, 13 (1979).

Massive blood replacement: Correlation of ionised calcium, citrate and hydrogen ion concentration. R.C.Kahn, D.Jascott, G.C.Carlon, O.Schweizer, W.S.Howland and P.L.Goldiner. Anesthesiol.Analg., 58, 274 (1979).

A factor V inhibitor. In vitro interference by calcium. M.C.Coots, A.F.Muhleman and H.I.Glueck. Amer.J.Haematol., 7, 173 (1979).

### 8.3.2 Sweat and saliva.

Determination of ionised calcium in parotid saliva. F.Lagerlöf. Clin.Chim.Acta, 102, 127 (1980).

Effect of Verapamil on ionised calcium excretion in human parotid saliva. H.Maier, G.Antonczyk, J.G.Schindler and A.Heidland. Eur.J. Clin.Pharm., 16, 7 (1979).

New ion-selective electrodes and their clinical and biological application (for $Na^+$ and $K^+$). D.Ammann, H.B.Jenny, P.C.Meier and W.Simon. Anal.Chem.Symp.Ser., 1979 (Published 1980). 2 (Electroanal. Hyg.,Environ.,Clin.Pharm.Chem.) page 3.

Diagnostic screening for cystic fibrosis with a combination chloride ion-selective electrode. G.J.Moody and J.D.R.Thomas in (Ed:W.Franklin Smyth) "Electroanalysis in Hygiene, Environmental, Clinical and Pharmaceutical Chemistry" Elevier, Amsterdam (1980), page 33.

### 8.3.3 Urine.

Renal fluoride excretion during and after enflurane anaesthesia. Dependency on spontaneous urinary pH variations. P.O.Järnberg, J.Ekstrand, L.Erestedt and J.Santesson. Acta Anaesthesiol.Scad., 24, 129 (1980).

Renal function and fluoride formation and excretion during enflurane anaesthesia. P.O.Järnberg, J.Ekstrand, L.Erestedt and J.Santesson. Acta Anaesthesiol.Scad., 23, 444 (1979).

Fluoride kinetics and renal function during enflurane anaesthesia. P.O.Järnberg, J.Ekstrand, L.Erestedt and J.Santesson. Acta Anaesthesiol.Scand.,Suppl., 71, 20 (1979).

Effects of enflurane anaesthesia on the function of ischemically damaged kidneys. I.Wickström. Acta Anaesthesiol.Scand.Suppl., 71, 15 (1979).

Lack of effect of fluoride on urinary (AMP excretion in rats.) R.H.Ophaug, K.M.Wong and L.Singer. J.Dent.Res., 58, 2036 (1979).

Determination of ammonia and urea in urine and of urea in blood by use of an ammonia selective electrode. J.Georges. Clin.Chem., 25, 1888 (1979).

Evaluation of an instrument (Nova-1) for direct potentiometric analysis of sodium and potassium in blood and their indirect potentiometric determination in urine. J.H.Ladenson. Clin.Chem.(Winston-Salem, N.C.) 25, 757 (1979).

### 8.3.4 Intracellular fluids.

Electrochemical profile for ion transport across the membranes of proxunal tubular cells. M.Fujimoto, K.Naito and T.Kubota. Membrane Biochem., 3, 67 (1980).

Sodium selective liquid ion-exchanger microelectrode for intracellular measurements. J.O'Doherty, J.F.Gareia-Diaz and W.McD. Armstrong. Science, 203, 1349 (1979).

A method for producing very low resistance micropipettes for intracellular measurements. S.L.Jacobson and G.A.R.Mealing. Electroencaphalog. and Neurophysiol., 48, 106 (1980).

Neutral carrier sodium ion-selective microelectrode for intracellular studies. R.A.Steiner, M.Oehme, D.Ammann and W.Simon. Anal.Chem., 51, 3 (1979).

The intracellular sodium activity of sheep heart Purkinje fibres; effect of local anaesthetics and tetrodotoxin. J.W.Deitmer and D.Ellis. J.Physiol., 300, 269 (1980).

Measurement of intracellular pH of bullfrog skeletal muscle and renal tubular cells with double barrelled antimony microelectrodes. Y.Matsumara, K.Kajina and M.Fujimoto. Membrane Biochem., 3, 99 (1980).

Intracellular pH in early Xenopus embryos: its effect on current flow between blastomeres. L.Turin and A.E.Warner. J.Physiol., 300, 489 (1980).

Intramyocardial pH as an index of myocardial metabolism during cardiac surgery. F.J.M.Walters, G.J.Wilson, D.J.Steward, R.J.Domenech and D.C.MacGregor. J.Thor.Card.Surg., 78, 319 (1979).

Intracellular chloride activity in quiescent cat papillary muscle. K.W.Spitzer and J.L.Walker. Amer.J.Physiol., 238, H487 (1980).

### 8.3.5 Miscellaneous.

Possibilities of using ISEs for studying some biological media. E.B.Nikol'skaya, E.A.Materova, A.L.Grekovich, N.V.Kopylo, A.V.Svyatkovskii, L.V.Ustimova and N.B.Shor. Ion.Obmen i Ionometriya, (2), 208 (1979).

ISEs in medicine and medical research. D.M.Band and T.Treasure in (Ed:A.K.Covington) Ion-Sel.Electrode Methodol., 2, 41 (1979).

X-ray fluorescence and potentiometry compared for determining iodine content of thyroid glands. F.Puttemans, F.Deconinck, M.Jonckheer, M.Vandeputte and D.L.Massart. Clin.Chem., 25, 1247 (1979).

Simultaneous recording of calcium ion and pH when studying the processes of calcium transport in the sarcoplasmic reticulum. V.A.Tugai and P.V.Usatyuk. Ukr.Biokhim.Zh., 51, 508 (1979).

The calibration and use of a calcium ISE for kinetic studies of mitochondrial calcium transport. R.K.Yamazaki, D.L.Mickey and M.Story. Anal.Biochem., 93, 430 (1979).

Improved technique for the measurement of the kinetics of $Ca^{2+}$ uptake by cells: The coupling of an amplifier with voltage regulator to a $Ca^{2+}$ selective electrode. U.Flora, R.Gennaro and D.Romeo. Anal. Biochem., 102, 77 (1980).

Comparative effects of calcium chloride and calcium gluceptate. L.J.Drop and D.J.Cullen. Brit.J.Anaesthesiol., 52, 501 (1980).

The state of accumulated calcium in mouse brain microsomes in vivo. D.Satomi J.Biochem., Japan, 85, 1439 (1979).

Nature of the calcium ion efflux induced in rat liver mitochondria by the oxidation of endogenous nicotinamide nucleotides. D.G.Nicholos and M.D.Brand. Biochem.J., 188, 113 (1980).

Internal effects of divalent cations on potassium permeability in molluscar neurones. A.L.F.Gorman and A.Herman. J.Physiol., 296, 393 (1979).

The state of accummulated calcium in mouse brain microsomes in vitro. D.Satomi. J.Biochem., 85, 1439 (1979).

Altered maternal calcium homeostasis in diabetic pregnancy. D.P. Cruikshank, R.M.Pitkin, W.A.Reynolds, G.A.Williams and G.K.Hargis. J.Clin.Endocrinol and Matab., 50, 264 (1980).

Effect of chronic parathyroidectomy on calcium metabolism in the lactating rat. J.M.Fry, D.H.Curnow, D.H.Gutteridge and R.W.Ketallach. J.Endocrinol., 82, 323 (1979).

Ion effects in measurement of ionised calcium with a calcium ISE. R.A.Kaufman and N.W.Tietz. Clin.Chem., 26, 640 (1980).

Improvements in and clinical utility of a continuous flow method for routine measurements of dialyzable (ultra filtrable) calcium. J.Toffaletti and G.N.Bowers. Clin.Chem., 25, 1939 (1979).

Chronic hypercalcaemia secondary to hyperparathyriodism: A risk factor during anaesthesia? M.A.Gunst and L.J.Drop. Brit.J.Anaesthesiol., 52, 507 (1980).

Ventricular pump performance during hypocalcaemia. P.M.Stulz, D.Scheideggar, L.J.Drop, E.Lowenstein and M.B.Laver. T.Thor.Card. Surg., 78, 185 (1979).

Effect of acylphosphates effects on calcium uptake by sarcoplasmic reticulum vesicles. G.Liguri, M.Stefani, A.Berti, P.Nassi and G.Ramponi. Arch.Biochem.Biophys., 200, 357 (1980).

Release of intracellular membrane bound calcium precedes the onset of stimulus induced exocytosis in platelets. M.B.Feinstein. Biochem. and Biophys.Res.Comm., 93,593 (1980).

Effects of glucagon and sodium on the control of extra mitochondrial free calcium by mitochondria from liver and heart. M.D.Brand and C.de Selincourt. Biochem.and Biophys.Res.Comm., 92, 1377 (1980).

Influence of posture on free calcium and related variables. B.W.Renoe, J.M.McDonald and J.H.Ladenson. Clin.Chem., 25, 1766 (1979).

The measurement of calcium ion movements in mitochondria. M.Compton and E.Carafoli. Methods Enzymol., 56, Biomembranes,(Part G) 338, (1979).

Pre-anesthetic cimetidine and gastric pH. K.Malinak and A.H.Vakil. Anesthesiol.Analg., 58, 309 (1979).

Selective ion-monitoring in clinical chemistry I.Bjoerkhem. CRC Crit.Rev.Clin.Lab.Sci., 11, 53 (1979).

Measurement of extracelluar ion activities ($[\text{potassium ion}]_e$, $[\text{sodium ion}]_e$, $[\text{calcium ion}]_e$) during acute coronary occlusion. M.Gueggi, M.Kessler, F.Greitschus, V.Wiegand and W.Meesmann. Front Biol.Energ. (Internat.Symp.) (Eds: P.L.Dutton, J.S.Leigh and A.Scarpo) Academic Press.New York. 2, 1427 (1978).

Study of a sodium ISE for medical - biological uses. E.A.Materova, Z.S.Alagova and G.I.Shumilova. Ion Obmen i Ionometriya, (2), 179 (1979).

Device for determining (potassium, sodium and chloride) ionic activity of components of liquid drops (of biological fluids). D.P.Hamblen, C.P.Glover and S.H.Kim. Canadian Patent, 1,071,305 (Cl. G01N27/26) 5 February 1980.

Possibilities of determining sodium ions in biological media using a membrane ISE. O.K.Stefanova, V.F.Gorshkova and E.A.Materova. Ion Obmen i Ionometriya, (2) 183 (1979).

The effects of local anesthetics on the potassium ISE. R.S.Greenwood. W.E.Dodson and S.Goldring. Brain.Res., 165, 171 (1979).

Potassium permeability of the mesothelium of the frog mesentery. J.F-Jensen and O.Christensen. Acta Physiol.Scand., 105, 228 (1979).

Membrane permeability during low potassium depolarisation in sheep cardiac Purkinje fibres. C.O.Lee and H.A.Fozzard. Amer.J.Physiol., 237, C156 (1979).

More on sodium determination by ISE versus flame photometry. W.W. Annan, N.A.Kirwan and W.S.Robertson. Clin.Chem., 25, 1865 (1979).

Peripheral vascular response to potassium administration during cardiopulmonery by-pass. A.J.Schwartz, T.J.Conahan, D.R.Jobes, R.W.Andrews, H.MacVaugh and A.J.Ominsky. J.Thor.Card.Surg., 79, 237 (1980).

Response of ion-selective sodium and potassium electrodes in the Beckman Astra 4 and Astra 8 Analyzers. P.Fievet, A.Truchard, J.Hersart and G.Glikmanas. Clin.Chem., 26, 138 (1980).

New ISEs and their clinical and biological application (for $H^+$, $Li^+$, $Na^+$, $K^+$, $Cl^-$ and $Ca^{2+}$). D.Ammann, H.B.Jenny, P.C.Meier and W.Simon. Anal.Chem.Symp.Ser. 1979 (Published 1980). 2 (Electroanal.Hyg.Envion. Clin.Pharm.Chem.,) page 3.

Determination of total protien in cerebrospinal fluid with a silver sulphide membrane electrode. Y.Morishita, K.Nokane and A.Kosaka. Rinsho Kensa, 23, 936 (1979).

A method for continuous measurement of ammonia in respiratory air-ways. T.V.Larson, D.S.Covert and R.Frank. J.Appl.Physiol., 46,603 (1979).

Hepatonecrosis caused by halothane and hypoxia in phenobarbital treated rats. W.T.Ross, B.P.Daggy and R.R.Cardell. Anesthesiol, 51, 327 (1979).

Pulmonary biotransformation of methoxyflurane. C.D.Blitt, B.R.Brown, B.J.Wright, A.J.Gandolfi and I.G.Sipis. Anesthesiol., 51, 528 (1979).

Uptake and bitransformation of Aliflurane in man. D.A.Holaday, M.C.Jardines and W.H.Greenwood. Anesthesiol., 51, 548 (1979).

Renal effects of enflurane anesthesia in Fischer 344 rats with pre-existing renal insufficiency, T.S.Sievenpiper, S.A.Rice, F.McClendon, J.C.Kosek and R.I.Mazze. J.Pharmacol.and Exp.Therap., 211, 36 (1979).

Toxic properties of an isomer of Halothane. H.F.Cascorbi and J.M. Verio. Der Anaesthetist, 29, 169 (1980).

Microsomal-thyroxine measurements with iodide-selective membrane electrodes. M.E.Meyerhoff and G.A.Rechnitz. Anal.Lett., Part A 12, 1339 (1979).

Analysis of hydrophobic amine antimalarials in biological fluids with the plastic ISE. D.W.Mendenhall, T.Higuchi and L.A.Sternson. J.Pharm. Sci., 68, 746 (1979).

Catheter-tip pH electrodes for continuous intravascular recording. S.M.Cobbe and P.A.Poole-Wilson. J.Meg.Eng.and Technol., 4, 122 (1980).

Tissue pH electrodes for chemical applications. D.K.Harrison and W.F.Walker. J.Med.Eng. and Technol., 4, 3 (1980).

## 8.4 Enzyme systems and bacteria

Potentionmetric electrode measurement of serum antibodies based on the complement fixation test. P.D'orazio and G.A.Rechnitz. Analytica Chim.Acta, 109, 25 (1979).

Antibody-selective membrane electrodes. Science, R.L.Solsky and G.A.Rechnitz. Science, 204, 1308 (1979).

Electrochemical biosensors in the assay of antibodies. M.Thompson, P.J.Worsfold, J.M.Holuk and E.A.Stubley. Analytica Chim.Acta, 104, 195 (1979).

Electrochemical method of investigation of antigen-antibody and enzyme-enzyme inhibitor reactions using chemically modified electrodes. N.Yamamoto, Y.Nagasawa, S.Shuto, M.Sawai, T.Sudo and H.Tsubomura. ChemLetts., 245, (1978).

ISE estimation of avidin and biotin using a lysozyme label. C.R. Gebauer and G.A.Rechnitz. Anal.Biochem., 103, 280 (1980).

Enzyme electrode for inhibitors determination - Urease - fluoride system. C.Tranminh and J.Blaux. Anal.Chem., 51, 91 (1979).

Enzyme kinetics and substrate stabilisation of detergent solubilized and membranerous ($Ca^{2+}$ + $Mg^{2+}$) - activated ATPase from sarcoplasmic reticulum. J.V.Møller, K.E.Lind and J.P.Anderson. J.Biol.Chem., 255, 1912 (1980).

Chemically modified electrodes bearing grafted enzymes. C.Bourdillon, J.P.Bourgeois and D.Thomas. Biotechnol.and Bioeng., 21, 1877 (1979).

Continuous-flow monitoring of lactose concentrations. B.Volesky and C.E.Mond. Biotechnol. and Bioeng., 21, 1251 (1979).

Determination of β-glucosidases, β-glucuronidase and alkaline phosphatase by an enzyme electrode sensitive to phenol. L.Macholan. Collect.Czech.Chem.Commun., 44, 3033 (1979).

Enzyme electrode-based kinetic assays of enzyme activities. C.R.Gebauer, M.E.Meyerhoff and G.A.Rechnitz. Anal.Biochem., 95, 479 (1979).

Electrode-based enzyme immunoassays using unrease conjugates (in conjunction with an $NH_3$- sensing electrode. M.E.Meyerhoff and G.A.Rechnitz. Anal.Piochem., 95, 483 (1979).

A rapid sensitive assay for starch phosphorylase and ADP glucose pyrophosphorylase. D.A.McCracken and W.M.Rutherford. Anal.Biochem., 101, 275 (1980).

Glucose analysis utilizing immobilized enzymes. M.H.Keyes, F.E. Semersky and D.N.Gray. Enzyme Microb.Technol., 1, 91 (1979).

Quantitation of glucose concentration using a glucose oxidase-catalase electrode by potentionmetric measurement. C.C.Lin, L.B.Wingard, S.K.Wolfson, S.J.Yao, A.L.Drash and J.G.Schiller. Bioelectrochem. Bioenerg., 6, 19 (1979).

The hydrogen peroxide sensing platinum anode as an analytical enzyme electrode (in conjunction with $H_2O_2$ - generating oxidases to measure substances of clinical interest in body fluids). L.C.Clark. Methods Enzymol., 56, (Biomembranes, Part G) 448 (1979).

Electrochemical application of oxireductases (with attention to enzyme electrodes and microflow immobilised enzyme reactor methods). G.Baum and H.H.Weetall. Methods Enzymol., 56 (Biomembranes, Part G), 479 (1979).

Electrochemical method for the determination of ethanol (with an ethanol-selective enzyme electrode). T.Araho. Japan Kokai Tokkyo Koho, 79 61,984 (C1.G01N27/52). 18 May 1979.

Amperometric determination of acetic acid with immobilized trichosporon brassicae. M.Hikuma, T.Kubo, T.Yasuda, I.Karu1 and S.Suzuki. Analytica Chim.Acta., 109, 33 (1979).

Bacterial membrane electrode for the determination of nitrate. R.K.Kobos, D.J.Rice and D.S.Flournoy. Anal.Chem., 51, 1122 (1979).

Microbial electrode sensor for alcohols. M.Hikuma, T.Kubo, T.Yasuka, I.Karube and S.Suzuki. Biotechnol. and Bioeng., 21, 1845 (1979).

Bioprobe electrode for phenol. H.Y.Neujahr and K.G.Kjellen. Biotechnol. and Bioeng., 21, 671 (1979).

Microbioassay of nystatin with a yeast electrode. I.Karube, T.Matsunga and S.Suzuki. Analytica Chim.Acta, 109, 39 (1979).

## 8.5 Medicinal and pharmaceutical preparations

Determination of phenylmercuric nitrate by potentiometric titration. R.W.Wood and H.L.Willes. J.Pharm.Science, 68, 1272 (1979).

Determination of aluminium in antacid formulations by Gran's plot titration with sodium fluoride. M.E.A.Cooper, J.Ballantine and A.D. Woolfson. J.Pharm Parmacol., 31, 403 (1979).

Determination of sodium in infusion solutions with an ISE. S.Zadeczky, D.Kuttel, J.Havas and L.Kecskes. Acta Pharm.Hung., 49, 54 (1979).

Rubidium ISE for assay of polyene antibiotics. R.F.Cosgrove and A.E.Beezer. Analytica Chim.Acta, 105, 77 (1979).

Determination of L-amino acids, alcohols and glucose with a tubular iodide selective electrode and oxidase enzymes. M.Mascini and G.Palleschi. Ann.Chim.(Roma), 69, 249 (1979).

Antidiuretic hormone specific electrode. S.Updike and I.Treichel, Anal.Chem., 51, 1643 (1979).

The potassium ISE as a tetraphenylborate sensor for quarternary ammonium salts analysis. S.Pinzauti and E.LaPorta. J.Pharm. and Pharmacol., 31, 573 (1979).

Differential potentiometric method for determining dissociation constants of very slightly water-soluble drugs applied to the sulphonamide diuretic chlorthalidone. H.L.J.Fleuren, C.A.M.van Ginneken and J.M.van Rossum. J.Pharm.Science, 68, 1056 (1979).

Liquid membrane electrode for direct potentiometry and potentiometric titration of strychnine. S.S.M.Hassan and M.B.Elsayes. Anal.Chem., 51, 1651 (1979).

Determination of chloramphenicol in pharmaceutical preparations by cadmium ISE spectrophotometry and aas. S.S.M.Hassan and M.H.Eldesonki. Talanta, 26, 531 (1979).

Determination of vitamin $B_1$ in pharmaceutical prepartions using a lead ISE and atomic absorption spectroscopy. S.S.M.Hassan, M.T.M. Zaki and M.H.Eldesouki. J.Ass.Offic. Anal.Chem., 62, 315 (1979).

Potentiometric determination of iodine in pharmaceutical preparations. M.Vandeputte, L.Dryon. L.de Hertogh and D.L.Massart. J.Pharm.Science, 68, 1416 (1979).

## 8.6 Beverages and foods

Fluoride content of sugar. R.G.Schamschula, H.A.Agus, K.H.Fong and G.G.Craig. J.Dent.Res., 58, 1915 (1979).

Microdiffusion and fluoride-specific electrode determination of fluoride in foods. R.W.Dabeka, A.D.McKenzie and H.B.S.Conacher. J.Ass.Offic.Anal.Chem., 62, 1065 (1979).

Fluoride content of tea leaves. W.Feldheim and S.O.Miehe. Z.Lebensm.-Unters.Forsch., 169, 453 (1979).

The determination of fluorine in coffee and tea using a microprocesser with a fluoride ion-selective electrode. G.J.Moody, B.Ong, K.Quinlan and J.D.R.Thomas. J.Food Technol., 15, 335 (1980).

Analytical methods for nitrate and nitrite in feeding-stuffs. II. Potentiometric determination of nitrate with ISE. K.Zimmermann and J.von Lengerken. Nahrang, 23, 523 (1979).

The application of ISEs to food analysis. J.Comer, Dev.Food Anal. Tech., 1, 197 (1978).

Fundamental studies on bio-electrochemical sensor. II. Amines (in meat extracts) sensor. S.Suzuki, I.Karube and I.Satch. Asahi Garasu Koyyo Gitutsu Shoreikoi Kenkyu Hokoku, 32, 355 (1978).

## 8.7 Vegetation, vegetables, fruits and juices

Rapid method for determining in vegetation using an ion-selective electrode. E.A.Villa. Analyst, 104, 545 (1979).

Determination of chlorine, fluorine and sulphur in plant leaves by the combustion method. M.T.Yu, C.F.Yu and C-Y Yang, Huan Ching K'o Hseuh, 5, 38 (1978).

Rapid method for determining fluoride in vegetation using an ion-selective electrode. A.E.Villa. Analyst, 104, 545 (1979).

Potassium and sodium determination with the help of ISEs and use of the method for technological evaluation of beets. R.S.Sova, S.N. Kalina, I.V.Zakharova and G.G.Rusin. Sakh.Prom-st., 43 (1979). Chem. Absts., 92, 24706 (1980).

## 8.8 Air and stack gases

Application of ISEs for determining trace amounts of (HF, HCl and HCN atmospheric pollutants). U.Fritze and C.Harzdorf. Gewaesser-schutz. Wasser, Abwasser, 39, 229 (1979).

Measurements of relative concentration of iodine in air by alkali treated filter paper method (by neutron activation and ISE). N.Fukuzaki, N.Moriyama, R.Sugai and T.Ohshina. Radioisotopes, 28, 371 (1979).

## 8.9 Natural water including sea water

Microdetermination of nitrate nitrogen in lake waters by potentiometry with an ISE. V.Simeonov, G.Andreev and A.Stoianov. Fres.Z. Anal.Chem., 297, 418 (1979).

Experiences with nitrate, fluoride and chloride electrodes in the laboratory (for water analysis). Gewaesserschutz, Wasser, Abwasser, 39, 161 (1979).

Microdetermination of nitrate nitrogen in lake waters by potentiometry with an ISE. V.Simeonov, G.Andreev and A.Stoianov. Fresenius' Z.Anal.Chem., 297, 418 (1979).

Ion-selective combination electrode for nitrate determination (in water and soil). V.Ebock. Chem.Tech.(Leipzig), 32, 91 (1980).

Determination of fluoride in river water. R.Gallego, J.L.Bernal and R.Pardo. Afinidad, 35, 333 (1978).

Direct determination of fluoride in water using acetylacetone as masking agent. H.Hashintani, H.Yoshida and T.Adachi. Bunseki Kagaku, 28, 680 (1979).

Direct potentiometric determination of fluoride in mineral, drinking and industrial waters without interference. H.Ballczo and M.Sager. Fresenius'Z.Anal.Chem., 298, 382 (1979).

Development of FACTS procedures for combined chlorine and ozone in aqueous solutions. N.M.Roscher, J.Liebermann, W.J.Cooper and E.P.Meier. U.S.N.T.I.S.

Practical determination of fluoride in water with ISEs. D.Weil and K.E.Quentin. Wasser Abwasser Forsch. 11, 133 (1978).

Determination of traces ($10^{-4}$ to $2 \times 10^{-6}$M) of iodine in natural water with use of ISEs. V.A.Nikashina and A.N.Krachak. Zh.Anal.Khim., 34, 2236 (1979).

Determination of thio-cyanate at the ppm level in waters using an ISE with liquid membrane: an automated procedure. T.Korenaga. Mikrochim.Acta, 2, (5-6), 455 (1979).

Application of calcium-selective electrodes for the rapid determination of macro-components of sea-water. A.L.Grekovich, D.E.Morachevskii and V.E.Yurinskaya. Ionnyi Obmen Ionometriya, 2, 221 (1979).

On the determination of $Ca^{2+}$ concentrations in sea-water. B.S.Krumgalz and R.Holzer. Limnol.and Oceanog., 25 , 367 (1980).

Use of ion-selective electrodes for rapid determination of trace components of sea-water. A.L.Grekovich, D.E.Morachevskii and V.E. Yurinskaya. Ion Obmen i Ionometriya, (2), 221 (1979).

Evaluation of (Orion) calcium ion-sensors in fresh and salt-water. G.K.Ward. Report, EPA/600/7-79/058, 39 pages,(1979).

Use of ISEs for determination of water hardness and the concentration of fluoride and chloride ion in water. D.Szyzlo. Rozpoznawanie Wod Podziemnych Potrzeb Gospod.Nar., 282 (1979).

Determination of total water hardness using ISEs. R.Perez Olmas. Ing.Quim., 10, 233 (1978).

Study of the possibility of using silver and calcium ion-selective membrane electrodes in viable closed ecological systems (for constant control of drinking water quality). V.S.Shterman, S.I.Petrov, I.I. Medvedev and L.T.Krasnoshchekova. Dokl.TSKLA, 238, 143 (1978).

Evaluation(for magnesium in marine applications) of the Orion divalent ISE. G.K.Ward. Report, EPA/600/7-70/059, 34 pages, (1979).

Copper ISE for determination of inorganic copper species in fresh waters. R.Stella and M.T.Ganzerlin Valentini. Anal.Chem., 51, 2148 (1979).

Test and evaluation of potassium sensors in fresh and salt water. G.K.Ward. Report, EPA/600/7-79/057, 65 pages, (1979).

ISEs: theory and application in water analysis. R.Sarin. Indian J. Environ.Health, 21, 244 (1979).

## 8.10 Rocks and soils

Fluorine contents in some standard (geological) samples. Y.Kato. Ganseki Kobutsu Kosho Gakkaishi, 74, 421 (1979).

Determination of fluorine in pyrite with a fluoride ISE. Human Inst. of Chem.Eng., Yen Hsi Hua Hsueh, 6, 279 (1978).

Direct potentiometric determination of fluoride in mineral raw materials of various compositions by using a fluoride ISE. G.I. Bebeshko, V.P.Roe and V.A.Khalizova. Zh.Anal.Khim., 34, 507 (1979).

Determination of fluoride in rocks, without preliminary distillation. by means of an ISE. I.(nee Dely) Szucs. Magy.Kem.Foly., 85, 481 (1979).

Fluorine determinations in minerals, rocks and raw materials. H.Ackermann. Interceram, 27, 404 (1978).

Determination of sulphide in pyritic soils and minerals with a sulphide ISE. D.L.Sorenson, W.A.Kneib and D.B.Porchella. Anal.Chem., 51, 1870 (1979).

Determination of chlorine in silicate rocks by ion-exchange chromatography and direct potentionmentry with an ISE. H.Akaiwa, H.Kawamotor and K.Hasegawa. Talanta, 26, 1027 (1979).

Sodium carbonate - potassium nitrate fusion and ISEs in direct potentiometric determination (in minerals and rocks) of fluorine and chlorine. K-H.Tau. Fen.Hsi Hua Hsueh, 6, 465 (1978).

Determination of iodine in phosphate ores by an iodide ISE. J-P.Li, M-C.Chow, Y-L.Chang and M-M.Chen. Fen Hsi Hua Hsueh, 6, 433 (1978).

Potentiometric determination of nitrate in soils using a membrane electrode. H.Knittel and G.Fischbeck. Z.Pflanzenernaehr.Bodenkd., 142, 669 (1979).

Determination of nitrate nitrogen in soil using nitrate ISE. V. Simeonov, I.Asenov and V.Dyadov. God. Sofii.Univ.Khim.Fak., 1975-76 70, Pt. 1, 11(Publ.1978).

An evaluation of pretreated natural zeolites for ammonium removal. J.R.Klieve and M.J.Semmens. Water Res., 14, 161 (1980).

Determination of silver in minerals with a silver ISE. M-S.Mo and Yu-T. Ho. Fen Hsi Hua Hsueh, 6, 274 (1978).

Determination of copper in minerals and soil by a copper ISE. P.H. Kuang, T-Y.Yang and K-K.Po. Fen Hsi Hua Hsueh, 7, 28 (1979).

8.11 Industrial

Potentiometric determination of sulphur in waters, chemicals, iron, steel and fly ash with a cadmium sulphide membrane electrode. D.Chakraborti and F.Adams. Analytica Chim.Acta, 109, 307 (1979).

Control of laboratory sulphidization with a sulphide ISE before flotation of oxidized lead-zinc-silver dump material. M.H.Jones and J.T.Woodcock. Int.J.Miner.Process., 6, 17 (1979).

Determination of nitrite ion in unused cutting fluids and cutting oils using a gas-sensing electrode method. F.R.S.Clark and H.B.MacPherson. Analyst, 104, 358 (1979).

Determination of nitrate radicals in filtration liquid for low temperature exchange methylation catalysts using a liquid membrane. S-M.Tung, T-Y.Pei and H-F.Chu. Fen Hsi Hua Hsueh, 7, 242 (1979).

Determination of chloride in high-purity waters in the range of 0-20 μg $dm^{-3}$ of chloride using ion selective membranes incorporating $Hg_2Cl_2$. G.B.Marshall and D.Midgley. Analyst 104, 55 (1979).

Application of fluoride ISE for determination of fluoride in zinc electrolyte (in zinc refining). A.Hulanicki, M.Trojanowicz and J.Sztandor. Chem.Anal. (Warsaw), 24, 617 (1979).

Fluoride ISE for control of chemical composition of slags and fluxes. (Determination of fluorine). V.N.Ioffe and T.D.Rukhtina. Zavod.Lab., 45, 969 (1979).

Potentiometric (barium ion-selective) electrode method for the rapid determination of sulphate in the presence of chromium(VI). R.Prasad. Analyst, 104, 164 (1979).

Application of and determination of anionic surfactant. K.L.Wang, S.F.Kao, M.H.Wang, J.F.Kao and A.L.Loshin. Ind.Eng.Chem.Prod.Res. Dev., 17, 186 (1978).

Two-phase potentiometric titration (of amines) with sodium dodecyl sulphate. I.A.Gur'ev, G.M.Lizunova and N.S.Bulanova. Zh.Anal.Khim., 34, (1979).

Liquid ISE for determining alkyl sulphates. G.L.Starobinets, G.A.Laevskaya, F.M.Rakhman'ko and A.K.Pirozhnikova. Zh.Anal.Khim., 35, 154 (1980).

(Review of) the role of ISEs in improving fermentation yields. D.B.Kell. Process Biochem., 15, 18 (1980).

ISEs and potentiometric methods for the determination of elements in phosphates, technical grade solutions, return - and waste water. (Review and practice of using ISEs). A.V.Dolgorev and R.S.Deinekina. Tr. NII Gornokhim. Syr'ya, (47), 71 (1978).

Application of silver ion electrode for indirect determination of phosphorous in niobium-ion alloys. Ta-M.Fung. Fen Hsi Hua Hsueh, 7, 130 (1979).

Potentiometric determination of sulphate in fertilizers with cadmium sulphide membrane electrode. D.Chakraborti and F.Adams. Fresenius' Z.Anal.Chem., 298, 397 (1979).

Continuous monitoring of industrial liquid effluents using ISEs. A.W.Nobrega and A.M.DaSilva. Quim.Nova, 2, 45 (1979).

Determination of sodium in aluminous materials with a sodium ISE. T.A.Palmer and J.M.Winkler. Analytica Chim.Acta, 113, 301 (1980).

Determination of copper in concentrated (acidic) electrolytes with a copper ISE. A.Hulanicki, K.T.vel Krawcznski and M.Trojanowicz. Chem. Anal.(Warsaw), 24, 435 (1979).

Volumetric determination of cobalt in cobalt-samarium-iron alloys using a silver ISE in aqueous di-methyl formamide and in water. A.Kosturiak and D.Kalavska. Collect.Czech.Chem.Commun., 44, 1742 (1979).

Application of polarography and ISEs in the photographic industry. E.C.Weller. Proc.Anal.Div.Chem.Soc., 16, 176 (1979).

Electroanalytical monitors for process control. P.L.Bailey. Kent Tech.Rev., 23, 14 (1978).

Industrial application for ISEs (a review). P.L.Bailey. Ion-Selective Electrode Revs., 1, 81 (1979).

## 8.12 Environmental and public health

Fluoride accummulation by long tailed field mice (Apodemus Sylvaticus L.) and field moles (Microtus Agrestic L.) from polluted environments. D.A.Wright, A.W.Davison and M.S.Johnson. Environment. Pollution, 17, 303 (1978).

ISEs and their application in the industrial hygiene field. R.Olmos Perez. Quim.Ind., 25, 361 (1979).

Development of a (hydrocarbon oxygenase) enzymic electrode as an oil in (sea) water sensor. A.M.Cundell and E.Findl. Report (Bio.Res.Inc. Farmingdale, NY, USA), ER - 013179, 29 pages (1979).

ISEs in environmental and toxicological analysis. J.D.R.Thomas in "Proceedings of the International Congress on Analytical Techniques in Environmental Chemistry". Editor: J.Albaiges, Pergamon, Oxford, 1980, page 543.

## 8.13 Application in non-aqueous media

Two-phase potentionmetric titration of phenols with use of an ISE. I.A.Gur'ev, E.A.Gushchina and E.N.Mitina. Zh.Anal.Khim., 34, 1184 (1979).

Volumetric determination of cobalt in cobalt-samarium-iron alloys using a silver ISE in aqueous dimethylformamide and in water. A.Kosturiak and D.Kalavska. Collect.Czech.Chem.Commun., 44, 1742 (1979).

Correction factors for the glass electrode in aqueous methanol. K.K.Agrawal. Talanta, 26, 599 (1979).

Effect of organic solvents and their dielectric constant on (lead and silver arsenite)ISE potential. S.N.K. Chaudhari and K.L.Cheng. Mikrochim.Acta, 2, 411 (1979).

Potentiometric determination of sulphate in white and green liquors using a lead ISE and acetone as solvent. E.J.Lokka. Pap.Puu, 60, 441 (1978).

## 8.14 Applications in chromatography

A dual ISE detector for selective detection and determination of atomic ratio in gas chromatography of chlorine - and bromine-containing compounds. T.Kojima, M.Ichise and Y.Seo. Asaki Garasu Kogyo Gijutsu Shoreikai Kenkyu Hokoku,32, 261 (1978).

Application of a copper electrode as a detector for high-performance liquid chromatography. C.R.Loscombe, G.B.Cox and J.A.W.Dalziel. J.Chromatogr., 166, 403 (1978).

Gas electrode detector for gas chromatography. Y.Seo. Bunseki Kagaku, 28, 334 (1979).

ISE detector for high pressure liquid chromatography. W.Ishibashi and A.Jyo. Asahi Garasu Kogyo Gijutsu Shoreikai Kenkyu Hokoku, 33, 47 (1978).

## 8.15 Applications in the analysis of organic compounds

Titrimetric determination of some analytically significant organic reagents using a cadmium(II) ISE Crytur. K.Vytras and M.Novotna-Horcicova. Collect.Czech.Chem.Commun., 44, 1126 (1979).

Organic analysis with a new silver-selective membrane electrode. V.V.Cosofret, C.Stefanescu and A.A.Banaciu. Talanta, 26, 1035 (1979).

ISEs in organic functional group analysis: Microdetermination of hydrazines with the copper electrode. S.S.M.Hassan and M.T.M.Zaki. Mikrochim.Acta, 1, 137 (1979).

Simple and specific microdetermination of organo iodine via oxygen flask combustion. F.W.Cheng. Microchem.J., 25, 86 (1980).

Determination of glutamic, aspartic and amino-acetic acids by potentiometric titration using a copper(II) ISE. I.Haiduc, Maria Mioscu and D.Cormos. Stud.Univ.Babes-Bolyai, Ser. Chem.,24, 59 (1979).

ISEs based on tetraphenylphosphonium 12-tungstosilicate and on crystal violet tetraphenylborate in the potentiometric titration of acid and basic dyes. A.G.Fogg and K.S.Yoo. Analytica Chim.Acta., 113, 165 (1980).

Nylon membrane electrode selective for high molecular weight alkyl sulphonates. S.H.Hoke, A.G.Collins and C.A.Reynolds. Anal.Chem., 51, 859 (1979).

Potentiometric semi microdetermination of some triazoles, diazoles and triazenes. W.Selig. Mikrochim.Acta, 1, 453 (1979).

Microdetermination of selenium in organic compounds by potentiometric selenile titration with lead(II) using a lead ISE. A.Campiglio. Mikrochim.Acta, 2, 245 (1979).

The potassium ISE as a tetraphenylborate sensor for quaternary ammonium salts analysis. S.Pinzauti and E.La Porta. J.Pharm.Parmacol., 31, 573 (1979).

Electrodes for organic ion concentration determination. K.Muramatsu. Japan Kokai Tokkyo Koho, 79,136,893 (Cl.G01N27/40) 24 October 1979.

## 8.16 Applications in inorganic analysis

Potentiometric titration of fluoride, sulphate, chromate, molybdate, tungstate, oxalate, phosphate, pyrophosphate and hexacyanoferrate(II) ions with lead(II) solutions and a fluoride ISE. C.E.Efstathiou and T.P.Hadjiioannou. Analytica Chim.Acta, 109, 319 (1979).

Potentiometric titration of selenium with a fluoride ISE. J.L.Bernal, E.Barrado and R.Pardo. Analytica Chim.Acta, 111, 71 (1979).

Determination of sulphate with fluoride ISE. A.F.Zhukov, A.V.Kopytin, G.K.Zhavoronkova and A.V.Gordievskii. Zavod.Lab., 45, 492 (1979).

Application of a fluoride ISE for determination of fluoride in zinc electrolyte. A.Hulanicki, M.Trojanowicz and J.Sztandor. Chem.Anal. (Warsaw) 24, 617 (1979).

Behaviour of iodide selective electrodes at low concentrations of iodide. A.Hulanicki, A.Lewenstam and J.Maj-Zurawska. Analytica Chim. Acta, 107, 121 (1979).

Determination of chlorine in concentrated sulphuric acid solution with a residual chlorine electrode. K.D.Brown and G.A.Parker. Anal. Chem., 51, 1332 (1979).

The use of iodate and periodate with iodine in the potentiometric titration of arsenite, sulphide and sulphite with mercury(II). H.Khalifa, N.T.Abdel-Ghani and M.S.Risk. Micochem.J., 24, 310 (1979).

Use of nitrate-selective liquid membrane electrode in high-halide concentration systems. A.Duca, E.Doniga and F.Matei. Rev.Chim. (Bucharest), 30, 916 (1979).

Potentiometric determination of beryllium using a salicylate ISE. M.Mikesova, J.Petr, J.Senkyr and M.Bartusek. Scr.Fac.Sci.Nat.Univ. Purkynianoe Brun., 9, 9 (1979).

Ultra-micro-simultaneous analysis of silver and cadmium by means of ISEs. P.Hartmann, O.Grunke and H.Berge. Z.Chemie, 19, 71 (1979).

## 8.17 Miscellaneous

Determination of copper in concentrated electrolytes with a copper ISE. A.Hulanicki, T.Krawczynski and M.Trojanowicz. Chem.Anal.(Warsaw) 24, 435 (1979).

ISE analysis (of chlorides in fluids containing bromides and iodides). A.Kumar. U.S.Patent 4,196,056 (Cl.240-IT; B01D59/40) 01 April 1980.

Direct potentiometric method for the determination of cyanide in biological materials. J.O.Egekeze and F.W.Oehme. J.Anal.Toxicol., 3, 119 (1979).

## 9. GLASS pH ELECTRODES: DEVELOPMENT, APPRAISAL, ETC.

A unified approach to glass electrode theory. W.E.Morf. Talanta, 26, 719 (1979).

Storage, ageing and regeneration of pH electrodes. H.Buehler. Chem. Tech.(Heidelberg), 8, 551 (1979).

Redetermination of the standard pH values for potassium hydrogen phthalate reference buffer solutions from 0 to 85$^{o}$C. H.P.Butikofer and A.K.Covington. Analytica Chim.Acta, 108, 179 (1979).

Manufacture of ion-selective electrodes with a solid membrane (by reacting metal or alloy with a sodium isopolycholiogenide). M.A. Llyuschenko, V.V.Bakanina, V.P.Gladyshev, A.I.Ozerov and E.P.Kireeva. U.S.S.R.Patent 712,749 (Cl.G01N27/30) 30 Jan 1980.

## 10. GAS-SENSING ELECTRODES: DEVELOPMENT AND NEW SYSTEMS

### 10.1 Gas sensors dependent on conventional ISEs

Response time characteristics of the p-carbon dioxide electrode. M.A.Jensen and G.A.Rechnitz. Anal.Chem., 51, 1972 (1979).

### 10.2 Other types of electrochemical gas sensors

The hydrogen electrode as pH indicator in hydrogen fluoride and superacid media. J.Devynck, A.B.Hadid and P.Fabre. J.Inorg.Nucl. Chem., 41, 1159 (1979).

Electrochemical oxygen analyser. V.N.Storozhenko, F.E.Dinkevich, V.V.Orlenko and V.Z.Moskovskii. Zavod.Lab., 45, 596 (1979).

An absolute electrode-active oxygen monitor (by corrosion equilibrium of Pd metal $\rightleftharpoons$ oxide). S.P.S.Badwal and H.J.DeBruin. Australas.Corros. Eng., 23, 23 (1979).

Preparation of semi conductor gas sensors. K.Ogawa, A.Abe, M.Nishikawa, S.Sekido and S.Hayakawa. Japan Kokai Tokkyo Koho, 80 27,954, (Cl.G01N27/00) 28 February 1980.

Thermoresistive sensitive component for gas analysis. Yu.I.Belgaev, V.S.Prokhorov, P.I.Stal'nov and M.V.Kulakov. U.S.S.R.Patent, 715,981 (Cl.G01N25/08) 15 February 1980.

Moisture sensor (electrodes of polymer electrolytes). K.Ebara. Japan Kokai Tokkyo Koho, 79,107,797, (Cl.G01N27/04) 23 February 1979.

Gas sensing elements (of coiled Pd fibres). H.Ohshima, K.Nakayama and H.Ohmura. Japan Kokai Tokkyo Koho, 79 62,893 (Cl.G01N27/12) 21 May 1979.

Determination of trace amounts of oxidizable gases such as hydrogen (in a concentration cell which simultaneously acts as a reactor). Japan Kokai Tokkyo Koho, 79 74,797 (Cl.GolN27/58) 15 June 1979.

Measuring cell (with Pt-Rh fibres or powder ) for determining oxygen (and CO by equilibrium with oxygen) concentration in a gas mixture. Fr.Demande, 2,390,7272 (Cl.G01N27/46) 8 December 1978.

Combustible (semiconductor) gas sensors (for isobutane). S.Kusanagi, H.Kawamura and S.Akiyama. Japan Kokai Tokkyo Koho, 79,130,996 (Cl. G01N27/12) 11 October 1979.

(Semiconductor) gas sensors (for H, CO, isobutane and alcohol). M.Katsura, T.Takashashi, T.Kaneda, H.Hiraki and M.Shiratori. Japan Kokai Tokkyo Koho, 79,145,192; 79,145,194; 79,145,199 and 79,145,200 all of (Cl.G01N27/12)13 November 1979.

Gas sensors (semiconductor type for $C_4 H_{10}$ and carbon dioxide). H.Ohkuma, T.Takahashi, M.Katsura and T.Kaneda. Japan Kokai Tokkyo Koho, 79,134,697 (Cl.G01N27/12) 19 October 1979.

Gas-sensitive device (of resistance film of ultra fine particles of metals or oxides for isobutane, water and ethanol). A.Abe, H.Ogawa, M.Nishikawa, S.Sekido and S.Hayakawa. Ger.Offen., 2,933,394, (Cl. G01N27/12) 06 March 1980.

Gas sensor (of iron(III) oxide coated with palladium for isobutane). M.Katsura, T. Takahashi, T.Kaneda, H.Hiraki and M.Shiratori. Japan Kokai Tokkyo Koho, 79,145,193 (Cl.G01N27/12) 13 November 1979.

Gas sensors (for carbon monoxide) using aluminium nitride as support (for calcined rhodium(III) nitrate). H.Taguchi, Y.Takahashi and C.Matsumoto. Funtai Oyobi Funmatsuyakin, 27, 6 (1980).

Gas sensors (of metal oxide semiconductors for carbon monoxide, ethanol, benzene and other hydro-carbons in air). L.Freitinger, R.Koepl and H.Pink. Ger.Offen., 2,821,267 (Cl.G01N27/12) 22 November 1979.

## 11. APPLICATIONS OF GAS SENSORS

Applications of a carbon dioxide selective electrode. I. Determination of meprobamate by decomposition with alkali. S.Tagami. Chem. and Pharm.Bull., 27, 1820 (1979).

A potentiometric analyser for carbon dioxide in gaseous mixture. E.Scarano, C.Calcagno and L.Cignoli. Annali di Chemica, 68, 747 (1979).

Rapid determination of blood ammonia by using Model $PNH_3$-1 ammonia-selective electrode. Ch-M.Liu, K.Chou and Po-M.Ho. Chung-hua I Hsueh Chien Yen Tsa Chih, 2, 114 (1979).

Determination of ammonia and urea in urine and of urea in blood by use of an ammonia-selective electrode. J.Georges. Clin.Chem., 25, 1888 (1979).

Automated determination of serum creatinine by an ammonia gas electrode. Y.Morishita and K.Nakane, Eisei Kensa, 28, 174 (1979).

A package containing a reference liquid for blood gas equipment. S.K.Sorenson and C.C.Holbek, British Patent, 1,555,626 (Cl.G01N33/16), 14 November 1979.

Continuous-flow analysis of ammonia in perchloric acid supernate of blood or plasma using an ammonia-selective electrode. G.C.Moses, R.J.Thibert and T.F.Draisey, J.Clin.Pathol., 31, 1207 (1978).

Automated determination of blood-plasma ammonia based on the ammonia gas membrane electrode. Y.Morishita, K.Nakane and A.Kosaka, Rinsho Kensa, 23, 393 (1979).

Determination of sub-micromolar concentrations of ammonia in natural waters by a standard-addition method using a gas-sensing electrode. C.Garside, G.Hull and S.Murray. Limnol.Oceanogr.,23, 1073 (1978).

Determination of ammonia in workphase air. R.Perez Olmos and M.Arnaiz. Quim.Ind., 25, 271 (1979).

Direct potentiometric determination of ammonia nitrogen in soil using an (ammonia) selective electrode. V.Simeonov, I.Asenov and V.Dyadov. God. Sofii.Univ., Kim.Fak., 1975-76. 70,Pt.1, 17 (Publ.1978).

Use of ammonia selective electrode for the microdetermination of organic nitrogen (Kjeldahl method). J.Varela Rial, J.Valdes Gonzalez and J.Gonzalez Carrero. An.Quim., 74, 1045 (1978).

Determination of nitrite ion in unused cutting fluids and cutting oils using a gas-sensing electrode method. F.R.S.Clark and H.B.MacPherson. Analyst, 104, 358 (1979).

Multiple measurement system for the electrochemical analysis (with ion-selective disc and gas sensors) of flowing liquids and gases. J.G.Schindler. Biomed.Tech., 22, 235 (1977).

Measurement of (ethanol and hydrogen sulphide) pollutant gases in air using (metal oxide) semiconducting sensors. M.Schulz, E.Bohn and G.Heiland. TM, Tech.Mess., 46, 405 (1979).

Direct determination of oxygen in molten steel. J.Skala. Freiberg. Forschungsh.B, 49 (1979).

Electrochemical determination of dissolved oxygen. T.K.Khamrakulov and V.V.Abramova. Zavod.Lab., 45, 99 (1979).

## 12. PIEZOELECTRIC SENSORS AND THEIR APPLICATION

Determination of micromolar concentrations of cyanide in solution with a piezoelectric detector. T.Nomura and O.Hattori. Analytica Chim. Acta, 115, 323 (1980).

Piezoelectricity - a review on materials, fundamental studies and applications. D.K.Dav-Grypte. Dig.Lit.Dielectr., 42, 420 (1978). (Published 1979).

Piezoelectric and electrolytic methods of moisture measurement in gases. S.D.Black. Anal,Instrum., 17, 187 (1979).

Use of a piezoelectric sensors as a continuous monitor of atmosphere pollutants. S.Cooke, T.S.West and P.Watts. Anal.Proc., 17, 2 (1980).

Detection of explosives with a coated piezoelectric quartz crystal. Y.Tomita, M.H.Ho and G.G.Guilbault. Anal.Chem., 51, 1475 (1979).

## 13. REFERENCE ELECTRODES

The dynamic hydrogen electrode. A practical reference in fluoro-sulphuric acid solutions. C.Pitti and M.Herlem. Anal.Lett., 12, 439 (1979).

Application of ISEs as reference electrodes. I.Sekerka and J.F.Lechner. Anal.Lett., Part A 12, 1239 (1979).

(Non-contaminating non-clogging) reference electrode. J.W.Ross and M.S.Frant. U.S.Patent. 4,166,021 (Cl.204,195F, G01N27/30) 28 August 1979.

Reference elements for ISEs. Anonymous. Res.Dicl., 190, 62 (1980).

Plate-shaped silver/silver halide reference electrode for determination of fluoride ion in microlitre solution with fluoride ISE. K.Chiba, K.Tsunoda, Y.Umezawa, H.Haraguchi, S.Fujiwara and K.Fuwa. Anal.Chem., 52, 596 (1980).

Reference electrode for determining values of electrochemical potentials under conditions of variable pressure and temperature. I.Janczarski, E.Lubaszka and A.Mazur. Polish Patent, 100,686 (Cl. G01N27/30) 31 January 1979.

Sterilizing reference electrodes and the like. J.Janata and P.T. McBride. U.S.Patent, 4,175,020 (Cl.204-195B; A61L1/00) 20 November 1978.

## 14. BOOKS AND GENERAL REVIEWS

Selective ion-sensitive electrodes. G.J.Moody and J.D.R.Thomas. Japanese edition. (Translation editors: K.Hiiro and H.Munemori), Kyoritsu Shuppon Co.Ltd., Japan, 1979, pp.148.

Electroanalysis in hygiene, environmental, clinical and pharmaceutical chemistry: Proceedings of a conference held at Chelsea College, London, April 17-20 1979. Editor: W.Franklin Smyth, Elsevier, Amsterdam, 1980. (Includes many papers on ISEs).

Metal ion selectivity of neutral carriers. K.Kimura and T.Shono. Kagaku (Kyoto), 35, 154 (1980).

Ion-selective electrodes. G.H.Fricke. Anal.Chem., 52, 259R (1980).

Analytical performance of ion-selective electrodes. M.A.Koupparis, K.N.Efstatiou, D.S.Papostathopoulos and T.P.Hadjiioanou. Chem.Cron., Genike Ekdase, 44, 45 (1979).

Ion-selective electrode methodology. Editor:A.K.Covington, CRC: Boca Raton, Fla., 1979. Volume 1 (257 pp) contains articles on
"Basic electrode types, classification and selectivity considerations" (A.K.Covington),
"Instrumentation for ISEs" (P.R.Burton),
"Practival techniques for ISEs" (R.J.Simpson),
"pX Standards" (A.K.Covington),
"Glass Electrodes" (A.K.Covington),
"Liquid ion-exchange types" (A.K.Covington and P.Davidson),
"PVC matrix membrane ISEs" (G.J.Moody and J.D.R.Thomas),
"Heterogeneous membrane, carbon support, and coated wire ISEs" (R.W. Cattrall),
"Crystalline and pressed powder solid membrane electrodes" (R.P.Buck).
Volume 2 (129 pp) contains articles on
"Gas-sensing probes" (M.Riley),
"Enzyme electrodes" (P.Vadgama),
"ISEs in medicine and medical research" (D.M.Band and T.Treasure),
"Analytical methods involving ISEs (including flow methods)" (E.Pungor, G.Nagy and K.Tóth).

Review of use of ion-selective electrodes in determination of halides. E.Pungor, Z.Feher, E.Linder, G.Nagy and K.Tóth. Z.Chem.,19, 367 (1979).

Modern applications for ion-selective electrodes. J.Comer. Lab.Equip. Dig., 17, 73 (1979).

Theory and applications of ion-selective electrodes Part III. J.Koryta. Analytica Chim.Acta, 111, 1 (1979).

Medical and biological applications of electrochemical devices. Editor: J.Koryta, J.Wiley and Sons, Chichester, 1980, Pp 331. Contains the following articles on ISEs:
"Liquid-membrane ISEs and their biomedical application" (P.C.Meier, D.Amman, W.E.Morf and W.Simon),
"Solid-state ISEs in clinical chemistry" (C.Fuchs),
"Single cell measurements with ISEs" (J.L.Walker),
"Determination of ion activity changes of excitable tissues with ion-selective microelectrodes" (P.Hnik, E.Syková, N.Kříž and F.Vyskočil),
"Use of enzyme electrodes in biomedical investigations" (G.G.Guilbault).

# SUBJECT INDEX

# AUTHOR INDEX